Photon-Counting Image Sensors

Special Issue Editors

Eric R. Fossum

Nobukazu Teranishi

Albert Theuwissen

David Stoppa

Edoardo Charbon

INTERNATIONAL
IMAGE SENSOR
SOCIETY

MDPI

Guest Editors
Eric R. Fossum
Dartmouth College,
USA

Nobukazu Teranishi
University of Hyogo and
Shizuoka University,
Japan

Albert Theuwissen
Harvest Imaging, Belgium and
Delft University of Technology,
The Netherlands

David Stoppa
Fondazione Bruno Kessler (FBK),
Italy

Edoardo Charbon
Delft University of Technology,
The Netherlands

Editorial Office
MDPI AG
St. Alban-Anlage 66
Basel, Switzerland

This edition is a reprint of the Special Issue published online in the open access journal *Sensors* (ISSN 1424-8220) in 2016 (available at: http://www.mdpi.com/journal/sensors/special_issues/PCIS).

For citation purposes, cite each article independently as indicated on the article page online and as indicated below:

Author 1; Author 2; Author 3 etc. Article title. *Journal Name.* **Year**. Article number/page range.

ISBN 978-3-03842-374-4 (Pbk)
ISBN 978-3-03842-375-1 (PDF)

Table of Contents

Chapter 1: Photon Counting with CMOS Image Sensors

Chapter 2: Photon Counting with Avalanche-Based Devices

Chapter 3: Other Devices, Materials and Applications for Photon Counting

Chapter 4: Image Reconstruction for Photon-Counting Image Sensors

About the Guest Editors

Eric R. Fossum received the B.S. degree in physics and engineering from Trinity College, Hartford, CT, and his Ph.D. degree in engineering and applied science from Yale University, New Haven, CT. He was a member of the Faculty of Electrical Engineering, Columbia University, New York, NY until he joined the NASA Jet Propulsion Laboratory (JPL), California Institute of Technology, Pasadena, CA. There he invented the modern CMOS active pixel sensor camera-on-a-chip technology which is used in billions of cameras each year. He co-founded Photobit Corporation to commercialize the technology and served in several top management roles, including CEO. He was later CEO of Siimpel Corporation, developing camera modules with MEMS-based autofocus and shutter functions for cell phones. He was a Consultant with Samsung Electronics engaged in various projects including 3D RGBZ ranging image sensors. Since 2010, he has been a Professor with the Thayer School of Engineering, Dartmouth College, Hanover, NH, USA focused on photon-counting image sensors. He co-founded the International Image Sensor Society (IISS) and served as its first President. He has published over 290 technical papers and holds over 160 U.S. patents. He is a Charter Fellow of the National Academy of Inventors, a member of the National Academy of Engineering and was inducted into the National Inventors Hall of Fame. In 2017 Dr. Fossum received the Queen Elizabeth Prize for Engineering for his invention and development of CMOS image sensor technology.

Nobukazu Teranishi received his B.S. degree and M.S. degree in physics from the University of Tokyo, Tokyo, Japan and is a professor at the University of Hyogo and Shizuoka University. Since 1978, he has developed image sensors at NEC Corporation (1978–2000) and at Panasonic Corporation (2000–2013) and different universities (2013–present). He and his group invented the pinned photodiode technology, vertical overflow structure, smear reduction structure, among others. They have developed image sensors for various applications, such as movie, digital still cameras, broadcast cameras, security, automobile, medical, scientific and space industries. They have also developed image sensors for infrared and X-ray use other than visible light. He has authored and co-authored 110 papers and has 46 Japanese patents and 21 US patents. Together with E. Fossum and A. Thuwissen, he founded the IISS (International Image Sensors Society), of which he is President. His leadership and image sensor technology development, including the pinned photodiode invention has been honored by government organizations as well as societies. He was awarded the National Invention Awards, Commendation by Minister of State for Science and Technology, Niwa-Takayanagi Award from the Institute of Image Information and Television Engineers (ITE), IEEE EDS J.J.Ebers Award, and is a Fellow of the ITE, and a Fellow of the IEEE. In 2017 Mr. Teranishi received the Queen Elizabeth Prize for Engineering for his invention and development of the pinned photodiode, widely used in CCD and CMOS image sensors.

Albert J.P. Theuwissen received his degree in electrical engineering from the Catholic University of Leuven (Belgium) in 1977. From 1977 to 1983, his work at the ESAT-laboratory of the Catholic University of Leuven focused on semiconductor technology for linear CCD image sensors. He received the Ph.D. degree in electrical engineering in 1983. In 1983, he joined the Micro-Circuits Division of the Philips Research Laboratories in Eindhoven (the Netherlands). In 1991 he became Department Head of the Division Imaging Devices, covering CCD as well as CMOS solid-state imaging activities. He is the author or coauthor of over 200 technical papers in the solid-state imaging field and has been issued several patents. In 1995, he authored a textbook "*Solid-State Imaging with Charge-Coupled Devices*" and in 2011 he co-edited the book "*Single-Photon Imaging*". In 1998, 2007 and 2015 he held the position of *IEEE* Electron Devices Society and Solid-State Circuits Society distinguished lecturer. He acted as general chairman of the International Image Sensor Workshop in 1997, 2003, 2009 and 2015. He was elected as the International Technical Program vice-chair and chair for respectively the ISSCC 2009 and ISSCC 2010. In March 2001, he was appointed as part-time professor at the Delft University of Technology, the Netherlands, where he teaches courses in solid-state imaging and coaches MSc and PhD students in their research on CMOS image sensors. In April 2002, he joined DALSA Corporation to act as the company's Chief Technology Officer. After he left DALSA in September 2007, he started his own company "Harvest Imaging", focusing on consulting, training, teaching and coaching in the field of solid-state imaging technology (www.harvestimaging.com). In 2006 he co-founded (together with his peers Eric Fossum and Nobukazu Teranishi) ImageSensors, Inc. (a Californian non-profit public benefit company) to address the needs of the image sensor community (www.imagesensors.org). In 2008, he received the SMPTE's Fuji gold medal for his contributions to the research, development and education of others in the field of solid-state image capturing. He is a member of the editorial board of the magazine "*Photonics Spectra*", an *IEEE* Fellow and member of SPIE. In 2011, he was elected as "Electronic Imaging Scientist of the Year"; in 2013 he received the Exceptional Service Award of the International Image Sensor Society; and in 2014 he was awarded with the SEMI Award.

David Stoppa received the Laurea degree in Electronics Engineering from Politecnico of Milan, Italy, in 1998, and his Ph.D. degree in microelectronics from the University of Trento, Italy, in 2002. He is the Head of the Integrated Radiation and Image Sensors research unit at FBK where he has been working as a research scientist since 2002, and as Group Leader of the Smart Optical Sensors and Interfaces group from 2010 to 2013. Since 2000 he has been teaching courses at the Telecommunications Engineering Faculty of the University of Trento on Analogue Electronics and Microelectronics. His research interests are mainly in the field of CMOS integrated circuits design, image sensors and biosensors. He has authored or co-authored more than 120 papers in international journals, and presentations at international conferences, and holds several patents in the field of image sensors. Since 2011 he has served as a program committee member of the 'International Solid-State Circuits Conference' (ISSCC) and the SPIE 'Videometrics, Range Imaging and Applications' conference, and was a technical committee member of the 'International Image Sensors Workshop' (IISW) in 2009, 2013, 2015 and 2017. He was a Guest Editor for *IEEE Journal of Solid-State Circuits* Special Issues on ISSCC'14 in 2015 and European Chair at ISSCC 2017. Dr. Stoppa received the 2006 European Solid-State Circuits Conference Best Paper Award.

Edoardo Charbon received his diploma from ETH Zürich, Switzerland, the M.S. from U.C. San Diego, and his Ph.D. from U.C. Berkeley in 1988, 1991, and 1995, respectively, all in electrical engineering and EECS. He was with Cadence Design Systems from 1995 to 2000, where he lead the company's initiative to develop information hiding for intellectual property protection. In 2000, he joined Canesta Inc., as Chief Architect, where he lead the development of time-of-flight 3D CMOS image sensors. From 2002 to 2008 he was assistant professor at EPFL; in 2008 he joined TU Delft as Chair of VLSI Design and, in 2015, EPFL as Chair of Advanced Quantum Architectures. He has authored and co-authored over 250 papers in journals, conference proceedings, magazines, and two books, and he holds 20 patents. Prof. Charbon was the initiator and coordinator of the EU projects Megaframe and SPADnet, which brought single-photon detectors to consumer and mainstream scientific and medical applications. His current research interests include 3D imaging, single-photon imaging, space-based detection, quantum-inspired circuits and systems, and cryo-CMOS technologies. Prof. Charbon has been a Guest Editor of numerous *IEEE* journals and is a member of the TPCs of ISSCC, ESSCIRC, ICECS, ISLPED, VLSI-SOC, and IEDM. He is a Distinguished Visiting Scholar with the W. M. Keck Institute for Space, California Institute of Technology, Pasadena, CA and a fellow of the Kavli Institute of Nanoscience Delft.

Preface to "Photon-Counting Image Sensors"

Papers presented at the 2015 International Image Sensor Workshop (IISW), in Vaals, the Netherlands, organized by the International Image Sensor Society (IISS), showed that photon-counting image sensors may represent the next step in the evolution of solid-state image sensors. In a photon-counting image sensor, it is possible to determine, with a high degree of accuracy, the number of photons that have struck a sensor photoelement during some interval of time. This is enabled by both high quantum efficiency and deep-sub-electron read noise. For some time, single-photon avalanche detectors (SPADs) have grown in performance and array size, with photoelement counts approaching 100,000 or more, today. Practical image capture at such array sizes can now be readily envisioned. SPADs also have an added advantage of allowing fine time resolution of photon arrival, enabling applications such as time-of-flight (TOF) range imaging and fluorescent lifetime imaging microscopy (FLIM). More recently, photoelements that do not require avalanche multiplication and instead use ultra-low capacitance to achieve voltage gain from captured photoelectrons, and/or use multiple sampling techniques to reduce read noise, have shown photon-counting capability. While not having the time resolution of SPADs, these devices offer the potential for large array formats and low power operation in a fabrication process consistent with modern backside-illuminated stacked CMOS image sensor manufacturing.

MDPI *Sensors* and the International Image Sensor Society (IISS) have joined together to create an all-invited special issue reviewing the status of photon-counting image sensors as well as recent developments. The issue includes papers on avalanche gain devices such as SPADs, CMOS image sensors with deep sub-electron read noise, and other devices offering photon-counting capability. Also included are papers on recent progress in image reconstruction, an important aspect in the practical application of photon-counting image sensors.

On behalf of the image sensor community, we extend our appreciation to the authors for accepting our invitation to contribute to the Special Issue. We hope that this collection of excellent papers on the topic of photon-counting image sensors is illuminating and useful in the years ahead.

Eric R. Fossum, Edoardo Charbon, David Stoppa, Nobukazu Teranishi and Albert Theuwissen
Guest Editors

Chapter 1:
Photon Counting with CMOS Image Sensors

sensors

MDPI

Article

Noise Reduction Techniques and Scaling Effects towards Photon Counting CMOS Image Sensors

Assim Boukhayma [1],*, Arnaud Peizerat [2] and Christian Enz [1]

[1]   Integrated Circuits Lab (ICLAB), École Polytechnique Fédérale de Lausanne (EPFL), Microcity,
      Rue de la Maladière 71, Neuchâtel 2000, Switzerland; christian.enz@epfl.ch
[2]   Laboratoire de l' Électronique et Technologies de l' Information (Leti),
      Commissariat a l' Énergie Atomique (CEA), Rue des Marthyrs 17, Grenoble 38000, France;
      arnaud.peizerat@cea.fr
*   Correspondence: assim.boukhayma@epfl.ch; Tel.: +4-121-695-4397

Academic Editor: Albert Theuwissen
Received: 25 January 2016; Accepted: 6 April 2016 ; Published: 9 April 2016

Abstract: This paper presents an overview of the read noise in CMOS image sensors (CISs) based on four-transistors (4T) pixels, column-level amplification and correlated multiple sampling. Starting from the input-referred noise analytical formula, process level optimizations, device choices and circuit techniques at the pixel and column level of the readout chain are derived and discussed. The noise reduction techniques that can be implemented at the column and pixel level are verified by transient noise simulations, measurement and results from recently-published low noise CIS. We show how recently-reported process refinement, leading to the reduction of the sense node capacitance, can be combined with an optimal in-pixel source follower design to reach a sub-0.3 e_{rms}^- read noise at room temperature. This paper also discusses the impact of technology scaling on the CIS read noise. It shows how designers can take advantage of scaling and how the Metal-Oxide-Semiconductor (MOS) transistor gate leakage tunneling current appears as a challenging limitation. For this purpose, both simulation results of the gate leakage current and $1/f$ noise data reported from different foundries and technology nodes are used.

Keywords: CMOS; image sensors; temporal read noise; $1/f$ noise; thermal noise; correlated multiple sampling; deep sub-electron noise

1. Introduction

The idea of an image sensor with photon counting capability is becoming a subject of interest for new applications and imaging paradigms [1–3]. Such a device must have an input-referred read noise negligible compared to a single electron. Among the state-of-the-art imaging devices, single photon detectors may appear to be the best candidate for such an application [4]. Historically, micro-electronics could not provide readout chains with noise levels as low as deep sub-electron. Hence, the solution was to introduce a gain at the level of the photon-electron conversion. In photomultipliers tubes (PMTs) and single photon avalanche photodiodes (SPADs), the electron generated by the incident photon is accelerated and multiplied to a number of electrons from a few hundred in PMTs to millions in SPADs. Such a signal level can be easily detected and quantized into two logic levels, since the number of incident photons during the period of detection is assumed to be much less than one. However, these devices present the following disadvantages [5]. First, they are limited to the case of single photon detection. In other words, the arrival of one photon and multiple photons are not distinguished. Second, these devices suffer from a dead time and after pulse following each photon detection, blinding the device for a certain time. The third limitation is related to the low resolution and fill factors of focal plane arrays using such devices. Additionally, they use high voltages, which are not compliant with standard CMOS image sensor (CIS) processes.

During the last decade, CISs have seen their performance increasing remarkably in terms of dynamic range, speed, resolution and power consumption. With a lower cost and better on-chip integration, CISs replaced progressively the charge coupled devices (CCDs) in many applications and enlarged the market of electronic imaging devices. In terms of sensitivity, the quantum efficiency has been improved to reach levels as high as 0.95 [3]. The fill factors have been constantly improved. The dark current in the pinned photodiodes (PPDs) has been reduced to levels making the process of electron-hole pair generation noiseless for integration times around tens of ms. The read noise has also been dramatically reduced to reach deep sub-electron levels [6–8]. Hence, CIS technologies are advanced enough to envisage the photon counting possibility.

Besides the quantum efficiency, this paper discusses the possibility of performing photon counting, with standard CIS, essentially from the read noise perspective. Starting from the analytical expressions of the input-referred noise, the noise reduction mechanisms at the circuit, device and process level are discussed and verified with simulation, measurements and data reported in recent works. The impact of the combination of different techniques is also analyzed, and the noise levels that can be reached with state-of-the-art technology in standard processes are quantified. This paper also shows how the technology downscaling can be used to reduce the read noise and how the gate leakage current could limit this advantage.

2. CMOS Image Sensors and Photon Counting Requirements

Figure 1 shows the schematic of a conventional low noise CIS readout chain. The corresponding timing diagram is shown in Figure 2. It also shows the potential profile across the PPD, the transfer gate (TX) and the sense node (SN) during the three phases of operation: the integration, the reset and the transfer phases. During the integration time, the PPD accumulates the electrons generated by the incident photons. During the readout, the pixel is connected to the column through the row selection switch (RS), then the reset switch (RST) is closed in order to set the SN voltage higher than the pinning voltage of the PPD. The voltage level at the SN after the reset is read with the in-pixel source follower (SF) and sampled at the end of the readout chain. The potential barrier between the PPD and the SN is controlled by the transfer gate (TX). When the barrier is lowered, the charges accumulated in the PPD are transferred to the SN. The SN voltage level after the transfer is sampled at the output of the readout chain. The reset and transfer samples are then differentiated. This operation is called correlated double sampling (CDS) [9].

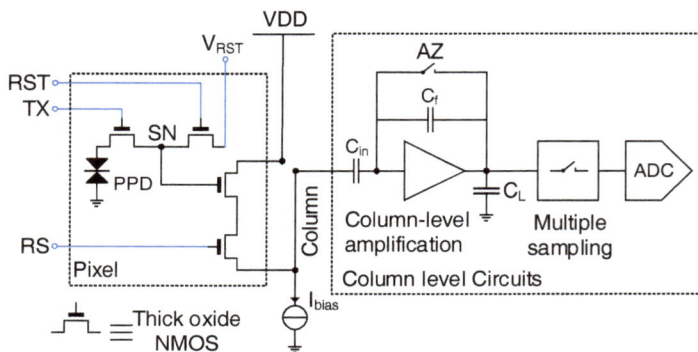

Figure 1. Schematic of a conventional low noise CMOS image sensor (CIS) readout chain. RST, reset switch; TX, transfer gate; RS, row selection switch; SN, sense node; PPD, pinned photodiode; AZ, auto-zero; VDD, supply voltage .

Figure 2 depicts also the different noise sources affecting the signal in the CIS apart from the photon shot noise. During the integration, the charge originating from the thermal generation of electron-hole pairs in the depleted region of the PPD (the dark current) can corrupt the signal. In state-of-the-art CIS, the dark current in PPDs has been reduced to a few e^-/s. Hence, for exposure times below hundreds of ms, the dark current can be neglected.

The reset of the SN leaves a kT/C noise charge held at the SN. This noise is as high as several electrons in the case of a SN capacitance of a few fF. However, for 4T pixels, it is canceled thanks to the CDS readout scheme, as depicted in the timing diagram of Figure 2.

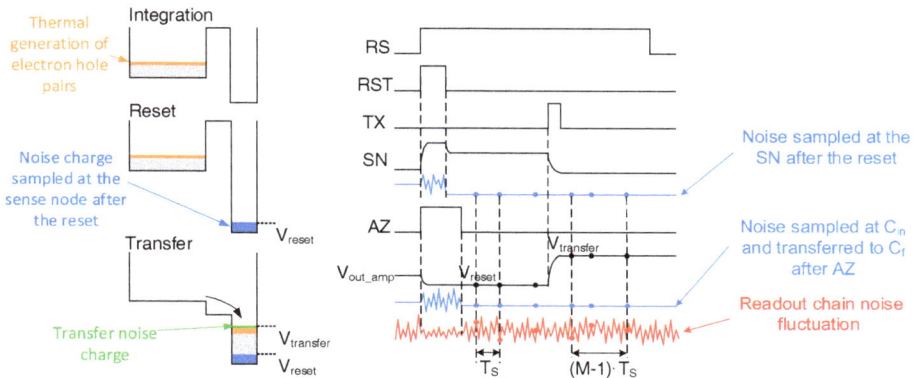

Figure 2. Timing diagram of the conventional CIS readout chain of Figure 1 with noise mechanisms affecting the signal at the PPD and the readout chain levels.

The charge transfer from the PPD to the SN can be affected by the noise related to the charge deficit due to incomplete transfer and lag [10,11]. Unlike the sampled reset kTC noise, this noise is not canceled by the CDS. The charge transfer noise has been extensively studied for CCDs [12,13] because an efficient charge transfer is crucial in such devices. In state-of-the-art CIS with 4T pixels, values of the lag as low as 0.1% have been reported. Thus, the lag can be neglected compared to the read noise in the low light context. The transient noise related to the lag is believed to behave as a shot noise [11], similarly to buried channel CCDs [13]. However, with a lag below 1%, this noise can be neglected in low light conditions. It is also believed that trapping mechanisms in the silicon oxide interface under the transfer gate also contribute to the transfer non-idealities [10,14–16], giving rise to a Random Telegraph Signal (RTS)-like noise.

Finally, the readout of the SN reset and transfer voltages is affected by random fluctuations due to the readout chain noise; starting with the in-pixel SF and noise coupling of the TX and RST lines with the SN, the power supply noise and ending with the column-level circuitry and analog-to-digital converters (ADCs). The column-level amplification is introduced in order to minimize the contribution of the next circuit blocks to the input-referred total noise, e.g., buffers, sample-and-holds and ADC. The column-level amplifier also limits the bandwidth in order to minimize the thermal noise [8]. A switched capacitor amplifier is usually used. An auto-zero (AZ) is performed in order to reset its feedback capacitor and to reduce its offset and $1/f$ noise [9]. When the AZ switch is opened, the noise is sampled at the integration capacitor and transferred to the output. This sampled noise is also canceled thanks to the CDS. Low noise CIS readout chains may also include correlated multiple sampling (CMS) that can be implemented with analog circuitry [17,18] or performed after the ADC [19]. CMS consists of averaging M samples after the reset and M other samples after the transfer with a sampling period T_S, then calculating the difference between the two averages.

With a careful design, the readout noise originating from the pixel and column-level amplifier is the dominant noise source in CIS. Figure 3 shows the calculated probability of a true photo-electron count and a single photo-electron detection as a function of the input-referred readout chain noise by assuming a Gaussian distribution of noise and using the error function. Based on Figure 3, 90% accuracy requires a read noise below 0.4 e_{rms}^- for single photo-electron detection and 0.3 e_{rms}^- for photo-electron count. Recently reported works are today closer than ever to these limits [7,8,20]. A detailed noise analysis of the readout noise is therefore necessary in order to determine the key design and process parameters that can be used for further noise reduction.

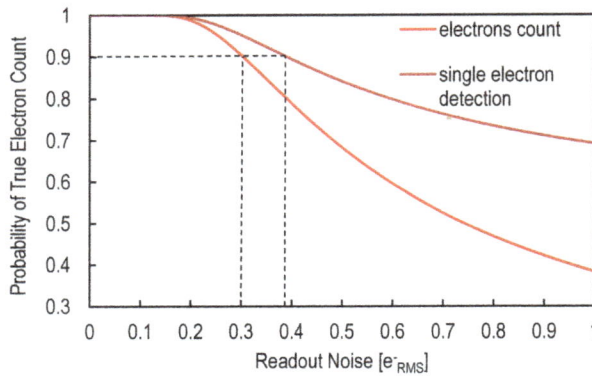

Figure 3. Probability of a true photo-electron count and single photo-electron detection as a function of the input-referred readout noise.

3. Read Noise in CIS

In a conventional CIS readout chain, three readout noise sources can be distinguished: thermal noise, $1/f$ noise and leakage current shot noise. For each noise source, the variance at the output of the readout chain is first calculated and then referred to the input as a noise charge. Hence, the pixel conversion gain is a key parameter in the noise analysis. The pixel conversion gain can be calculated using a small-signal analysis of the pixel. It is crucial to take into account the effect of parasitic capacitances. Figure 4 presents a schematic of a 4T pixel section view showing all of the parasitic capacitances connected to the sense node. These include the overlap capacitances of the transfer and reset gates, C_{Tov} and C_{Rov}, respectively, the sense node junction capacitance, C_J, and the parasitic capacitance related to the metal wires, C_W. These capacitances are independent of the in-pixel SF. Their sum is defined as:

$$C_P = C_{Tov} + C_{Rov} + C_J + C_W \tag{1}$$

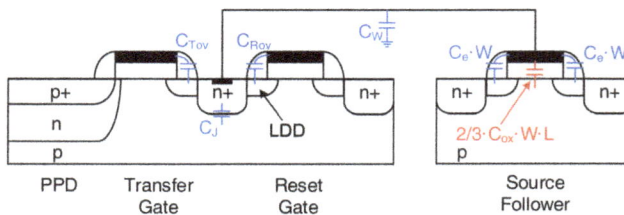

Figure 4. Cross-section of a conventional 4T pixel showing the different parasitic elements contributing to the sense node capacitance.

Figure 5 presents a simplified small-signal schematic of the CIS readout chain of Figure 1. This small-signal schematic is used to calculate the conversion gain together with the noise and signal transfer functions. Based on the detailed analytical calculation presented in [8], the conversion gain of a conventional CIS 4T pixel can be expressed as:

$$A_{CG} = \frac{\frac{1}{n}}{C_P + C_e \cdot W + (1 - \frac{1}{n})(C_e \cdot W + \frac{2}{3}C_{ox} \cdot W \cdot L)} \tag{2}$$

Here, n is the slope factor of the in-pixel SF [21] defined as G_{ms}/G_m, where G_m and G_{ms} are the SF gate and source transconductances, respectively. C_e is the extrinsic capacitance per unit width of the in-pixel source follower transistor. It includes the overlap and fringing capacitances as depicted in Figure 4. C_{ox} is the SF oxide capacitance per unit area.

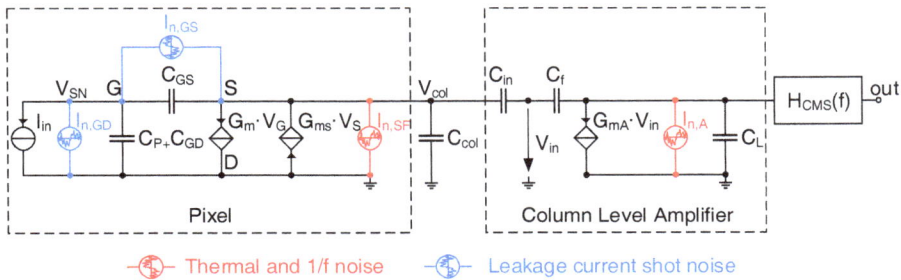

Figure 5. Small-signal analysis of the CIS readout chain depicted in Figure 1 showing the different readout noise sources considered in the analysis.

3.1. 1/f Noise

Under the long-channel approximation, the gate-referred $1/f$ noise power spectral density (PSD) of a MOS transistor operating in the saturation region is commonly expressed as:

$$S_{V_g}(f) = \frac{K_F}{C_{ox}^2 \cdot W \cdot L \cdot f} \tag{3}$$

Here, W and L are the gate width and length; C_{ox} is the oxide capacitance per unit area; and K_F is a $1/f$ noise process and bias-dependent parameter. This empirical model is easy to use for hand calculation and remains valid even for advanced CMOS technologies for adequate gate widths and lengths [22]. The parameter K_F can be expressed as [21,23]:

$$K_F = K_G \cdot k \cdot T \cdot q^2 \cdot \lambda \cdot N_t \tag{4}$$

where k is the Boltzmann constant, T is the absolute temperature, q is the electron charge, λ is the tunneling attenuation distance ($\simeq 0.1$ nm) [24], N_t is the oxide trap density and K_G is a bias-dependent parameter. It has been shown in [21] that K_G is close to unity when the transistor is operating in the weak and moderate inversion regime.

Most analog circuit simulators use the Berkeley Short-channel Model (BSIM) to predict the $1/f$ noise behavior of circuits. It is important to establish a relationship between the parameters used by the simulator and the simple equation used for hand calculations in order to best exploit the noise calculation results. The oxide trap density is the key process-dependent parameter. In the BSIM model, it is referred to as the noise parameter A (noiA) [25].

It is well known that the $1/f$ noise PSD is inversely proportional to the gate area. In low noise CIS readout chains, the transistors located outside the pixels array can be designed with gate dimensions much larger than the in-pixel source follower transistor. In this case, the latter becomes the dominant

$1/f$ noise source in the readout chain, and the other $1/f$ noise sources can be neglected. Based on the small-signal schematic of Figure 5 and the calculation detailed in [8], the input-referred $1/f$ noise can be expressed as:

$$\overline{Q^2_{1/f}} = \alpha_{1/f} \cdot \frac{K_F(C_P + 2C_e \cdot W + \frac{2}{3}C_{ox} \cdot W \cdot L)^2}{C^2_{ox} \cdot W \cdot L} \tag{5}$$

where $\alpha_{1/f}$ is a unitless circuit design parameter reflecting the impact of the CMS noise reduction on the $1/f$ noise. Based on the detailed analytical calculation [26], it can be expressed as:

$$\alpha_{1/f} = \int_0^\infty \frac{1}{f} \cdot \frac{4}{M^2} \frac{sin^4(\pi \cdot M \cdot T_S \cdot f)}{sin^2(\pi \cdot T_S \cdot f)} \cdot \frac{1}{1 + \left(\frac{f}{f_c}\right)^2} df \tag{6}$$

where f_c is the cutoff frequency of the column-level amplifier, which is assumed to be lower than the SF stage bandwidth. T_S is the sampling period of the correlated sampling. $\alpha_{1/f}$ is calculated numerically and plotted as a function of $f_c.T_S$ in Figure 6. It shows that $\alpha_{1/f}$ is weekly dependent on T_S when M is higher than two. In this case, $\alpha_{1/f}$ ranges between three and four.

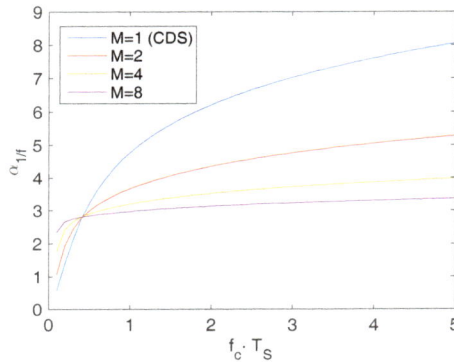

Figure 6. Numerical calculation of the parameter $\alpha_{1/f}$ from Equation (6) as a function of the ratio between the cutoff frequency of the readout chain and the sampling frequency of the correlated sampling T $f_c \cdot T_S$ for a simple CDS and CMS with different orders M.

3.2. Thermal Noise

The thermal noise of a MOS transistor operating in saturation is modeled by a drain current source that adds to the signal. The drain current noise PSD is commonly expressed as [21]:

$$S_{I_d}(f) = 4 \cdot k \cdot T \cdot \gamma \cdot g_m \tag{7}$$

where g_m is the gate transconductance of the transistor and γ is the excess noise factor given by $\frac{2n}{3}$, for a long-channel transistor biased in strong inversion [21].

In a conventional CIS readout chain, besides the power supply and bias voltage noise, there are two dominant thermal noise sources: the in-pixel SF transistor operating in saturation and the column-level amplifier. The latter makes the noise sources from the next stages (ADC, CMS, etc.) negligible when enough gain is provided. The two dominant noise sources are uncorrelated; thus, their noise PSDs add. We assume that the bandwidth of the in-pixel SF stage is limited by the column-level amplifier. We consider that the column-level gain is provided by a closed-loop operational transconductance amplifier (OTA). Using the small-signal analysis of the SF stage and the column-level amplifier [8], the thermal noise voltage variance at the output of the column-level amplifier is calculated. It is then referred to the input using the column-level and conversion gain Equation (2), resulting in:

$$\overline{Q_{th}^2} = \alpha_{th} \cdot \frac{kT}{A_{col} \cdot C} \left(\frac{\gamma_{SF} G_{m,A} (\frac{2}{3} C_{ox} \cdot W \cdot L + 2C_e \cdot W + C_P)^2}{G_{m,SF}} + \frac{\gamma_A}{A_{CG}^2} \right) \tag{8}$$

where $C = C_L + \frac{C_{in}}{A_{col}+1}$. Here, C_L and C_{in} are the integration and load capacitances of the column-level amplifier. γ_{SF} and γ_A are the noise excess factors corresponding to the in-pixel source follower transistor and the OTA of the column-level amplifier, respectively. $G_{m,SF}$ and $G_{m,A}$ are the transconductances of the in-pixel SF stage and column-level OTA, respectively. α_{th} is a unitless circuit design parameter dependent on the circuit or processing techniques used after the column-level amplification stage. In the case of CMS, α_{th} is given by [26]:

$$\alpha_{th} = \frac{1}{\pi f_c} \int_0^\infty \frac{4}{M^2} \frac{sin^4(\pi \cdot M \cdot T_S \cdot f)}{sin^2(\pi \cdot T_S \cdot f)} \cdot \frac{1}{1 + \left(\frac{f}{f_c}\right)^2} df \simeq \frac{2}{M} \tag{9}$$

Note that for proper settling of the signal between sampling instants, $2\pi \cdot f_c \cdot T_S$ has to be typically larger than five, and under such conditions, α_{th} can simply be approximated by $\frac{2}{M}$.

3.3. Leakage Current Shot Noise

During the readout, the charge transferred to the SN may be corrupted by all of the leakage currents through the junctions and gate oxide due to tunneling. Since these leakage currents are due to barrier control processes, they give rise to shot noise. As shown in the small-signal schematic of Figure 5, the leakage current shot noise can be modeled by two noise current sources: $I_{n,GD}$ and $I_{n,GS}$. $I_{n,GD}$ represents the shot noise of all of the leakage currents flowing between the SN and the ground, which includes the SN junction leakage and the SF gate oxide tunneling current that sinks into the bulk and the drain. $I_{n,GS}$ represents the shot noise associated with part of the SF gate oxide tunneling current that flows to the source. The unilateral PSD of the current shot noise can be expressed as [27]:

$$S_{I_L}(f) = 2 \cdot q \cdot I_L \tag{10}$$

where I_L is the mean value (DC current) of the total leakage current. It can be shown that both shot noise components $I_{n,GD}$ and $I_{n,GS}$ have the same transfer function magnitude, between the noise current source and the output of the column level amplifier. The leakage current shot noise PSD at the output of the column level amplifier can therefore be simplified as:

$$S_{L,Amp}(f) = S_{I_L}(f) \cdot \frac{A_{CG}^2 \cdot A_{col}^2}{(2\pi f)^2} \cdot \frac{1}{1 + \left(\frac{f}{f_c}\right)^2} \tag{11}$$

Note that I_L is the sum of all of the sense node leakage currents. The noise PSD after the CMS, taking into account the impact of aliasing, can be expressed as:

$$S_{L,CMS}(f) = sinc(\pi f T_S)^2 \cdot \frac{4}{M^2} \frac{sin^4(\pi \cdot M \cdot T_S \cdot f)}{sin^2(\pi \cdot T_S \cdot f)} \cdot \sum_{n=-\infty}^{+\infty} S_{L,Amp}(f - nT_S) \tag{12}$$

Figure 7a shows a plot of the input-referred shot noise PSD, normalized to $2 \cdot q \cdot I_L \cdot T_S$. It can be noticed that due to the $1/f^2$ term in Equation (11), the PSD is independent of f_c, and the area of the PSD increases with M. It can be shown that the input-referred charge variance due to the total leakage current's shot noise can be expressed as:

$$\overline{Q_L^2} = 2 \cdot \alpha_{shot} \cdot q \cdot I_L \cdot T_S \tag{13}$$

with:

$$\alpha_{shot} = \int_0^\infty \frac{1}{\pi} \cdot \frac{sin(M \cdot x)^4}{(x \cdot sin(x))^2} \cdot \frac{1}{1 + \left(\frac{x}{f_c \cdot T_S}\right)^2} dx \simeq \frac{M}{3} \ (for \ M \geq 2) \tag{14}$$

Note that the shot noise current sources feature a white PSD. However, when integrated in the SN capacitance, they give rise to a Wiener process [28]. The variance of this noise is thus expected to rise with the readout time. In order to evaluate the impact of the CMS on the leakage current shot noise, α_{shot} is calculated numerically and plotted in Figure 7b as a function of M. In the case of a simple CDS, α_{shot} is equal to 0.5; hence, the shot noise variance is given by $q \cdot I_L \cdot T_S$, which corresponds to a typical case of a Wiener process [28]. Figure 7b also shows that, in the general case, the leakage current shot noise increases linearly with $T_S \cdot M$.

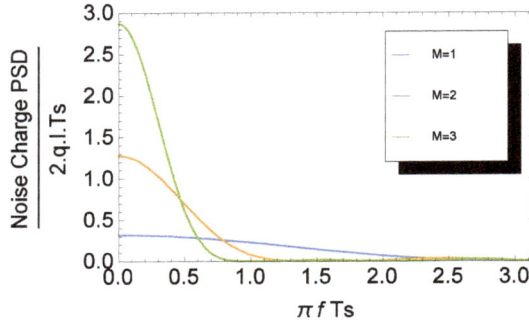

(a)

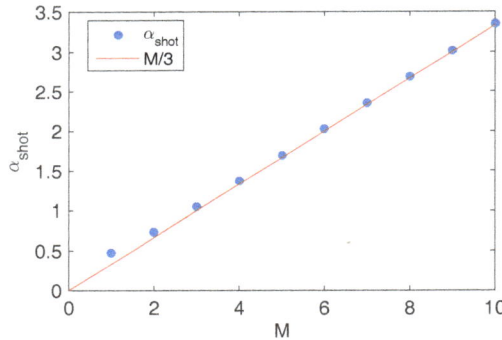

(b)

Figure 7. Input-referred shot noise PSD (**a**) and variance (**b**), normalized to $2 \cdot q \cdot I_L \cdot T_S$ as a function of the correlated multiple sampling (CMS) order M.

4. CIS Read Noise Reduction Techniques

4.1. Column-Level Techniques

Based on Equation (8), the parameters that can be used to reduce the readout thermal noise independently of the pixel design are the column-level gain A_{col}, the capacitance C acting on the bandwidth of the column-level amplifier and the CMS order M that determines the value of the parameter α_{th}. For thermal noise, the column-level gain A_{col}, the capacitance C and the CMS order M all have the same impact on the input-referred noise. In order to validate this result, transient noise simulations [29] have been performed on a conventional CIS readout chain with a 4T pixel using a standard thick oxide NMOS source follower transistor, a column-level amplifier based on an OTA with a feedback capacitance C_f and the passive CMS circuit presented in [17]. Figure 8a shows the impact of the column-level gain and bandwidth control on the input-referred noise when a simple correlated double sampling is used after the column-level amplifier. Figure 8b shows the impact of the correlated multiple sampling on the input-referred thermal noise for different column-level gains. These simulation results show that the thermal noise can be reduced drastically using only the column-level parameters.

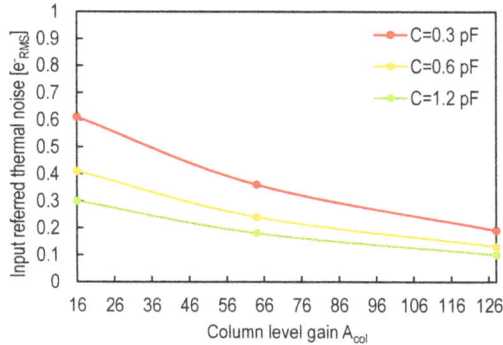

(a)

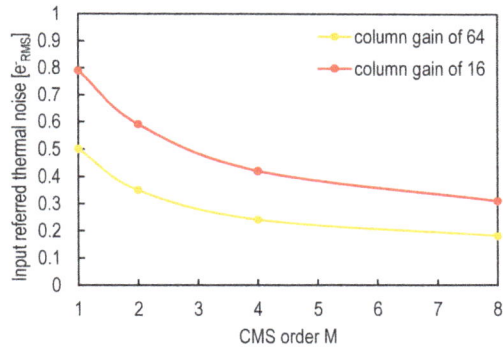

(b)

Figure 8. Input-referred thermal noise, obtained from transient noise simulations, of a CIS readout chain, with a 4T pixel (standard NMOS source follower) with a conversion gain of 85 $\mu V/e^-$, column amplification (closed loop gain with the operational transconductance amplifier (OTA)) and CMS implemented with the analog circuit presented in [17], as a function of: (**a**) the column-level gain A_{col} for different values of C and a simple CDS ($M = 1$); (**b**) the CMS order M for different values of the column-level gain A_{col} and $C = 0.2$ pF.

Based on Equation (5), $\alpha_{1/f}$ is the only parameter in the input-referred $1/f$ noise expression that is independent of the pixel-level device and process parameters. As shown in Section 3.1, $\alpha_{1/f}$ decreases with the CMS order. Figure 9 shows transient noise simulations of the input-referred $1/f$ noise for readout chains with different in-pixel SF types as a function of the CMS order. Figure 9 demonstrates that even if the CMS comes with some $1/f$ noise reduction, the impact of the device parameters (SF type) is much more significant.

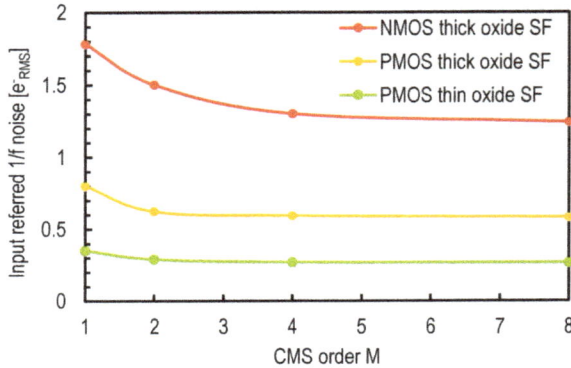

Figure 9. Input-referred $1/f$ noise of a CIS readout chain, with 4T pixel, column amplification and CMS [17], obtained with transient noise simulations, as a function of the CMS order M for different in-pixel source follower transistor types. The pixels with thick oxide NMOS and PMOS SFs feature a conversion gain of about 85 $\mu V/e^-$, while the thin oxide SF based pixel features a conversion gain of 185 $\mu V/e^-$.

4.2. Pixel-Level Techniques

The thermal noise can be reduced to extremely low levels by implementing column-level circuit techniques as shown in Figure 8a,b. Consequently, further reduction of the thermal noise at the pixel level is less efficient, and noise optimization at the pixel level should be mostly focused on reducing the remaining and dominant $1/f$ noise. Equation (5) is the starting point for the $1/f$ noise optimization and suggests different approaches, including proper device selection, design optimizations, as well as process improvements.

4.2.1. Reduce the Capacitance C_P and the Source Follower Transistor Overlap Capacitance

This point remains an active research topic. Careful layout is not enough to significantly decrease the contributions of the wiring parasitic capacitances, the transfer and reset overlap capacitances, as well as the junction capacitance of the floating diffusion. For this purpose, process improvements are necessary. Many recent works presenting sub-electron readout noise CIS actually focused on this point. In [30], different process-level techniques have been presented leading to the reduction of C_P. It has been shown that the omission of the low doped drains (LDDs) used in standard CMOS transistors reduces effectively the gate overlap capacitances. Furthermore, increasing the depletion depth under the floating diffusion by reducing the doping concentration reduces the junction capacitance. The combination of these techniques led to a C_P reduction of about 47%. In [7], the capacitance C_P has been reduced by using an idea called "virtual phase", well known in CCDs, consisting of creating a potential profile that isolates the floating diffusion from the transfer gate. In this way, the overlap capacitance between the transfer gate and the floating diffusion (denoted C_{Tov} in Figure 4) is dramatically reduced. Furthermore, the channel width of the reset transistor is reduced by controlling the doping profile in order to reduce the overlap between the reset gate and the floating diffusion (denoted C_{Rov} in Figure 4). However, this was obtained at the cost of a low pixel full-well capacity and a relatively higher lag. In [20], C_{Tov} is reduced by introducing a special implant isolating the transfer gate from the SN, and C_{Rov} is reduced by omitting the reset transistor. However, this requires the reset to be performed with a high voltage clock of 25 V connected directly to an implant close to the SN. Figure 10 shows the impact of the C_P reduction through a plot of the calculated input-referred $1/f$ noise as a function of the gate width and length, based on Equation (5), for a C_P of 0.75 fF, corresponding to a standard process, and a C_P of 0.25 fF, corresponding to the one that could be obtained through advanced process refinements [30]. Figure 10 shows the effectiveness of this C_p reduction, which leads to a reduction of the input-referred $1/f$ noise from 0.4 to 0.3 e^-_{rms}.

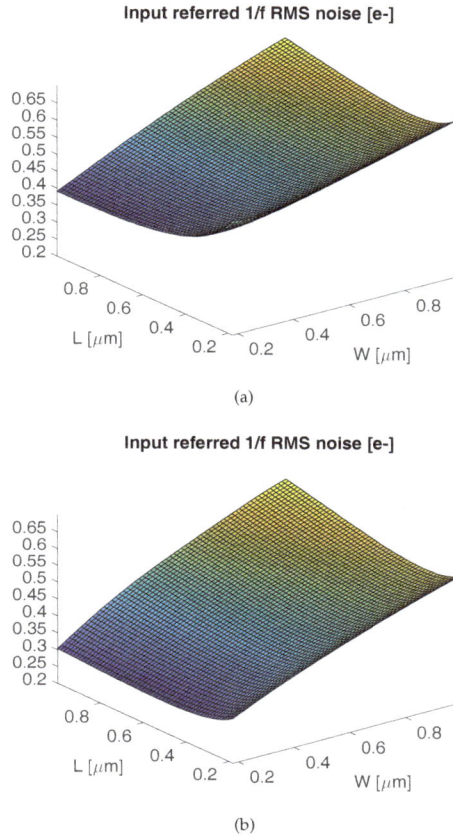

Input referred 1/f RMS noise [e-]

(a)

Input referred 1/f RMS noise [e-]

(b)

Figure 10. The calculated input-referred $1/f$ noise, based on Equation (5), as a function of the in-pixel source follower width W and length L for a thin oxide transistor in a 180-nm technology where $C_{gse} = C_{gde} = 0.95$ fF/μm, $C_{ox} = 9.5$ fF/μm^2, $K = 10^{-11}$ F^{2}V^2/m^2, $\alpha_{CMS} = 3$, and in (a) $C_P = 0.75$ fF, it corresponds to the SN capacitance in a standard process, (b) $C_P = 0.25$ fF corresponds to an SN capacitance reduced with process-level optimization.

4.2.2. Reduce the $1/f$ Noise Process Parameter K_F (the Oxide Trap Density N_t)

This point can be addressed through design choices and technological improvements. It is known that buried channel devices have a lower $1/f$ noise by featuring a lower K_F parameter. This is likely due to the fact that the charge carriers are kept away from the silicon oxide interface [31]. It has been shown that using buried channel NMOS source followers leads to sub-electron noise performance [19]. From a design aspect, thick-oxide transistors that operate at voltages as high as 3.3 V are commonly used in CIS pixels. Figure 11 shows the oxide trap density of PMOS and NMOS thick oxide transistors from different foundries and technology nodes. It shows that PMOS transistors feature, generally, a K_F parameter lower than NMOS transistors. Using an in-pixel PMOS source follower transistor also led to a sub-electron noise performance [32]. The drawback of in-pixel PMOS transistor is the reduction of the fill factor due to the spacings imposed by the layout design rules and the possible quantum efficiency reduction if the PMOS n-well is too close to the PPD.

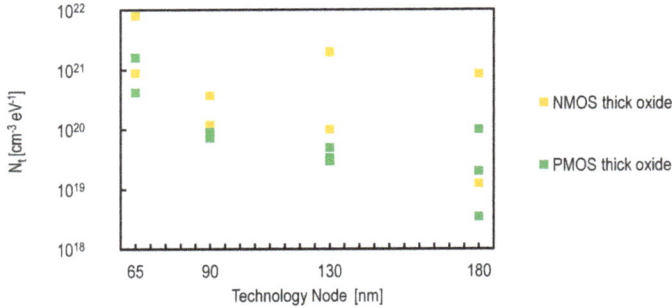

Figure 11. The oxide trap density N_t, of PMOS and NMOS thick oxide transistors, as a function of the technology node based on measurement results reported in design kits from different foundries.

4.2.3. Increase the Oxide Capacitance per Unit Area C_{ox}

Based on Equation (5), the $1/f$ noise can also be reduced by increasing the oxide capacitance per unit area of the in-pixel SF. From a design perspective, this corresponds to the selection of a thin oxide SF instead of the traditional thick oxide transistor. In most CIS processes, all of the gates included in the pixel feature thick oxides, since the transfer gate and the reset gate are controlled by high voltages (3.3 V); the SF is also chosen as a thick oxide transistor to exploit a high dynamic range. In [8], it has been shown how a thin oxide transistor can be implemented in a CIS pixel without degrading the dynamic range or dramatically reducing the fill factor at the benefit of a much reduced input-referred $1/f$ noise.

4.2.4. Use a Minimum Gate Width and an Optimal Length

Based on Equation (5), it can be shown analytically [33,34] or numerically using the a plot of Equation (5) *versus* the gate width and length that the lowest input-referred $1/f$ noise corresponds to the minimum gate width and an optimal length generally slightly higher [33]. Figure 10 illustrates this principle on a practical example. It shows a plot of the input-referred $1/f$ noise as a function of the gate width and length for a pixel based on a thin oxide PMOS SF of a 180-nm CMOS process for two different values of C_P. It shows how, for both C_P values, the input-referred $1/f$ noise can be reduced by choosing a minimum SF gate width and a slightly larger length. Note that the reduction of the SF gate size might increase the probability of RTS noise occurrence [35]. However, the amplitude of the RTS noise is inversely proportional to the gate area [36]; thus, it would be also reduced by using a minimum SF gate width.

4.2.5. Thin Oxide Source Follower: A Good Match

From the designer's perspective, the points mentioned in Sections 4.2.3 and 4.2.4 can both be addressed by using a thin oxide SF. Indeed, A thin oxide SF with a minimum gate width features also lower overlap capacitances. It is important to verify that the choice of a thin oxide transistor does not come at the cost of a negative impact on the $1/f$ noise process parameter K_F. A thinner oxide is expected to come with a better control of the gate over the channel and, therefore, a lower K_F. The oxide trap density of thin oxide PMOS transistors and thick oxide NMOS transistors of different foundries and technology nodes has been compared in [8] based on data reported in design kits. This comparison showed that the thin oxide PMOS transistors generally feature a lower oxide trap density. Thus, using a thin oxide PMOS source follower addresses the points of Sections 4.2.2, 4.2.3 and 4.2.4 at once. Figure 12 shows the schematic of a 4T pixel based on a thin oxide PMOS SF. In order to validate this idea, a transient noise simulation is performed on three readout chains based respectively on a standard thick oxide NMOS, a thick oxide PMOS and a thin oxide PMOS source follower in a 180-nm CIS process. The compared readout chains share the same column level amplification and CMS circuit presented in [17]. Figure 9 shows the impact of the in-pixel source follower transistor type, as well as the CMS order M on the input-referred $1/f$ noise. The PMOS SF-based readout chain features a lower $1/f$ noise than the NMOS-based one thanks to the increase of C_{ox} and the reduction of the parameter K_F. The readout chain based on the thin oxide PMOS SF features the lowest input-referred noise because its SF cumulates a lower K_F, a higher oxide capacitance per unit area and a lower minimum width.

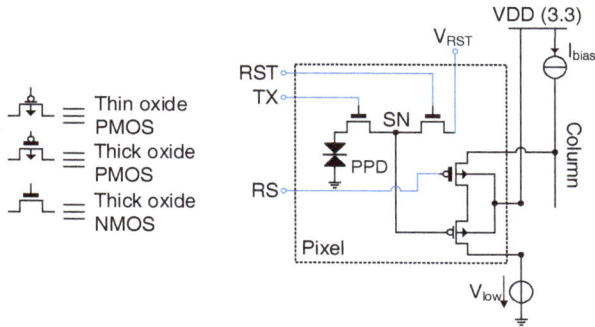

Figure 12. Schematic of the recently-proposed pixel [8] based on a thin oxide PMOS source follower.

It is important to verify that these techniques are not harmful in terms of the thermal noise. Equation (8) shows that the thermal noise is reduced by increasing the conversion gain. Thus, using a thin oxide SF with smaller gate dimensions is also expected to reduce the input-referred thermal noise.

The benefits of using a thin oxide SF transistor have been confirmed with measurement results. A test chip comparing pixels based on thin oxide PMOS source followers with state-of-the-art pixels with thick oxide buried channel NMOS source followers has been presented in [8]. Figure 13 shows the measured average total input-referred noise of the two different pixels for two column-level gain values and with a simple CDS. Figure 13 shows how both thermal (at low column gain) and $1/f$ noise (at high column gain) are dramatically reduced thanks to the implementation of the source follower with a thin oxide PMOS transistor.

The tested pixel was designed using a standard CIS process and fulfilling the standard design rules. There was therefore no possibility to exploit the impact of the reduction of C_P through process optimization. In order to predict the impact of the combination of thin oxide SF with process optimizations reducing the sense node capacitance, the parameter C_P of Equation (5) is replaced by the measurement results from [30]. The starting point corresponds to the result of 0.4 e_{rms}^- obtained in [8] and corresponding to the case of a thin oxide SF-based pixel design with standard rules. Then, the values of C_P based on [30] are used to predict the evolution of the input-referred noise for each additional technique used to reduce the the SN capacitance. The result is plotted in Figure 14. It shows the expected input-referred noise, at the optimal gate width and length, for each value of C_P. Figure 14 shows that the 0.3 e_{rms}^- limit can be crossed if the thin oxide PMOS SF is combined with process optimizations reducing C_P in the 180-nm process used in [8].

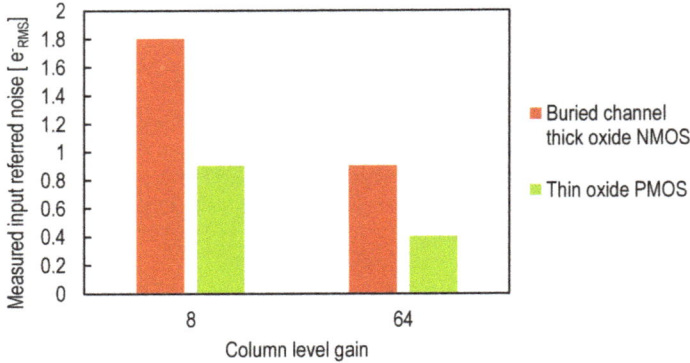

Figure 13. The measured input-referred total noise of a CIS readout chain, with 4T pixel and column amplification, for two column level gains A_{col}, for an in-pixel buried channel thick oxide NMOS source follower and a thin oxide PMOS source follower from [8].

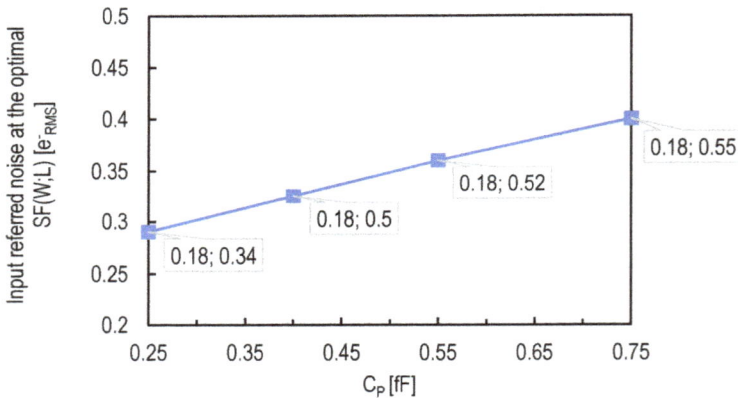

Figure 14. The input-referred $1/f$ noise as a function of the capacitance C_P determined by the floating diffusion, wiring and parasitic capacitances independent from the source follower transistor. $C_P = 0.75$ fF corresponds to the case of a conventional sense node junction, transfer gate and transistors. $C_P = 0.55$ fF corresponds to the case of transfer and reset gates without low doped drains [30]. $C_P = 0.4$ fF corresponds to the case of transfer and reset gates without low doped drains and a sense node without channel stop underneath [30]. $C_P = 0.25$ fF corresponds to the case of more advanced process refinements as [7,30]. The numbers in labels correspond to the minimum width and optimum length as discussed in Section 4.2.4.

5. CIS Read Noise and Technology Downscaling

Since their first development, CIS pixels have always been designed with thick oxide transistors compatible with high voltages (3.3 V). The device parameters of thick oxide transistors do not follow the scaling rules as the thin oxide transistors. The impact of the technology downscaling on these devices is rather limited. Moreover, it appears that the oxide trap density of thick oxide transistors tends to increase with the technology downscaling, as shown in Figure 11.

It has been demonstrated in [8] that a thin oxide SF can be used together with a conventional PPD for a low noise performance. Thin oxide transistors, on the other hand, take full advantage of technology downscaling. Thus, it is interesting to investigate the impact of technology downscaling on the input-referred noise. The starting point for analyzing the impact of the technology downscaling on the input-referred noise of a readout chain based on a thin oxide SF is Equations (5) and (8). The conclusions can be made based on how the technology downscaling affects the different process and device parameters. Table 1 shows the scaling factor corresponding to the relevant device parameters [37]. The technology downscaling allows a higher oxide capacitance per unit area, a lower gate width and lower overlap and parasitic capacitances. Hence, the input-referred $1/f$ noise variance is supposed to decrease with κ^2, assuming that the oxide trap density N_t remains constant with the technology downscale. The thermal noise is expected to decrease with κ^2 and, hence, would remain negligible. The International Technology Roadmap for Semiconductors (ITRS) expects the oxide trap density to decrease with the technology downscaling [22]. Figure 15 shows the N_t values, for thin oxide transistors, reported in design kits of three foundries for different technology nodes. It shows that the oxide trap density follows the ITRS roadmap when downscaling from 180 nm to 130 nm. For more advanced technologies, the data are not conclusive and must be verified by measurements. The $1/f$ noise of NMOS transistors does not increase dramatically. On the contrary, PMOS transistors appear to show a higher N_t for advanced technologies. In bulk CMOS, the buried channel conductance of the PMOS transistors is likely the reason for their lower $1/f$ noise. While deep submicron PMOS transistors are expected to behave as surface channel devices, which explains the fact that their $1/f$ noise becomes comparable to the one of NMOS transistors.

Table 1. Impact of technology downscaling on the parameters of the input-referred $1/f$ noise in Equation (5).

Parameter	Scaling Factor
W	$\frac{1}{\kappa}$
L	$\frac{1}{\kappa}$
C_{ox}	κ
C_P	$\frac{1}{\kappa}$
$C_e \cdot W$	$\frac{1}{\kappa}$

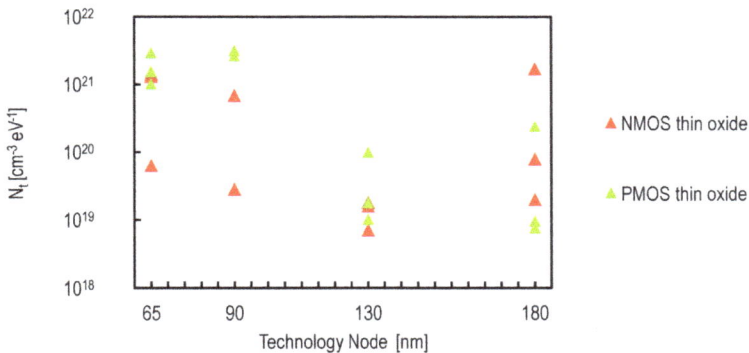

Figure 15. The evolution of the oxide trap density N_t, as a function of the technology node based on measurement results reported in design kits from different foundries.

The measurement results presented in [8,38] explore indirectly the impact of the technology downscaling on the noise reduction. A pixel with a thin oxide SF transistor have been compared to a thick oxide SF based one. For the 180-nm process used in [8,38], the thick oxide transistor features an oxide capacitance per unit area of 5 fF$/\mu$m^2 compared to 9.55 fF$/\mu$m^2 for the thin oxide transistor. In addition, the minimum width determined by the design rules is 0.4 μm for the thick oxide compared to 0.22 μm for the thin oxide transistor. Consequently, using a thin oxide source follower transistor instead of a thick oxide has the same effect as a technology downscaling with a scaling factor of two. Based on this observation, the input-referred noise is expected to decrease by a factor of two, which matches the measurement results shown in Figure 13.

Besides the read noise originating from the $1/f$ and thermal noise, the gate leakage current shot noise has been up to now neglected due to the extremely low levels of the leakage currents achieved in the used technology. It is important to investigate the evolution of this noise when using more advanced technologies. Indeed, the gate leakage current increases by several orders of magnitude when downscaling from 180-nm to 65-nm technologies [37]. Based on Equation (13), the shot noise associated with the gate leakage current is hence expected to increase significantly. In order to evaluate its impact, simulations have been performed with transistors having a minimum gate width and length from technologies between 180 nm and 65 nm. The corresponding leakage current shot noise RMS is given by the square root of the total number of electrons crossing the gate in a time interval of 10 µs (enough to read two samples). The results are plotted in Figure 16. Figure 16 also shows how the input-referred noise is expected to decrease by only taking advantage of the technology downscaling based on Equation (5) and the assumption of constant oxide trap density for deep submicron technologies. The starting point corresponds to the input-referred noise obtained using a thin oxide SF in a 180-nm CMOS process [8]. It can be noticed that the $0.3\ e^-_{rms}$ limit can be crossed if a CIS process is developed with a technology node under 130 nm and a thin oxide transistor is used as a SF. However, for technologies under 90 nm, the gate leakage current appears to be a severe problem starting to dominate the total noise. Hence, the optimal technology node is between 130 nm and 90 nm, unless process improvements are applied to reduce the gate leakage current. Figure 16 shows also the impact on the technology node when the process refinements reducing the SN capacitance C_P are applied. The noise levels for each technology node are obtained using Equation (5) and the scaling rules. The starting point corresponds to noise expected when combining the thin oxide PMOS SF with a C_P of 0.25 fF, as shown in Figure 14. The latter shows that an input-referred read noise under $0.2\ e^-_{rms}$ could be possible with a technology node between 130 nm and 90 nm, a thin oxide SF and process level C_P reduction.

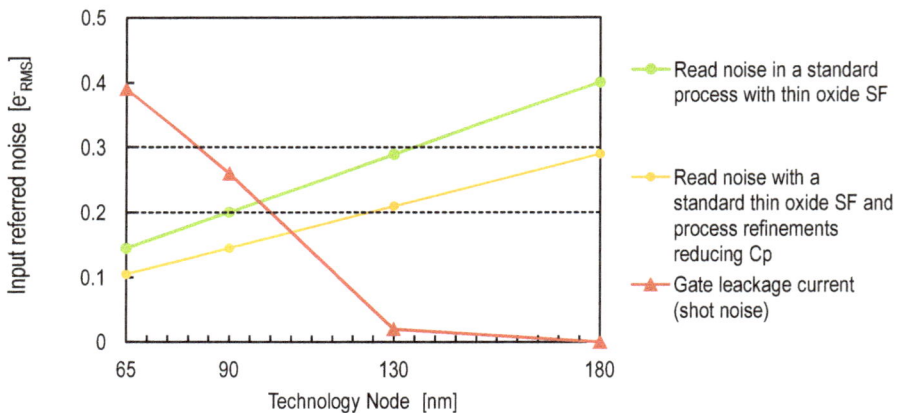

Figure 16. The expected evolution, with technology downscale, of the input-referred $1/f$ noise of CIS designed with standard CMOS process with thin oxide in-pixel SF.

RTS noise may also be a concern with the technology downscaling. In sate-of-the-art low noise CMOS image sensors, it may result in a dramatically high input referred-noise value of about several e^-_{rms}, but it is only present in a minority of pixels (the tail of the noise histogram). Therefore, RTS noise was not accounted for in this work, including in the extrapolation towards downscaled technologies, because we limited the latter to 65 nm, where leakage is much more an issue. A further investigation of input-referred noise for such 4T pixels in deep submicron technology would definitely require one to account for RTS noise. Unfortunately, RTS noise is not modeled in the most common simulators, and the complexity of this phenomena still impedes an analytically- or empirically-precise expression of its occurrence.

6. Conclusions

The capability of performing photo-electron counting, with an accuracy higher than 90%, using conventional CIS readout chains requires a total read noise level below 0.3 e_{rms}^-. This read noise is mainly composed of the $1/f$ noise originating from the in-pixel SF, the thermal noise originating from the pixel- and column-level saturated transistors and the shot noise associated with the leakage current at the level of the SN. The latter is negligible in the technology nodes used currently (above 100 nm).

The thermal noise can be drastically reduced, to extremely low levels, by combining column-level gain, bandwidth control and CMS. The $1/f$ noise becomes then the dominant noise source. The reduction of the $1/f$ noise can involve process-, device- and circuit-level optimizations. The process-level refinements include the sense node total capacitance and the SF K_F parameter reduction. At the device level, the input-referred noise can be reduced by using an in-pixel SF with a higher oxide capacitance per unit area, a minimum gate width and an optimal gate length. The implementation of an in-pixel thin oxide PMOS SF-instead of a thick oxide NMOS presents a practical example of how this device level optimization can be performed in a standard process. At the circuit level, the $1/f$ noise can be slightly further reduced using the CMS.

Based on measurement results reported in recent works and the analytical expressions of the input-referred noise, the combination of a standard thin oxide PMOS SF with the process refinement reducing the SN capacitance is expected to decrease the total read noise of a conventional CIS below 0.3 e_{rms}^-.

The input-referred thermal and $1/f$ noise are expected to decrease with the technology downscaling to levels below 0.2 e_{rms}^-. For technologies below 90 nm, the SF gate oxide leakage current is expected to increase dramatically. Therefore, unless the leakage current is reduced by some other means at the process level, the optimum technology node ranges between 90 nm and 130 nm.

Acknowledgments: Commissariat à l'Énergie Atomique (CEA) and Délégation générale de l'armement (DGA) are acknowledged for the financial support of this work.

Author Contributions: Assim Boukhayma carried out the data analysis, the theoretical and experimental work. Assim Boukhayma, Arnaud Peizerat and Christian Enz contributed to write this paper. All the authors have read and approved the final manuscript.

Conflicts of Interest: The authors declare no conflict of interest.

References

1. Seitz, P.; Theuwissen, A.J.P. *Single-Photon Imaging*; Springer: Berlin, Germany, 2011.
2. Fossum, E. The quanta image sensor (QIS): Concepts and challenges. *Imaging Appl. Opt. Opt. Soc. Am.* **2011**, doi:10.1364/COSI.2011.JTuE1.
3. Teranishi, N. Required conditions for photon-counting image sensors. *IEEE Trans. Electron Devices* **2012**, *59*, 2199–2205.
4. Bronzi, D.; Villa, F.; Tisa, S.; Tosi, A.; Zappa, F. SPAD Figures of merit for photon-counting, photon-timing, and imaging applications: A review. *IEEE Sens. J.* **2016**, *16*, 3–12.
5. Charbon, E. Single-photon imaging in complementary metal oxide semiconductor processes. *Philos. Trans. R. Soc. Lond. A Math. Phys. Eng. Sci.* **2014**, *372*, doi:10.1098/rsta.2013.0100.
6. Wakashima, S.; Kusuhara, F.; Kuroda, R.; Sugawa, S. A linear response single exposure CMOS image sensor with 0.5 e- readout noise and 76 ke- full well capacity. In Proceedings of the 2015 Symposium on VLSI Circuits, Kyoto, Japan, 17–19 June 2015; pp. C88–C89.

7. Ma, J.; Starkey, D.; Rao, A.; Odame, K.; Fossum, E. Characterization of quanta image sensor pump-gate jots with deep sub-electron read noise. *IEEE J. Electron Devices Soc.* **2015**, *3*, 472–480.
8. Boukhayma, A.; Peizerat, A.; Enz, C. Temporal readout noise analysis and reduction techniques for low-light CMOS image sensors. *IEEE Trans. Electron Devices* **2016**, *63*, 72–78.
9. Enz, C.; Temes, G. Circuit techniques for reducing the effects of op-amp imperfections: Autozeroing, correlated double sampling, and chopper stabilization. *Proc. IEEE* **1996**, *84*, 1584–1614.
10. Bonjour, L.E.; Blanc, N.; Kayal, M. Experimental analysis of lag sources in pinned photodiodes. *IEEE Electron Device Lett.* **2012**, *33*, 1735–1737.
11. Fossum, E.R. Charge transfer noise and lag in CMOS active pixel sensors. In Proceedings of the 2003 IEEE Workshop on Charge-Coupled Devices and Advanced Image Sensors, Elmau, Germany, 15 May 2003; pp. 1–6.
12. Thornber, K. Noise suppression in charge transfer devices. *Proc. IEEE* **1972**, *60*, 1113–1114.
13. Janesick, J.R. *Scientific Charge-Coupled Devices*; SPIE: Bellingham, WA, USA, 2001.
14. Wang, X.; Rao, P.R.; Theuwissen, A.J.P. Fixed-pattern noise induced by transmission gate in pinned 4T CMOS image sensor pixels. In Proceeding of the 36th European Solid-State Device Research Conference, Montreux, Switzerland, 19–21 September 2006; pp. 331–334.
15. Goiffon, V.; Virmontois, C.; Magnan, P. Investigation of dark current random telegraph signal in pinned photodiode CMOS image sensors. In Proceeding of the 2011 IEEE International Electron Devices Meeting (IEDM), Washington, DC, USA, 5–7 December 2011; pp. 8.4.1–8.4.4.
16. Fowler, B.; Liu, X.C. Charge Transfer Noise in Image Sensors. In Proceeding of the 2007 International Image Sensor Workshop (IISW), Ogunquit, ME, USA, 7–10 June 2007; pp. 51–54.
17. Boukhayma, A.; Peizerat, A.; Enz, C. A correlated multiple sampling passive switched capacitor circuit for low light CMOS image sensors. In Proceeding of the 2015 International Conference on Noise and Fluctuations (ICNF), Xi'an, China, 2–6 June 2015; pp. 1–4.
18. Suh, S.; Itoh, S.; Aoyama, S.; Kawahito, S. Column-Parallel correlated multiple sampling circuits for CMOS image sensors and their noise reduction effects. *Sensors* **2010**, *10*, 9139–9154.
19. Chen, Y.; Xu, Y.; Chae, Y.; Mierop, A.; Wang, X.; Theuwissen, A. A 0.7 e^-_{rms} temporal-readout-noise CMOS image sensor for low-light-level imaging. *IEEE Int. Solid-State Circuits Conf. Dig. Tech. Pap.* **2012**, *55*, 384–386.
20. Seo, M.; Kawahito, S.; Kagawa, K.; Yasutomi, K. A 0.27 e^-_{rms} Read Noise 220 µV/e^- Conversion Gain Reset-Gate-Less CMOS Image Sensor With 0.11-µm CIS Process. *IEEE Electron Device Lett.* **2015**, *36*, 1344–1347.
21. Enz, C.; Vittoz, E. *Charge Based MOS Transistor Modeling: The EkV Model For Low-Power and RF IC Design*; Wiley: Hoboken, NJ, USA, 2006.
22. Nemirovsky, Y.; Corcos, D.; Brouk, I.; Nemirovsky, A.; Chaudhry, S. 1/f noise in advanced CMOS transistors. *IEEE Instrum. Meas. Mag.* **2011**, *14*, 14–22.
23. Hung, K.; Ko, P.; Hu, C.; Cheng, Y. A physics-based MOSFET noise model for circuit simulators. *IEEE Trans. Electron Devices* **1990**, *37*, 1323–1333.
24. Ghibaudo, G. Low-frequency noise and fluctuations in advanced CMOS devices. *Proc. SPIE* **2003**, *5113*, 16–28.
25. Celik-Butler, Z. Low-frequency noise in deep-submicron metal-oxide-semiconductor field-effect transistors. *IEEE Proc. Circuits Devices Syst.* **2002**, *149*, 23–31.
26. Enz, C.; Boukhayma, A. Recent trends in low-frequency noise reduction techniques for integrated circuits. In Proceeding of the 2015 International Conference on Noise and Fluctuations (ICNF), Xi'an, China, 2–6 June 2015; pp. 1–6.
27. Scholten, A.; Tiemeijer, L.; van Langevelde, R.; Havens, R.; Zegers-van Duijnhoven, A.; Venezia, V. Noise modeling for RF CMOS circuit simulation. *IEEE Trans. Electron Devices* **2003**, *50*, 618–632.
28. Papoulis, A. *Probability, Random Variables, and Stochastic Processes*; McGraw-Hill: New York, NY, USA, 1965.
29. Bolcato, P.; Tawfik, M.; Poujois, R.; Jarron, P. A new efficient transient noise analysis technique for simulation of CCD image sensors or particle detectors. In Proceedings of the IEEE Custom Integrated Circuits Conference, San Diego, CA, USA, 9–12 May 1993; pp. 14.8.1–14.8.4
30. Fumiaki, K.; Shunichi, W.; Satoshi, N.; Rihito, K.; Shigetoshi, S. Analysis and Reduction of Floating Diffusion Capacitance Components of CMOS Image Sensor for Photon-Countable Sensitivity. In Proceedings of the International Image Sensors Workshop (IISW), Vaals, The Netherlands, 8–11 June 2015; pp.120–123.

31. Chen, Y.; Wang, X.; Mierop, A.; Theuwissen, A. A CMOS image sensor with in-pixel buried-channel source follower and optimized row selector. *IEEE Trans. Electron Devices* **2009**, *56*, 2390–2397.
32. Lotto, C.; Seitz, P.; Baechler, T. A sub-electron readout noise CMOS image sensor with pixel-level open-loop voltage amplification. In Proceedings of the 2011 IEEE International Solid-State Circuits Conference Digest of Technical Papers (ISSCC), San Francisco, CA, USA, 20–24 February 2011; doi:10.1109/ISSCC.2011.5746370.
33. Boukhayma, A.; Peizerat, A.; Dupret, A.; Enz, C. Design optimization for low light CMOS image sensors readout chain. In Proceedings of the IEEE 12th International New Circuits and Systems Conference (NEWCAS), Trois-Rivieres, QC, Canada, 22–25 June 2014; pp. 241–244.
34. Boukhayma, A.; Peizerat, A.; Dupret, A.; Enz, C. Comparison of two optimized readout chains for low light CIS. *Proc. SPIE* **2014**, *9022*, 90220H-90220H-11.
35. Martin-Gonthier, P.; Magnan, P. Novel readout circuit architecture for CMOS image sensors minimizing RTS noise. *IEEE Electron Device Lett.* **2011**, *32*, 776–778.
36. Kwon, H.M.; Han, I.S.; Bok, J.D.; Park, S.U.; Jung, Y.J.; Lee, G.W.; Chung, Y.S.; Lee, J.H.; Kang, C.Y.; Kirsch, P.; *et al.* Characterization of random telegraph signal noise of high-performance p-MOSFETs with a high-k dielectric/metal gate. *IEEE Electron Device Lett.* **2011**, *32*, 686–688.
37. Lewyn, L.; Ytterdal, T.; Wulff, C.; Martin, K. Analog circuit design in nanoscale CMOS technologies. *Proc. IEEE* **2009**, *97*, 1687–1714.
38. Boukhayma, A.; Peizerat, A.; Enz, C. A 0.4 e^-_{rms} Temporal Readout Noise 7.5 μm Pitch and a 66% Fill Factor Pixel for Low Light CMOS Image Sensors. In Proceedings of the International Image Sensors Workshop (IISW), Vaals, The Netherlands, 8–11 June 2015; pp.365–367.

sensors

MDPI

Article

Reduction of CMOS Image Sensor Read Noise to Enable Photon Counting

Michael Guidash [1],*, Jiaju Ma [2], Thomas Vogelsang [1] and Jay Endsley [1]

[1] Rambus Inc., Sunnyvale, CA 94089, USA; tvogelsang@rambus.com (T.V.); jendsley@rambus.com (J.E.)
[2] Thayer School of Engineering, Dartmouth College, Hanover, NH 03755, USA; jiaju.ma@dartmouth.edu
* Correspondence: mguidash@rambus.com; Tel.: +1-585-802-1532

Academic Editor: Albert Theuwissen
Received: 1 February 2016; Accepted: 31 March 2016; Published: 9 April 2016

Abstract: Recent activity in photon counting CMOS image sensors (CIS) has been directed to reduction of read noise. Many approaches and methods have been reported. This work is focused on providing sub 1 e^- read noise by design and operation of the binary and small signal readout of photon counting CIS. Compensation of transfer gate feed-through was used to provide substantially reduced CDS time and source follower (SF) bandwidth. SF read noise was reduced by a factor of 3 with this method. This method can be applied broadly to CIS devices to reduce the read noise for small signals to enable use as a photon counting sensor.

Keywords: CMOS; image sensor; photon counting; read noise

1. Introduction

1.1. Read Noise Reduction for CIS Devices

In the past several years there has been a substantial amount of work directed to the use of CMOS Image Sensors (CIS) for single photon detection and photon counting [1–5]. A key requirement and development area for photon counting CIS devices is low read noise [1–5]. It has been shown that CIS read noise should be reduced to 0.15 electrons (e^-) or less [2]. The state of the art for high volume consumer application small pixel CIS is in the 1.2 e^- to 2.0 e^- range. Recent work on CIS read noise reduction has been directed to increasing conversion gain (CG) [6–9], correlated multiple sampling (CMS) [9,10], source follower (SF) transistor structure [11], and SF accumulation [12]. Results of these papers are summarized in Table 1. Sub 1 e^- rms read noise was achieved with results in the range of 0.28 e^- rms to 0.86 e^- rms.

Table 1. Sub 1 e^- SF read noise results from various references.

Ref #	Noise Reduction Approach	Conversion Gain (µV/e⁻)	Analog Gain	Number of Reads	Read Noise (µV rms)	Read Noise (e⁻ rms)	Pixel Size (µm)	Process Node (nm)
[6]	CS AmpHigh CG	300	10	1	258	0.86	11	180
[7]	High CG	240		1	120	0.50	5.5	180
[8]	High CG	426		1	137	0.28	1.4	65
[8]	High CG	256		1	97	0.32	1.4	65
[9]	CMS		64	4		0.70[1]	10.0	180
[10]	CMS	110	16	5	73	0.66	1.1	
[11]	Bch SF	185	64	1	74	0.40	7.5	180
[12]	CMSInver. Cycling	~400		1600	136	0.34	25	180

[1] This reference provided total read noise only (SF read noise was not determined).

1.2. New Method for Read Noise Reduction for Photon Counting CIS Devices

This paper addresses a different approach to read noise reduction for photon counting CIS devices. For single bit photon counting CIS devices, the pixel output signal readout will be binary (*i.e.*, no signal or >1 photon signal) [1]. In this case the signal readout path needs to be designed for a signal range that corresponds to 1 e$^-$ with some headroom (e.g., 5 e$^-$). There is no reason to measure or precisely know the output signal value above this maximum signal. For conversion gains in the range of 200 μV/e$^-$ to 500 μV/e$^-$ this is a maximum signal swing of 1–2.5 mV. For multi-bit photon counting CIS devices the pixel output signal swing needs to be precisely known only for signal levels corresponding to the maximum number of electrons to be counted per readout (e.g., 20 e$^-$) [2,3]. As a result the maximum output signal to be precisely determined is <10 mV. In both cases this is substantially less than the maximum signal swing for a conventional CIS device, which is typically on the order of 0.5–1.0 V. The signal readout path can be designed and optimized for this. Since the maximum signal level that needs to be accurately quantified is small compared to a conventional CIS, a shorter Correlated Double Sample (CDS) time (t_{CDS}) and reduced source follower (SF) bandwidth (BW) can be used for a photon counting CIS compared to a conventional CIS.

Referring to Figure 1, the t_{CDS} is defined as the time between the falling edge of sample-and-hold reset (SHR) pulse to the falling edge of sample-and-hold signal (SHS) pulse. SF read noise is limited by 1/f noise and Random Telegraph Signal (RTS) noise, [13,14]. Reduction of t_{CDS} and SF BW will have an attendant reduction on 1/f and Johnson or thermal noise of the SF readout [13,15]. However, it is difficult to achieve significantly reduced t_{CDS} and SF BW due to the limitations of transfer gate (TG) feed-through (FT) to the floating diffusion (FD). This is shown in Figure 1.

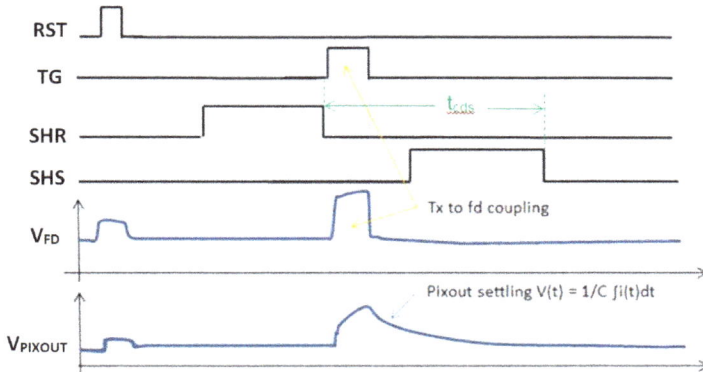

Figure 1. Conventional pixel readout timing diagram and output waveform. TG feed-through limits small signal settling time.

The coupling capacitance from TG to FD (C_{tgfd}) and reset gate (RG) to FD (C_{tgrg}) is shown in Figure 2. The FD node will see a FT signal that follows the pulses from TG and RG signals. The magnitude of the FT signal (ΔV_{FT}) is given by:

$$\Delta V_{FT} = \Delta V_{tg} * \left(C_{tgfd} / C_{fd} \right) \qquad (1)$$

where ΔV_{tg} is the voltage swing of the TG pulse and C_{fd} is the total capacitance of the FD node.

The pixel output settling time is dominated by the TG FT to the FD for small signal levels. This is especially true for high conversion gain pixels where the C_{tgfd} can be a larger percentage of C_{fd}.

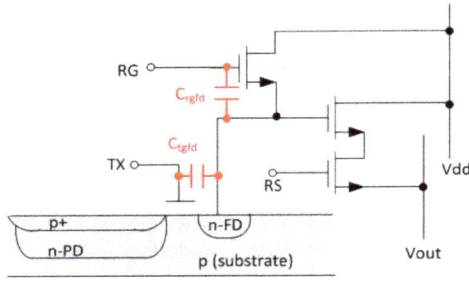

Figure 2. Pixel schematic with TG and RG coupling capacitances to FD.

1.3. A New Timing Method for CIS Read Noise Reduction

In order to substantially reduce the t_{CDS} and SF BW we have devised a new readout timing method where TG feed-through is compensated and the small signal settling time is dramatically reduced. This is shown at a high level in Figure 3a,b below. A signal tg_null is used to null (*i.e.*, cancel or compensate) the TG feed-through. This signal can be provided as an additional and separate signal wire with an attendant decoder/driver. This null signal line is preferably row based, to match skew and droop over the array of the TG signals that are to be compensated. There are many possible approaches to provide the tg_null signal including use of existing pixel signal lines. One such approach is described in Section 2 of this paper.

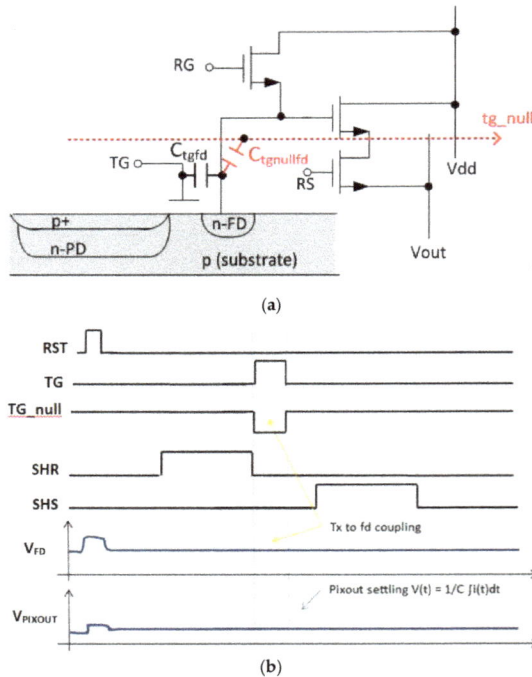

(a)

(b)

Figure 3. (a) New pixel schematic showing tg_null signal line used to compensate TG feed-through. (b) New pixel readout timing diagram and output waveform; TG feed-through is compensated.

Figure 3a is a pixel schematic showing the case of an additional row based signal line "tg_null". There is a coupling capacitance from the tg_null signal line to the FD, C_{tg_nullfd}. The magnitude of the nulling pulse designed to cancel the TG feed-through will depend on the value of C_{tg_nullfd}. The compensation or feed-through cancellation signal does not have to be perfectly aligned with the TG signal in order to provide a substantially reduced pixel output settling time. For example, when referring to Figure 3b, the edges of the tg_null signal do not have to be exactly aligned with the TG signal pulse. In addition the product of the voltage swing and coupling capacitance (C_{tg_nullfd}), of the tg_null pulse does not need to be exactly the same as that of the TG signal.

One of the advantageous effects of the feed-through compensation is the elimination of the trade-off between conversion gain (CG) and settling time. CG is the conversion factor of e^- to volts in the readout of the pixel. This is determined by C_{fd} according to Equation (2):

$$CG = q/C_{fd} \qquad (2)$$

If the feed-through to the floating diffusion is not compensated, the ΔV_{FT} will increase as the conversion gain is increased (*i.e.*, conversion gain is increased by decreasing C_{fd}, so if coupling capacitance remains the same, the ΔV_{FT} is larger). This larger feed-through then causes a longer FD and V_{out} settling time and increases t_{CDS}. By compensating the FD feed-through, one can increase CG without increasing settling time, and thus further reduce input referred read noise.

As mentioned above, an alternate approach to use of an additional null signal and signal line is to use existing signals and structures in the pixel for TG to FD feed-through signal compensation. This does not add capacitance to the FD and as a result does not reduce CG. One such approach was used on an existing sensor. The details of this method and attendant results are described in the next sections.

2. Materials and Methods

2.1. Sensor Description

TG to FD feed-through compensation timing was implemented on an existing prototype sensor with programmable timing to investigate the effect of reduced t_{CDS} and SF BW on the sensor read noise. The chip photograph is shown in Figure 4.

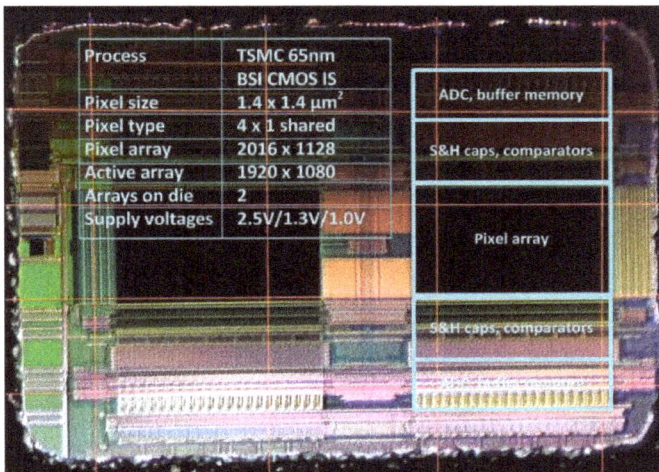

Process	TSMC 65nm BSI CMOS IS
Pixel size	1.4 x 1.4 μm²
Pixel type	4 x 1 shared
Pixel array	2016 x 1128
Active array	1920 x 1080
Arrays on die	2
Supply voltages	2.5V/1.3V/1.0V

Figure 4. Sensor die photograph.

The chip contains two 1920 by 1080 pixel arrays. Each array contains different pixel architectures. One half of one the arrays is a 4-shared amplifier 4T pinned photodiode pixel architecture. The pixel size is 1.4 μm. The chip was fabricated in 65 nm BSI CIS process technology. A simplified schematic of the 4-shared unit pixel cell and the array readout path is shown in Figure 5.

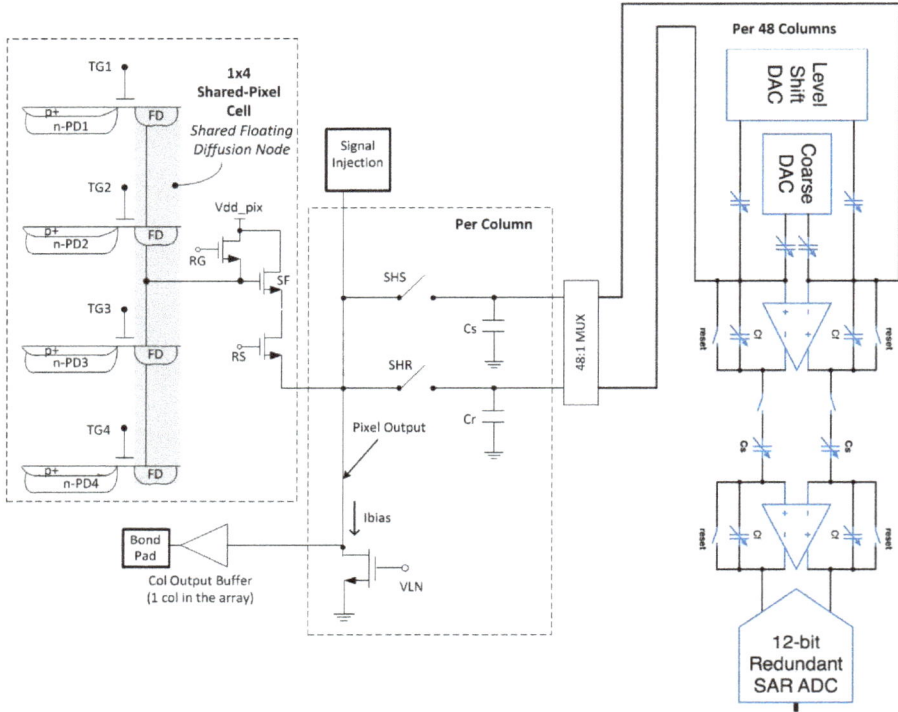

Figure 5. Block diagram of pixel array and readout signal path.

The unit pixel cell is one column by four rows (1 × 4). The SF has a width of 0.28 μm and a length of 0.7 μm. The row select transistor has a width of 0.28 μm and a length of 0.29 μm. Per column sample and hold capacitors are used to store the reset and transfer signal levels for CDS readout. A switched capacitor programmable gain amplifier (PGA) and 12 bit SAR ADC are shared by 48 columns. The PGA and ADC layout is split into two banks, one at the top and one at the bottom of the array. Adjacent groups of four columns are routed to the top and bottom ADC banks. This architecture was chosen for fast readout. The PGA has a selectable gain of 2×, 4× or 8×. A gain of 8× was used for the noise measurements. One column output line is connected to an analog output buffer to view the pixel output waveform. An injection point was included at the input of the PGA to determine the electrons per Data Number (DN) of the readout path. The SF Ibias current is programmable by an external master current and on-chip current mirror. The sensor readout timing and control is implemented on a FPGA external to the sensor, and is fully programmable.

The pixel output lines have a total resistance of 1261 Ω and capacitance of 906 fF. The sample and hold capacitors are ~400 fF. The total capacitance (C_{pixout}) of the column readout is ~1.3 pF (400 fF + 906 fF). The pixel output bandwidth is limited by the transconductance (g_m) of the pixel source follower which is 12 to 55 μs depending on the SF bias current used in this experiment. The dominant time constant due to g_m ($\tau D = C_{pixout}/g_m$), is ~22 ns at the baseline SF Ibias condition of 8 μA. The conversion gain of the 1×4 pixel is 75.6 μV/e⁻.

2.2. New Readout Method Details for Reduction of Read Noise

As discussed in the Introduction section, in conventional CIS timing and readout, the t_{CDS}, is limited by the TG FT settling time, especially for small signals. In addition, for photon counting CIS devices, the t_{CDS} and SF BW can be reduced given the maximum output signal swing to accurately measure is very small compared to that of a conventional CIS device. We have modified the CIS timing to compensate or cancel the TGFT. By canceling the TGFT, the output signal settling time is reduced. A variety of timing approaches can be implemented depending on the pixel architecture and the row decoder/driver design details. The timing we intended to use is shown in Figure 6. The t_{CDS} is the time between the falling edge of the SHR signal pulse to the falling edge of the SHS signal pulse during the readout phase. The various TG signal levels are indicated by name. Vtg_off is the TG off level used during integration, and is typically a negative voltage in order to reduce TG dark current. Vtg_mid1 level is typically used during readout of the pixel, and is less negative or 0 V in order to avoid any gate induced drain leakage (GIDL) on the FD during readout. Vtg_on is the signal level used to provide lag free transfer from the photodiode (PD) to FD.

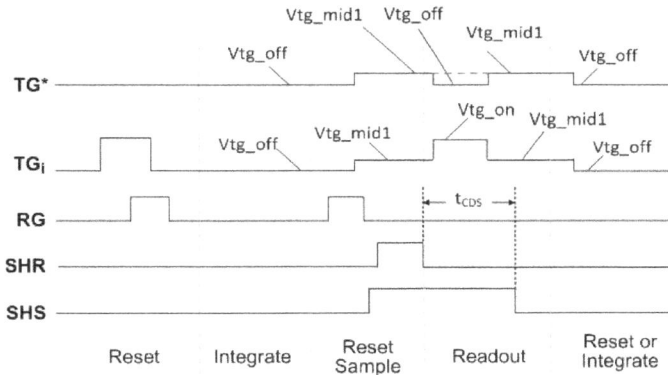

Figure 6. Intended pixel timing diagram for TG feed-through compensation: TGi is the row being readout; TG* is 3 "other" TG's in the 1×4 pixel cell.

Referring to Figure 6, for any given row being read out in the 1×4 unit cell, the other 3 TG's are used to compensate the TG FT for the pixel being read out. With this approach, the compensating voltage of the three TG's is in a sub-threshold range and will not cause charge transfer. Conventional TG timing is shown by the dotted red line for the TG* signal. Since the local overlap capacitance of the TGs to FD is well matched, this method will provide a very small residual FT signal, and the timing skew across the array will be very well matched for the TG* and TGi signals. Based on the row decoder/driver design of our sensor, we had to use the RG signal to compensate the rising edge of TG, and 3 TG's to compensate the falling TG edge. This timing diagram is shown in Figure 7.

2.3. New Readout Method Measured Timing and Waveforms

Pixel output waveforms were captured at the column analog output buffer to verify operation of the timing and cancellation of the TG FT. These waveforms are shown in Figure 8.

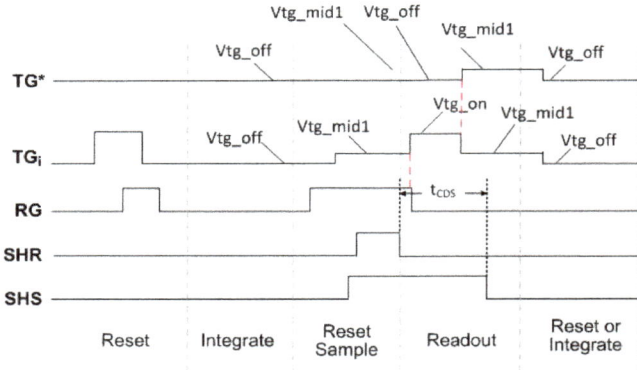

Figure 7. Pixel timing diagram used in this experiment due to limitations with sensor row decoder design.

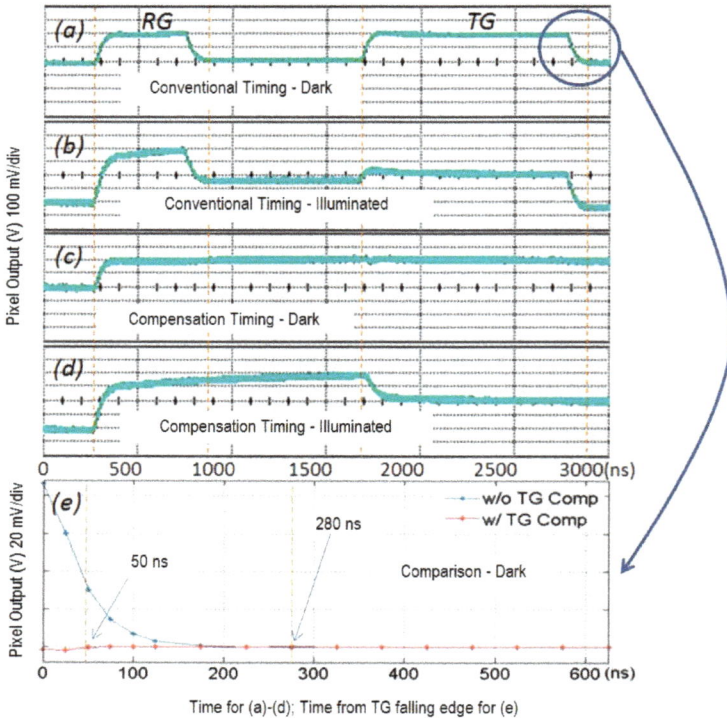

Figure 8. Analog output waveforms *vs.* time from TG falling edge.

Pixel output waveforms for conventional timing and our new timing are shown for both dark and illuminated conditions. The outputs for conventional timing for dark and illuminated conditions are shown in waveforms (a) and (b). The outputs for TG compensation timing for dark and illuminated conditions are shown in waveforms (c) and (d). The signal level for the illuminated condition is ~64 mV (~1000 e⁻). Comparing waveforms (a) and (c), it is evident that the compensation signals cancel the TG FT and the output settles much faster for both dark and illuminated signals. The zoomed in section of dark condition waveforms for conventional and compensation timing is shown in plot (e). This shows the dark level settling time is reduced from 280 ns to 50 ns by the TG FT compensation method. Note also that the RG falling edge feed-through is compensated by the TG rising edge. We briefly examined the variation in the residual feed-through for the single column of pixels that could be observed. The variation was very small, and we attribute this to the local matching of C_{tgfd} using this cancellation method. Further work is required to quantify this variation for the whole column and for an array.

Figure 9 below shows the measured ADC output *vs.* TG rising edge to SHS falling edge time for two signal levels when using TG compensation timing. The settling time is 150 ns for a signal of 230 e⁻ and 100 ns for a signal level of 25 e⁻. The 150 ns settling time is less than the dark settling time of 280 ns for conventional timing. The measured settling times are in reasonable agreement with simulation results of 86 ns and 137 ns, respectively.

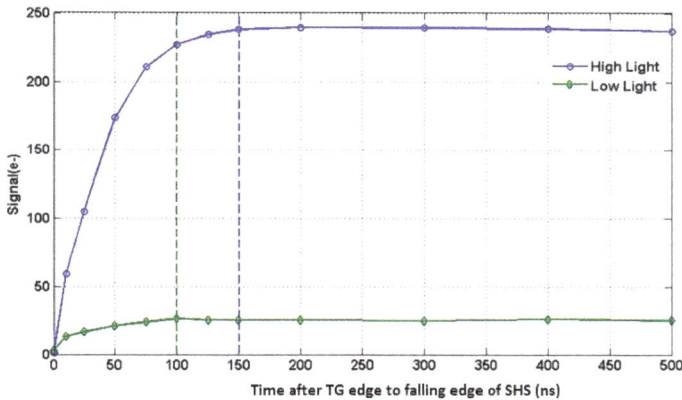

Figure 9. ADC output signal *vs.* time from edge of TG to falling edge of SHS.

These simulation results include the calculation of the number of settling time constants ($N\tau$) that are required for small and quantized signals. For conventional CIS readout where it is required to convert the maximum signal level to n-bits, the required number of setting time constants is given by Equation (3):

$$N\tau = \ln(2 * (1 - \text{slewp}) * (V_{max}/V_{lsb})) \qquad (3)$$

where V_{max} is the full signal swing, V_{lsb} is the lsb voltage and slewp is the slew percentage of the full signal swing. Assuming a signal swing of 500 mV to 1 V, a slew percentage of 70% and a 12 bit ADC for a conventional CIS device, this would yield $N\tau$ of 7–9 for conventional CIS devices. For photon counting devices the Vmax is only a few electrons (e.g., 1–20 e⁻). In this case $N\tau$ will be 0.7 to 3.7.

The t_{CDS} is the time between the falling edge of the SHR signal to the falling edge of the SHS signal as shown in Figures 6 and 7. With TG compensation timing and attendant reduced settling time, the t_{CDS} can be reduced from the baseline time of 750 ns. In addition to t_{CDS}, the SF load current is also programmable. The SF load current (Ibias) was adjusted to change the τD of the readout. In conventional CIS, reduced BW can preclude readout of a full signal swing, but can be used with a photon counting CIS as previously discussed. CDS times of 750, 250, 100 and 50 ns were implemented with SF Ibias values of 8.0, 0.8 and 0.4 µA.

100 dark frames were captured for each operating condition, at room temperature with an integration time of 16.5 µs. The frame rate was 56 frames-per-second (frame time of 17.8 ms). Total

sensor temporal noise was measured at the ADC output in the dark as a function of t_{CDS} and SF Ibias with TG compensation timing implemented. Since the total sensor read noise was measured at the ADC output, this included the SF, PGA and ADC read noise. Noise measurements were then made at the ADC output by overlapping the SHR and SHS pulses during readout to determine the read noise of only the PGA and ADC. This is referred to as base noise in the rest of the paper. The SF noise was then calculated by an rms subtraction of the base noise from the total noise (Equation (4) below):

$$\sigma_{sf} = sqrt(\sigma_{tot}^2 - \sigma_{base}^2) \tag{4}$$

3. Read Noise Results

3.1. Read Noise Histograms and Average Read Noise

Half of one of the imaging arrays was used since this contained the baseline 4T pixel. The data from one bank of ADCs was used to avoid any differences in noise related to layout, routing or timing skew details of the two banks. A histogram of total read noise *vs.* t_{CDS} is shown in Figure 10. The Ibias value shown in the legend of the graph is the master Ibias current. The source follower load current is supplied through a current mirror with a reduction ratio of 12.5 (*i.e.*, 100 μA master current is 8 μA source follower load current). The baseline t_{CDS} and SF Ibias were 750 ns and 8 μA, respectively.

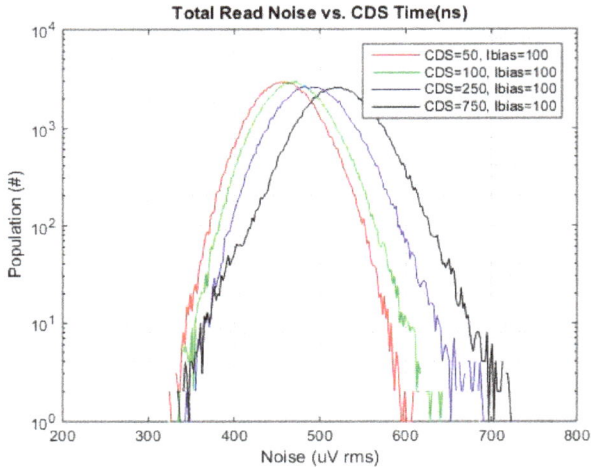

Figure 10. Total read noise for each t_{CDS} and SF Ibias of 8 μA (100 μA master current).

A histogram of base read noise *vs.* t_{CDS} and SF Ibias is shown in Figure 11. As expected t_{CDS} and SF Ibias do not have an effect on the PGA + ADC read noise, and base read noise distribution is Gaussian.

Referring to Figure 10, at the baseline condition of 750 ns, a tail in the histogram is clearly evident. This tail is due to the pixel source follower given this tail is not evident in the base read noise histogram. Such a tail is typical and is attributed to pixels with higher 1/f and RTS noise [14]. As the CDS time is reduced, the tail of the distribution is also reduced. This general trend is expected since the reduced CDS time will reject low frequency 1/f noise, [13,15]. The specific results that are obtained are dependent on the specific thermal noise and 1/f noise magnitude, and specific 1/f noise characteristics of the sensor, [13]. This will be foundry and process specific.

Figure 12a–d are total noise histograms for t_{CDS} of 750, 250, 100 and 50 ns each with SF Ibias of 8 μA, 0.8 μA and 0.4 μA. For each t_{CDS}, the total noise is reduced as SF Ibias is reduced from 8 μA to 0.8 μA. There is not much of a change as the SF Ibias is reduced from 0.8 μA to 0.4 μA.

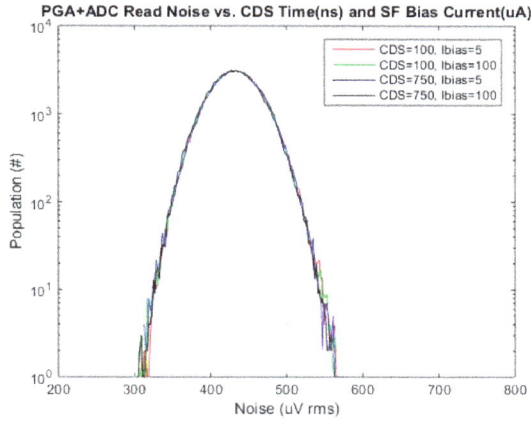

Figure 11. Base read noise histogram for selected t_{CDS} and SF Ibias.

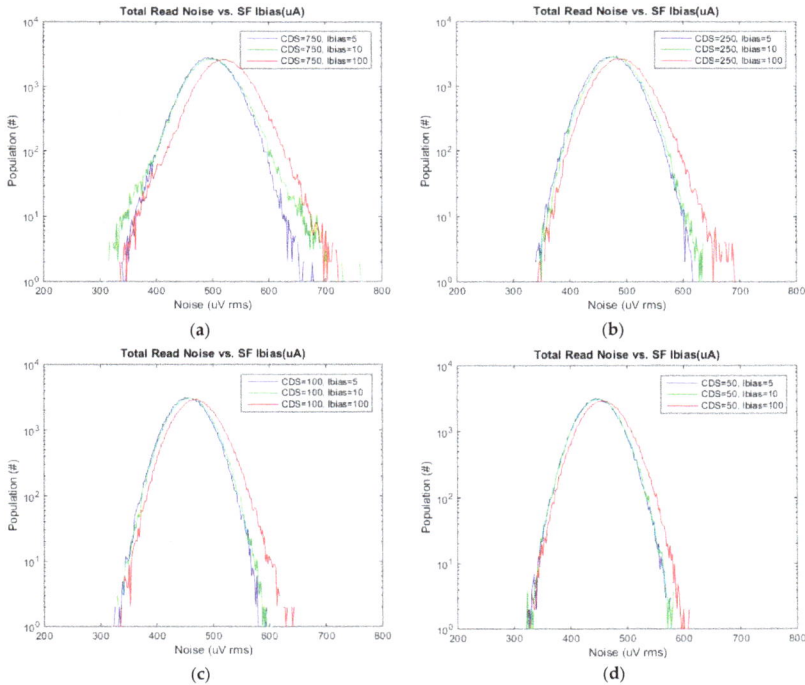

Figure 12. Total read noise histograms for each CDS time *vs.* SF Ibias. t_{CDS}: (**a**) 750 ns, (**b**) 250 ns, (**c**) 100 ns and (**d**) 50 ns.

Figure 13 is a histogram of total noise for selected t_{CDS} and SF Ibias. Based on measurements and circuit simulations, 100 ns t_{CDS} and SF Ibias current of 0.8 μA was selected as a practical minimum operating condition to be able to handle a signal swing of 2 e⁻ (simulated to be 95 ns).

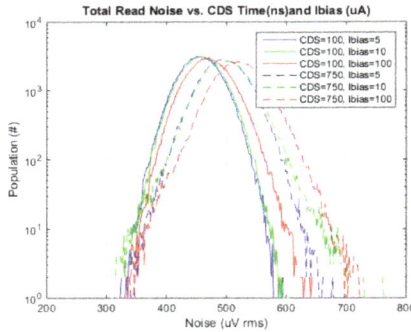

Figure 13. Total read noise histogram for selected CDS times and SF Ibias.

A summary of the average SF read noise vs bias condition is shown in Table 2 below.

Table 2. Measured average SF read noise (μVrms) for selected t_{CDS} (ns) and SF Ibias (μA).

CDS Time (ns)	Ibias (μA)		
	8	0.8	0.4
750	246	217	205
250	189	160	149
100	105	79	78

This SF read noise was calculated by an rms subtraction of the average base noise from the average total read noise. The average read SF read noise is reduced by a factor of 3.1 (from 246 μVrms to 79 μVrms; or 3.2 e⁻rms to 1.0 e⁻rms), for the baseline condition of 750 ns and 8 μA compared to 100 ns and 0.8 μA.

These results are compared to the expected reduction in 1/f and thermal noise based on reduction of τD and t_{CDS}, [15], and based on the baseline thermal and 1/f noise components provided by the process design kit (PDK) for our test sensor. The expected results are shown Table 3 below.

Table 3. Expected average SF read noise (μVrms) as a function of t_{CDS} (ns) and SF Ibias (μA).

CDS Time (ns)	Ibias (μA)		
	8	0.8	0.4
750	246	191	179
250	191	148	130
100	157	98	79

There is reasonable agreement with most of operating conditions, and very good agreement with the noise reduction factor observed from the 100 ns and 0.40 μA compared to the baseline value. An exact agreement would not be likely since the analysis in [15] assumes all 1/f noise has the same slope, and not all of the pixels in the tail of the histogram of our sensor are known to have, nor likely will have identical 1/f noise behavior [13]. A transient noise simulation was also performed for the sensor with the standard noise models provided with the PDK of the 65 nm CIS process. These simulation results predicted SF read noise of 200 µV$_{rms}$ for 750 ns, 8 µA operating point and 85 µV$_{rms}$ for 100 ns, 0.8 µA operating point. This is also in reasonable agreement with the observed results.

3.2. Investigation of Individual Pixels vs. CDS Time and SF Ibias

Several pixels were selected from various points on the read noise histogram, the mode, the tail and selected points in between. Plots of pixel value *vs.* frame #, and histograms of pixel values for the 100 frames, are provided for each of these selected pixels. These are shown in Figures 14–17 below.

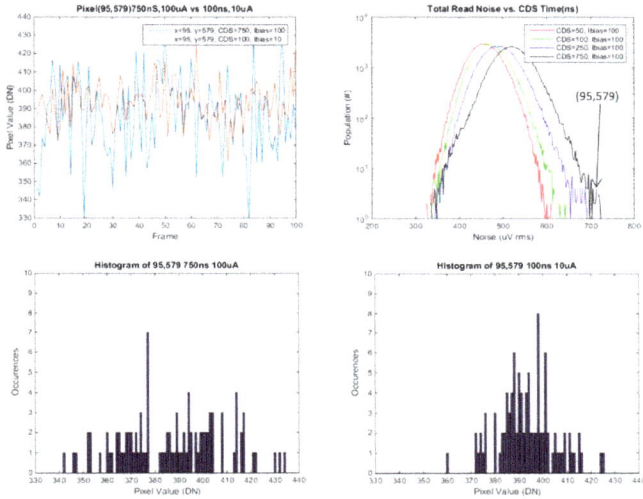

Figure 14. Pixel value *vs.* frame, and histogram of pixel values for pixel 95,579 (from the tail of the noise histogram).

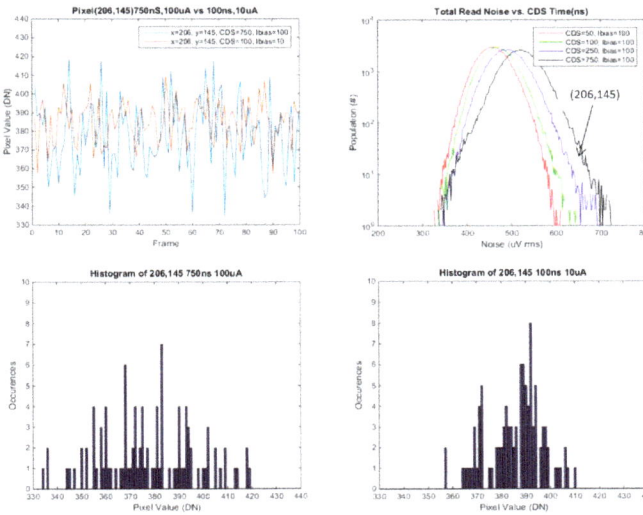

Figure 15. Pixel value *vs.* frame, and histogram of pixel values for pixel 206,145 (from the shoulder of the noise histogram).

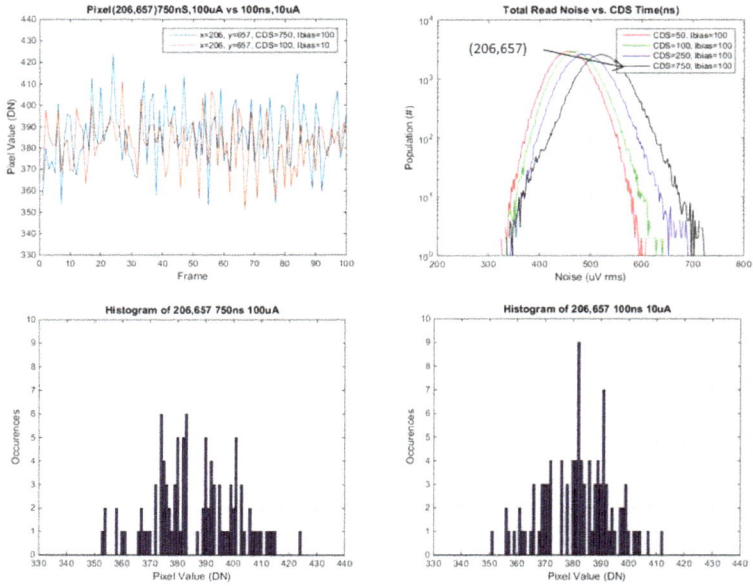

Figure 16. Pixel value *vs.* frame, and histogram of pixel values for pixel 206,765 (from the mode of the noise histogram).

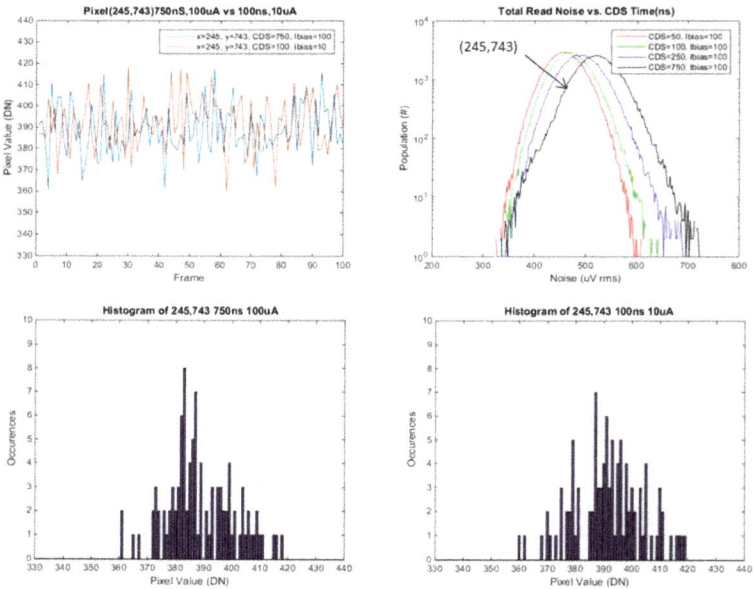

Figure 17. Pixel value *vs.* frame, and histogram of pixel values for pixel 245,743 (from below the mode of the noise histogram).

In general the histograms appear not to be multi-modal which would be indicative of RTS pixels [16], although 100 samples may not be enough in order to see this behavior. It is evident that from comparing histograms of higher noise pixels at 750 ns t_{CDS} and 8 μA Ibias *vs.* 100 ns t_{CDS} and 0.8 μA Ibias, that the histogram is a significantly tighter distribution for 100 ns CDS time and 0.8 μA Ibias. For lower noise pixels near to or less than the mean of the baseline distribution, there is little or no change in the histograms of the two operating conditions. The low noise histograms may likely appear to be more Gaussian if more samples were taken.

A summary of the ratio of total noise reduction is provided in Table 4 below. Since only 100 samples were used, only general trends can be observed. For a high noise pixel from the tail of the baseline distribution, there is close to a factor of 2 reduction in the total read noise. For pixels near or below the mean there is little or no change in the ratio of the total read noise.

Table 4. Normalized total noise for selected pixels *vs.* t_{CDS} (ns) and SF Ibias (μA).

Pixel (location in histogram)	CDS time (ns), SF Ibias (μA)			
	750, 8.0	250, 8.0	100, 8.0	100, 0.8
95, 579 (tail)	1	0.82	0.67	0.52
206, 145 (shoulder)	1	0.74	0.74	0.57
206, 657 (mean)	1	1.02	0.85	0.82
245, 743 (< mean)	1	1.07	1.07	1.02

In order to examine this further, the base noise for each pixel was averaged over the 400 frames captured for the baseline noise measurement (100 frames each for t_{CDS} of 750 ns and 100 ns and Ibias of 8 μA and 0.4 μA). An rms subtraction of the average base noise from the average total noise for each selected pixel was done to determine the SF read noise for each of the selected pixels. This result is shown in Table 5.

Table 5. SF read noise for selected pixels *vs.* t_{CDS} (ns) and SF Ibias (μA).

Pixel (locationin histogram)	SF noise (μV) rms
95, 579 (tail)	187
206, 145 (shoulder)	151
206, 657 (mean)	88
245, 743 (< mean)	62

The results from Table 3 are now as expected given pixels in the tail and shoulder of the histogram have high SF read noise and in general will therefore have higher 1/f noise, and will be impacted more by reduced t_{CDS} and SF Ibias [13,15]. In contrast pixels 206,657 and 245,743 have low SF read noise, and are likely dominated by thermal noise, and as a result will not change much with reduced t_{CDS} and SF Ibias, and may increase slightly due to the reduced SF g_m at lower SF Ibias.

4. Discussion

The maximum voltage swing for readout is much lower for a photon counting CIS device than that of a conventional CIS device. As a result the CDS time and dominant time constant of the SF readout can be reduced significantly. We have experimentally shown that reduced CDS time and dominant time constant of the SF readout can provide significant read noise reduction. The ratio of noise reduction will depend on the baseline characteristics of the CIS device. We achieved a factor of three reduction in the average SF read noise for the device used in this study (246 μV$_{rms}$ to 79 μV$_{rms}$). This was a reduction in input referred SF read noise from 3.2 e$^-$ to 1.0 e$^-$ (CG of 75.6 μV/e$^-$). This was is reasonable agreement with the transient noise simulation results completed using the PDK and noise models provided by the foundry, 200 μV$_{rms}$ to 85 μV$_{rms}$.

Sensors **2016**, *16*, 517

The conversion gain for the sensor in this study was not optimized, and in general the 1×4 shared pixel architecture will have a lower conversion gain than an unshared or 2×2 shared pixel. For unshared or 2×2 shared pixel architectures, conversion gains in the range of 100 $\mu V/e^-$ to 300 $\mu V/e^-$ have been reported, [8,11,17]. For a device with similar SF device characteristics, but a higher conversion gain (e.g., 200 $\mu V/e^-$), the reduced t_{CDS} and SF Ibias would provide a SF input referred read noise of 0.38 e^-. In addition the SF noise for this sensor (200 μV_{rms} to 250 μV_{rms}), was high by state of the art standards (<100 μV_{rms}).

Further work is planned to look at noise characteristics of individual pixels and the subsequent effects of reduced t_{CDS} and SF BW in more detail. For our test sensor, this will require many more frames (1000–1500) in order to have sufficient statistical accuracy in the rms subtraction of total read noise from base read noise.

Acknowledgments: The authors would like to acknowledge Eric Fossum for valuable discussions and the opportunity to publish this work, Daniel VanBlerkom for valuable discussions and assistance in circuit and noise simulations, Frank Armstrong and Craig Smith for image capture collection and assistance with Matlab code.

Author Contributions: Michael Guidash co-conceived the TG feed-through compensation method, co-designed the specific methods, co-conceived and designed the experiments, analyzed the data and wrote the paper. Jiaju Ma co-designed the timing method, co-designed the experiments, measured and analyzed the data. Thomas Vogelsang co-conceived the TG feed-through compensation method, co-conceived and designed the experiments, and analyzed the data. Jay Endsley co-conceived the TG feed-through compensation method, co-conceived and designed the experiments.

Conflicts of Interest: The authors declare no conflict of interest.

References

1. Fossum, E.R. What to do with sub-diffraction limit (SDL) pixels?—A proposal for gigapixel digital film sensor (DFS). In Proceedings of the IEEE Workshop on CCDs and Advanced Image Sensors, Nagano, Japan, 8 June 2005; pp. 214–217.
2. Vogelsang, T.; Guidash, M.; Xue, S. Overcoming the Full Well Capacity Limit: High Dynamic Range Imaging Using Multi-Bit Temporal Oversampling and Conditional Reset. In Proceedings of the International Image Sensor Workshop (IISW), Snowbird, UT, USA, 12 June 2013.
3. Fossum, E.R. Modeling the performance of single-bit and multi-bit quanta image sensors. *IEEE J. Electron Devices Soc.* **2013**, *1*, 166–174. [CrossRef]
4. Dutton, N.A.W.; Parmesan, L.; Holmes, A.J.; Grant, L.A.; Henderson, R.K. 320 × 240 oversampled digital single photon counting image sensor. In Proceedings of the 2014 Symposium on VLSI Circuits Digest of Technical Papers, Honolulu, HI, USA, 10–13 June 2014; pp. 1–2.
5. Dutton, N.; Parmesan, L.; Gnecchi, S.; Henderson, R.K.; Calder, N.J.; Rae, B.R.; Grant, L.A.; Henderson, R.K. Oversampled ITOF imaging techniques using SPAD-based quanta image sensors. In Proceedings of the International Image Sensor Workshop (IISW), Vaals, The Netherlands, 8–11 June 2015; pp. 170–173.
6. Lotto, C.; Seitz, P.; Baechler, T. A Sub-Electron Readout Noise CMOS Image Sensor with Pixel-Level Open-Loop Voltage Amplification. In Proceedings of the 2011 IEEE International Solid-State Circuits Conference Digest of Technical Papers (ISSCC), San Francisco, CA, USA, 20–24 February 2011; pp. 402–403.
7. Wahashima, S.; Kusuhara, F.; Kuroda, R.; Shigetoshi, S. A Linear Response Single Exposure CMOS Image Sensor with 0.5 e^- Readout Noise and 76 ke^- Full Well Capacity. In Proceedings of the Symposium on VLSI Circuits (VLSI Circuits), Kyoto, Japan, 17–19 June 2015; pp. C88–C89.
8. Ma, J.; Fossum, E. Quanta Image Sensor Jot with Sub 0.3 e^- r.m.s. Read Noise and Photon Counting Capability. *IEEE Electron Device Lett.* **2015**, *36*, 926–928. [CrossRef]
9. Chen, Y.; Xu, Y.; Chae, Y.; Mierop, A.; Wanf, X.; Theuwissen, A. A 0.7 e^-_{rms} Temporal-Readout-Noise CMOS Image Sensor for Low-Light-Level Imaging. In Proceedings of the IEEE International Solid-State Circuits Conference Digest of Technical Papers (ISSCC), San Francisco, CA, USA, 19–23 February 2012; pp. 385–386.
10. Yeh, S.; Chou, K.; Tu, H.; Chao, C.; Hsueh, F. A 0.66 e^-_{rms} Temporal-Readout-Noise 3D-Stacked CMOS Image Sensor with Conditional Correlated Multiple Sampling (CCMS) Technique. In Proceedings of the Symposium on VLSI Circuits (VLSI Circuits), Kyoto, Japan, 17–19 June 2015; pp. C84–C85.

11. Boukhayma, A.; Peizerat, A.; Enz, C. A 0.4 e$^-$$_{rms}$ Temporal Readout Noise 7.5 µm Pitch and a 66% Fill Factor Pixel for Low Light CMOS Image Sensors. In Proceedings of the International Image Sensor Workshop (IISW), Vaals, The Netherlands, 8–11 June 2015; pp. 365–368.

12. Yao, Q.; Dierickx, B.; Dupont, B.; Ruttens, G. CMOS image sensor reaching 0.34 e$^-$$_{rms}$ read noise by inversion-accumulation cycling. In Proceedings of the International Image Sensor Workshop (IISW), Vaals, The Netherlands, 8–11 June 2015; pp. 369–372.

13. Janesick, J.; Elliott, T.; Andrews, J.; Tower, J. Fundamental Performance Differences of CMOS and CCD imagers: Part VI. In Proceedings of the SPIE Optics and Photonics, San Diego, CA, USA, 9–13 August 2015.

14. Martin-Gonthier, P.; Magnan, P. RTS Noise Impact in CMOS Image Sensors Readout Circuit. In Proceedings of the 16th IEEE ICECS 2009, Hammamet, Tunisia, 13–16 December 2009.

15. Kansy, R. Response of a Correlated Double Sampling Circuit to 1/f Noise. *IEEE J. Solid-State Circuits* **1980**, *15*, 373–375. [CrossRef]

16. Wang, X.; Rao, P.; Mierop, A.; Theuwissen, A. Random Telegraph Signal in CMOS Image Sensor Pixels. In Proceedings of the Electron Devices Meeting, IEDM'06, San Francisco, CA, USA, 11–13 December 2006.

17. Kusuhara, F.; Wakashima, S.; Nasuno, S.; Kuroda, R.; Sugawa, S. Analysis and Reduction of Floating Diffusion Capacitance Components of CMOS Image Sensor for Photon-Countable Sensitivity. In Proceedings of the International Image Sensor Workshop (IISW), Vaals, Netherlands, 8–11 June 2015; pp. 120–123.

sensors

MDPI

Article

Analysis of Subthreshold Current Reset Noise in Image Sensors

Nobukazu Teranishi [1,2]

[1]   Research Institute of Electronics, Shizuoka University; 3-5-1 Johoku, Naka-ku, Hamamatsu 432-8011, Japan;
      teranishi@idl.rie.shizuoka.ac.jp; Tel.: +81-53-478-1313
[2]   Laboratory of Advanced Science and Technology for Industry, University of Hyogo; 1-1-2 Koto, Kamigori,
      Ako-gun, Hyogo 678-1205, Japan

Academic Editor: Eric R. Fossum
Received: 26 January 2016; Accepted: 4 May 2016; Published: 10 May 2016

Abstract: To discuss the reset noise generated by slow subthreshold currents in image sensors, intuitive and simple analytical forms are derived, in spite of the subthreshold current nonlinearity. These solutions characterize the time evolution of the reset noise during the reset operation. With soft reset, the reset noise tends to $\sqrt{mkT/2C_{PD}}$ when $t \to \infty$, in full agreement with previously published results. In this equation, C_{PD} is the photodiode (PD) capacitance and m is a constant. The noise has an asymptotic time dependence of t^{-1}, even though the asymptotic time dependence of the average (deterministic) PD voltage is as slow as $\log t$. The flush reset method is effective because the hard reset part eliminates image lag, and the soft reset part reduces the noise to soft reset level. The feedback reset with reverse taper control method shows both a fast convergence and a good reset noise reduction. When the feedback amplifier gain, A, is larger, even small value of capacitance, C_P, between the input and output of the feedback amplifier will drastically decrease the reset noise. If the feedback is sufficiently fast, the reset noise limit when $t \to \infty$, becomes $\frac{mkT(C_{PD}+C_{P1})^2}{2q^2 A(C_{PD}+(1+A)C_P)}$ in terms of the number of electron in the PD. According to this simple model, if $C_{PD} = 10$ fF, $C_P/C_{PD} = 0.01$, and $A = 2700$ are assumed, deep sub-electron rms reset noise is possible.

Keywords: CMOS image sensor; 3-transistor scheme; reset noise; subthreshold current; hard reset; soft reset; feedback reset; tapered reset

1. Introduction

Four-transistor (4-Tr) complementary metal-oxide-semiconductor (CMOS) image sensors [1] are widely used in various applications, such as mobile phone cameras, digital still cameras, security, industrial, medical equipment, *etc.* They have significant advantages compared with three-transistor (3-Tr) CMOS image sensors. Firstly, the 4-Tr scheme can use pinned photodiodes (PPDs) [2–6] to reduce the dark current. Secondly, the complete charge transfer by the PPD [2] realizes "first reset, later signal" and correlated double sampling (CDS) [7], which eliminates both the reset noise at the floating diffusion node and the low frequency noise at the source follower amplifier. Thirdly, the capacitance of the floating diffusion can be decreased by fine processing technology and a large conversion gain can be obtained, which increases the signal-to-noise ratio. Fourthly, the shared transistor technology [8,9] reduces the number of transistors per pixel. The minimum reported transistor number per pixel is 1.375 transistors/pixel [10], which is much smaller than that of 3-Tr scheme.

The 3-Tr scheme is now being used for large pixel CMOS image sensors. One example is its use in medical X-ray image sensors. The typical pixel size is around 100 μm. There are several reasons why the 3-Tr scheme is being used. The first one is that it is difficult to achieve a complete charge transfer of the PPD with such large pixels or PDs. Another one lies in the fact that X-ray image sensors usually suffer from photon shot noise and the readout noise of the 3-Tr scheme is acceptable. A third reason is that the fabrication process for the 3-Tr scheme is simpler than that of the 4-Tr scheme. A fourth reason is the fact that the 3-Tr scheme can be operated in non-destructive readout mode, and can realize dose sensing during radiation or auto exposure control (AEC) using fast-frame-rate skip mode [11]. Finally, a fifth reason is that the 3-Tr scheme can reach a higher number of saturation electrons at the PD than that of the 4-Tr scheme.

If the 3-Tr scheme were able to achieve low readout noise, it could be used in more applications, in particular, elevated image sensors, or photosensitive material hybrid image sensors, which cannot use the PPD complete charge transfer scheme. Some organic photoconductive films have larger absorption coefficients than that of silicon, and smaller photosensitive layer thicknesses can provide enough sensitivity. Crosstalk could then be reduced even for small pixel size, and elevated image sensors with organic photoconductive films would become candidates for small pixel image sensors [12,13]. Elevated image sensors can have sensitivities beyond the silicon sensitive wavelength range, well within the ultraviolet (UV) and infrared (IR) range. For example, crystal selenium (c-Se) has a 1.74 eV bandgap and is a good sensitive material for both UV and visible light [14]. Germanium (Ge) and indium-gallium-arsenide (InGaAs) have 0.8 eV and 0.36–1.43 eV direct bandgaps, respectively, and are good photosensitive materials for near IR [15,16]. These hybrid image sensor developments might be accelerated by recent advances in 3D and hybrid technology.

A 3-Tr pixel consists of an N-type PD, a reset transistor (RST) to reset PD, a source follower amplifier (SF) which picks up the PD voltage and sends the voltage signal to the column circuit, and a select transistor (SEL) which activates the selected row, as shown in Figure 1. The reset noise of the PD is the dominant noise source in the 3-Tr scheme. The original reset method is hard reset. Its noise variance is calculated as kTC [17,18], where k is the Boltzmann constant, T the absolute temperature, and C the detection capacitance. This noise is therefore called "kTC noise". Various other reset methods have been proposed to reduce the reset noise and will be discussed later soft reset [19,20], feedback reset [21,22], feedback reset with taper control [13,23–26]. Feedback reset has realized a reset noise level as small as 2.9 e^- · rms (electrons rms) [13]. While those approaches aim to reduce the reset noise itself, other approaches to the problem have been attempted; one of them is to reduce effective detection capacitance, thus increasing signal voltage. For this purpose, a charge sensitive amplifier or capacitive transimpedance amplifier is introduced [27,28]. Another approach is to introduce in-pixel CDS [29].

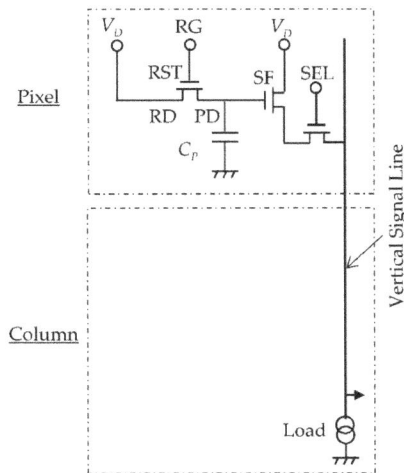

Figure 1. Pixel and column schematic for 3-Tr scheme CMOS image sensor.

In this paper, we will discuss the reset noise reduction itself. A fundamental time-domain analysis of various reset methods is presented, and the reset noise is studied in detail. In the next section, our reset noise analysis technique is introduced. In Sections 3–6 the hard reset, soft reset, tapered reset, and feedback reset with reverse taper control methods will be analyzed. In Section 7, the possibility of photon counting by the 3-Tr scheme is discussed.

2. Reset Noise Analysis Technique

To discuss the various reset methods, a reset noise analysis technique must be prepared, preferably one capable of providing an intuitive and simple analytical solution without numerical or Monte Carlo simulations in spite of the nonlinearity of the subthreshold current. The subthreshold current causes a slow reset operation, therefore, the time dependence of the reset noise during the reset operation period needs to be evaluated, from the initial condition to the final state.

A frequency domain analysis has previously been published, where the estimated reset noise was compared with measurement result [26]. The steady-state noise (final stage noise) was calculated using a resistor instead of the reset transistor. A time domain analysis was proposed, using effectively-second-order differential equation [25]. To derive a closed form expression, a fixed resistance was also used instead for the reset transistor. Another time domain method was proposed for soft reset analysis, directly treating the subthreshold current nonlinearity and assuming the existence of a shot noise in the subthreshold current [20]; it obtained a soft reset noise of $kT/2C$, which agrees well with the measurements. However, it would be desirable that improved reset methods such as feedback reset with taper control could also be analytically treated.

In the rest of this section, our reset noise analysis is introduced. The PD node voltage $V_{PD}(t)$ is decomposed into a deterministic (or average) part $V_{PDa}(t)$ and a stochastic (or noise) part $v_{PD}(t)$. Naturally:

$$V_{PD}(t) = V_{PDa}(t) + v_{PD}(t) \tag{1}$$

To derive the analytical form of the reset noise variance $< v_{PD}(t)^2 >$ three steps are needed in this analysis:

Step 1: The equation for the average part $V_{PDa}(t)$ is derived, and the solution is obtained.
Step 2: The equation for the noise part $v_{PD}(t)$ is derived, and $v_{PD}(t)$ is obtained explicitly.
Step 3: The variance $< v_{PD}(t)^2 >$ is calculated.

This approach is straightforward and logically simple. In the following sections, this analysis is applied to hard reset, soft reset, tapered reset and feedback reset with taper control.

3. Hard Reset

Hard reset is originally applied in the reset of the floating diffusion of CCD (Charge coupled device), and is the original reset method of 3-Tr CMOS image sensor. Its timing diagram is shown in Figure 2a. The hard reset noise variance was derived as kTC using frequency domain analysis [17] and time domain analysis [18]. The same result will be derived here.

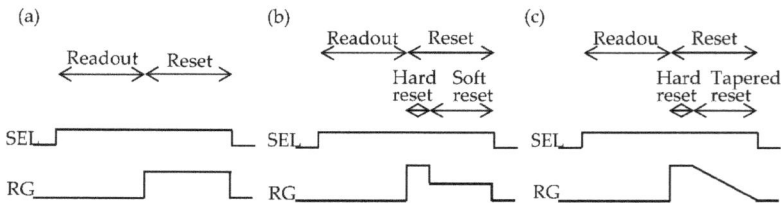

Figure 2. Timing diagram for one pixel. (**a**) Hard reset and soft reset; (**b**) Flush reset; and (**c**) Tapered reset.

With hard reset, the RST channel can be regarded as a pure resistance R, because the RST operates in the linear region. Resistances generate Johnson noise or thermal noise, from which the reset noise arises. To simplify the model, it is also assumed that there is no dark current or no incident light during the reset phase. This assumption is also used in Sections 4–6. The continuity equation is:

$$C_{PD} \frac{dV_{PD}(t)}{dt} = \frac{V_{RD0} - V_{PD}(t)}{R} + i_n(t) \tag{2}$$

where C_{PD} is the PD capacitance, V_{RD0} is the reset transistor drain (RD) voltage, and $i_n(t)$ is the thermal noise associated with resistance R, whose autocorrelation is:

$$< i_n(t_1) i_n(t_2) > = \frac{2kT}{R} \delta(t_1 - t_2) \tag{3}$$

When applied at Step 1, the equation of continuity becomes:

$$C_{PD} \frac{dV_{PDa}(t)}{dt} = \frac{V_{RD0} - V_{PDa}(t)}{R} \tag{4}$$

The solution is obtained as:

$$V_{PDa}(t) = V_{PDa}(0) e^{-\frac{t}{\tau_{HR}}} + V_{RD0} \left(1 - e^{-\frac{t}{\tau_{HR}}} \right) \tag{5}$$

where time constant, τ_{HR} is given by:

$$\tau_{HR} \equiv C_{PD} R \tag{6}$$

If the parameters of a typical RST are assumed, with a 0.4 μm channel width, 0.55 μm channel length, 6 nm thick gate oxide, $V_{GS} = 3.3$ V, and $V_{RD0} = 3.3$ V, we will have R ≈ 10 kΩ. For $C_{PD} = 10$ fF, τ_{HR} becomes 100 ps, which is much smaller than the typical reset period, 1 μs. When t → ∞, $V_{PDa}(t)$ converges to V_{RD0}.

For Step 2, we substitute Equation (5) into Equation (2), to obtain the equation for $v_{PD}(t)$ as:

$$C_{PD} \frac{dv_{PD}(t)}{dt} = \frac{v_{PD}(t)}{R} + i_n(t) \tag{7}$$

The solution of this equation is:

$$v_{PD}(t) = \frac{1}{C_{PD}} \int_0^t dt' e^{\frac{t'-t}{\tau_{HR}}} i_n(t') + e^{-\frac{t}{\tau_{HR}}} v_{PD}(0) \tag{8}$$

Finally, for Step 3, we square Equation (8) to obtain:

$$v_{PD}(t)^2 = \frac{1}{C_{PD}{}^2} \int_0^t \int_0^t dt_1 dt_2 e^{\frac{t_1-t}{\tau_{HR}} + \frac{t_2-t}{\tau_{HR}}} i_n(t_1) i_n(t_2) + \frac{2e^{-\frac{t}{\tau_{HR}}} v_{PD}(0)}{C_{PD}} \int_0^t dt_1 e^{\frac{t_1-t}{\tau_{HR}}} i_n(t_1) + e^{-\frac{2t}{\tau_{HR}}} v_{PD}{}^2(0) \tag{9}$$

Averaging Equation (9) and using Equation (3), the hard reset noise variance is obtained as:

$$< v_{PD}(t)^2 > = \frac{kT}{C_{PD}} \left(1 - e^{-\frac{2t}{\tau_{HR}}} \right) + < v_{PD}(0)^2 > e^{-\frac{2t}{\tau_{HR}}} \tag{10}$$

The first term is caused by thermal noise, and the second term comes from the initial condition. Because of the exponential decay, the reset noise variance $< v_{PD}(t)^2 >$ is sufficiently settled within the reset period. When $t \to \infty$:

$$< v_{PD}(\infty)^2 > = \frac{kT}{C_{PD}}. \tag{11}$$

The well-known *kTC* noise is therefore produced.

4. Soft Reset

The soft reset method was introduced to reduce the reset noise. Even though the timing diagram is the same as that of the hard reset, the RG (RST gate) on voltage is smaller. With the soft reset method, the RST is operated first in the saturation region, and then in the subthreshold region. The signal charge transfer from the PD to the RD in the saturation region is smooth and the period is as small as a few nanoseconds, it does not substantially contribute to the reset noise, when compared with the following subthreshold region period. The reset noise will therefore be calculated neglecting the saturation period, using only subthreshold region period; $t = 0$ in this analysis corresponds to the moment when RST enters this region.

The equation of continuity then becomes:

$$C_{PD}\frac{dV_{PD}(t)}{dt} = I_a(t) + i_n(t) = I_0 e^{-\beta V_{PD}(t)} + i_n(t) \tag{12}$$

where $I_a(t)$ is the average drain current, I_0 is a constant, $\beta \equiv q/mkT$, $m \equiv 1 + C_D/C_G$, C_D is the depletion-layer capacitance and C_G is the gate capacitance. Typically, m is slightly above 1. The subthreshold current has shot noise with autocorrelation:

$$< i_n(t_1)\, i_n(t_2) > = qI_a(t)\, \delta(t_1 - t_2) \tag{13}$$

For Step 1, the continuity equation for the average voltage $V_{PDa}(t)$ is given by:

$$C_{PD}\frac{dV_{PDa}(t)}{dt} = I_0 e^{-\beta V_{PDa}(t)} \tag{14}$$

Even though this equation is nonlinear, it has an analytical solution, which can be obtained with the variation of parameters method. The solution [2] is:

$$V_{PDa}(t) = \frac{1}{\beta}\log\left[e^{\beta V_{PDa}(0)} + \frac{t}{\tau}\right] \tag{15}$$

where:

$$\tau \equiv C_{PD}/\beta I_0 \tag{16}$$

The existence of this analytical solution is essential for the subthreshold current reset noise analyses. When $t \to \infty$, $V_{PDa}(t)$ diverges slowly as a logarithmic function. The soft reset has no finite limit, even though the hard reset has V_{RD0} as a limit. This is an important characteristic for the soft reset.

For Step 2, we substitute Equation (15) into Equation (12) and obtain the equation for $v_{PD}(t)$:

$$C_{PD}\frac{dv_{PD}(t)}{dt} = -I_0 e^{-\beta V_{PDa}(0)}\frac{1 - e^{-\beta v_{PD}(t)}}{1 + \frac{t}{\tau}e^{-\beta V_{PDa}(0)}} + i_n(t) \tag{17}$$

Considering that $\beta v_{PD}(t) \ll 1$, the approximation, $e^{-\beta v_{PD}(t)} \approx 1 - \beta v_{PD}(t)$ can be used. Equation (17) then becomes a linear equation:

$$C_{PD}\frac{dv_{PD}(t)}{dt} = -\frac{\beta I_0 e^{-\beta V_{PDa}(0)}}{1 + \frac{t}{\tau}e^{-\beta V_{PDa}(0)}}v_{PD}(t) + i_n(t) \tag{18}$$

Its solution is given by:

$$v_{PD}(t) = \frac{1}{C_{PD}}\int_0^t dt\prime\frac{1 + \frac{t\prime}{\tau}e^{-\beta V_{PDa}(0)}}{1 + \frac{t}{\tau}e^{-\beta V_{PDa}(0)}}i_n(t\prime) + \frac{1}{1 + \frac{t}{\tau}e^{-\beta V_{PDa}(0)}}v_{PD}(0) \tag{19}$$

For Step 3, squaring Equation (19), averaging and using Equation (13), the soft reset noise variance can be obtained:

$$< v_{PD}(t)^2 >= \frac{mkT}{2C_{PD}}\left(1 - \frac{1}{\left(1 + \frac{t}{\tau}e^{-\beta V_{PDa}(0)}\right)^2}\right) + < v_{PD}(0)^2 > \frac{1}{\left(1 + \frac{t}{\tau}e^{-\beta V_{PDa}(0)}\right)^2} \quad (20)$$

Using the fact that $I(0) = I_0 e^{-\beta V_{PDa}(0)}$, Equation (20) can be rewritten as:

$$< v_{PD}(t)^2 >= \frac{mkT}{2C_{PD}}\left(1 - \frac{1}{\left(1 + \frac{qI_a(0)t}{mkTC_{PD}}\right)^2}\right) + < v_{PD}(0)^2 > \frac{1}{\left(1 + \frac{qI_a(0)t}{mkTC_{PD}}\right)^2} \quad (21)$$

The first term is caused by shot noise, and the second term results from the initial condition. When $t \rightarrow \infty$, the asymptotic form and the limit are obtained as:

$$< v_{PD}(t)^2 > \approx \frac{mkT}{2C_{PD}}\left(1 - \left(\frac{mkTC_{PD}}{qI_a(0)t}\right)^2\right) + < v_{PD}(0)^2 > \left(\frac{mkTC_{PD}}{qI_a(0)t}\right)^2 \quad (22)$$

$$< v_{PD}(t)^2 > \rightarrow \frac{mkT}{2C_{PD}} \quad (23)$$

It should be noted that the asymptotic time dependence of the noise standard deviation $\sqrt{< v_{PD}(t)^2 >}$ behaves as t^{-1} although the asymptotic time dependence of the average PD voltage, $V_{PDa}(t)$ behaves as $\log t$ (as shown in Equation (15)), which is much slower than t^{-1}. In the hard reset case, $V_{PDa}(t)$ and $\sqrt{< v_{PD}(t)^2 >}$ have the same exponential time dependence (with $e^{-t/\tau_{HR}}$). The determinant time constant in Equations (21) and (22), $\tau_{SR} \equiv mkTC_{PD}/qI(0)$, is calculated for a typical case, as follows. Assuming that $C_{PD} = 10$ fF, $I_a(0) = 0.5$ µA, $v_{th} \equiv kT/q = 26$ mV (at 300 K), $m = 1$, we have that $\tau_{SR} = 0.52$ ns. It is small enough when compared with the typical reset period, 1 µs. The limit at $t \rightarrow \infty$ is $mkT/2C_{PD}$, which fits the results obtained in previous works [19,20,30–33].

To alleviate the image lag problem of soft reset image sensors [2], the flushed reset method was proposed [20,22,33]. In this method, during one reset period, a hard reset is first carried out to eliminate vestige of the previous signal, and a soft reset is then performed to reduce the reset noise. The timing chart for a simple case of the flushed reset method is shown in Figure 2b. The reset noise variance after the hard reset is kT/C_{PD}, as given as Equation (11), and this becomes the initial condition for the soft rest period. Substituting $< v_{PD}(0)^2 >= kT/C_{PD}$ into Equation (21), the flushed reset noise can be derived as:

$$< v_{PD}(t)^2 >= \frac{mkT}{2C_{PD}}\left(1 + \left(\frac{2}{m} - 1\right)\frac{1}{\left(1 + \frac{qI(0)t}{mkTC_{PD}}\right)^2}\right) \quad (24)$$

If the reset period is enough long, $< v_{PD}(t)^2 >$ becomes:

$$< v_{PD}(t)^2 > \rightarrow \frac{mkT}{2C_{PD}} \quad (25)$$

The hard reset part eliminates image lag, and the soft reset part reduces the reset noise to the soft reset level.

5. Tapered Reset

To improve the convergence at the soft reset method, tapered reset is proposed. In this method, the RST gate voltage is gradually decreased to 0 V during the soft reset period, as shown in Figure 2c. The continuity equation becomes:

$$C_{PD}\frac{dV_{PD}(t)}{dt} = I_a(t) + i_n(t) = I_0 e^{-\beta V_{PD}(t) - \beta a t} + i_n(t) \tag{26}$$

where a is a positive constant characterizing the slope of the RST taper, in unit of V/s.

For Step 1, the continuity equation for average voltage $V_{PDa}(t)$ is:

$$C_{PD}\frac{dV_{PDa}(t)}{dt} = I_0 e^{-\beta V_{PDa}(t) - \beta a t} \tag{27}$$

Its solution is:

$$V_{PDa}(t) = \frac{1}{\beta}\log\left[e^{\beta V_{PDa}(0)} + \frac{I_0}{C_{PD}a}\left(1 - e^{-\beta a t}\right)\right] \tag{28}$$

When $t \to \infty$, $V_{PDa}(t)$ converges to:

$$V_{PDa}(\infty) = \frac{1}{\beta}\log\left[e^{\beta V_{PDa}(0)} + \frac{I_0}{C_{PD}a}\right] \tag{29}$$

As seen, while $V_{PDa}(t)$ for the soft reset diverges slowly as logarithmic function, that of the tapered reset converges exponentially to a constant; this happens because the drain current is extinguished as a consequent of the taper control.

For Step 2, substituting Equation (28) into Equation (26), the equation for $v_{PD}(t)$ can be obtained as:

$$C_{PD}\frac{dv_{PD}(t)}{dt} = -I_0 e^{-\beta V_{PDa}(0) - \beta a t}\frac{1 - e^{-\beta v_{PD}(t)}}{1 + \frac{I_0 e^{-\beta V_{PDa}(0)}}{C_{PD}a}\left(1 - e^{-\beta a t}\right)} + i_n(t) \tag{30}$$

Considering that $\beta v_{PD}(t) \ll 1$, the approximation $e^{-\beta v_{PD}(t)} \approx 1 - \beta v_{PD}(t)$ can be used. Equation (30) then becomes a linear equation:

$$C_{PD}\frac{dv_{PD}(t)}{dt} = -\frac{\beta I_0 e^{-\beta V_{PDa}(0) - \beta a t}}{1 + \frac{I_0 e^{-\beta V_{PDa}(0)}}{C_{PD}a}\left(1 - e^{-\beta a t}\right)}v_{PD}(t) + i_n(t) \tag{31}$$

Its solution is written as:

$$v_{PD}(t) = \frac{1}{C_{PD}}\int_0^t dt\prime\frac{1 + \frac{I_0 e^{-\beta V_{PDa}(0)}}{C_{PD}a}\left(1 - e^{-\beta a t\prime}\right)}{1 + \frac{I_0 e^{-\beta V_{PDa}(0)}}{C_{PD}a}\left(1 - e^{-\beta a t}\right)}i_n(t\prime) + \frac{1}{1 + \frac{I_0 e^{-\beta V_{PDa}(0)}}{C_{PD}a}\left(1 - e^{-\beta a t}\right)}v_{PD}(0) \tag{32}$$

For Step 3, squaring Equation (32), averaging and using Equation (13), the tapered reset noise variance can be derived:

$$< v_{PD}(t)^2 >= \frac{mkT}{2C_{PD}}\left(1 - \frac{1}{\left(1 + \frac{I_a(0)}{C_{PD}a}\left(1 - e^{-\beta a t}\right)\right)^2}\right) + < v_{PD}(0)^2 > \frac{1}{\left(1 + \frac{I_a(0)}{C_{PD}a}\left(1 - e^{-\beta a t}\right)\right)^2} \tag{33}$$

If a is so small that $\beta a t \ll 1$, $e^{-\beta a t} \approx 1 - \beta a t$, Equation (33) then becomes identical to that of the soft reset case, Equation (22). On the other hand, if a is enough large, $e^{-\beta a t}$ decays so fast that the reset noise variance $< v_{PD}(t)^2 >$ cannot reach the soft reset level. Therefore, the tapered reset shown in Figure 2c is not useful for noise reduction, although the average voltage $V_{PDa}(t)$, converges exponentially.

6. Feedback Reset with Reverse Taper Control (FRRT)

The feedback reset method was also proposed as a mean to reduce the reset noise [13,20–26]. In this method, during the reset period, the noisy PD voltage is detected and the resulting negative feedback forces the PD voltage to approach the reference level. A bidirectional current is needed at the RST for effective feedback, even though the subthreshold current is essentially unidirectional [13,24]. The concept of an "unidirectional current" means, in this context, that if the feedback (relaxation) times for upper and lower fluctuations are very different because of the current nonlinearity, the current looks unidirectional from the feedback point of view. One solution to overcome this contradiction is to constantly inject electrons into the PD; these injected electrons can then effectively play the role of a current flowing in the opposite direction. There are a couple of methods to perform this injection; one is to slowly ramp the RST gate toward the on-direction or positive direction [23,24], in contrast with what is done in the tapered reset method discussed in Section 5. Another method is to ramp the RST source voltage toward the on-direction or negative direction, as will be explained in detail in this section. It should be noted that electrons flow from the RD to the PD in both cases, regardless of the name of "drain". Therefore, this method can be named feedback reset with reverse taper control (FRRT).

There is another important point to be considered when discussing reset noise reduction; excrescent noise should not be generated when the RST is turned off at the end of the reset period. If there are electrons at the RST channel just before it is turned off, these electrons are partitioned to the PD and the RD. This partitioning has a stochastic nature, and generates the partition noise [34]. Subthreshold operation tends to reduce this effect, because the electron number at the RST channel is smaller and unidirectional current is involved.

Figure 3a shows a schematic diagram of the pixel and the related column-based feedback circuits for the FRRT to be analyzed in this section. The feedback is applied through the RD. The pixel structure is the same as the conventional 3-Tr scheme, as shown in Figure 1. The exception is that the RD wiring is prepared separately from the SF drain line. Other feedback circuits and ramp circuits are column-based. The vertical signal line, transferring the SF output voltage, is connected to the negative input of a column-based differential amplifier together with the load transistor and the following signal circuits. The positive input is connected with a ramp generator, $V_{Ref} = a_0 - at$. The output of the amplifier is connected to the RD through the additional vertical RD line. Both the parasitic capacitance between PD and RD, C_{P1}, and the parasitic capacitance between the vertical signal line and the RD line, C_{P2}, are included in the analysis. It is assumed that C_{PD} does not include C_{P1}. The timing chart for this structure is shown in Figure 4. After one row is selected by SEL, signal is read out in a fashion similar to the one of the conventional 3-Tr scheme at first. During the reset period, a hard reset is carried out in front to eliminate vestiges of previous signals, by setting Flush to ON. Subsequently, FRRT is executed turning FB ON and gradually decreasing V_{Ref}.

Figure 3b shows the simplified schematic diagram for noise modeling; the SF is merged with the high-gain differential amplifier, of gain A. The amplifier is also assumed to be faster than the reset motion. The parasitic capacitances and parasitic resistances (which delay the feedback) for both the vertical signal line and the RD line are neglected because fast feedback is assumed. The parasitic capacitances C_{P1} and C_{P2} are included because they have an important role in the feedback connecting the amplifier's input and output. Only reset noise or RST channel noise is considered here; noises from the SF, the differential amplifier, SEL and wiring resistances are not included, because the reset noise is dominant.

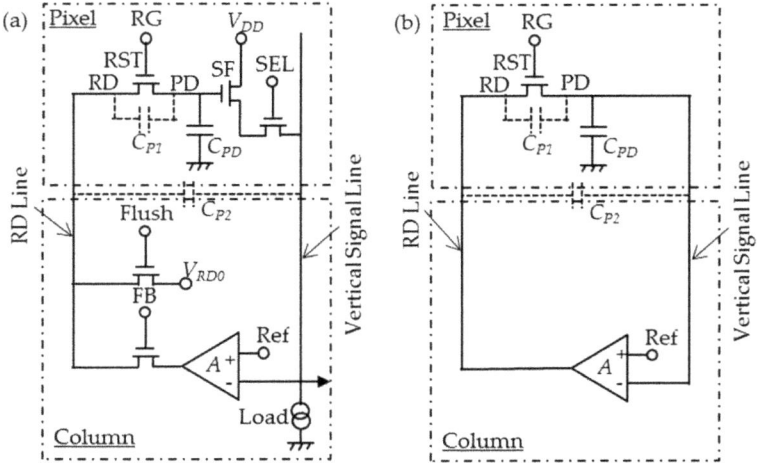

Figure 3. Feedback reset with reverse tapered control (FRRT); (a) Schematic; (b) Simplified model for noise analysis.

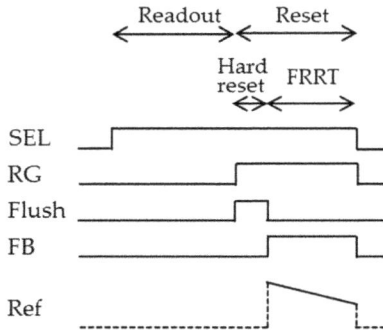

Figure 4. Timing diagram for one pixel in the case of feedback reset with reverse tapered control (FRRT).

The continuity equation becomes:

$$(C_{PD} + C_{P1} + C_{P2})\frac{dV_{PD}(t)}{dt} - (C_{P1} + C_{P2})\frac{dV_{RD}(t)}{dt} = -I_a(t) - i_n(t) = -I_0 e^{-\beta V_{RD}(t)} - i_n(t) \quad (34)$$

$$V_{RD}(t) = A\left(V_{Ref}(t) - V_{PD}(t)\right) = A(a_0 - at - V_{PD}(t)) \quad (35)$$

Substituting Equation (35) into Equation (34), the equation for $V_{PD}(t)$ is obtained as:

$$C_T \frac{dV_{PD}(t)}{dt} = -AC_P a - I_0 e^{-\beta A(a_0 - at - V_{PD}(t))} - i_n(t) \quad (36)$$

where $C_T \equiv C_{PD} + (1 + A)C_P$ and $C_P \equiv C_{P1} + C_{P2}$.

For Step 1, the continuity equation for average voltage $V_{PDa}(t)$, is:

$$C_T \frac{dV_{PDa}(t)}{dt} = -AC_P a - I_0 e^{-\beta A(a_0 - at - V_{PDa}(t))} \quad (37)$$

Equation (37) can be transformed to eliminate the constant term, as follows:

$$C_T \frac{d(V_{PDa}(t) + \frac{AC_{pa}}{C_T}t)}{dt} = -I_0 e^{-\beta A(a_0 - a\frac{C_{PD}+C_P}{C_T}t - (V_{PDa}(t) + \frac{AC_{pa}}{C_T}t))} \tag{38}$$

Its solution is:

$$V_{PDa}(t) = -\frac{AC_{pa}}{C_T}t - \frac{1}{\beta A}\log\left[e^{-\beta AV_{PDa}(0)} + \frac{I_0 e^{-\beta Aa_0}}{a(C_{PD}+C_P)}\left(e^{t/\tau_{FRRT}}-1\right)\right] \tag{39}$$

where:

$$\tau_{FRRT} \equiv \frac{1}{\beta Aa}\frac{C_T}{C_{PD}+C_P} \tag{40}$$

when $t \to \infty$, $V_{PDa}(t)$ approaches the asymptotic form exponentially:

$$V_{PDa}(t) \to -at \tag{41}$$

This divergence is reasonable because of the substantial charge injection to the PD.
For Step 2, substituting Equation (39) into Equation (36) we obtain the equation for $v_{PD}(t)$ as:

$$C_T \frac{dv_{PD}(t)}{dt} = \frac{I_a(0)\,e^{t/\tau_{FRRT}}\left(1 - e^{\beta Av_{PD}(t)}\right)}{1 + \frac{I_a(0)}{a(C_{PD}+C_P)}\left(e^{t/\tau_{FRRT}}-1\right)} - i_n(t) \tag{42}$$

Assuming that $\beta Av_{PD}(t) \ll 1$, one can use the approximation:

$$e^{\beta Av_{PD}(t)} \approx 1 + \beta Av_{PD}(t) \tag{43}$$

The validity of this assumption will be discussed later. Equation (42) then becomes a linear equation as:

$$C_T \frac{dv_{PD}(t)}{dt} = -\frac{\beta AI_a(0)\,e^{t/\tau_{FRRT}}v_{PD}(t)}{1 + \frac{I_a(0)}{a(C_{PD}+C_P)}\left(e^{t/\tau_{FRRT}}-1\right)} - i_n(t) \tag{44}$$

The solution can be written as:

$$v_{PD}(t) = -\frac{1}{C_T}\int_0^t dt' \frac{1 + \frac{I_a(0)}{a(C_{PD}+C_P)}\left(e^{t'/\tau_{FRRT}}-1\right)}{1 + \frac{I_a(0)}{a(C_{PD}+C_P)}\left(e^{t/\tau_{FRRT}}-1\right)}i_n(t') + \frac{1}{1 + \frac{I_a(0)}{a(C_{PD}+C_P)}\left(e^{t/\tau_{FRRT}}-1\right)}v_{PD}(0) \tag{45}$$

For Step 3, squaring Equation (45), averaging and using Equation (13), the noise variance can be obtained as:

$$< v_{PD}(t)^2 > = \frac{mkT}{2AC_T}\left(1 - \frac{1}{\left(1 + \frac{I_a(0)}{a(C_{PD}+C_P)}\left(e^{t/\tau_{FRRT}}-1\right)\right)^2}\right) + < v_{PD}(0)^2 > \frac{1}{\left(1 + \frac{I_a(0)}{a(C_{PD}+C_P)}\left(e^{t/\tau_{FRRT}}-1\right)\right)^2} \tag{46}$$

The first term is caused by shot noise, and the second term results from the initial condition. When $C_P = 0$, the asymptotic form and the limit for $t \to \infty$ are obtained as:

$$< v_{PD}(t)^2 > \approx \frac{mkT}{2AC_{PD}}(1 - e^{-2\beta Aat}) + < v_{PD}(0)^2 > e^{-2\beta Aat} \tag{47}$$

$$< v_{PD}(\infty)^2 > = \frac{mkT}{2AC_{PD}} \tag{48}$$

When $C_P \neq 0$, the asymptotic form and the limit for $t \rightarrow \infty$ become:

$$< v_{PD}(t)^2 > \approx \frac{mkT}{2A(C_{PD} + (1+A)C_P)}(1 - e^{-2t/\tau_{FRRT}}) + < v_{PD}(0)^2 > e^{-2t/\tau_{FRRT}} \tag{49}$$

$$< v_{PD}(\infty)^2 > = \frac{mkT}{2A(C_{PD} + (1+A)C_P)} \tag{50}$$

Considering that the detection capacitance is $C_{PD} + C_{P1}$, the reset noise variance in electron numbers at the PD, $< n_{PD}(\infty)^2 >$, is derived as:

$$< n_{PD}(\infty)^2 > = \frac{mkT(C_{PD} + C_{P1})^2}{2q^2 A(C_{PD} + (1+A)C_P)} \tag{51}$$

where q denotes the electronic elementary charge.

When $C_P = 0$, the limit of the noise variance $< v_{PD}(\infty)^2 >$ is $1/A$ times smaller than the soft reset noise variance, $mkT/2C_{PD}$, according to Equation (48). The reset noise is therefore much reduced. When $C_P \neq 0$, $< v_{PD}(\infty)^2 >$ becomes even smaller than that when $C_P = 0$, because C_P couples the output of the amplifier to the PD directly, which contributes as a capacitive feedback [21], in addition to the feedback path through the RST. However, large values of C_P have some drawbacks as well; if the vertical signal line capacitance and the RD line capacitance are large, the feedback speed is limited. If C_{P1} is large, the conversion gain is decreased. Figure 5a,b show the reset noises for $C_{PD} = 10$ fF and 1 fF, respectively. The horizontal axis denotes the amplifier gain A, the left vertical axis represents the reset noise voltage $\sqrt{< v_{PD}(\infty)^2 >}$, and the right vertical axis represents the reset noise in number of electrons, $\sqrt{< n_{PD}(\infty)^2 >}$. The parameter for the curves is C_P/C_{PD}. Here, $C_{P1} = 0$ fF is assumed for simplicity. According to this simple model, the reset noise decreases as $A^{-1/2}$ when $C_P/C_{PD} = 0$. When $C_P/C_{PD} \neq 0$, the reset noise decreases also as $A^{-1/2}$ for large values of A. It should be noted that even small values of C_P/C_{PD} will drastically decrease the reset noise when A is larger. For example, when $C_P/C_{PD} = 0.01$, the reset noise is decreased to 30% at $A = 1000$. If $C_{PD} + C_{P1}$ is smaller, reset noise in number of electrons becomes smaller while the reset noise in voltage becomes larger. It is important to reduce the detection capacitance, $C_{PD} + C_{P1}$, as is also the case with the 4-Tr. scheme.

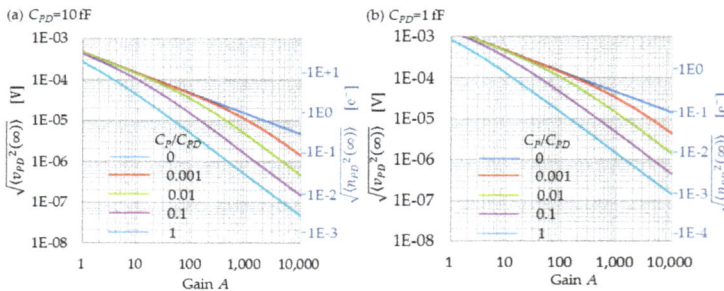

Figure 5. FRRT reset noise, with $C_{P1} = 1$ fF. (**a**) $C_{PD} = 10$ fF; (**b**) $C_{PD} = 1$ fF. $< v_{PD}(\infty)^2 >$ is the reset noise variance in voltage at the PD as expressed by (50), and $< n_{PD}(\infty)^2 >$ is the reset noise variance in number of electrons at the PD as expressed by (51).

Even though FRRT uses a subthreshold current mode, $< v_{PD}(t)^2 >$ still converges with fast exponential decay. Figure 6 shows the time constant, $\tau_{FRRT}/2$. In the figure, the horizontal axis represents the gain A, assuming that $a = 0.1$ V/μs. When $C_P = 0$, the time constant decreases linearly with $1/A$. When $C_P \neq 0$, $\tau_{FRRT}/2$ decreases linearly with $1/A$ at $AC_P << C_{PD}$, while it becomes a constant with the value $\frac{C_P}{2\beta a(C_{PD}+C_P)}$ for $AC_P \gg C_{PD}$. In the extremely unfavorable case of $C_P/C_{PD} = 10$, $\tau_{FRRT}/2$ is in practice sufficiently small (as small as 0.12 μs) for $A > 20$.

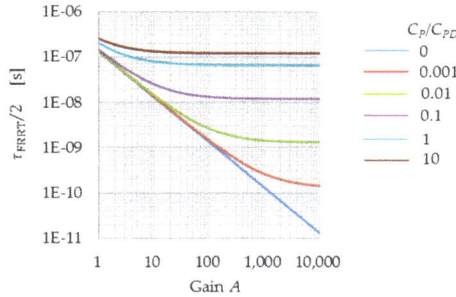

Figure 6. Time constant of the FRRT process (τ_{FRRT}), for $a = 0.1$ V/μs. $\frac{\tau_{FRRT}}{2} \equiv \frac{1}{2\beta Aa}\frac{C_{PD}+(1+A)C_P}{C_{PD}+C_P}$.

Using FRRT, the reset voltage is reduced by aT_{FRRT}, where T_{FRRT} is FRRT period. This decreases the saturation of the PD. If a is increased τ_{FRRT} decreases as $1/a$, and T_{FRRT} can be decreased in the same manner. If T_{FRRT} is adjusted properly, a will not affect the reset voltage reduction, because the reset voltage reduction dependence on a is given as $aT_{FRRT} \sim a^0$.

According to this simple model, if $A \to \infty$, $< v_{PD}(\infty)^2 > \to 0$. Limitations to this ideal case should be discussed below.

Firstly, the approximation:

$$e^{\beta Av_{PD}(t)} \approx 1 + \beta Av_{PD}(t)$$

is examined. If 30% of error is allowed, $\beta Av_{PD}(t)$ is limited by:

$$1 \leqslant \frac{e^{\beta Av_{PD}(t)}}{1 + \beta Av_{PD}(t)} < 1.3 \tag{52}$$

which is always larger than 1. This means that $\beta Av_{PD}(t)$ should be smaller than 0.91. Figure 7a,b shows the exponent $\beta A\sqrt{< v_{PD}(\infty)^2 >}$, substituting $\sqrt{< v_{PD}(\infty)^2 >}$ to $v_{PD}(t)$ and using $m = 1$ and $1/\beta = mkT/q = 0.026$ V. When $C_P = 0$, the upper limit of A is obtained as 2,700 for $C_{PD} = 10$ fF and as 270 for $C_{PD} = 1$ fF, respectively. When $C_P \neq 0$, and because Equation (50) < Equation (48), the range within which approximation Equation (43) can be used becomes larger. For example, when $C_P/C_{PD} = 0.01$, the upper limit becomes more than 10,000 for both $C_{PD} = 10$ fF and $C_{PD} = 1$ fF.

If the approximation Equation (43) becomes invalid, the quantitative discussion is difficult. However, the feedback effect becomes rather larger because the first term of right hand side at Equation (42) has larger negative value than that of Equation (44).

Secondly, the assumption that the feedback is faster than the reset motion should be discussed. Both the differential and SF amplifiers have finite output impedances, and both the vertical signal and RD lines have parasitic resistances and capacitances; this means that the feedback has a finite time constant. It increases as A increases. If the feedback becomes slow compared with the reset motion, the reset noise is increased in reverse. Therefore, the reset noise has a minimum at some value of A.

The dimensionless factor $I_a(0)/a(C_{PD} + C_P)$, at Equation (46) represents the ratio between the PD voltage change, $I_a(0)/(C_{PD} + C_P)$, and the taper slope, a. Assuming typical parameters: $I_a(0) = 0.5$ μA, $C_{PD} = 10$ fF, $C_P/C_{PD} = 0.01$, and $a = 0.1$ V/μ s, we obtain $I_a(0)/a(C_{PD} + C_P) = 0.5$, which is in the order of 1 and does not affect the convergence.

As discussed above, it can be said that FRRT reduces both reset noise and the convergence time constant. In fact, 2.5 e$^-$·rms reset noise and 2.9 e$^-$ rms readout noise have been reported, using organic photoconductive film CMOS image sensor with 3 μm pixel, 5 μs reset period and $A = 100$ [13].

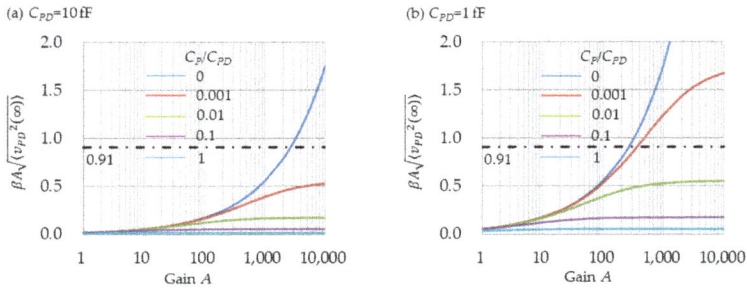

Figure 7. Depnedence of $\beta A \sqrt{\langle v_{PD}{}^2 (\infty) \rangle}$ on gain A and the ratio C_P/C_{PD} If $\beta A \sqrt{\langle v_{PD}{}^2 (\infty) \rangle}$ is smaller than 0.91, the approximation Equation (43) is permitted with an error of 30%.

7. Electron Counting Possibility

Photon counting imaging is one of the grand targets for image sensor development. It requires two conditions; electron counting and high quantum efficiency [35]. In this section, the possibility of electron counting using FRRT is discussed, leveraging the good properties of the 3-Tr scheme discussed in Section 1. Before that, the statuses of other approaches, such as single-photon avalanche diode (SPAD) image sensors [36,37] and 4-Tr CMOS image sensors are reviewed.

The Geiger mode avalanche in SPADs creates a sharp spike signal from one original photon-generated electron-hole pair. The spike signal is so large that the in-pixel circuitry detects it as a digital signal and subsequent stages do not add any noises. The sharp spike also realizes time stamp, which is important for various applications such as time of flight, or fluorescence lifetime imaging microscopy (FLIM). Its weak points are the large dark count, the after pulse, and small fill factor because of the necessity of a guardring.

The 4-Tr CMOS image sensor saw some progress in 2015 [38–42], obtained mainly by reducing the detection capacitance or floating diffusion capacitance. It brings large conversion gain (as large as 426 $\mu V/e^-$) [41] and a readout noise as small as 0.27 $e^-\cdot$rms [42]. If the readout noise is less than 0.3 $e^-\cdot$rms, it can be said that elctron counting is possible with the 90% confidence level [35]. This method has a rather small dark current and a lager fill factor—even for small pixels—than SPADs. It also does not suffer from the after pulse. The 4-Tr scheme is not convenient for time stamping, because the pixel has to wait for a photon after holding the reset level for CDS and a longer period between reset sampling signal sampling makes the CDS $1/f$ noise reduction less effective. In contrast, the 3-Tr scheme is operated in a "signal-first, reset-later" mode, which is suitable for time stamping.

With the FRRT simple model shown in the previous section, the reset noise, $\sqrt{< n_{PD} (\infty)^2 >}$, becomes 0.10 $e^-\cdot$rmsfor C_{PD} = 10 fF, C_P/C_{PD} = 0.01, and A = 2,700, and becomes 0.29 $e^-\cdot$rms for C_{PD} = 1 fF, C_P/C_{PD} = 0.01, and A = 270. In those cases, the possibility of electron counting exists if the other noises are small enough.

The readout noise for the 3-Tr scheme is constituted by the reset noise, SF thermal noise, SF $1/f$ noise, column circuit noise, and ADC quantization noise. The most effective method to reduce these noises is to reduce the detection capacitance and to increase the conversion gain as done in the 4-Tr scheme; the noises will then be reduced in terms of the number of electrons at the PD. There is, however, a sharp tradeoff between the reduction of the detection capacitance and sensitivity, because using a smaller PD area to decrease the detection capacitance originates also a small sensitivity. Therefore, it is much difficult to achieve an electron counting capability with a 3-Tr scheme than with a 4-Tr scheme. One possible circuit-based approach is to combine a capacitive transimpedance amplifier [27,28] with the FRRT.

Another possibility is to reduce the SF thermal noise, SF $1/f$ noise, column circuit noise, and ADC quantization noise themselves. The SF thermal noise, column circuit noise, and ADC quantization noise could be reduced by circuit technologies. Although various methods have been reported to reduce the SF $1/f$ noise, additional improvements are needed to perform the electron counting with a 3-Tr scheme.

8. Conclusions

To discuss the reset noise generated by a slow subthreshold current, intuitive and simple analytical forms are derived in spite of the subthreshold current nonlinearity, which characterize the time evolution of the reset noise during the reset operation.

For soft reset, the reset noise limit when $t \to \infty$, $\sqrt{< v_{PD}(\infty)^2 >}$, is given by $\sqrt{mkT/2C_{PD}}$, which agrees with previous published works. The asymptotic time dependence of the noise, $\sqrt{< v_{PD}(t)^2 >}$, decreases with t^{-1}, even though the asymptotic time dependence of the average PD voltage, $V_{PDa}(t)$, is as slow as $\log t$. The flush reset method is effective because the hard reset part eliminates the image lag, and soft reset part reduces the noise to the soft reset noise level.

The tapered reset method achieves exponential convergence, but the reset noise reduction is insufficient.

Finally, the FRRT shows both a fast convergence and a good reset noise reduction. When A is large, even small values of C_P/C_{PD} can drastically decrease the reset noise. If the feedback is sufficiently fast, the reset noise limit when $t \to \infty$, $\sqrt{< n_{PD}(\infty)^2 >}$, becomes $\frac{mkT(C_{PD}+C_{P1})^2}{2q^2 A(C_{PD}+(1+A)C_P)}$. Assuming that $C_{PD} = 10$ fF, $C_P/C_{PD} = 0.01$ and $A = 2700$, $\sqrt{< n_{PD}(\infty)^2 >}$ becomes 0.10 e^-·rms according to this simple model. Achieving an electron counting capability with this architecture requires a challenging $1/f$ noise reduction, even if the reset noise can be decreased.

Conflicts of Interest: The authors declare no conflict of interest.

Abbreviations

The following abbreviations are used in this manuscript:

Tr	Transistor
CMOS	Complementary metal oxide semiconductor
PPD	Pinned photodiode
CDS	Correlated double sampling
AEC	Auto exposure control
PD	Photodiode
UV	Ultraviolet
IR	Infrared
Ge	Germanium
InGaAs	Indium-gallium-arsenide
RST	Reset transistor
SF	Source follower
SEL	Row select transistor
k	Boltzmann constant
T	Absolute temperature
C	Capacitance
e^-	Electron
CCD	Charge coupled devices
RD	Reset transistor drain
FRRT	Feedback reset with reverse taper control
FB	Feedback
Ref	Reference
SPAD	Single photon avalanche diode
FLIM	Fluorescence lifetime imaging microscopy
rms	Root mean square

References

1. Mendis, S.K.; Kemeny, S.E.; Fossum, E.R. A 128 × 128 CMOS Active Pixel Image Sensor for Highly Integrated Imaging Systems. In Proceedings of the IEEE International Electron Devices Meeting (IEDM), Washington, DC, USA, 5–8 December 1993; pp. 583–586.
2. Teranishi, N.; Kohono, A.; Ishihara, Y.; Oda, E.; Arai, K. No image Lag Photodiode Structure in the Interline CCD Image Sensor. In Proceedings of the IEEE International Electron Devices Meeting (IEDM), San Francisco, CA, USA, 13–15 December 1982; pp. 324–327.
3. Lee, P.P.K.; Gee, R.C.; Guidash, R.M.; Lee, T.; Fossum, E.R. An Active Pixel Sensor Fabricated Using CMOS/CCD Process Technology. In Proceedings of the of IEEE Workshop on CCD and Advanced Image Sensors, Dana Point, CA, USA, 20–22 April 1995.
4. Inoue, I.; Nozaki, H.; Yamashita, H.; Yamaguchi, T.; Ishiwata, H.; Ihara, H.; Miyagawa, R.; Miura, H.; Nakamura, N.; Egawa, Y.; *et al.* New LV-BPD (Low Voltage Buried Photo-Diode) for CMOS Imager. In Proceedings of the Technical Digest International Electron Devices Meeting (IEDM), Washington, DC, USA, 5–8 December 1999; pp. 883–886.
5. Yonemoto, K.; Sumi, H.; Suzuki, R.; Ueno, T. A CMOS Image Sensor with a Simple FPN Reduction Technology and Hole Accumulated Diode. In Proceedings of the IEEE International Solid-State Circuits Conference (ISSCC) Digest of Technical Papers, San Francisco, CA, USA, 9 February 2000; pp. 102–103.
6. Fossum, E.R.; Hondongwa, D.B. A Review of the Pinned Photodiode for CCD and CMOS Image Sensors. *IEEE J. Electron Devices Soc.* **2014**, *2*, 33–43. [CrossRef]
7. White, M.H.; Lampe, D.R.; Blaha, F.C.; Mack, I.A. Characterization of Surface Channel CCD Image Arrays at Low Light Levels. *IEEE J. Solid State Circuit* **1974**, *SC-9*, 1–13. [CrossRef]
8. Takahashi, H.; Kinoshit, M.; Morita, K.; Shirai, T.; Sato, T.; Kimura, T.; Yuzurihara, H.; Inoue, S. A 3.9 μm pixel pitch VGA format 10 b digital image sensor with 1.5-transistor/pixel. In Proceedings of the IEEE International Solid-State Circuits Conference (ISSCC) Digest of Technical Papers, San Francisco, CA, USA, 15–19 February 2004; pp. 108–109.
9. Mori, M.; Katsuno, M.; Kasuga, S.; Murata, T.; Yamaguchi, T. A 1/4 in 2M pixel CMOS Image Sensor with 1.75 Transistor/Pixel. In Proceedings of the IEEE International Solid-State Circuits Conference (ISSCC) Digest of Technical Papers, San Francisco, CA, USA, 15–19 February 2004; pp. 110–111.
10. Itonaga, K.; Mizuta, K.; Kataoka, T.; Yanagita, M.; Ikeda, H.; Ishiwata, H.; Tanaka, Y.; Wakano, T.; Matoba, Y.; Oishi, T.; *et al.* Extremely-Low-Noise CMOS Image Sensor with High Saturation Capacity. In Proceedings of the IEEE International Electron Devices Meeting (IEDM), Washington, DC, USA, 5–7 December 2011; pp. 171–174.
11. Korthout, L.; Verbugt, D.; Timpert, J.; Mierop, A.; de Haan, W.; Maes, W.; de Meulmeester, J.; Muhammad, W.; Dillen, B.; Stoldt, H.; *et al.* A wafer-scale CMOS APS imager for medical X-ray applications. In Proceedings of the International Image Sensors Workshop, Bergen, Norway, 22–28 June 2009.
12. Ihama, M.; Inomata, H.; Asano, H.; Imai, S.; Mitsui, T.; Imada, Y.; Hayashi, M.; Gotou, T.; Suzuki, H.; Sawaki, D.; *et al.* CMOS Image Sensor with an Overlaid Organic Photoelectric Conversion Layer: Optical Advantages of Capturing Slanting Rays of Light. In Proceedings of the International Image Sensors Workshop, Hokkaido, Japan, 8–11 June 2011; pp. 153–156.
13. Ishii, M.; Kasuga, S.; Yazawa, K.; Sakata, Y.; Okino, T.; Sato, Y.; Hirase, J.; Hirose, Y.; Tamaki, T.; Matsunaga, Y.; *et al.* An Ultra-low Noise Photoconductive Film Image Sensor with a High-speed Column Feedback Amplifier Noise Canceller. In Proceedings of 2013 Symposium on VLSI Circuits Digest of Technical Papers, Kyoto, Japan, 10–13 June 2013; pp. C8–C9.
14. Imura, S.; Kikuchi, K.; Miyakawa, K.; Ohtake, H.; Kubota, M.; Okino, T.; Hirose, Y.; Kato, Y.; Teranishi, N. High Sensitivity Image Sensor Overlaid with Thin-Film Crystalline-Selenium-based Heterojunction Photodiode. In Proceedings of the 2014 IEEE International Electron Devices Meeting, San Francisco, CA, USA, 15–17 December 2014; pp. 88–91.
15. Aberg, I.; Ackland, B.; Beach, J.V.; Godek, C.; Johnson, R.; King, C.A.; Lattes, A.; O'Neill, J.; Pappas, S.; Sriram, T.S.; *et al.* A Low Dark Current and High Quantum Efficiency Monolithic Germanium-on-Silicon CMOS Imager Technology for Day and Night Imaging Applications. In Proceedings of the 2010 IEEE International Electron Devices Meeting (IEDM), San Francisco, CA, USA, 6–8 December 2010; pp. 344–347.

16. De Borniol, E.; Guellec, F.; Castelein, P.; Rouvié, A.; Robo, J.; Reverchon, J. High-Performance 640 × 512 Pixel Hybrid InGaAs Image Sensor for Night Vision. In Proceedings of the Infrared Technology and Applications XXXVII, SPIE, Baltimore, MD, USA, 23 April 2012; Volume 8353. [CrossRef]

17. Carnes, J.E.; Kosonocky, W.F. Noise Source in Charge-coupled Devices. *RCA Rev.* **1972**, *33*, 327–343.

18. Barbe, D.F. Imaging Devices Using the Charge-Coupled Concept. *IEEE Proc.* **1975**, *63*, 38–67. [CrossRef]

19. Pain, B.; Yang, G.; Ortiz, M.; Wrigley, C.; Hancock, B.; Cunningham, T.J. Analysis and Enhancement of Low-Light-Level Performance of Photodiode Type CMOS Active Pixel Imagers Operated with Subthreshold Reset. In Proceedings of the IEEE Workshop on Charge Coupled Devices and Advanced Image Sensors, Karuizawa, Japan, 9–11 June 1999; pp. 140–143.

20. Tian, H.; Fowler, B.; el Gamal, A. Analysis of Temporal Noise in CMOS Photodiode Active Pixel Sensor. *IEEE J. Solid State Circuits* **2001**, *36*, 92–101. [CrossRef]

21. Takayanagi, I.; Fukunaga, Y.; Yoshida, T.; Nakamura, J. A Four-Transistor Capacitive Feedback Reset Active Pixel and Its Reset Noise Reduction Capability. In Proceedings of the IEEE Workshop on Charge Coupled Devices and Advanced Image Sensors, Lake Tahoe, CA, USA, 7–9 June 2001.

22. Pain, B.; Yang, G.; Cunningham, T.J.; Wrigley, C.; Hancock, B. An Enhanced-Performance CMOS Imager with a Flushed-Reset Photodiode Pixel. *IEEE Trans. Electron Devices* **2003**, *50*, 48–56. [CrossRef]

23. Fowler, B.; Godfrey, M.D.; Balicki, J.; Canfield, J. Low Noise Readout Using Active Reset for CMOS APS. *SPIE Proc.* **2000**, *3965*, 126–135.

24. Pain, B.; Cunningham, T.J.; Hancock, B.; Yang, G.; Seshadri, S.; Ortiz, M. Reset Noise Suppression in Two-Dimensional CMOS Photodiode Pixels through Column-Based Feedback-Reset. In Proceedings of the IEEE International Electron Devices Meeting (IEDM), San Francisco, CA, USA, 8–11 December 2002; pp. 809–812.

25. Fowler, B.; Godfrey, M.D.; Mims, S. Analysis of Reset Noise Suppression via Stochastic Differential Equations. In Proceedings of the 2005 IEEE Workshop on Charge Coupled Devices and Advanced Image Sensors, Karuizawa, Japan, 9–11 June 2005; pp. 19–22.

26. Kozlowski, L.J.; Rossi, G.; Blanquart, L.; Marchesini, R.; Huang, Y.; Chow, G.; Richardson, J.; Standley, D. Pixel Noise Suppression via SoC Management of Tapered Reset in a 1920 × 1080 CMOS Image Sensor. *IEEE J. Solid State Circuits* **2005**, *40*, 2766–2776. [CrossRef]

27. Yang, J.; Fife, K.G.; Brooks, L.; Sodini, C.G.; Mudunuru, A.B.P.; Lee, H.-S. A 3Mpixel Low-Noise Flexible Architecture CMOS Image Sensor. In Proceedings of the 2006 IEEE International Solid State Circuits Conference—Digest of Technical Papers, San Francisco, CA, USA, 6–9 February 2006.

28. Lotto, C. Energy-Sensitive Single-photon X-ray and Particle Imaging. In *Single-Photon Imaging*; Seitz, P., Theuwissen, A.J.P., Eds.; Springer-Verlag: Berlin, Germany; Heidelberg, Germany, 2011; pp. 255–271.

29. Merrill, R.B. *kTC* Noise Cancellation Pixel. In Proceedings of the 2001 IEEE Workshop on Charge-Coupled Devices and Advanced Image Sensors, Lake Tahoe, CA, USA, 7–9 June 2001; pp. 118–121.

30. Thornber, K.K. Theory of Noise in Charge-Transfer Devices. *Bell Syst. Tech. J.* **1974**, *53*, 1211–1262. [CrossRef]

31. Emmons, S.P.; Buss, D.D. Noise Measurements on the Floating Diffusion Input for Charge-Coupled Devices. *J. Appl. Phys.* **1974**, *45*, 5303–5306. [CrossRef]

32. Shioyama, Y.; Hatori, F.; Matsunaga, Y. Analysis of Random Noise in STACK-CCD. *ITE Techn. Rep.* **1993**, *17*, 1–6. (In Japanese)

33. Nakamura, N.; Shioyama, Y.; Ohsawa, S.; Sugiki, T.; Matsunaga, Y. Random Noise Generation Mechanism for a CCD Imager with an Incomplete Transfer-Type Storage Diode. *IEEE Trans. Electron Devices* **1996**, *43*, 1883–1889. [CrossRef]

34. Teranishi, N.; Mutoh, N. Partition Noise in CCD Signal Detection. *IEEE Trans. Electron Devices* **1986**, *ED-33*, 1696–1701. [CrossRef]

35. Teranishi, N. Required Conditions for Photon Counting Image Sensors. *IEEE Trans. Electron Devices* **2012**, *59*, 2199–2205. [CrossRef]

36. Charbon, E.; Fishburn, M.W. Monolithic Single-Photon Avalanche Diodes: SPADs. In *Single-Photon Imaging*; Seitz, P., Theuwissen, A.J.P., Eds.; Springer-Verlag: Berlin, Germany; Heidelberg, Germany, 2011; pp. 123–157.

37. Dutton, N.A.W.; Gyongy, I.; Parmesan, L.; Gnecchi, S.; Calder, N.; Rae, B.R.; Pellegrini, S.; Grant, L.A.; Henderson, R.K. A SPAD-Based QVGA Image Sensor for Single-Photon Counting and Quanta Imaging. *IEEE Trans. Electron Devices* **2016**, *63*, 189–196. [CrossRef]

38. Kusuhara, F.; Wakashima, S.; Nasuno, S.; Kuroda, R.; Sugawa, S. Analysis and Reduction of Floating Diffusion Capacitance Components of CMOS Image Sensor for Photon-Countable Sensitivity. In Proceedings of the International Image Sensor Workshop, Vaals, the Netherlands, 8–11 June 2015; pp. 120–123.

39. Boukhayma, A.; Peizerat, A.; Enz, C. A 0.4 e-rms Temporal Readout Noise, 7.5 μm Pitch and a 66% Fill Factor Pixel for Low Light CMOS Image Sensors. In Proceedings of the International Image Sensor Workshop, Vaals, the Netherlands, 8–11 June 2015; pp. 365–368.

40. Yao, Q.; Dierickx, B.; Dupont, B. CMOS Image Sensor Reaching 0.34 e^- RMS Read Noise by Inversion-Accumulation Cycling. In Proceedings of the International Image Sensor Workshop, Vaals, the Netherlands, 8–11 June 2015; pp. 369–372.

41. Ma, J.; Fossum, E.R. Quanta Image Sensor Jot with Sub 0.3 e^- r.m.s. Read Noise and Photon Counting Capability. *IEEE Electron Device Lett.* **2015**, *36*, 926–928. [CrossRef]

42. Seo, M.-W.; Kawahito, S.; Kagawa, K.; Yasutomi, K. A 0.27 e^- rms Read Noise 220-μV/e^- Conversion Gain Reset-Gate-Less CMOS Image Sensor with 0.11- CIS Process. *IEEE Electron Device Lett.* **2015**, *36*, 1344–1347.

sensors

MDPI

Article

Particle and Photon Detection: Counting and Energy Measurement

James Janesick [1,*] and John Tower [2]

[1]   SRI-Sarnoff, 4952 Warner Avenue, Suite 300, Huntington Beach, CA 92649, USA
[2]   SRI-Sarnoff, 201 Washington Road, Princeton, NJ 08540, USA; john.tower@sri.com
*   Correspondence: CMOSCCD@aol.com; Tel.: +1-714-377-6223

Academic Editor: Eric R. Fossum
Received: 22 March 2016; Accepted: 4 May 2016; Published: 12 May 2016

Abstract: Fundamental limits for photon counting and photon energy measurement are reviewed for CCD and CMOS imagers. The challenges to extend photon counting into the visible/nIR wavelengths and achieve energy measurement in the UV with specific read noise requirements are discussed. Pixel flicker and random telegraph noise sources are highlighted along with various methods used in reducing their contribution on the sensor's read noise floor. Practical requirements for quantum efficiency, charge collection efficiency, and charge transfer efficiency that interfere with photon counting performance are discussed. Lastly we will review current efforts in reducing flicker noise head-on, in hopes to drive read noise substantially below 1 carrier rms.

Keywords: ultra low noise; CCD; CMOS; imagers

---

1. Introduction

Silicon CCD and CMOS imagers have been demonstrated to be exceptional detectors for particle counting and energy measurement for some time. The spectral range where photon counting is possible covers an extensive wavelength range from 0.1 to 1000 nm (1.24 to 12,400 eV), *i.e.*, nIR, visible, UV, EUV and soft X-ray. At the beginning of the EUV range (10 eV) photon energy absorbed by the imager can be determined by using the simple relation [1],

$$E(\text{eV}) = 3.65 n_i \qquad (1)$$

where 3.65 is an experimentally determined constant for silicon (eV/carriers) and n_i is measured quantum yield (carriers generated/interacting photon). The equation is applicable to photon energies greater than ~10 eV. The formula is not useful for energies less than this because the constant 3.65 wildly fluctuates. Besides photons, this equation is also useful for any particle that ionizes silicon atoms (electrons, protons, muons, *etc.*). The uncertainty in energy measurement is limited by the detector's read floor and Fano noise. Fano noise, the variation of charge generated per photon, is found by,

$$FN = (F n_i)^{0.5} \qquad (2)$$

where F is the Fano factor (~0.1 for silicon) and n_i is the quantum yield (carriers generated per photon). Physically Fano noise arises within the silicon where a small amount of thermal energy is lost to the silicon lattice (phonons) instead of creating electron-hole pairs. The variation in the loss from pixel to pixel is the Fano noise generated and represents a fundamental noise source in determining the energy of high energy particles [2].

Imagers where Fano noise is greater than the sensor's read noise are referred to as "Fano noise limited" [2]. Figure 1 plots Fano noise as a function of photon energy (eV) and wavelength (μm) showing the Fano limited range that can be covered by an imager for a given read noise floor. For example, Figure 2 presents a histogram taken from a Fano noise limited 8 um 3T NMOS pinned photo diode (PPD) pixel array showing multiple energy lines from a basalt target fluoresced with 5.9 keV Mn X-rays. The sensor's read noise is slightly less than 2 electrons (e^-) allowing Fano limited performance to cover the entire soft X-ray range (0.12 nm–10 nm). This spectral range has been particularly fruitful for CCD soft X-ray imaging spectrometers used in scientific and space applications. The width of each spectral line revealed in Figure 2 is a measurement of the amount of Fano noise generated. The spectral range for this imager can be further extended into the EUV range (10 nm–124 nm) if only photon counting is desirable.

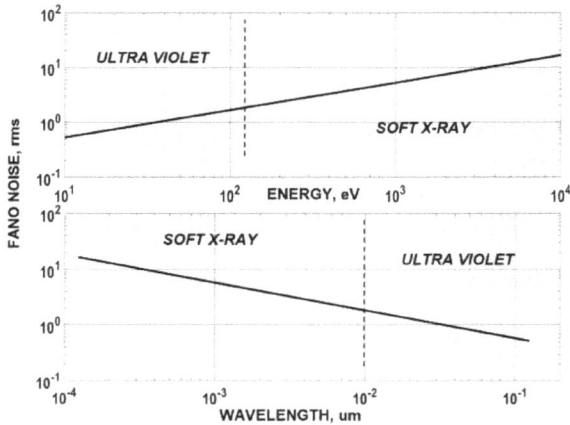

Figure 1. The plots above are used to determine the "Fano noise limit" for an imager with a given read noise. For example, a read noise of one carrier rms will cover the soft X-ray and extend into the UV at wavelength of 0.03 um (~40 eV).

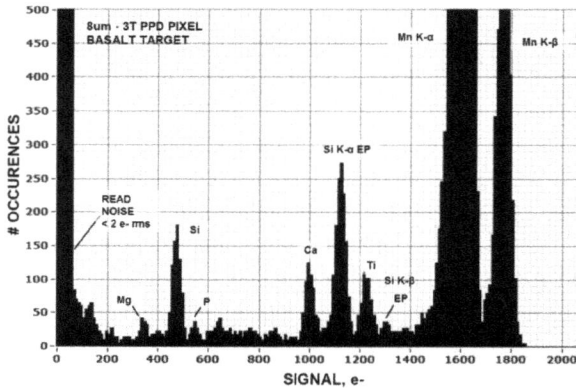

Figure 2. Photon counting and energy histogram generated by a 3T PPD CMOS pixel imager demonstrating Fano-noise limited performance over the entire soft X-ray regime.

The photon energy for visible (400 nm–700 nm) and nIR (700 nm–1100 nm) wavelengths is only able to generate one electron-hole pair/photon, limiting sensing to only photon counting. But when leaving the visible range into the UV multiple carriers per photon are generated allowing their energy to be determined. The challenge left today is to extend energy measurement into the UV and provide photon counting in the visible/nIR wavelengths by reducing the sensor's read noise floor.

The average noise floor for high performance CCD and CMOS imagers is typically shy of achieving "one" carrier of noise (this excludes EMCCD and SPAD detectors). For example, Figure 3 presents a photon transfer (PT) [2,3] curve generated by a 512 um × 512 um × 8 um PPD PMOS CMOS imager with a read noise floor of 1.08 holes (h+)at room temperature. It is extraordinary that the CCD/CMOS imaging community for the most part has achieved approximately "one" carrier of noise considering the multitude of solid state phenomena at play at world wide fabrication foundries for several years (decades in the case of CCDs). But why "one"? Is this apparent final outcome simply coincidental? One also wonders further why "one" carrier of noise along with "one" carrier of signal forces the minimum detection limit (MDL) of the detector to be "one" (*i.e.*, S/N = "one"). The conundrum continues with why today's imagers are very close but yet so far from "routinely" counting single visible photons consistently across full imaging arrays.

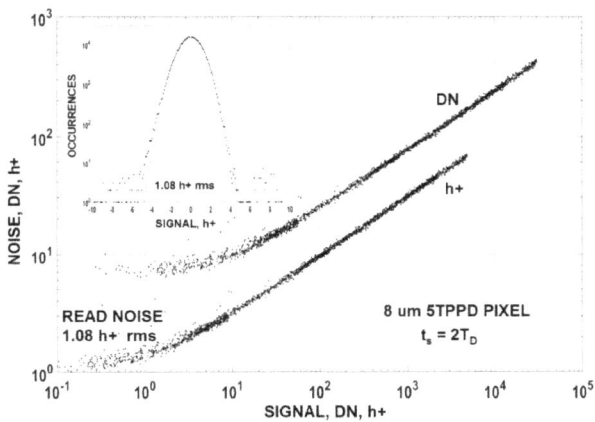

Figure 3. Photon Transfer curves demonstrating a read noise floor of ~1 hole rms. Most high performance CCD and CMOS imagers are close to this noise level.

To appreciate the challenge, Figure 4 displays computer simulated histograms for different read noise levels for an average signal level of "one" interacting photon/pixel and quantum yield of "one". These plots show that it is necessary to have a read noise floor of <0.3 e⁻ before the histograms appear "quantized", which in turn sets the stage for visible/nIR photon counting. Also notice from these plots that the net signal-to-noise (S/N) determined directly from the simulated plots hardly changes as read noise is lowered from 0.3 to 0.1 carriers because the presence of photon shot noise. Nonetheless, read noise improvement in this range is especially critical to future visible/nIR photon counting imagers and applications.

Occasionally we do run into an individual low noise pixel that is capable (but barely) of photon counting allowing us to see the Poisson distribution profile that governs photon counting statistics. The top histogram of Figure 5a is generated by an exceptionally low noise "cherry picked" NMOS 3T PPD pixel with a 0.35 e⁻ noise floor taken under cooled conditions (~−80 °C). An average signal level of 0.8 e⁻ is adjusted for the experiment (*i.e.*, 0.8 interacting photons/pixel). The lower histogram shown in Figure 5b is a computer simulation showing that data and theory match up quite well. Figure 6 is a different NMOS 3T pixel illustrating how 4.5 e⁻ photon shot noise folds into the 0.78 e⁻ noise floor as the light source slowly turns off. These low noise pixels imply that the "one" carrier noise level that manufacturers have produced may not represent a fundamental limit but simply related to the noise level being "good enough". This stance is reasonable to assume because for normal imagery the S/N is always less than "one" regardless of how much lower the read noise is beyond "one". Hence, imager manufactures may not feel obligated to aggressively lower read noise any further (especially monetarily). This position produces an illusion that there may be a physical barrier of some kind holding us to "one" noise carrier. Fairly recently (2015) it has become more apparent that the "one" noise barrier has been a deception for groups are now reporting sub carrier performance down to 0.23 e⁻ for selected pixels [4,5] allowing for visible photon counting. The future objective for sensors with sub carrier read noise floors is to have all pixels contained on an imager exhibit close to the same low noise level and not just a few while at the same time achieve a reasonable saturation level for high dynamic range.

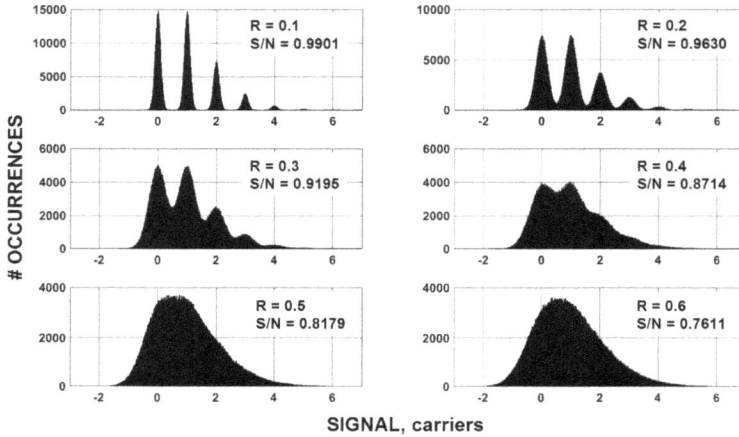

Figure 4. Simulated histograms for an average signal level of "one" carrier and six different read noise (R) levels. The histogram shows that a read noise <0.3 carriers rms is required for precise photon counting results.

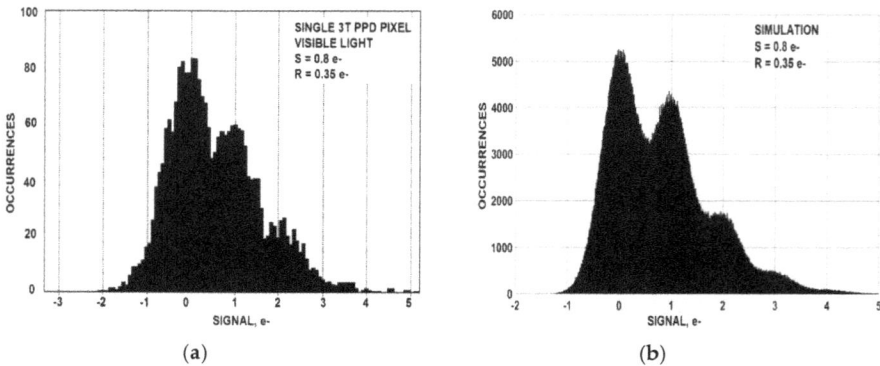

(a)   (b)

Figure 5. Experimental (**a**) and simulated (**b**) histograms for a low noise 3T PPD CMOS pixel showing a Poisson distribution profile. Read noise for this pixel is 0.35 e⁻ rms.

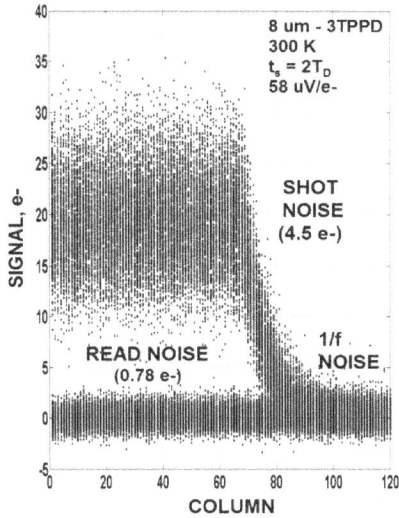

Figure 6. Raw video response for a sub carrier noise 3T PPD pixel in response to a light source that slowly turns off. Read noise for this pixel is 0.78 e⁻ rms.

2. Read Noise Reduction

2.1. Flicker and Random Telegraph Noise

Flicker and random telegraph noise (RTN) generated by the pixel's source follower (SF) MOSFET are the dominant noise sources that limit read noise performance for an imager. It is generally believed that RTN is associated with traps within the gate oxide. Free carriers from the SF channel tunnel in and out of these traps at different rates depending on the tunneling distance to the trap and the barrier height encountered. The local "on" and "off" potential in the localized region of a trapped carrier modulates the SF channel current in a digital manner and is seen as RTN output voltage fluctuation (with varying amplitude and frequency from pixel to pixel depending on the trap time constant and location). NMOS imagers have a great deal more RTN pixels than PMOS devices do. For instance, Figure 7 compares the read noise for NMOS and PMOS pixels showing very few RTN pixels for the PMOS imager. The background flicker noise levels are different for the plots because the sense node (SN) conversion gains are not the same. However, it is observed that when the sensitivity is the same the background flicker noise level (rms carriers) is nearly identical for PMOS and NMOS pixels.

Other sources of pixel noise have been observed that mimic SF RTN. For example, a poor SN metal contact exhibits RTN behavior. Leakage currents related to the pixels transfer gate (TG) and reset gate that are in close proximity to the SN can also contribute noise for some local pixels. However, one can usually differentiate these noise sources from SF noise by varying the voltages to these gates and the SN reset bias level and noting the affect.

The reason for the NMOS/PMOS RTN population difference is holes are much less likely to tunnel into oxide traps than electrons because the barrier height is greater. On these grounds we do not need to consider RTN as a fundamental barrier to lowering read noise. Therefore, the problem of beating down SF flicker noise has become the only fundamental noise source in reducing the read noise floor for PMOS imagers. However, for NMOS devices we must use other strategies to reduce RTN. For example buried channel MOSFETs can be employed which to some degree isolates the SF current carrying channel from the surface. In turn this increases the barrier to the traps and lowers RTN. We will briefly discuss this technology below.

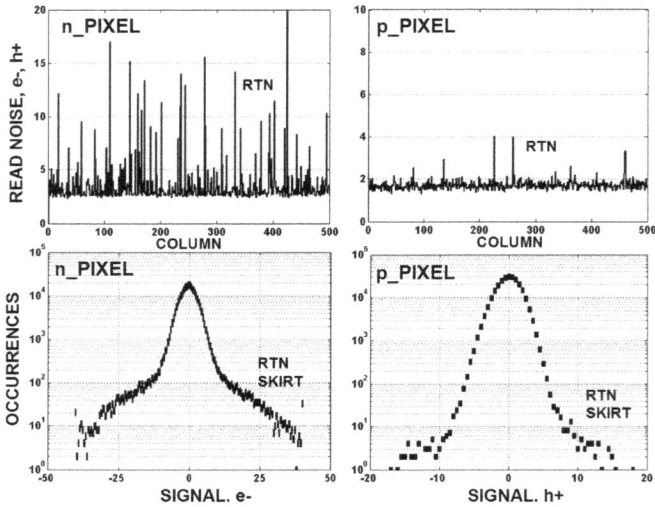

Figure 7. Random telegraph noise (RTN) population comparison for NMOS (n_pixel) and PMOS (p_pixel) pixels.

Two primary theories describe flicker noise generation in MOSFETs: Hooge's mobility fluctuation and McWhorter's carrier density (or number) fluctuation models [6,7]. The latter mechanism is based on the random trapping and release of conduction band carriers located at the Si-SiO$_2$ interface of the SF MOSFET (*i.e.*, generation recombination (GR) noise). The former is related to conductance or resistance fluctuations within the channel of the SF MOSFET. There are several possible causes for channel mobility noise. For example, fluctuations may be associated with a lower surface mobility compared to bulk silicon because carriers scatter off the Si-SiO$_2$ interface. Another possibility could be linked to the SF channel's implanted impurities that are randomly imbedded in the silicon lattice. The depleted ions in the SF channel produce small individual potential barriers that "fluctuate" current flow thus producing noise [8].

To help identify where $1/f$ noise sources are located, pixels have been irradiated with high energy radiation sources to deliberately increase interfacial Si-SiO$_2$ surface states. In general when such experiments are performed we find that the read noise level does not change. For example, Figure 8a shows that the read noise remains constant after irradiating an unbiased NMOS CMOS imager to different dose levels of high energy electrons. The noise data shown was generated using a very short exposure time (~10 μs) such that the increase in dark current shot noise due to radiation damage was negligible. However, the read noise for 10 Mrd exposure appears to be higher because thermally generated dark current was high enough to become an issue. Cooling the pixels slightly as displayed in Figure 8b lowers the noise down to the same level as other irradiated imagers.

The same sensors irradiated did exhibit charge transfer efficiency (CTE) degradation related to the TG demonstrating that surface states were in fact being generated. Some flat-band shift due to positive fixed charge buildup in the oxide induced by the radiation was also measured again indicating some damage took place but without a read noise increase. Other research groups have experienced the same surprising outcome [6]. Although far from being proven these findings begin to bias us to believe that the background flicker noise may not be dominantly associated with the oxide but instead within the current carrying channel of the SF.

We have had some past success in reducing read noise using buried channel NMOS MOSFETs. Others working with this approach have realized the same benefit [9]. For example, Figure 9a,b shows the temporal noise difference for surface and buried channel MOSFETs. Notice that a significant RTN noise reduction takes place with buried use but does not change the background $1/f$ noise. Figure 10 also shows an unchanged background noise level for a row of 500 pixels where the rms noise level is determined for each pixel. The outcome is reasonable for RTN because the barrier increase provided by buried channel should diminish electron tunneling into the oxide. But why not a $1/f$ noise reduction? Possibly the buried channel employed here did not provide a sufficient barrier or shield deep enough for bulk generated $1/f$ noise. Or perhaps the $1/f$ noise is being generated in another location or mechanism other than just traps.

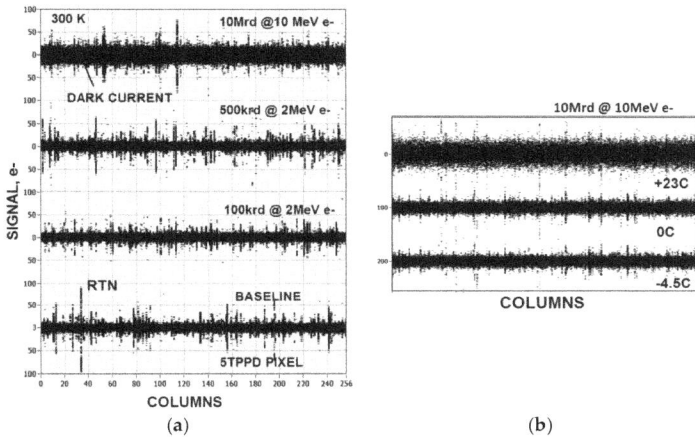

(a)   (b)

Figure 8. High energy electron damage source follower (SF) noise responses taken from a CMOS 5TPPD image. As the radiation dose level dramatically increases there is no significant increase in read noise even though gate oxide damage is taking place. The lower plot of (**b**), which is taken at −4.5 °C, shows that read noise at 23 °C for the 10 Mrd exposure is limited by dark current noise. Cooling the detector eliminates this dark current noise source resulting in the SF read noise level.

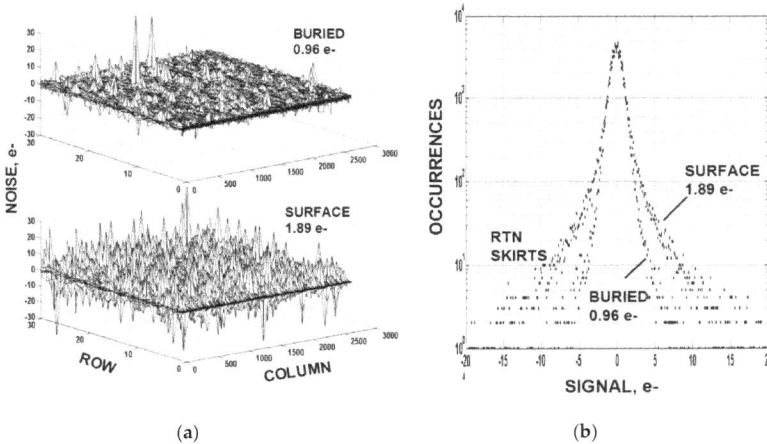

(a)   (b)

Figure 9. Noise performance comparison for surface and buried channel SF MOSFETs. The data shows that the buried channel device has reduced RTN but not reduced $1/f$ noise.

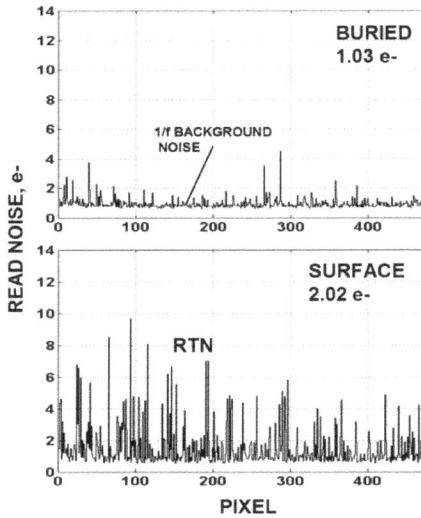

Figure 10. Buried (**top**) and surface (**bottom**) channel SF data showing more clearly that $1/f$ background noise is unaffected by buried channel use.

2.2. Correlated Double Sampling

Flicker noise and correlated double sampling (CDS) share a fascinating relationship. Analysis and measurements show that optimum flicker noise reduction takes place when:

$$t_s = 1/(8B) \tag{3}$$

where t_s is the time difference between samples before and after charge is transferred to the SN and B is the net equivalent noise bandwidth for the sensor and CDS system [10]. Equation (3) can also be expressed as $t_s = 2\tau_D$ where τ_D is the system dominate time constant. Figure 11 plots CDS output noise as a function of t_s for various τ_D for a flicker noise input. For a specified τ_D the output noise increases with t_s because $1/f$ noise correlation is being lost between the two samples. For very long t_s time the CDS output noise will eventually level off to $2^{0.5}$ times the input noise because the samples are fully uncorrelated and differenced. It is important to point out from this plot that CDS output noise is constant with t_s given that $t_s = 2\tau_D$ is fixed (*i.e.*, the dotted horizontal line shown on the plot). Under this timing boundary condition $1/f$ noise decreases with τ_D at the same rate as $1/f$ noise increases with t_s. This is the central reason why it's not feasible to eliminate $1/f$ noise by CDS processing. It should be mentioned that sampling less than $2\tau_D$ results in a signal gain loss and lower S/N performance (the reason why the plots are bounded by the dotted horizontal line). Also, sampling an exponentially changing video waveform can produce camera instabilities (e.g., offset control).

To demonstrate the claim above, stacked raw clamped dark video plots for a number of PMOS pixels is presented in Figure 12a. The response shown is how the noise would appear on a high persistence oscilloscope before it enters an analog-to-digital converter with internal sample and hold. The magnified plot presented in Figure 12b shows that timing starts by resetting and clamping the pixels' video. The white noise seen in the clamp period is associated with the clamp switch (~0.1 h+). As one would expect $1/f$ noise slowly increases after the clamp release due to lack of correlation. Figure 13 plots this read noise buildup for sixteen of the pixels. Ideally the video should be sampled as soon as the clamp is released after the required video τ_D settling time is satisfied for lowest noise performance (*i.e.*, $t_s = 2\tau_D = 7$ μs in this case). This special timing condition is illustrated by the "squares" on the plot. An average noise level of ~0.9 h+ is measured for a row of 500 pixels of which the sixteen pixels are contained. The data points with "stars" include the pixel's TG overhead time where a higher noise is measured (1.7 h+) because the $1/f$ noise is less correlated. It should be also mentioned that averaging multiple samples in Figure 13 does not help to reduce the read noise floor. This is because the samples are semi correlated and the noise within the sample increases for each new sample taken. As a result the noise increases with sample number at the same rate as the noise reduction offered by averaging (*i.e.*, $N^{1/2}$). It is very difficult to circumvent flicker noise presence for any signal processing scheme employed.

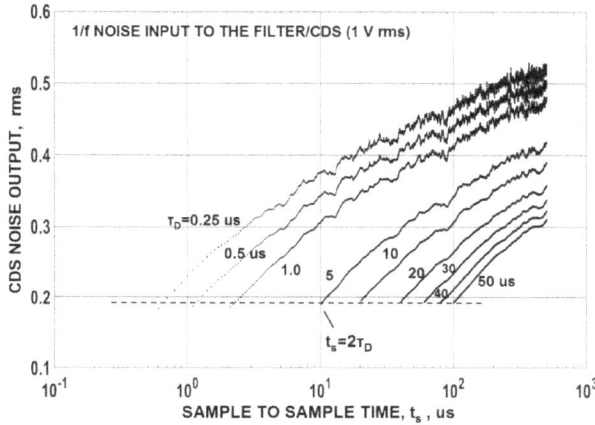

Figure 11. Computer simulated correlated double sampling (CDS) output noise in response of $1/f$ noise [9]. The lack of correlation on $1/f$ noise between samples is the cause for increase.

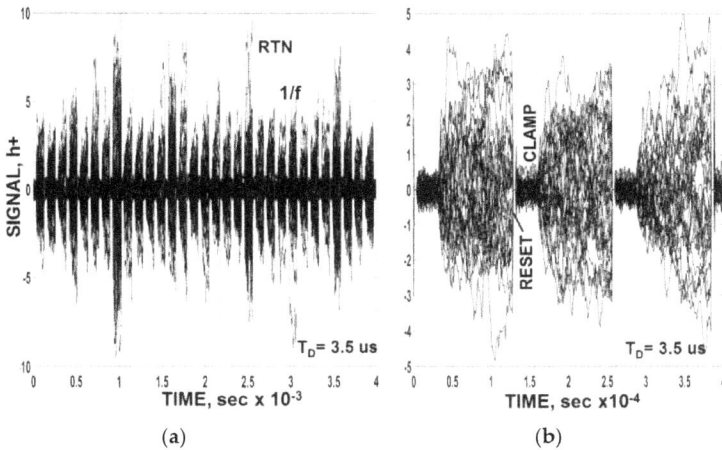

Figure 12. Raw video showing how noise increases after clamp is released for several pixels. (**a**) stacked raw clamped dark video plots; (**b**) magnified plot.

It is interesting to point out that using a 3T pixel for some applications can achieve a slightly lower noise than a 4T pixel simply because the 3T does not need to contend with TG overhead settling time. This assumes that the pixels are individually read out one by one by undergoing the entire exposure and CDS timing process (reset, clamp, flash expose and sample) before moving onto the next pixel. This approach is often used to test CMOS imagers including charge transfer pixels. Many figures in this paper are generated this manner to avoid dark current problems at room temperature.

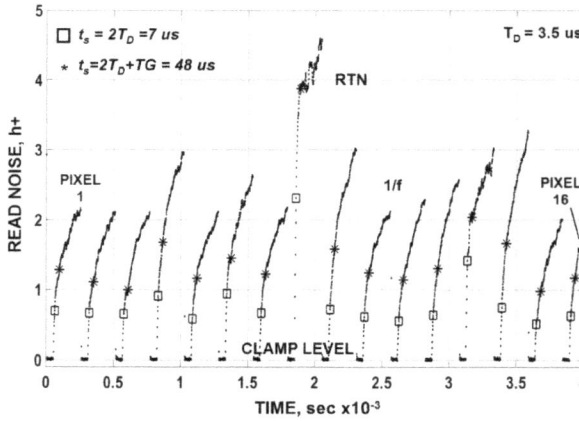

Figure 13. Measured noise for the raw clamped video in Figure 12. The "square" points represent the ideal sampling time after clamp is released (*i.e.*, $t_s = 2\tau_D$). The "star" points include TG overhead and settling time.

Although not conventional, one can shorten the TG overhead time by sampling the video on the leading edge of the TG instead of the lagging edge. In fact sampling can be performed at any time throughout the TG clock period as long as the $t_s = 2\tau_D$ settling time is satisfied. Figure 14 presents raw video traces taken under dark and light conditions. The traces show the reset clock feedthrough, clamp period, TG clock feedthrough and the downward change in video level to the light level applied. Figure 15 is the corresponding noise measured for Figure 14. Note after the clamp is released that the read noise increases due to correlation loss. The presence of a large TG clock feedthrough does not influence the noise measured because it only represents a changing offset which can be readily removed by a computer.

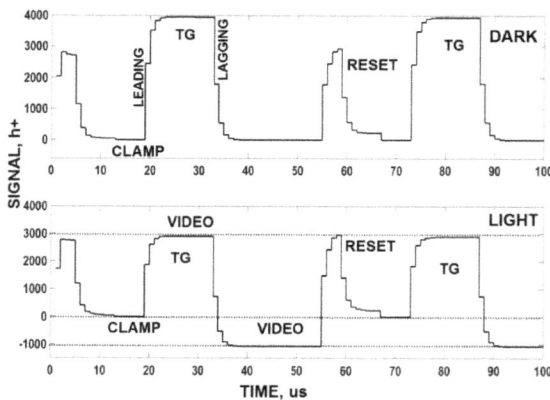

Figure 14. Dark (**top**) and light (**bottom**) raw video responses showing the pixel's clock feedthroughs and the video level.

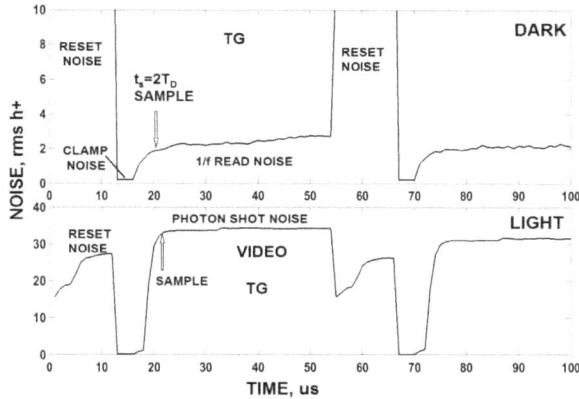

Figure 15. Noise levels for signal shown in Figure 14. The top dark plot shows how the read noise increases throughout the TG time period. No excess TG clock feed-through noise is observed.

2.3. Conversion Gain

The conversion gain (V/carrier) depends on the various parasitic capacitances attached to the pixel's floating diffusion SN besides the SF gate capacitance. It is our experience that the highest conversion gain realized does not bear the lowest average noise as long as the SF gate capacitance dominates. Instead an optimum SF size exists for lowest noise. Also in general, $1/f$ and SN sensitivity scale together for a large range of SF width and lengths. This strong relationship between $1/f$ and V/carrier has clearly been experienced when CCD and CMOS read noise performance is compared. The former technology typically has a considerably lower conversion gain than the latter yet both technologies in essence are limited to the same read noise floor (*i.e.*, ~1 carrier rms). This outcome is because flicker noise is proportional to SF gate capacitance while the SN conversion gain is inversely proportional to it. SF MOSFET size and gate capacitance have grown smaller as imager technology improves but without a significant reduction in read noise.

2.4. Nondestructive Readout

Nondestructive floating gate readout schemes have worked in a straightforward fashion in producing sub carrier noise performance assuming there are no restrictions involving frame time requirements and operating temperature. The technology coined floating gate "Skipper" long ago can theoretical achieve any desired noise level as the read noise decreases by the square-root of the number of samples taken for each pixel [2]. Figure 16 presents the read noise for a Skipper CCD where ~0.4 e^- is achieved with 64 samples beginning with a 3.25 e^- noise floor for a single sample [1]. Unfortunately no attempt was made to count single visible photons with this imager since it was busy counting X-ray photons for a soft X-ray flight mission called Cosmic Unresolved Background Instrument (CUBIC). The SN conversion factor is quite low for the device (3 uV/ e^-) because of large parasitic SN and SF gate capacitances. More recent Skipper CMOS pixels have shown that the conversion gain for a floating gate and a floating diffusion are about the same. Hence, it may be possible for a 1 carrier noise Skipper device to achieve 0.3 e^- with nine samples using today's fabrication technology without reducing $1/f$ noise. These Skipper CMOS pixels were fabricated using a multi transfer buried channel gate process developed with Jazz Semiconductor for CMOSCCDs. No serious attempt has been made to characterize these pixels for photon counting use as of yet (only single sample data has been taken). The Skipper noise reduction approach was employed in the far past when the SN conversion was very low as sub micron fab technology did not exist. Today one can achieve the same 0.3 e^- noise floor with one sample with much higher conversion gain [5].

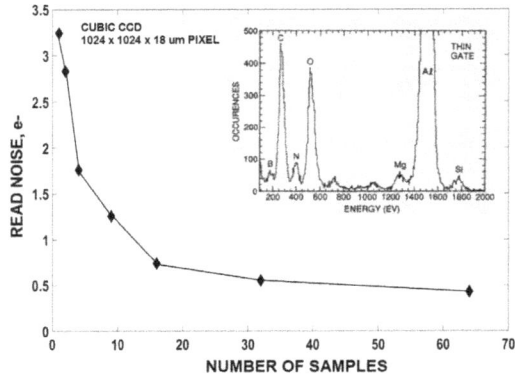

Figure 16. Nondestructive readout of the CUBIC imager showing how the read noise decreases with the number of samples averaged [1]. Embedded in the figure is a low energy soft X-ray histogram showing B, C, N and O lines with Fano-noise limited performance.

3. Additional Counting Issues

Besides read noise, quantum efficiency (QE), CTE and charge collection efficiency (CCE) interfere with photon counting precision. We will take a look at these problems in this section.

3.1. Quantum Efficiency

QE is never ideal and hence sensors will fall short in precisely counting the number of incoming photons. Photons can be reflected, absorbed by non-active regions of the pixel or entirely transmitted through the active silicon in lowering QE. QE is dependent on silicon wafer thickness, photon wavelength and the pixel technology utilized (e.g., backside *versus* frontside illumination). Fortunately a photon miscount due to QE is probably not that critical to most future applications given that today's imagers exhibit fairly high QE performance.

3.2. Charge Transfer Efficiency

CTE is rarely perfect either (3T CMOS pixels are not vulnerable to CTE problems since charge is not transferred). A CTE issue usually translates to deferred carriers rather than their absolute disappearance. Seldom do signal carriers recombine but are instead trapped and released at later time. For CCDs trapped charge is spatially and time deferred whereas carriers are only time deferred for CMOS pixels. Also CCD CTE degrades as signal level decreases whereas CMOS CTE typically increases. Lastly, CMOS usually involves only a single transfer within the pixel where CCDs must transfer charge hundreds or thousands of times in taking charge to the output amplifier which only exasperates other CTE issues. Also the number of transfers is especially critical when a sensor operates in a damaging high energy radiation environment. For these reasons CMOS imagers are highly favored over CCD for single photon detection work.

Nonetheless, a few CMOS CTE issues still can hamper photon counting today. For example, the pixel's TG often has a front edge built-in thermal potential barrier that slows charge carrier transit time to the SN region. The barrier height observed is a complex function of PPD and TG implants (dose, energy and alignment), silicon resistively, specific details with the layout of the pixel and the fabrication process employed. Even so, trial and error design and processing can sufficiently reduce the barrier to an acceptable level for a given application. Also unlike buried channel CCDs the TG usually runs surface channel. That is, charge is transferred at the Si-SiO$_2$ surface where traps abound. However, if charge transfers very quickly through the TG the probability that a carrier meeting up with trap can be kept very low yielding imagers with immeasurable CTE for small charge packets (our measurement limit is CTE ~0.999). It is more demanding when large charge packets are transferred because CTE degrades with signal as collapsing fields between the photo diode and SN slows transit time increasing the chances of trap interaction. Few sensors have been tested where CTE is next to perfect as full well is approached. Figure 17 demonstrates a high performance 10 um PMOS pixel imager without a low light deferred CTE problem. Notice as the light level decreases across the chip the signal slowly disappears into the read noise floor (1.5 h+ for this imager).

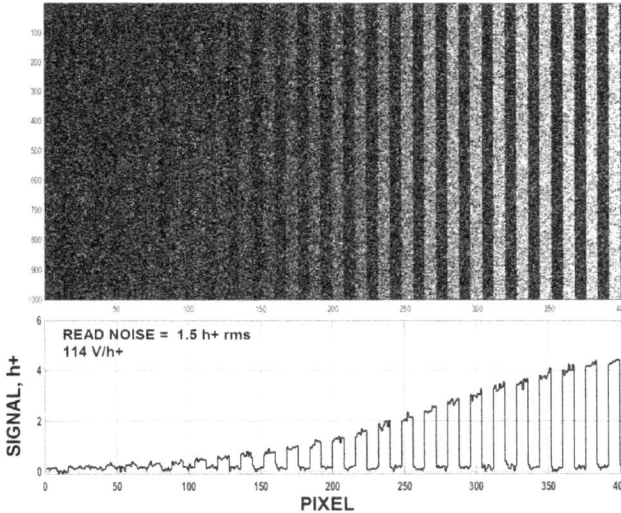

Figure 17. Low light image for a 10 um 5T pinned photo diode (PPD) PMOS pixel imager.

Either inverting a buried channel or accumulating surface channel poly gate with free carriers from the substrate has been important clocking scheme to achieving low dark current generation for both CCD and CMOS technologies [10]. For our PMOS pixels accumulating a surface channel TG takes place at 4.7 V for a p_doped TG and 3.7 V for n_doped TG assuming a substrate voltage of 3.3 V. However, doing so promotes an inconspicuous but adverse CTE problem that is particularly important to night vision and future photon counting applications. For the case of a PMOS pixel, accumulation (ACC) takes place when electrons from the substrate diffuse and collect within TG surface state traps. This in turn arrests TG dark current. Also any presence of trapped holes in the gate oxide in the form of image lag (deferred charge) is completely annihilated by the presence of these electrons. However, this action can deceive the user thinking that perfect CTE without image lag is being achieved when in fact a loss of charge is taking place. When the TG is clocked low to transfer signal charge the majority of these electrons return back to the substrate. However, some electrons remain trapped which can now recombine with the signal holes being transferred at the same time. It is difficult to determine the amount of absolute loss that is taking place as it is similar to experiencing a QE loss at very low light. The percentage of charge that is lost typically increases with decreasing signal and can be significant without the user even realizing it (tens of percent have been experienced). Low energy X-rays can be used for an acid test to determine absolute CTE loss. But a rather easier method to find loss in a relative sense is by measuring the signal transferred as the TG barrier voltage is varied.

For example, Figure 18 presents a response for single CMOS 4T pixel as the light to it is turned on and off as indicated. TG not accumulated (non ACC) and accumulated (ACC) responses are presented as the pixel is switched between these two modes. This particular pixel has a CTE problem associated with the TG front edge barrier discussed above and is seen in Figure 18 as deferred charge tails after the light is turned off for the non ACC mode. Charge loss is observed when TG is biased into ACC as trapped electrons recombine with signal holes. Also the deferred charge tail disappears as electrons recombine with trapped holes.

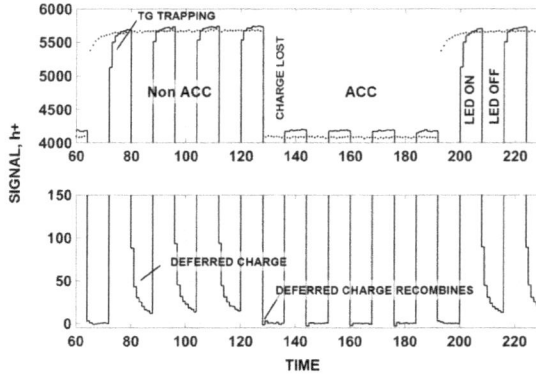

Figure 18. Response for a troubled imager when the transfer gate (TG) is switched in and out of accumulation. Deferred and recombination loss are clearly seen for the device. The top trace shows how charge is trapped whereas the lower trace shows how trapped charge is released and deferred in time.

In general, the amount of recombination loss and deferred charge increase as TG charge transfer transit time increases. We find that a transit time of ~100 ns is sufficient to avert these problems. For example, Figure 19 compares ACC and non ACC settings for a single 10 um PMOS pixel that is transferring single carriers to the SN on the average. The SN is not reset for this test allowing charge to build on the SN in a linear fashion as shown. The light source (LED) is turned "on" and "off" as SN integration takes place on the SN. We can safely conclude that no recombination is taking place when the ACC and non ACC curves are compared. Figure 20 is taken from the same imager biased in the non ACC mode and exposed and stimulated with a high light level source. There is no deferred charge (*i.e.*, lag) detected for the imager and yield a CTE >0.999 over its dynamic range (full well is 15,000 h+). This imager would be ideally suited for photon counting if the read noise was lower (1.4 h+ read noise is measured).

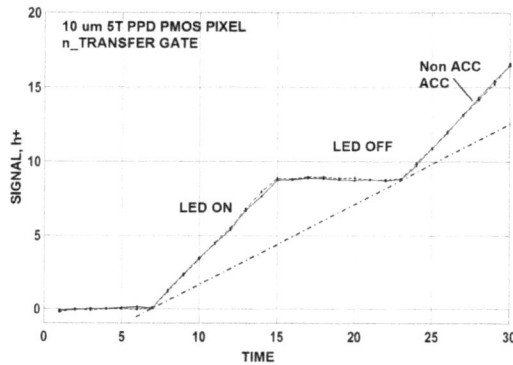

Figure 19. Ultra-low light accumulation (ACC) and non ACC responses without recombination loss.

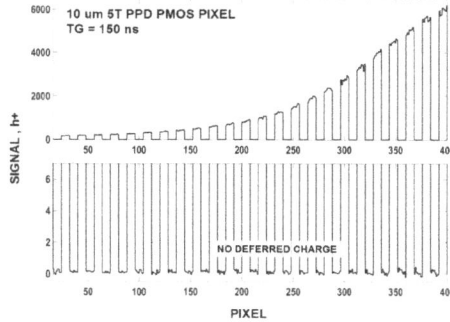

Figure 20. Response from the same imager as Figure 19 showing no deferred charge up to 6000 h+. Charge transfer efficiency (CTE) performance loss for this imager is immeasurable.

3.3. Charge Collection Efficiency

CCE is by no means perfect. Ideally carriers generated by a photon should only be collected by the target pixel without sharing with its neighbors. MTF, OTF, point spread and edge response measure CCE performance and signify the degree of sharing that takes place. Pixel size, silicon thickness/resistivity and the potential difference across the active silicon all play a role on CCE. A CCE problem is especially influential on high energy photon detection applications since a charge packet can diffuse and divide with neighboring pixels (referred to as "split" events). For instance, Figure 21a shows 5.9 keV Mn X-ray events generated by an 8 um 3T PPD pixel fabricated on 25 um intrinsic epi silicon. One can see many split events where the 1620 e^- Mn charge packets diffuse to other pixels instead of being fully contained in the target pixel. Figure 21b presents a response when the same pixel is fabricated on 15 um intrinsic epi silicon where CCE performance is much improved. Neighboring pixels must be summed with the target pixel to determine the energy of the photon. However, this off-chip summation results in an increase in read noise (by the square-root of the number pixels summed) and a corresponding loss of energy resolution. Also split events may experience some recombination loss which leads to an absolute error in energy measurement. This occurs when photons interact with highly doped regions within the pixel (e.g., CCD channel stops, CMOS p and $n\_$wells and at epitaxial/substrate interface). Also backside illuminated imagers often have a finite amount of recombination loss associated with the surface accumulation layer when dealing with photons with a very short absorption length. Fortunately the lack of perfect CCE is not as detrimental to visible photon counting in that the problem only translates to a loss of spatial information.

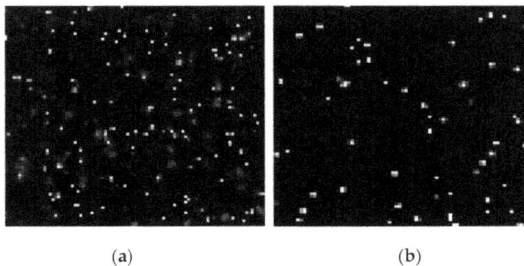

(a)   (b)

Figure 21. Mn (Fe-55) soft X-ray responses taken from an 8 um 3TPPD pixel image fabricated on 25 and 15 μm epitaxial silicon. Charge collection efficiency (CCE) performance for the thicker imager (**a**) is inferior compared to the thinner device (**b**).

4. Conclusions and Future Development

It could be tough going in lowering the "average" read noise substantially below 1 carrier rms because of the $1/f$ noise wall. CDS processing has been pushed to its limit in reducing noise from this source. Averaging multiple samples doesn't help us since $1/f$ noise increases with time as a function of sample count increases. Using buried channel MOSFETS have been successful for reducing RTN but not for $1/f$ noise. Increasing the SN conversion gain is also restricted for the given design and process rules that must be followed. Although our SN conversion is presently limited to 160 uV/h+ there is still some wiggle room left in reducing parasitic capacitances associated with the TG and reset MOSFET that couple to the SN. However, the reduction won't be that significant in that the SF capacitance currently dominates. Consequently, it appears the only means to lower read noise for us is to reduce $1/f$ noise head on.

Hence, future development will focus on $1/f$ reduction without knowing where it is coming from. Various trial and error approaches will be employed on a new runs currently being fabed (be it for mobility or GR reasons). For example one lot run contains deeper buried channel MOSFETs in hopes to see a $1/f$ noise change. Special attention is being given to punch-through issues which usually emerge with buried channel use. Also in order to maximize the surface barrier potential for a buried device the SN must be reset to a SF gate bias voltage that is close as possible to substrate potential. Doing so will reduce dynamic range but the foremost objective behind these experiments is to locate the $1/f$ noise source without much regard to using the pixel in practice.

Experiments involving implants associated with the wells which contain the pixel MOSFETs are being fabricated. We can do this because the pixel implants are derived from custom reticules and independent from the standard n and p_wells used for support circuitry. As mentioned above reducing the doping concentration near the surface could lead to a more uniform current flow with less fluctuation. For example, experimental pixels being fabricated will include a single "minimum implant dose" SF MOSFET. Possibly there will be a $1/f$ noise change since this implant recipe is considerably different than the 3 implant process being used now that is intended for high speed digital MOSFETs. Also further thermal annealing is being incorporated to better activate the pixel wells. The sub carrier noise pixels in Figures 5 and 6 were fabricated with a past 0.25 μm CMOS process which is known to produce lower $1/f$ noise than the 0.18 um process applied today. The 0.25 μm process used a longer thermal anneal cycle. In addition a very long and greater temperature anneal at the very start of the process is being tried to make sure the implants are fully activated.

Native MOSFETs of various geometries are also being fabricated. The native MOSFET is known to have lower $1/f$ noise given that the SF channel doping is the epitaxial silicon [8]. Our epi silicon resistivity varies widely from 10 to 1000 ohm-cm, and hence, we should see a $1/f$ noise change. The experiment will also include baseline MOSFETs with the same geometries as the natives in order to compare $1/f$ noise levels.

Lastly we are trying a proprietary non-imaging gate oxide process offered by Jazz Semiconductor that claims to lower MOSFET $1/f$ RTN noise by 4 to 5 times compared to 0.18 um processing that is used now. All $1/f$ noise reduction approaches above will need to determine the optimum SF geometry for lowest noise performance.

Acknowledgments: The authors are grateful to Tom Elliott also of SRI International for his help in generating some of the data presented. Special thanks also to Richard Mann of Jazz Semiconductor for processing the imagers.

Author Contributions: James Janesick conceived and designed the experiments, defined the fabrication process, generated and analyzed the data and interpreted the results. John Tower supported the work performed and helped interpret the results.

Sensors **2016**, *16*, 688

Abbreviations

The following abbreviations are used in this manuscript:

ACC	accumulated
CCD	charge collection efficiency
CTE	charge transfer efficiency
E	photon energy (eV)
F	Fano factor
FN	Fano noise (rms carriers)
non ACC	not accumulated
PPD	pinned photo diode
PT	photo transfer
QE	quantum efficiency
RTN	random telegraph noise
S/N	signal to noise
SF	source follower
SN	sense node
TG	transfer gate
τ_D	aCDS dominant time constant
t_s	clamp to sample time

References

1. Canfiled, L.; Vest, R.; Korde, R.; Schmidtke, H.; Desor, R. Absolute Silicon Photodiodes for 160 to 254 nm Photons. *Metrologia* **1998**, *35*, 329–334. [CrossRef]
2. Janesick, J. *Scientific Charge-Coupled Devices*; SPIE Press: Bellingham, WA, USA, 2001.
3. Janesick, J. *Photon Transfer λ → DN*; SPIE Press: Bellingham, WA, USA, 2007.
4. Ma, J.; Starkey, D.; Rao, A.; Odame, K.; Fossum, E.R. Characterization of Quanta Image Sensor Pump-Gate Jots with Deep Sub-electron Read Noise. *IEEE J. Electron Devices Soc.* **2015**, *3*, 472–480. [CrossRef]
5. Seo, M.-W.; Kawahito, S.; Kagawa, K.; Yasutomi, K. A 0.27 e_{rms}^- Read Noise 220-µV/e^- Conversion Gain Reset-Gate-Less CMOS Image Sensor with 0.11-µm CIS Process. *IEEE Electron Device Lett.* **2015**, *36*, 1344–1347.
6. Vandamme, L.K.J.; Li, X.; Rigaud, D. 1/*f* Noise in MOS Devices, Mobility or Number Fluctuations? *IEEE Trans. Electron Devices* **1994**, *41*, 1936–1945. [CrossRef]
7. Vandamme, L.K.J.; Hooge, F.N. What Do We Certainly Know about 1/*f* Noise in MOSTs? *IEEE Trans. Electron Devices* **2008**, *55*, 3070–3085. [CrossRef]
8. Miller, D.A. Random Dopants and Low-Frequency Noise Reduction in Deep-Submicron MOSFET Technology. Ph.D. Thesis, Oregon State University, Corvallis, OR, USA, 2011.
9. Levinzon, F.A.; Vandamme, L.K.J. Comparison of 1/*f* noise in JFETs and MOSFETs with several figures of merit. *Fluct. Noise Lett.* **2011**, *10*, 447–465. [CrossRef]
10. Janesick, J.; Elliott, T.; Andrews, J.; Tower, J. Fundamental Performance Differences of CMOS and CCD Imagers: Part VI. *Proc. SPIE* **2015**, *9591*. [CrossRef]

Review

sensors

MDPI

The Quanta Image Sensor: Every Photon Counts

Eric R. Fossum *, Jiaju Ma, Saleh Masoodian, Leo Anzagira and Rachel Zizza

Thayer School of Engineering at Dartmouth, Dartmouth College, Hanover, NH 03755, USA;
Jiaju.Ma.TH@dartmouth.edu (J.M.); Saleh.Masoodian.TH@dartmouth.edu (S.M.);
Leo.Anzagira.TH@dartmouth.edu (L.A.); rzizza111@gmail.com (R.Z.)
* Correspondence: eric.r.fossum@dartmouth.edu; Tel.: +1-603-646-3486

Academic Editor: Nobukazu Teranishi
Received: 24 April 2016; Accepted: 2 August 2016; Published: 10 August 2016

Abstract: The Quanta Image Sensor (QIS) was conceived when contemplating shrinking pixel sizes and storage capacities, and the steady increase in digital processing power. In the single-bit QIS, the output of each field is a binary bit plane, where each bit represents the presence or absence of at least one photoelectron in a photodetector. A series of bit planes is generated through high-speed readout, and a kernel or "cubicle" of bits (x, y, t) is used to create a single output image pixel. The size of the cubicle can be adjusted post-acquisition to optimize image quality. The specialized sub-diffraction-limit photodetectors in the QIS are referred to as "jots" and a QIS may have a gigajot or more, read out at 1000 fps, for a data rate exceeding 1 Tb/s. Basically, we are trying to count photons as they arrive at the sensor. This paper reviews the QIS concept and its imaging characteristics. Recent progress towards realizing the QIS for commercial and scientific purposes is discussed. This includes implementation of a pump-gate jot device in a 65 nm CIS BSI process yielding read noise as low as 0.22 e− r.m.s. and conversion gain as high as 420 µV/e−, power efficient readout electronics, currently as low as 0.4 pJ/b in the same process, creating high dynamic range images from jot data, and understanding the imaging characteristics of single-bit and multi-bit QIS devices. The QIS represents a possible major paradigm shift in image capture.

Keywords: photon counting; image sensor; quanta image sensor; QIS; low read noise; low power

1. Introduction

The Quanta Image Sensor (QIS) was conceived in 2004 and published in 2005 [1–4] as a forward look at where image sensors may go in the 10 to 15-year future as progress in semiconductor device technology would allow sub-diffraction limit (SDL) pixels to be readily implemented, and advancement in circuit design and scaling would permit greater pixel throughput at reasonable power dissipation levels. Active research began in 2008 at Samsung (Yongin, Korea) [5] but was short lived due to economic pressure in that period. Research began anew at Dartmouth in 2011 and was supported from 2012 to the present by Rambus Inc. (Sunnyvale, CA, USA). Since 2011, progress has been made in pixels, readout circuits, and image formation [6–26]. In this paper, the QIS concept and progress to date is reviewed.

In the QIS, SDL pixels (e.g., 200 nm–1000 nm pitch) are sensitive to single photoelectrons, so that the presence or absence of one electron will result in a logical binary output of 0 or 1 upon readout. The specialized pixel, called a "jot", (Greek for "smallest thing") needs only the smallest full-well capacity (FWC). It is envisioned that a QIS will consist of hundreds of millions or perhaps billions of jots read out at perhaps 1000 fields per second, resulting in a series of bit planes, each corresponding to one field. The bit data can be thought of as a jot data cube, with two spatial dimensions (x and y) with the third dimension being time.

Output image pixels are created by locally processing the jot data cube to create a representation of local light intensity received by the QIS. Since this processing occurs post-capture, great flexibility is afforded in choosing the effective spatial dimensions of a pixel as well as its temporal dimension (e.g., digital integration time). Conceptually, if the bit data is an accurate representation of the collection and counting of photoelectrons, the combining of jot data is noiseless, allowing functionality such as time-delay-and-integration (TDI) to be performed post-capture on the data in an arbitrary track direction. In fact, different tracks can be used in different portions of the image. Indeed, even relative motion of objects within the field of view can be determined and refined iteratively to optimize the image generation process. Spatial and temporal resolution can also be adjusted for different portions of the image.

After the QIS concept was introduced in 2005, the concept was applied for use with single-photon avalanche detectors (SPADs) by the group at the University of Edinburgh [27–30] as published starting in 2014 as part of their SPAD research program. Other work on SPADs published in 2005 shows concurrent conception of some of the same ideas [31], and in 2009 a "gigavision camera" using binary pixels was proposed by researchers from École Polytechnique Fédérale de Lausanne (EPFL) [32–36]. Some characteristics of the QIS can be traced to photographic film as reported in 1890 [37] and other characteristics were observed in photon-counting devices implemented using vacuum tubes and solid-state devices [38–43]. Progress in other devices achieving sub-electron and deep sub-electron read noise has mainly been made in the past few years [44–51].

Possible applications of the QIS include scientific low-light imaging such as in life sciences (e.g., microscopy), defense and aerospace, professional and consumer photography and cinematography, multi-aperture imaging, cryptography (quantum random number generation), direct detection of low-energy charged particles, and others.

To realize the QIS in a convenient form, several theoretical and technological issues require exploration. These are: (1) image formation algorithms that yield high quality images; (2) understanding the imaging characteristics of QIS devices; (3) the implementation of pixels (jots) that enable photon counting; (4) low-power readout of high volumes of data (readout of a 1 Gjot sensor at 1000 fps yields a data throughput rate of 1 Tb/s) and (5) on-focal-plane processing to reduce the data volume. Exploration of these issues of the course of the last several years has led to significant advancement in the first four areas with the fifth just being explored now. This paper reviews this progress. Additional details may be found in the cited references.

2. Creating Images from Jots

Readout of jots results in a bit cube of data, with two dimensions representing spatial dimensions of the field-of-view of the sensor, and the third representing time. Each bit-plane slice is a single readout field. For single-bit QIS devices, the jot data cube is binary in nature. For multi-bit QIS, the jot data cube consists of words of bit length corresponding the readout quantization bit depth described in more detail below.

In perhaps the simplest form, image pixels can be created from the sum of a small x-y-t "cubicle" of bits from the jot data cube (see Figure 1). The dimensions of the cubicle determine the spatial and temporal resolution of the output image, and all jot values within the cubicle are weighted equally. The maximum signal-to-noise ratio (SNR) of the image pixel (assuming an ensemble of pixels created from the same illumination and readout conditions) is determined by the size of the cubicle. For example, a cubicle of size $16 \times 16 \times 16$ of single-bit QIS jots summed together would have a maximum value of 4096 and a maximum SNR of $\sqrt{4096} = 64$. Illustration of image formation from simulated jot data is shown in Figure 2. Note that cubicles do not have to have equal dimensions in x, y and t. Furthermore, all image pixels need not be formed from the same sized cubicles, and their cubicles may overlap. These choices can be made post-capture and on an output frame-by-frame basis to optimize particular imaging aspects such as the trade-off between image SNR and spatial-temporal resolution. An example of this processing was recently demonstrated using a QIS-like device with

SPAD-based jots [28] where the pixels near the edges of rotating fan blade were processed differently than either the slowly varying body of the blades or the static background.

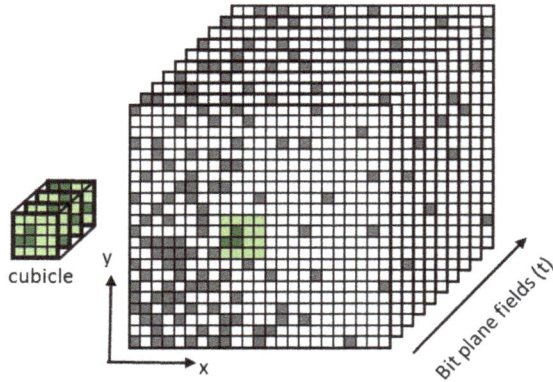

Figure 1. Conceptual illustration of a jot data cube and a 4 × 4 × 4 cubicle subset.

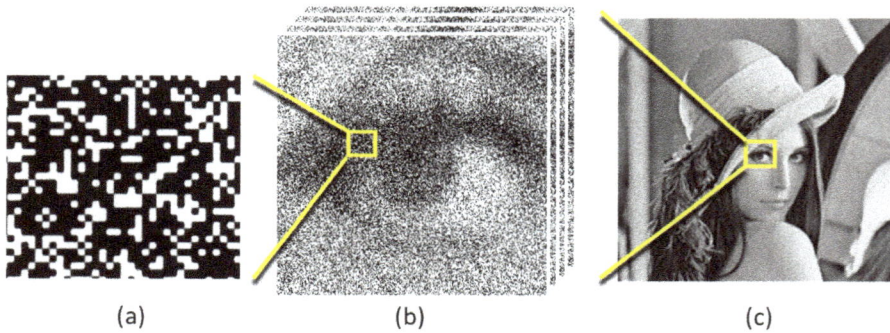

(a)   (b)   (c)

Figure 2. Simulation of (**a**) raw jot data; (**b**) same at lower magnification; (**c**) after processing cubicles to form grey scale image. From [16].

Using cubicles that have a non-orthogonal trajectory in the time dimension can be used for performing operations analogous to time-delay and integration (TDI) but along an arbitrary track direction [6]. Different objects in the scene with different motion trajectories could be processed with independent tracks to improve SNR. The utility of such processing is enhanced by deep sub-electron read noise in the QIS and quantizer, so that the noise is inherently low and does not accumulate like $\sqrt{M}$, where M is the ensemble size of jots summed together.

Measuring photon flux with a single-pixel photon-counting photomultiplier tube was reported as early as 1968 [38]. The generation of an image with improved SNR from a series of readout fields goes back to 1985 when photon-counting detectors were used this way for astronomy purposes [39,40]. The technique was also applied to CCDs with built-in avalanche multiplication [43]. Similar operations with CMOS image sensors (CIS) were envisioned as early as 1998 [52] and proposed as a digital integration sensor (DIS) [7]. The technique was also applied in SPAD arrays [31]. SPAD arrays operating as QIS devices were demonstrated in [30]. Following the introduction of the early QIS concept, the "gigavision" binary image sensor was proposed in 2009 [33,34]. Image formation from bits in the binary sensor was explored mathematically by EPFL [36] and by Harvard [53]. These investigations concerned a cubicle that was essentially $1 \times 1 \times N$, that is, one pixel read out N times.

QIS devices use both spatial and temporal sampling to create one image pixel using sub-diffraction limit pixel sizes [2], although binning multiple pixels to improve SNR at the expense of spatial resolution dates back to at least the early days of CCDs in the 1970's. Uniform weighting and other weight distributions applied to jots of the bit cube to form image pixels have been explored by Zizza [16]. It was found that there was little apparent impact on image quality between non-uniformly and uniformly weighted cubicles, and modulation transfer function (MTF) and SNR were also not significantly impacted. It was also found that the EPFL and Harvard algorithms for creating images from jots worked well for single, static images, but when processing time and latency are considered for continuous image acquisition, simple summation of cubicle data is preferred.

3. Imaging Characteristics

3.1. Hurter-Driffield Characteristic Response (D-LogH)

In the QIS, the statistical nature of the arrival of photons and the photoelectrons they produce are well described by Poisson arrival statistics. For convenience we define the quanta exposure H as the average number of photoelectrons collected by a jot over an integration period, which depends on factors such as the incident photon flux, effective jot area, quantum efficiency, carrier collection efficiency and integration time. The probability of there being k photoelectrons is given by the Poisson mass function:

$$\mathbb{P}[k] = \frac{e^{-H} H^k}{k!} \tag{1}$$

The probability that there are no photoelectrons is $\mathbb{P}[0] = e^{-H}$, and the probability of at least one photoelectron is $\mathbb{P}[k > 0] = 1 - \mathbb{P}[0] = 1 - e^{-H}$. In the single-bit QIS, reading out no photoelectrons is a logic "0" and reading out one or more photoelectrons is set as a logic "1". Essentially, the full-well capacity (FWC) of the single-bit QIS is 1 e−. The bit density D of jots (fraction per bit ensemble that have logic value "1") gives rise to an S-shaped curve if D is plotted as a function of $\log(H)$, reflecting the relationship [3,9]:

$$D = 1 - e^{-H} \tag{2a}$$

This characteristic QIS-response curve is shown in Figure 3a. This behavior is similar to the behavior of an ensemble of avalanche detectors in a Si photomultiplier and also observed in SPADs and in subsequent analysis of binary sensors [33,34,41]. The standard deviation of D (or noise, σ) is given by [9]:

$$\sigma = \sqrt{e^{-H} \left[1 - e^{-H} \right]} \tag{2b}$$

as shown in Figure 4.

The statistical nature of photoelectron counting (or essentially photon counting if the efficiency factors above are close to unity) is the same as that which gives rise to the D-$\log(H)$ nature of film exposure reported by Hurter and Driffield in 1890 [37] due to the statistical exposure of film grains, as shown in Figure 3b. *Henceforth, the asymptotic response for Figure 3a,b is referred to as the Hurter-Driffield response curve in their honor, whether referring to film, SPADs or QIS jots.* In the case of Ag-I film grains, about 3 photoelectrons are required to result in exposure leading to slightly different slopes [9,33,42]. In fact, many photographers and cinematographers desire this non-linear behavior that results in a response curve with good overexposure latitude [54]. The curve shape is determined by the underlying statistical physics of exposure and threshold for creating jots with value "1", and is not influenced strongly by circuit or device performance. Further, the exposure-referred signal-to-noise ratio (SNR_H) as a function of exposure is well-behaved [9], unlike that which is found in conventional image sensors operating in high dynamic range mode, and which suffer from one or more large dips in SNR_H as the exposure is increased (e.g., [7]).

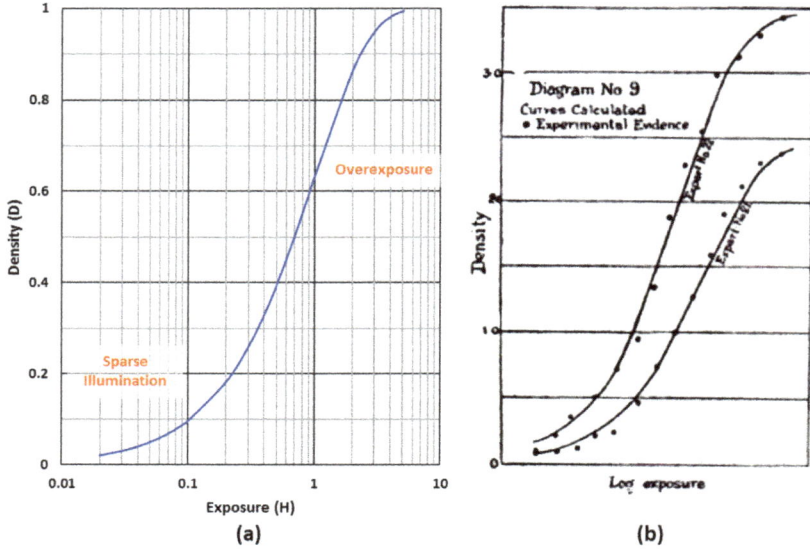

Figure 3. (a) Single-bit QIS bit density as a function of quanta exposure, in D-log(H) format. Sparse and over exposure regions are labelled [9]; (b) D-log(H) curve for photographic plates as reported by Hurter and Driffield in 1890 [37].

3.2. Flux Capacity

An important figure of merit for QIS devices is *flux capacity*. Flux capacity ϕ_w is defined as the nominal maximum photon flux that results in $H = 1$. It is dependent on the density of jots j in the image sensor (jots/cm^2), the readout field rate f_r, the shutter duty cycle δ and the effective quantum efficiency $\bar{\gamma}$ according to [14]:

$$\phi_w = jf_r/\delta\bar{\gamma} \qquad (3)$$

For photography and cinematography, high flux capacity is required so that the QIS does not saturate under normal imaging conditions. Note that the QIS can handle exposures for $H > 1$ without saturating, typically $5\times$ higher due to its overexposure latitude, but $H = 1$ is taken for convenience. For example, with a jot pitch of 500 nm, 1000 fps field rate, unity duty cycle and 50% avg. QE, the flux capacity is 8×10^{11} photons/cm^2/s which at F/2.8, QE = 50%, lens T = 80%, scene R = 20%, corresponds to ~400 lux at the scene (yielding $H = 1$ at the sensor). It can be seen that high jot density and high field readout rate are driven by flux capacity and not necessarily by improved spatial nor temporal resolution of the final image, although these are additional benefits. The sub-diffraction jot pitch requires use of advanced-node processes that are expensive and difficult to access today. The high readout rate creates challenges in controlling power dissipation in the readout circuit.

3.3. Multi-Bit QIS

To further improve flux capacity, the multi-bit QIS was proposed. In the multi-bit QIS, the readout result can result in 2^n states, where n is the readout bit depth (a single-bit QIS is, in essence, a special case of a multi-bit QIS with $n = 1$). For example, if $n = 2$ then 4 possible states can be considered, (1) no photoelectron, $\mathbb{P}[0] = e^{-H}$ as before, now coded logically as "00"; (2) one photoelectron, $\mathbb{P}[1] = He^{-H}$ now coded logically as "01"; (3) two photoelectrons, $\mathbb{P}[2] = H^2e^{-H}/2$ coded logically as "10"; and (4) 3 or more photoelectrons, $\mathbb{P}[k > 2] = 1 - e^{-H} - He^{-H} - H^2e^{-H}/2$, coded logically as "11". During readout of the signal from the jot, each of these states must be discriminated, such as by using a 2b analog-to-digital converter (ADC). The multi-bit QIS has a *FWC* given by $2^n - 1$ and the flux capacity ϕ_{wn} is increased to:

$$\phi_{wn} = jf_r\left(2^n - 1\right)/\delta\bar{\gamma} \tag{4}$$

The Hurter-Driffield response is modified by multi-bit QIS readout resulting in higher saturation signal, less non-linearity and less overexposure latitude [9,14]. The expected number of electrons read out from a multi-bit QIS jot is given by *<k>* where:

$$< k > = \sum_{k=0}^{FWC} k \cdot \mathbb{P}\left[k\right] + \sum_{k=FWC+1}^{\infty} FWC \cdot \mathbb{P}\left[k\right] = FWC\left[1 - \sum_{k=0}^{FWC}\left(1 - \frac{k}{FWC}\right) \cdot \mathbb{P}\left[k\right]\right] \tag{5}$$

The variance in the number of electrons is:

$$\sigma^2 = < k^2 > - < k >^2 \tag{6}$$

where $< k^2 >$ is given by:

$$< k^2 > = \sum_{k=0}^{FWC} k^2 \cdot \mathbb{P}\left[k\right] + \sum_{k=FWC+1}^{\infty} FWC^2 \cdot \mathbb{P}\left[k\right] = FWC^2\left[1 - \sum_{k=0}^{FWC}\left(1 - \frac{k^2}{FWC^2}\right) \cdot \mathbb{P}\left[k\right]\right] \tag{7}$$

The multi-bit signal summed over an ensemble of *M* jots is $M < k >$, and the noise (standard deviation) is $\sqrt{M\sigma^2}$.

Consider an ensemble of 4096 jots. For a single-bit QIS, the maximum signal obtained by adding together the logical readout signal from each jot (0 or 1) is 4096. For the 2-bit QIS, the maximum signal is increased by three-fold ($2^n - 1$) to 12,288 and the noise characteristic rolls off more steeply. The summed signal of an ensemble of 4096 multi-bit jots is shown in Figure 4. The predicted signal and noise vs. exposure relationship for a single-bit QIS was first experimentally verified by Dutton et al. [29].

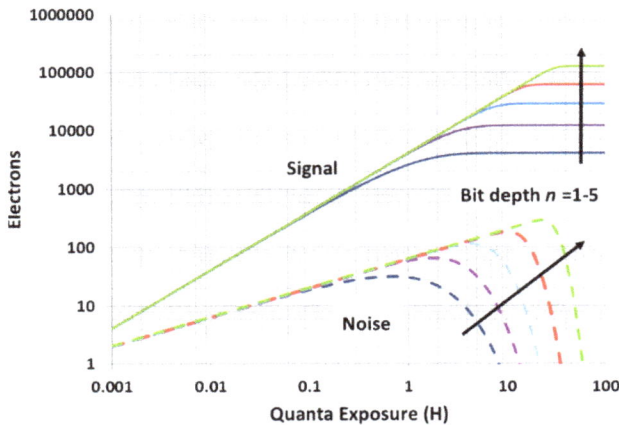

Figure 4. Log signal and noise as a function of log exposure for multi-bit QIS jots with varying bit depth. The signal is the expected sum over 4096 jots (e.g., 16 × 16 × 16). Saturation signal is 4096 ($2^n - 1$) (From [14]).

3.4. Signal-to-Noise Ratio (SNR) and Dynamic Range (DR)

The use of exposure-referred *SNR* is useful for non-linear devices, especially when intrinsic signal noise drops near saturation. The SNR_H for normal readout of a single-bit QIS is given by [9]:

$$SNR_H = \sqrt{M}\frac{H}{\sqrt{e^H - 1}} \tag{8}$$

where M is the number of jots in the ensemble or cubicle used for the read signal sum. This assumes the read noise is low enough that the readout bit error rate (BER) [9,14] does not significantly affect the sum—a reasonable assumption for the QIS target read noise of less than 0.15 e− r.m.s. [9,21]. For normal readout, the SNR_H reaches a maximum value at $H \cong 1.6$ and the maximum value of SNR_H is approximately $0.8\sqrt{M}$. For multi-bit QIS, maximum SNR_H increases as $\sim\sqrt{FWC}$.

Dynamic range (DR) for the QIS is defined as the range between low signal H_{min} where $SNR_H = 1$ (essentially where the ensemble of read out jots has a total sum of one photoelectron) and high signal H_{max} where SNR_H drops back down to unity and lower due to saturation [9]. The DR depends on the size of the ensemble—more jots, higher dynamic range, and the DR scales approximately as M, where M is the number of jots in the ensemble. The DR and the maximum value of SNR_H are shown in Figure 5 as a function of ensemble size, calculated using the expressions derived in [9]. At $M = 4096$, for example, the DR is 95 dB and the maximum value of SNR_H is 34 dB. This can be compared to a conventional CIS with FWC of 4096 e− and read noise of 1.5 e− r.m.s. The CIS would have a DR of 68 dB and maximum SNR of 36 dB assuming linear response. Increasing the bit depth does not substantially increase the DR due to the reduction in non-linearity (overexposure latitude) as the bit depth is increased. For example, going from single-bit to 2b QIS (3× increase in FWC) only increases the DR by approximately 3 dB.

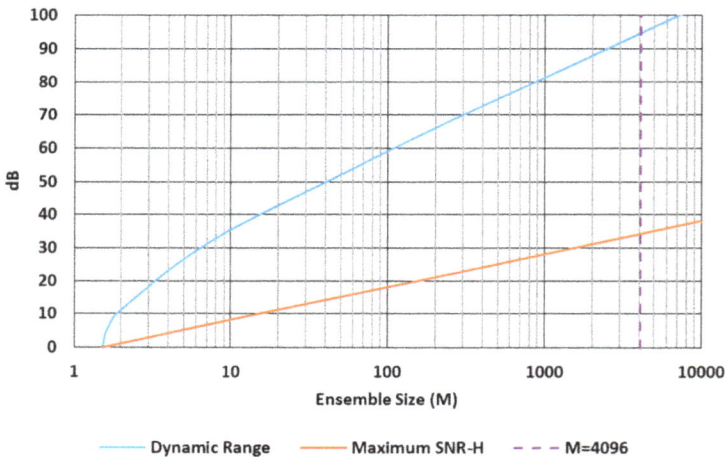

Figure 5. Single-bit QIS normal readout dynamic range (blue) and maximum SNR_H as a function of ensemble or cubicle size M. For convenience, $M = 4096$ is highlighted by the purple dashed line. Note that M is confined to integer values despite the continuous nature of the curves in this figure.

The response of the single-bit QIS is only linear for $H \lesssim 0.1$ with increasing non-linearity above this exposure level. The non-linearity is a desirable feature for photography and cinematography, but not as much for some other applications, and undesirable for most scientific photon-counting applications. In the latter case, the linear response can be retrieved from the non-linear signal or the exposure must be kept so $H \lesssim 0.1$. Multi-bit QIS has a larger range of linearity as can be readily seen in Figure 4. As was noted in [14] multi-bit signals can be easily transformed into lower bit depth signals, or single-bit signals, by post-readout digital signal processing. This permits some flexibility in the response curve by trading FWC or flux capacity for linearity.

3.5. High Dynamic Range (HDR)

Since the QIS consists of multiple fields of jot data that may be combined in a cubicle ensemble, it is possible to have a different electronic shutter duty cycle or "speed" for each field or time slice. Thus some time slices can have high flux capacities allowing capture of brighter portions of scenes without saturation. A similar idea has been used in CMOS image sensors for many years [55] for high dynamic range (HDR) imaging, although it can suffer from imaging artifacts due to relative motion of the scene between captured fields. The higher field readout rate of the QIS will ameliorate some of those artifacts. The Hurter-Driffield response characteristics of the QIS help reduce SNR "dips" caused by the fusion of multiple fields of data taken with different shutter speeds [9]. Multi-bit QIS devices can also be operated in HDR mode. For example, consider 2b-QIS output formed from a 16 × 16 × 16 cubicle. In normal readout mode summing all 16 fields, each with 100% shutter duty cycle, the *DR* is approximately 98 dB. In an HDR mode, with summing a cubicle where 13 fields are exposed with 100% shutter duty cycle, 1 field at 20% duty cycle, 1 field at 4% duty cycle, and 1 field at 0.8% duty cycle, the dynamic is extended to approximately 135 dB as illustrated in Figure 6. Figure 6 shows the log signal vs. log exposure characteristic of a 1b QIS, 2b QIS and their attendant SNR_H when the 16 fields in the cubicle all have 100% duty cycle and are summed (green and blue respectively). For the 2b QIS, the maximum signal is 3 e− per jot leading to a maximum sum of the cubicle of 3 × 4096 = 12,288. Also shown in the figure are the 2b QIS cubicle sums vs. exposure for the 4 component fields. S1 shows the sum of 13 fields taken with 100% shutter duty cycle. The number of fields should be large in order to capture low light detail in the image with good SNR. The three following fields' cubicle sums S5, S25 and S125 are taken with 20%, 4%, and 0.08% duty cycles respectively—essentially 1/5, 1/25, and 1/125 relative shutter speeds. The sum of all these fields of the cubicle is the 2b HDR signal vs. exposure characteristic (red) along with its attendant SNR_H. The latter has one large SNR_H dip from its peak of ~37 dB to a plateau of 27 dB in the extended range. While on a log scale, the HDR curve (red) looks similar to the normal readout curve (blue), the inset shows a linear-linear plot of the extended range, showing significant contrast for the HDR response. Different transfer curves can be generated by varying the relative duty cycles and number of fields and this particular set of signal vs. exposure curves are just one example set.

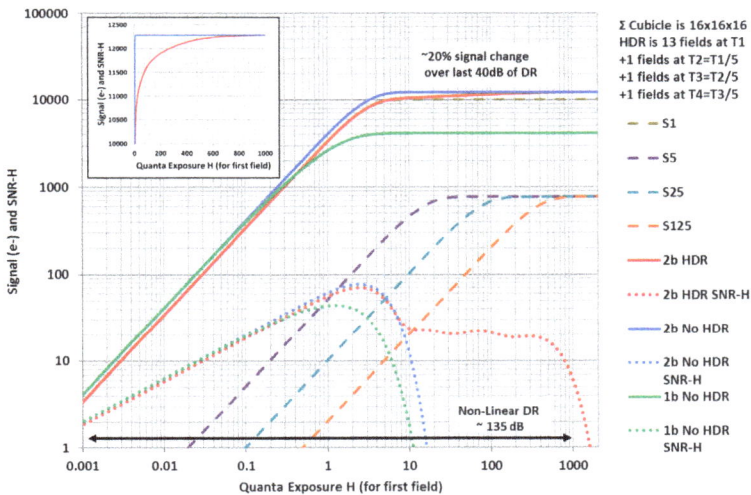

Figure 6. Log signal vs. log exposure for a 2b QIS operated in normal readout mode (blue) and high dynamic range mode (red), showing extension of the dynamic range by approximately 40 dB. Inset shows linear-scale response curves for the extended exposure range.

4. Read Noise and Counting Error Rates

4.1. Read Noise and Readout Signal Probability

Counting photon or photoelectrons requires deep sub-electron read noise (DSERN), that is, read noise less than 0.50 e− r.m.s. It has been suggested that 0.30 e− r.m.s. read noise is sufficient for many photon-counting applications [47,56,57] however accurate counting with low error rate under low exposures (e.g., $H < 0.2$) requires read noise less than 0.15 e− r.m.s. [21].

When both read noise and conversion gain variation is considered for an ensemble of jots that are read out, the probability distribution of readout voltages is given by:

$$P[U] = \sum_{k=0}^{\infty} \frac{\mathbb{P}[k]}{\sqrt{2\pi\sigma_k^2}} exp \left[-\frac{(U-k)^2}{2\sigma_k^2} \right] \tag{9}$$

where U is the readout signal normalized by mean conversion gain (in electron number), u_n is the read noise (in e− r.m.s.), and where σ_k is given by:

$$\sigma_k \triangleq \sqrt{u_n^2 + (k\sigma_{CG}/\overline{CG})^2} \tag{10}$$

and σ_{CG} is the standard deviation of conversion gain in the ensemble, and $\overline{CG}$ is the mean conversion gain in the ensemble.

An example of the distributions arising from different levels of read noise is shown in Figure 7 Electron number quantization is seen for read noise in the deep sub-electron range. Using a fine resolution ADC, plots of frequency of occurrence vs. readout voltage can be made for experimental jot devices under illumination and are called photon-counting histograms (PCH). The ratio of valley amplitude to peak amplitude, called valley-peak modulation (VPM) can be used to experimentally determine read noise from the PCH, the peak spacing can be used to determine conversion gain, and the relative peak heights can be used to determine exposure [22].

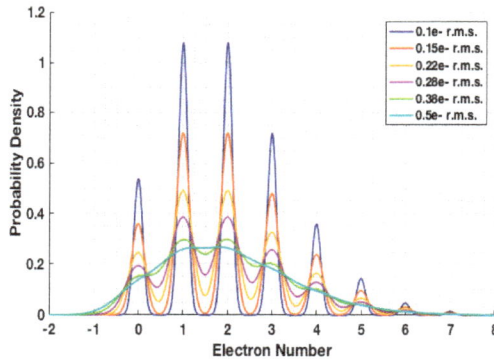

Figure 7. Readout signal probability density as a function readout signal for a quanta exposure $H = 2$ and for various read noise levels. Examples with CG variation are shown in [21,22].

4.2. Quantization and Bin Counts

In a single-bit QIS, the threshold level U_{T1} for setting the output to a logic "1" is $U_{T1} = 0.5$, so that the digital output is "1" for $U > 0.5$ and otherwise "0". For multi-bit QIS, additional thresholds are set at integer increments above U_{T1} (e.g., $U_{T2} = 1.5$, $U_{T3} = 2.5$, etc.). All read out signals lying between two adjacent thresholds (a bin) result in the count for that bin C_N being incremented by one. False positive counts are generated when a read out signal is misquantized into the wrong bin due to noise and conversion gain variation. These false positives (and their corresponding false negatives) give rise to an error in the total count.

An ensemble of M jots results in a total of M counts spread across the 2^n bins of a multi-bit QIS. The total count in each bin can be used to determine the expected total number of photoelectrons collected by the ensemble N_{TOT} such that [21]:

$$N_{TOT} = M \sum_{N=0}^{\infty} N \cdot C_N \tag{11}$$

For $\sigma_k \lesssim 0.50$ e– r.m.s., the expected count in each bin for a single jot is given by:

$$C_N \cong \sum_{k=N-1}^{N+1} \frac{1}{2} \mathbb{P}[k] \left[\mathrm{erf}\left(\frac{N + \frac{1}{2} - k}{\sigma_k \sqrt{2}} \right) - \mathrm{erf}\left(\frac{N - \frac{1}{2} - k}{\sigma_k \sqrt{2}} \right) \right] \tag{12}$$

except for bin 0 and last bin $2^n - 1$ where in the former the bin extends to $U = -\infty$, and in the latter to $U = +\infty$, so that:

$$C_0 = \frac{1}{2} \sum_{k=0}^{\infty} \mathbb{P}[k] \left[\mathrm{erf}\left(\frac{\frac{1}{2} - k}{\sigma_k \sqrt{2}} \right) + 1 \right] \tag{13}$$

and:

$$C_{2^n - 1} = \frac{1}{2} \sum_{k=0}^{\infty} \mathbb{P}[k] \left[1 - \mathrm{erf}\left(\frac{(2^n - 1) - \frac{1}{2} - k}{\sigma_k \sqrt{2}} \right) \right] \tag{14}$$

In [21] it was found that for higher quanta exposures ($H > 0.2$), the count was not significantly affected by (deep sub-electron) read noise nor conversion gain variation in the ensemble, assuming M was sufficiently large, and the non-linear Hurter-Driffield response dominated counting error in a predictable way. For lower quanta exposure, systematic count error was introduced by read noise. Essentially under sparse illumination conditions ($H < 0.1$), even a small amount of read noise can cause excess counting by the occasional misquantization of the dominant "0" signal as "1". This systematic count error can result in counting rate error of 34% for $H = 0.1$ and read noise of 0.30 e– r.m.s. yet nearly no error at a read noise of 0.20 e– r.m.s. The systematic counting rate error increases dramatically for lower exposures, strongly indicating that for applications requiring accurate photon counting in this realm, read noise should be 0.15 e– r.m.s. or smaller.

Count vs. quanta exposure is shown in Figure 8. Ideally the count should be equal to the quanta exposure leading to a linear relationship shown by the diagonal gray line (mostly obscured). For a 4b QIS with read noise of 0.15 e– r.m.s., (purple solid line), the count is nearly indistinguishable from the linear relationship. However, for higher read noise levels, significant systematic departure from the ideal behavior under sparse illumination conditions can be observed, independent of bit depth. At higher exposures, the Hurter-Driffield response dominates the non-linear behavior independently of read noise. In all cases, the impact of conversion gain variation in the ensemble is negligible if M is sufficiently large, and if not, then photoresponse non-uniformity can be an issue as in conventional CIS devices.

The expected count in Equations (11)–(14) can be used to estimate the count for an ensemble of M jots. For example, consider a single-bit QIS array of jots with pitch of 1 μm. An ensemble of 100 jots, formed from $10 \times 10 \times 1$ cubicle would cover an area of size 10 μm $\times$ 10 μm. For a quanta exposure $H = 0.01$, the ideal expected count from the ensemble would by $N_{TOT} = 1$. Using Equations (11)–(14) or Figure 8 with $H = 0.01$, and read noise of 0.15 e– r.m.s., one obtains an expected count of $100 \times 0.0104 = 1.04$ e–, but for a higher read noise of 0.30 e– r.m.s., one obtains an expected count of $100 \times 0.0568 = 5.68$ e–. It is noted that this systematic error is different in nature than a typical manifestation of read noise which leads to a correct average value over a large number of samples but with some standard deviation or noise. In this case, the average readout value itself is offset.

Sub-electron (voltage) quantizer resolution (e.g., 0.05 e−) may be used to provide more accurate counting in the presence of higher read noise, by computing the mean signal of a larger number of samples and converting to electrons, as is done conventionally and which was used to calibrate the horizontal axis of Figure 8. However this requires a more accurate ADC, higher power, and likely slower field readout rate.

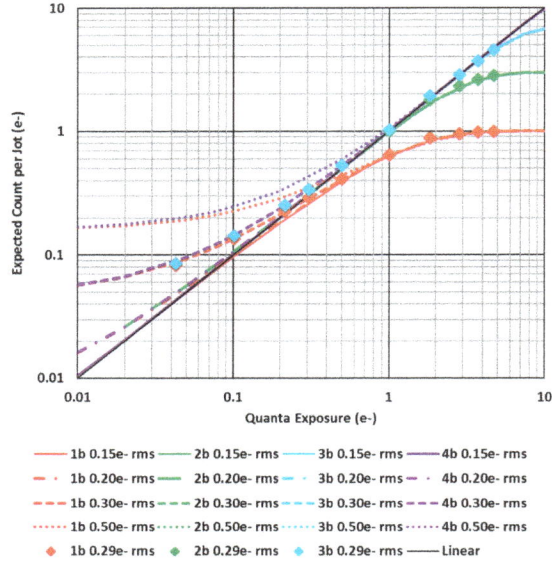

Figure 8. Expected count vs. quanta exposure for various bit depths and read noise levels. Experimental data is shown by the diamond symbols (after [21]).

5. Jot Device

5.1. Background and Motivation

The ultimate goals of a jot device include small pitch size (200 nm–500 nm), low read noise (<0.15 e− r.m.s.), low dark current (<1 e−/s), small FWC (1–100 e−) and strong compatibility with a CIS fabrication line. One big difference between a jot and a conventional CIS pixel is its deep sub-electron read noise and photoelectron counting capability. A conventional CIS often has voltage read noise higher than 100 µV r.m.s. and CG lower than 100 µV/e−, yielding read noise higher than 1 e− r.m.s. Higher CG and lower voltage noise reduce input-referred read noise. As a possible candidate for a jot device, SPADs are widely used for photon counting [see this Special Issue]. Through the avalanche multiplication effect, it can provide a higher CG (>1 mV/photoelectron) and low read noise (<0.15 e− r.m.s.). It has been used to demonstrate the QIS concept and showed interesting results [27–30]. Unfortunately, its relatively large size (typically 5–10 µm pitch) limits flux capacity and resolution [32], high electric fields result in high dark count rate (~1000 counts/s/pix), low fill factor and dead time reduce photon detection efficiency, and manufacturing yield is lower than conventional CIS process on a pixel-by-pixel basis. Electron-multiplying CCD (EMCCD) technology provides high CG by an electron multiplication process and is able to achieve 0.45 e− r.m.s. average read noise [49]. But similar to SPAD arrays, it has a high dark current due to thermal generation of carriers under high electric fields. EMCCDs also have relatively low frame rates as the signal is read out by CCD circuitry. We have considered these devices and other devices such as floating-base bipolar transistors, as candidates [11].

In consideration of fabrication feasibility, we started the jot design based on a conventional intra-pixel charge transfer approach similar to conventional CIS 4T pixels with a "pinned photodiode" due to its mature fabrication process, low dark current and high quantum efficiency, which also provides the jot device with good compatibility to many techniques developed in CIS, such as BSI, shared readout and stacked process. For example, in a BSI device, fill factor is very high, nearly unity, and backside treatments to reduce reflection losses are well known from the CCD era. Carrier collection efficiency can also high, depending on detailed device design and the funneling of carriers to the storage well. The readout introduces minimal dead time compared to SPADs, and may be as low as 0.004% in a gigajot sensor. A typical CIS readout chain includes an in-pixel source follower (SF), a correlated double sampling (CDS) circuitry, a high analog-gain amplifier and an ADC. Voltage noise is added to the voltage signal by each readout component before the signal is digitized in the ADC, and in standard practice, it is best to add gain earlier in the signal chain to ameliorate the impact of downstream noise components. Generally, in a low-noise (1 e− r.m.s. to 1.5 e− r.m.s.) CIS, the in-pixel SF contributes most of the input-referred voltage noise, typically 100–200 μV r.m.s. An in-pixel common source amplifier can provide a higher than unity gain and suppress latter noise sources without increasing the pixel size, but it also generates high gain variation [45], which can be detrimental for multi-bit QIS application. The major noise components in an in-pixel SF are $1/f$ noise and random-telegraph signal (RTS), and both appear to be related to the carrier capture and emission process of surface interface traps, either at the gate oxide-semiconductor interface, or due to shallow trench isolation sidewalls [58], although other sources of $1/f$ noise such as turbulent flow have been suggested [59,60]. Buried-channel SF and correlated multiple sampling (CMS) techniques [61] were applied to reduce the SF noise, and 35 μV r.m.s. voltage noise was achieved, but with a relatively low CG (46 μV/e−) yielding 0.76 e− r.m.s. average read noise [46]. The CMS technique was also explored with low temperature (because of dark current considerations) and achieved limited photoelectron counting capability [44,47,51]. However, the CMS technique with a large number of samples is not feasible for QIS application due to its relatively low speed.

Our approach to achieve deep sub-electron read noise is to improve CG and reduce SF transistor noise. Since our first report of success with this approach [14,17] other groups have also reported success at achieving deep sub-electron read noise (and photon counting) without the use of avalanche gain [50,51]. Improvement of CG was also reported in [62] leading to 0.46 e− r.m.s. read noise, just short of what is needed to demonstrate photoelectron counting.

The photoelectron signal is converted to a voltage signal for readout using the capacitance of the floating diffusion (FD) node. The voltage signal generated by one photoelectron is given by:

$$CG = \frac{q}{C_{FD}} \tag{15}$$

where q is the elementary charge of one electron and C_{FD} is the node capacitance of FD that includes several major components: depletion capacitance between FD and substrate, overlap capacitance between FD and the transfer gate (TG), overlap capacitance between FD and reset gate (RG), SF effective gate capacitance, and inter-metal capacitance. To improve CG, the capacitance of FD needs to be reduced. Note that the reduction of FD capacitance may lead to reduction of FWC in conventional CIS, but it is not a concern for a jot device as the required FWC is very small.

5.2. High CG Pump-Gate Jot Devices

Depending on the process feature size and layout design, the overlap capacitance between FD and TG in a CIS pixel can be 0.3 fF or higher, especially in a pixel with a shared readout structure. A pump-gate (PG) technique was developed by our group to eliminate the overlap capacitance between FD and TG without affecting complete charge transfer [13]. A cross-section doping profile of a PG jot device is shown in Figure 9a. A distal FD is formed with no spatial overlap with TG. With different doping concentration in PW, PB and VB regions, two built-in electrostatic potential steps are formed, as shown in Figure 9b. The photoelectrons accumulate in SW during the integration period. During this period, dark current generated directly under TG at the Si-SiO$_2$ interface, is blocked from flowing to SW by a barrier, and instead dark current flows to FD. As a result of SW being an n-region fully surrounded in 3D by single crystal p-type silicon, dark current is extremely low. For readout, FD is reset and sampled, and then integrated carriers in SW are transferred to the PW region under TG as TG is turned "on" by a positive bias, and then transferred to FD in a "pump" action when TG is turned "off", since a built-in barrier prevents their return to SW. With the transferred charge, FD is sampled a second time for correlated double sampling (CDS). The PG jot device has a FWC of about 200 e− and can achieve lag-less charge transfer [18].

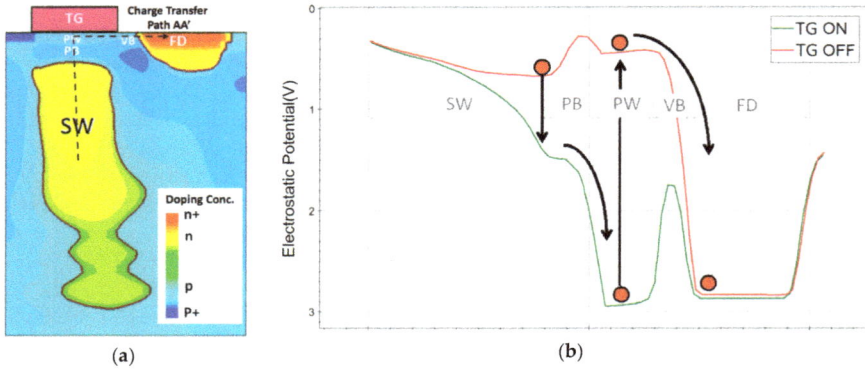

Figure 9. (**a**) PG jot cross-section doping profile from TCAD simulation; (**b**) Electrostatic potential curve along charge transfer path AA′. Both are presented in [13].

A tapered RG technique was developed to reduce the overlap capacitance between FD and RG, which uses STI to shrink the width of reset transistor on the FD end. The use of the tapered RG (aka tapered PG (TPG) jot) significantly increased conversion gain from 250 μV/e− to over 400 μV/e− and helped reduce read noise from approximately 0.33–0.45 e− r.m.s. range to the 0.22–0.35 e− r.m.s. range as shown in Figure 10. The variation in read noise may be due to fluctuations in the energy levels of traps in the readout transistor, the total number of traps, and other random factors.

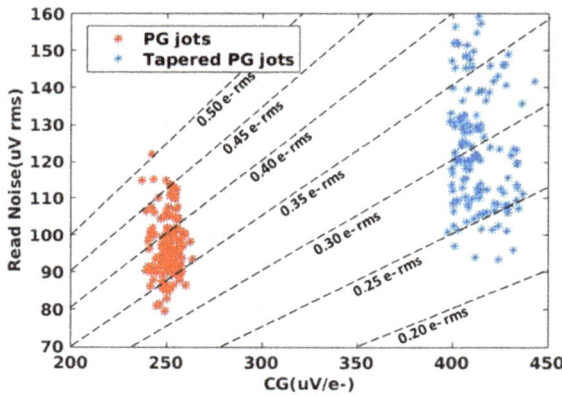

Figure 10. Scatter plot of voltage read noise vs. CG for PG jots and TPG jots. The read noise in e− r.m.s. levels are shown with dashed lines. Presented in [18].

The pump-gate technique enables the implementation of shared readout structure (shared PG jot) without adding overlap capacitance due to the distal FD, and the 3D TCAD model of a 4-way shared readout PG jot is depicted in Figure 11. The shared readout jot has a more compact layout design with 1 μm pitch, but since FD needs to be connected to the SF, more inter-metal parasitic capacitance is added to FD, which yields a mildly lower CG.

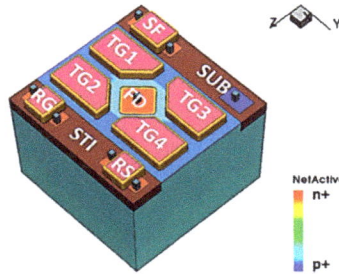

Figure 11. TCAD 3D model of a 4-way shared readout jot, from [20].

Both PG jot and TPG jot (PG jot with tapered RG) were designed and fabricated in the TSMC BSI 65 nm process. The fabrication followed baseline process with implantation modifications, and no extra mask was required. The TPG jot pitch is 1.4 μm and has 410 μV/e− CG (0.39 fF FD capacitance), the non-shared PG jot pitch is 1.4 μm and has 250 μV/e− CG (0.64fF FD capacitance), and the 4-way shared PG jot pitch is 1 μm and has 230 μV/e− CG (0.7 fF FD capacitance). As expected, extremely low SW dark current (0.1 e−/s at RT) was measured and almost lag-less (<0.1 e−) charge transfer was achieved. The measured characteristics of jot devices are listed in tab:sensors-16-01260-t001.

Table 1. Summary of characterization results of PG jot devices.

Quantity	TPG Jot	Non-Shared PG Jot	Shared PG Jot
CG	410 μV/e−	250 μV/e−	230 μV/e−
Read Noise	0.29 e− r.m.s. (129 μV r.m.s.)	0.38 e− r.m.s. (95.3 μV r.m.s.)	0.48 e− r.m.s. (110 μV r.m.s.)
SF Size	0.2×0.2 μm^2	0.2×0.4 μm^2	0.2×0.4 μm^2
Dark Current @ RT	0.09 e−/s (0.73 pA/cm^2)	0.12 e−/s (0.98 pA/cm^2)	Not measured
Dark Current @ 60 °C	1.29 e−/s (10.5 pA/cm^2)	1.26 e−/s (10.2 pA/cm^2)	0.71 e−/s (11.4 pA/cm^2)
Lag @ RT	<0.1 e−	<0.1 e−	<0.12 e−

5.3. Photoelectron Counting Capability

Both the PG and TPG jots are demonstrated to have deep sub-electron read noise, and the PCH-VPM method was used to characterize their photoelectron counting capability [17,18]. The jots in each 32 × 32 array were readout by single CDS under room temperature (RT). TPG jots have an average read noise of 0.29 e− r.m.s., or 129 μV r.m.s. voltage noise, and a "golden" TPG jot achieved 0.22 e− r.m.s. read noise. The PCH of the "golden" TPG jot is depicted in Figure 12a. It was the first time that a CIS pixel without avalanche gain achieved deep sub-electron read noise and photoelectron counting capability. PG jots have an average read noise of 0.38 e− r.m.s., or 95.3 μV r.m.s. voltage noise. Shared readout PG jots have an average read noise of 0.48 e− r.m.s., or 110 μV r.m.s. voltage noise [20]. The PCHs of these jots are also shown in Figure 12b,c.

A more straightforward method was also used to illustrate the photoelectron counting capability of TPG jot. The TPG jot was kept in an integration state under a low illumination and the FD voltage was read out continuously. The quantized voltage steps generated by photoelectrons (and possibly by some thermally generated electrons) can be clearly seen in Figure 13, in which the FD voltage (*y*-axis) is normalized by CG. This is a very basic electrical engineering demonstration of putting one electron on a capacitor and seeing a step in the voltage, but we have not found many prior examples of such an elementary measurement in the literature. It is possible the unfiltered noise in Figure 13 is related to RTS but detailed exploration of this noise has not yet been performed.

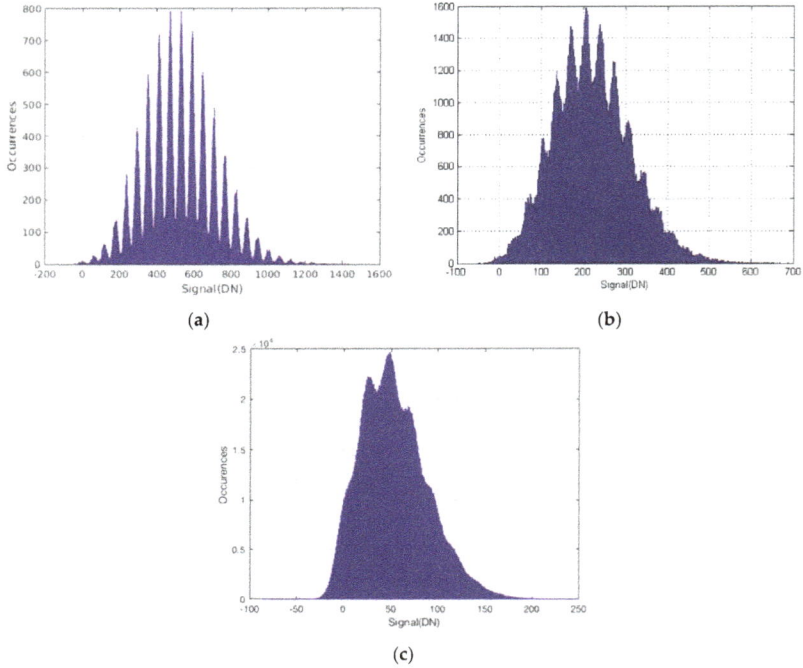

(a)

(b)

(c)

Figure 12. (**a**) PCH of a "golden" TPG jot with 0.22 e− r.m.s. read noise for a quanta exposure of 9. Presented in [18]; (**b**) PCH of a PG jot with 0.32 e− r.m.s. for a quanta exposure of 6.5. Presented in [17]; (**c**) PCH of a shared readout PG jot with 0.42 e− r.m.s. read noise for a quanta exposure of 2.4. Presented in [20].

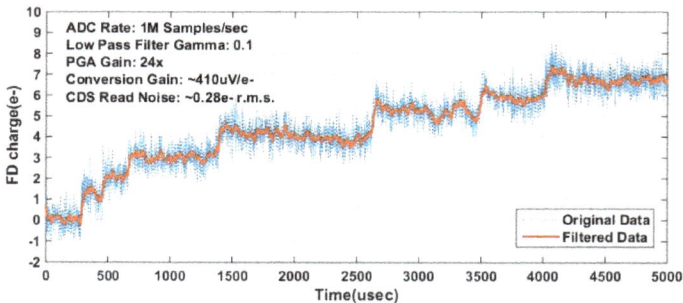

Figure 13. Illustration of photoelectron counting. The signal is the continuously sampled FD voltage from a TPG jot (with 0.28 e− r.m.s. read noise when operated in a CDS mode.) The FD voltage was changed by photoelectrons from SW (and possibly dark generated electrons.) Each single electron generates a fixed voltage jump on FD, and with deep sub-electron read noise, the electron quantization effect is visible.

5.4. Jot Device with JFET SF

It was noticed that although the TPG jot yielded a lower read noise than the PG jot, it actually had higher voltage noise. This effect is believed to be caused by a smaller SF gate area in the TPG jot. As the SF gate capacitance can dominate the total FD capacitance in the PG jot, smaller SF area can provide a higher CG, but also makes the SF more susceptible to the random fluctuation caused by interface traps and leads to increased $1/f$ noise and RTS. With this tradeoff between gate capacitance (that is, CG) and SF voltage noise, further reduction of read noise becomes challenging. The scatter plot in Figure 10 suggests that further reduction of SF size will not allow us to achieve 0.15 e− r.m.s. read noise even with CG of 1 mV/e−. Generally, to achieve the ultimate goal of high accuracy photoelectron counting, more innovation is needed for the jot device to reduce noise or increase conversion gain.

A one-transistor single-electron field effect transistor (SEFET) was proposed as a possibly jot device by earlier work at Samsung [5]. This device used direct collection of photoelectrons in the gate of a junction field effect transistor (JFET) to modulate the current flow of the transistor with CDS performed by resetting the gate back to a fully depleted state. The goals were both small jot size and use of a JFET with high CG and low channel noise to implement low read noise. Only preliminary simulations were performed on the device before the QIS work was abandoned at Samsung.

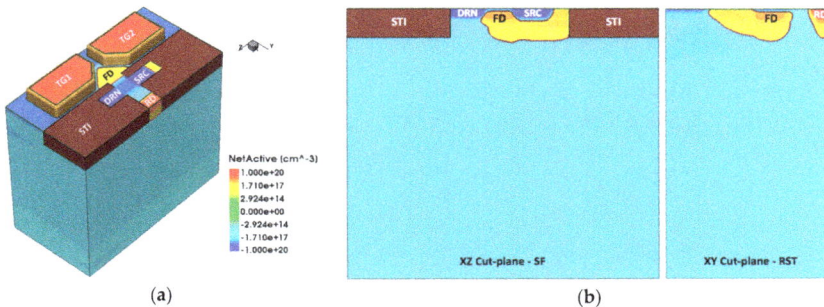

(a)   (b)

Figure 14. (**a**) 3D doping profile of JFET jot device from TCAD simulation; (**b**) Cross-section doping profiles of JFET SF region.

A jot device with an in-jot JFET SF has been explored with TCAD to address the dilemma in PG jots [23,26]. The doping profile of this device is shown in Figure 14. In this device, FD is the n-type doping well located underneath a p-type shallow channel in the JFET SF, and it also functions as the gate of SF. As photoelectrons are transferred from SW to FD, the potential change in FD modulates the depletion region width in the channel, so as to affect the effective channel depth. With the JFET working as a SF, the source (SRC) is biased by a current source and the drain (DRN) connected to ground. Working in saturation mode, the source voltage would follow the gate (FD) voltage. In the PG jot device with a MOSFET SF, FD is connected to the gate of SF through metal wire, and in order to form an Ohmic contact, FD needs to be heavily doped. In this device, since FD is merged with the gate of SF, no metal connection is needed, so the doping concentration of FD can be much lower, which helps reduce the depletion capacitance between FD and substrate. Also, in a MOSFET SF the gate capacitance is relatively large as a result of the extremely thin gate oxide, but in the JFET SF it is replaced by a much smaller junction capacitance between gate and channel. To further reduce the FD node capacitance, a punch-through reset diode is used in this device. Under this mechanism, FD would be reset when a positive pulse is applied on reset drain (RD). Comparing to CIS pixels with punch-through reset [51,63] taking advantage of the small FWC needed for QIS application, the reset state RD voltage can be much lower (e.g., 2.5 V), and FD would be reset to about 1 V to provide enough FWC. As a result of the reduction in FD capacitance, the JFET jot yields a CG of 1400 µV/e−

according to TCAD simulation. Similar to a conventional JFET, this device gate does not interact with channel on the surface interface, which could lead to reduced $1/f$ noise and RTS. Other JFET-based readout devices are also under investigation. Generally, the features of high CG and potentially low noise makes this device a promising candidate to achieve the desired 0.15 e− r.m.s. or less read noise.

5.5. Color and Polarization Filters

For many applications, color filter arrays (CFAs) are needed to enable color imaging. In this case, the bit planes can be separated by color into groups and processed independently, and then re-fused for a full color image. Generally, cubicle sizes for the color groups need not be the same and may facilitate particular improvements in image quality. Color processing could also be performed for each combined bit plane followed by cubicle processing. The options are certainly broad but mostly unexplored.

For SDL jots, one can consider microlenses and color filters that cover multiple jots (e.g., 2 × 2) since diffraction will likely result in optical resolution lower than the jot pitch [1,64]. Color crosstalk was analyzed in [20] for example, and a new color filter array pattern to ameliorate the impact of color crosstalk was proposed and analyzed in [12]. Polarization filter gratings can also be applied to jots, or groups of jots to select particular polarization of photons [24]. For example, 4 polarization filters formed by gratings, corresponding to 0°, 45°, 90°, and 135° polarization selection angles can each be placed over a group of jots, e.g., 4 × 4 jots under each filter. Color filters can also be adjacent to the polarization filters to form a 3 × 3 super-kernel of filters for polarization and color as shown in the inset to Figure 15. Thus, both color and polarization information can be obtained from the 12 × 12 × t super-cubicle of jots, with accuracy dependent on exposure and cubicle size.

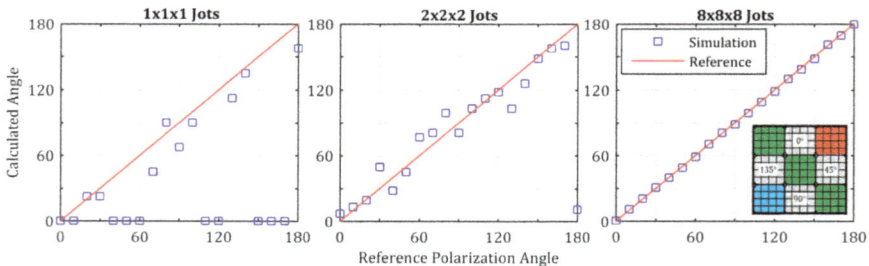

Figure 15. Illustration of polarization angle extraction from various sized cubicles of jots under each polarization-angle filter, from one Monte-Carlo simulation iteration ($H = 1$) for each reference angle. Inset shows 4 × 4 jots under both polarization-angle filters and color filters, to create a 3 × 3 super-kernel to extract polarization and color information from a single-bit QIS. From [24].

6. Low-Power and High-Speed Readout Circuits

The principal challenge addressed in this section is the design of internal high-speed and low-power addressing and readout circuitry for the QIS. A QIS may contain over a billion jots, each producing just 1 mV/e− of signal, with a field readout rate 10–100 times faster than conventional CMOS image sensors.

6.1. Readout Circuits for Single-Bit QIS

To implement the single-bit QIS ADC, the inherent random offset in a comparator and latch circuit must be overcome to permit practical use of a 500 μV comparator threshold voltage. This traditionally requires additional gain and concomitant power dissipation. For low power, a charge-transfer amplifier (CTA) approach was taken [10]. Minimizing the power dissipation was achieved by using a 4-stage charge-transfer amplifier (CTA) as a gain stage in the analog readout signal chain. Use of CTA technique implemented in pathfinder test chips have resulted in a significant improvement in an energy-per-bit figure of merit (FOM) compared to previous work, although detailed comparison is complicated.

In the first test chip, low-power readout circuits based on the CTA were implemented in a 1000 fps megapixel binary imager [15,19]. The architecture of the 1 Mpixel pathfinder image sensor is shown in Figure 16a. The 1376 (H) × 768 (V) pixel image sensor uses a partially-pinned photodiode, 3.6 μm 3T pixel, and readout architecture implemented in the X-FAB 0.18 μm process. The sensor is operated in a single-row rolling-shutter mode so true correlated double sampling (CDS) can be utilized. This means that when a particular row is accessed, it is first reset, allowed to briefly integrate a signal, and then read out before moving to the next row. However, to achieve 1000 fps, this leads to extremely short integration times (i.e., <1 μs), useful only in the lab. To characterize the pixels, lower frame rates were used.

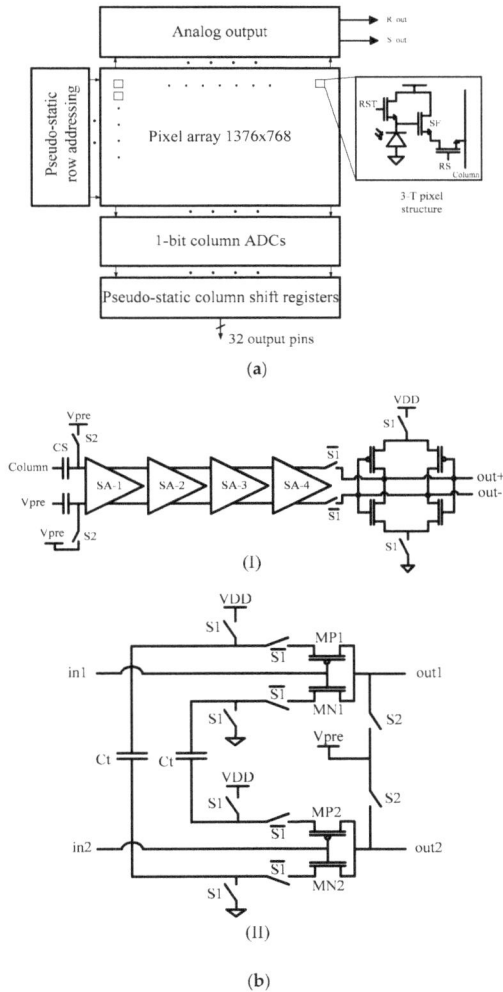

Figure 16. (a) Architecture of the 1Mpixel pathfinder image sensor; (b) (I) 1-b ADC based on a cascade of sense amplifiers and a single D-latch comparator; (II) Schematic of each sense amplifier that is implemented as a differential charge transfer amplifier.

A column-parallel single-bit ADC using a CTA-based design detects a minimum 0.5 mV output swing from the pixel (Figure 16b). The ADC is capable of sampling at speeds of 768 kSa/s. The sensor operates at 1000 fps, which corresponds to a row time of 1.3 μs, a signal integration time, T_{int}, of 0.9 μs, and an output data rate of 1 Gb/s.

The final specifications of the image sensor are shown in tab:sensors-16-01260-t002. The power consumption of the entire chip (including I/O pads) is 20 mW. Total power consumption of the ADCs is 2.6 mW which corresponds to 1.9 μW per column. The row addressing circuits including the buffers consume 0.73 μW per row, whereas the column shift registers dissipate 2.3 μW per column. The ADCs working in tandem with digital circuits consume an average power of 6.4 mW. It is also noted that in the QIS, input offset at 3σ must be less than 1/2 VLSB (=0.5 mV for this chip) which requires additional power dissipation. The FOM of the pathfinder chip is 2.5 pJ/b.

Table 2. Specifications of the 1 Mpixel binary image sensor.

	Process	X-FAB, 0.18 μm, 6M1P (Non-Standard Implants)
	VDD	1.3 V (Analog and Digital), 1.8 V (Array), 3 V (I/O pads)
	Pixel type	3T-APS
	Pixel pitch	3.6 μm
	Photo-detector	Partially pinned photodiode
	Conversion gain	119 μV/e−
	Array	1376 (H) × 768 (V)
	Column noise	2 e−
	Field rate	1000 fps
	ADC sampling rate	768 KSa/s
	ADC resolution	1 bit (VLSB = 1 mV)
	Output data rate	32 (output pins) × 33 Mb/s = 1 Gb/s
	Package	PGA with 256 pins
	Pixel array	8.6 mW
	ADCs	2.6 mW
Power	Addressing	3.8 mW
	I/O pads	5 mW
	Total	20 mW

The second test sensor explores the low-power readout circuits needed for a 1040 fps gigapixel binary image sensor [23,25]. Due to limited available area on the die, only 32 of the columns (12,000 pixels in each column) and 16 1b-ADCs were implemented in this test chip. Since the column parallel architecture is used, the power consumption of a column can be multiplied by 2 × 42,000 to estimate the expected total power consumption of a gigajot QIS. This imager was implemented in a 65 nm BSI CIS process. Pixel pitch is 1.4 μm pitch, and 4-way-shared PPD pixels are used in the imager. The same structures of the sense-amplifier and 1b-ADC (size of the transistors and capacitors are scaled down) are implemented in this test chip.

The average power consumption per column (biasing a column with 24,000 pixels and a sense-amplifier and a 1b-ADC) is 68 μW. It is estimated that the power consumption of a gigapixel QIS imager, (ADCs and column biasing) would be approximately 2.85 W. The FOM of the sense-amplifiers and ADC is 0.4 pJ/b. Comparing the power consumption of the ADC that is used in the first single-bit chip (FOM = 2.5 pJ/b) with the ADC in this work, shows that using more advanced technology node (65 nm in this work and 0.18 μm in the first test chip) yields 6× improvement in FOM.

6.2. Readout Circuits for Multi-Bit QIS

Conceptually, once input-referred read noise is low enough to count a single photoelectron reliably, counting multiple photoelectrons with the same photodetector and readout structure and a low-bit-depth ADC also becomes practical, allowing implementation of a multi-bit QIS. The ADC digital value is the number of photoelectrons in the jot.

Increasing the bit depth of a jot from single-bit to *n* bits allows the field readout rate to be reduced while maintaining constant flux capacity. Thus, while ADC energy per readout is increased by increasing the jot bit depth, the power dissipation increase is mitigated or negated by the reduced field readout rate. The multi-bit QIS approach also addresses the column limited bandwidth issue, where in single-bit QIS imager, since the integration time is shorter than the integration time in multi-bit QIS, imaging throughput is limited. We have explored several variations of multi-bit QIS architectures, including single-slope, cyclic, and successive approximation ADCs implemented in 180nm CIS process [14]. Results are promising and will be reported in a future publication.

6.3. Stacked QIS

A stacked QIS addresses the limited bandwidth problem of the source-follower amplifiers in the pixels or jots. In the stacked QIS approach, more than one substrate or layer could be used to implement the readout circuits. These layers are stacked over each other with bonding interconnections. To readout the jots, the readout and image processing circuits are implemented on the separate substrates.

A stacked QIS may consist of a billion jots which are organized as an array of M row and N column jots [23]. A cluster of jots is defined as a sub-array of m rows and n columns of jots. Figure 17 shows one example of a simplified schematic of a cluster of jots, their analog readout circuits and chip-level signal or image processing units. In each cluster, the RS switches turn on and off sequentially and only one RS switch is connected to the column bus in a cluster at a time. During the selection of one jot, the reset and signal voltage levels are stored on the correlated double sampling (CDS) unit. A differential CTA amplifies the signals stored in the CDS on the level which is bigger than input referred offset and input referred noise of the ADC. All the clusters function in parallel. ADC can be single-bit or multi-bit, based on the readout structure of the entire image sensor system. After quantization of the signal by the ADC, simple digital processing is done on the digital signal by image processor (IP1) and the output is saved in a memory. The simple digital process can be an adder or a digital convolver. The next ADC output, which is the quantized output of the subsequent jot, is summed or convolved with the value stored in the memory. This process continues until all the jots in the cluster have been readout. At this moment, the value stored in the memory, and all other clusters memories are transferred to a chip-level image processor for further processing. After reading one cluster of jots, the clusters readout is re-done for the next frame. By using this method, the bandwidth of the columns in clusters are wide enough to produce thousands of frames per second while consuming very low-power.

As an example, in a gigajot, 1000 fps QIS with 16:9 aspect ratio, with cluster size of 32 (m) × 32 (n), there are 42,000 columns (N) and 24,000 (M) rows of jots and 984,750 clusters as 750 row and 1313 column.

In this system there are 984,750 current sources, CDSs, SAs, ADCs, IP1s, 256-bit memories and one chip-level image processor. The sampling rate of the CDS, SA, ADC, IP1 and memory is 1 MSa/s. Considering 2 W as the power budget for entire chip, 0.5 W may be consumed in chip-level image processing and pad frame, and the rest of 1.5 W budget provides almost 1.5 μW per cluster. Using a more advanced CMOS process such as a 45 nm technology node, charge transfer circuits in the analog domain and sub-threshold regime operation in the digital domain, we estimate it is possible to design the blocks for each cluster to consume less than 1.5 μW power.

It should be mentioned that by using a digital kernel and memory, the output data rate can be significantly reduced, although post-readout processing flexibility is reduced. In the above example, if no image processing was implemented on-chip then the output data rate is about 1 Tb/s; whereas by using simple digital kernels in each cluster, the output data rate could be reduced, for example, to about 8 Gb/s. Using a 3rd stacking layer for chip-level image processing could reduce the output data rate to similar data rates as in conventional cameras.

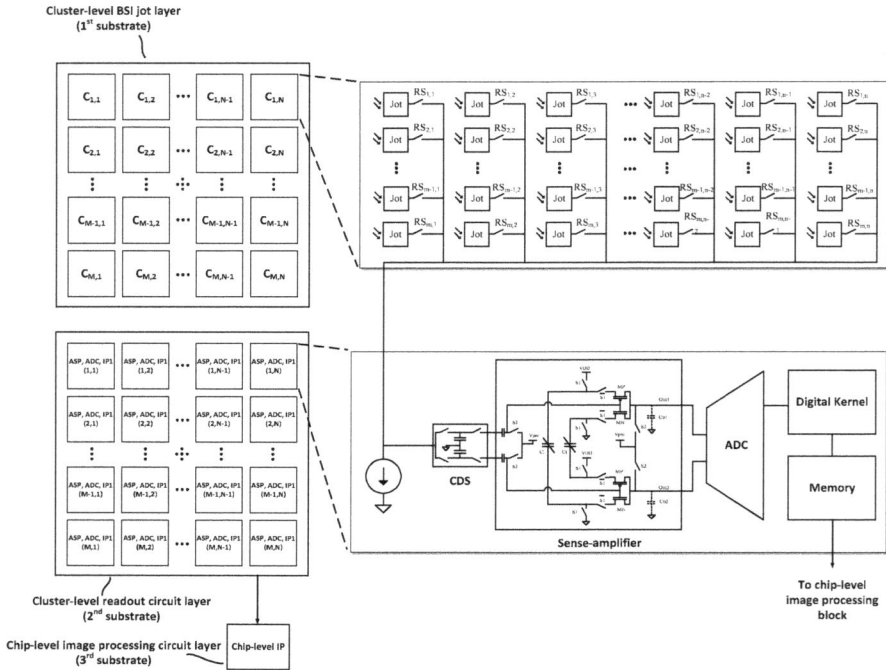

Figure 17. Block diagram of jot clusters, readout circuits and image processing layers.

7. Conclusions

This paper has presented a review of progress to date on Quanta Image Sensor made by the group at Dartmouth and others, as well as a brief review of related activity. Much progress has been made since 2012 when work started in earnest at Dartmouth. Implementation of all the critical elements of the QIS has been demonstrated, including image formation, photon-counting jots, and low-power readout electronics. Demonstration of megajot QIS arrays is possible over the next year or two.

Acknowledgments: This work was sponsored in part by Rambus Inc. (Sunnyvale, CA, USA) The authors appreciate discussion with the students and faculty at Dartmouth as well as with visiting scientists and colleagues in academic and industry, including Forza Silicon (Pasadena, CA, USA), TSMC Ltd. (Hsinchu, Taiwan), XFAB (Erfurt, Germany) and Rambus. In particular, we would like to thank Kofi Odame at Dartmouth, Barmak Mansoorian at Forza Silicon, and Jay Endsley and Michael Guidash at Rambus Inc. We would also like to especially thank graduate students Donald Hondongwa, Arun Rao, Yue Song, Song Chen, and Dakota Starkey for their contributions.

Conflicts of Interest: The authors declare no conflict of interest.

Abbreviations

The following abbreviations are used in this manuscript:

ADC	analog to digital converter
BSI	backside illumination
CCD	charge-coupled device
CDS	correlated double sampling
CFA	color filter array
CG	conversion gain
CIS	CMOS image sensor

CMOS	complementary metal oxide semiconductor
CMS	correlated multiple sampling
CTA	charge transfer amplifier
DR	dynamic range
DRN	drain
DSERN	deep sub-electron read noise
EMCCD	electron-multiplying CCD
EPFL	École Polytechnique Fédérale de Lausanne
FD	floating diffusion
FOM	figure of merit
FWC	full-well capacity
HDR	high dynamic range
I/O	input-output
JFET	junction field effect transistor
MOSFET	metal oxide semiconductor field effect transistor
MTF	modulation transfer function
PCH	photon-counting histogram
PG	pump gate
PW	p-type well
QE	quantum efficiency
QIS	quanta image sensor
RG	reset gate
r.m.s.	root-mean-square
RT	room temperature
RTS	random telegraph signal
SDL	sub-diffraction limit
SEFET	single-electron field effect transistor
SF	source-follower
SNR	signal to noise ratio
SNR-H	exposure-referred SNR
SPAD	single photon avalanche detector
SRC	source
STI	shallow trench isolation
SW	storage well
TCAD	technology computer-aided design
TDI	time delay integration
TG	transfer gate
TPG	tapered reset-gate pump gate
VPM	valley peak modulation

References

1. Fossum, E.R. Image Sensor Using Single Photon Jots and Processor to Create Pixels. U.S. Patent 8,648,287, 11 February 2014.
2. Fossum, E.R. Some Thoughts on Future Digital Still Cameras. In *Image Sensors and Signal Processing for Digital Still Cameras*; Nakamura, J., Ed.; CRC Press: Boca Raton, FL, USA, 2005; pp. 305–314.
3. Fossum, E.R. What to do with sub-diffraction-limit (SDL) pixels?—A proposal for a gigapixel digital film sensor (DFS). In Proceedings of the 2005 IEEE Workshop on Charge-Coupled Devices and Advanced Image Sensors, Karuizawa, Japan, 9–11 June 2005.
4. Fossum, E.R. Gigapixel Digital Film Sensor. In *Nanospace Manipulation of Photons and Electrons for Nanovision Systems*, Proceedings of the 7th Takayanagi Kenjiro Memorial Symposium and the 2nd International Symposium on Nanovision Science, University of Shizuoka, Hamamatsu, Japan, 25–26 October 2005.

5. Fossum, E.R.; Cha, D.-K.; Jin, Y.-G.; Park, Y.-D.; Hwang, S.-J. High Sensitivity Image Sensors Including a Single Electron Field Effect Transistor and Methods of Operating the Same. U.S. Patent 8,546,901, 1 October 2013. U.S. Patent 8,803,273, 12 August 2014.

6. Fossum, E. The Quanta Image Sensor (QIS): Concepts and Challenges. In Proceedings of the 2011 OSA Topical Mtg on Computational Optical Sensing and Imaging, Toronto, ON, Canada, 10–14 July 2011.

7. Chen, S.; Ceballos, A.; Fossum, E.R. Digital integration sensor. In Proceedings of the 2013 International Image Sensor Workshop, Snowbird, UT, USA, 12–16 June 2013.

8. Hondongwa, D.; Ma, J.; Masoodian, S.; Song, Y.; Odame, K.; Fossum, E.R. Quanta Image Sensor (QIS): Early Research Progress. In Applied Industrial Optics: Spectroscopy, Imaging and Metrology, Proceedings of the 2013 Optical Social America Topical Meeting on Imaging Systems, Arlington, VA, USA, 24–27 June 2013.

9. Fossum, E.R. Modeling the performance of single-bit and multi-bit quanta image sensors. *IEEE J. Electron Devices Soc.* **2013**, *1*, 166–174. [CrossRef]

10. Masoodian, S.; Odame, K.; Fossum, E.R. Low-power readout circuit for quanta image sensors. *Electron. Lett.* **2014**, *50*, 589–591. [CrossRef]

11. Ma, J.; Hondongwa, D.; Fossum, E.R. Jot devices and the quanta image sensor. In Proceedings of the 2014 IEEE International Electron Devices Meeting (IEDM) on Technical Digest, San Francisco, CA, USA, 15–17 December 2014; pp. 247–250.

12. Anzagira, L.; Fossum, E.R. Color filter array patterns for small-pixel image sensors with substantial cross talk. *J. Opt. Soc. Am. A* **2015**, *32*, 28–34. [CrossRef] [PubMed]

13. Ma, J.; Fossum, E.R. A pump-gate jot device with high conversion gain for quanta image sensors. *IEEE J. Electron Devices Soc.* **2015**, *3*, 73–77. [CrossRef]

14. Fossum, E.R. Multi-bit Quanta Image Sensors. In Proceedings of the 2015 International Image Sensor Workshop (IISW), Vaals, The Netherlands, 8–11 June 2015.

15. Masoodian, S.; Rao, A.; Ma, J.; Odame, K.; Fossum, E.R. A 2.5 pJ Readout Circuit for 1000 fps Single-Bit Quanta Image Sensors. In Proceedings of the 2015 International Image Sensor Workshop (IISW), Vaals, The Netherlands, 8–11 June 2015.

16. Zizza, R. Jots to Pixels: Image Formation Options for the Quanta Image Sensor. M.S. Thesis, Thayer School of Engineering at Dartmouth College, Hanover, NH, USA, July 2015.

17. Ma, J.; Fossum, E.R. Quanta image sensor jot with sub 0.3 e− r.m.s. read noise and photon counting capability. *IEEE Electron Device Lett.* **2015**, *36*, 926–928. [CrossRef]

18. Ma, J.; Starkey, D.; Rao, A.; Odame, K.; Fossum, E.R. Characterization of Quanta image sensor pump-gate jots with deep sub-electron read noise. *IEEE J. Electron Devices Soc.* **2015**, *3*, 472–480. [CrossRef]

19. Masoodian, S.; Rao, A.; Ma, J.; Odame, K.; Fossum, E.R. A 2.5 pJ/b binary image sensor as a pathfinder for Quanta image sensors. *IEEE Trans. Electron Devices* **2016**, *63*, 100–105. [CrossRef]

20. Ma, J.; Anzagira, L.; Fossum, E.R. A 1 μm-pitch quanta image sensor jot device with shared readout. *IEEE J. Electron Devices Soc.* **2016**, *4*, 83–89. [CrossRef]

21. Fossum, E.R. Photon counting error rates in single-bit and multi-bit quanta image sensors. *IEEE J. Electron Devices Soc.* **2016**, *4*, 136–143. [CrossRef]

22. Starkey, D.A.; Fossum, E.R. Determining conversion gain and read noise using a photon-counting histogram method for deep sub-electron read noise image sensors. *IEEE J. Electron Devices Soc.* **2016**, *4*, 129–135. [CrossRef]

23. Fossum, E.R. Photon counting without avalanche multiplication-progress on the quanta image sensor. In Proceedings of the Image Sensors 2016 Europe, London, UK, 15–17 March 2016.

24. Anzagira, L.; Fossum, E.R. Application of the quanta image sensor concept to linear polarization imaging—A theoretical study. *J. Opt. Soc. Am. A* **2016**, *33*, 1147–1154. [CrossRef] [PubMed]

25. Masoodian, S.; Fossum, E.R. A 32 × 12,000, 1040 fp/s binary image sensor with 0.4 pJ/b readout circuits fabricated in 65 nm backside-illuminated CIS process as a path-finder for 1 Gpixel 1040 fps binary image sensor. to be published.

26. Ma, J.; Fossum, E.R. TCAD simulation of a quanta image sensor jot device with a JFET source follower. to be published.

27. Dutton, N.A.W.; Parmesan, L.; Holmes, A.J.; Grant, L.A.; Henderson, R.K. 320 × 240 oversampled digital single photon counting image sensor. In Proceedings of the 2014 IEEE Symposium on VLSI Circuits Digest of Technical Papers, Honolulu, HI, USA, 10–13 June 2014; pp. 1–2.

28. Gyongy, I.; Dutton, N.; Parmesan, L.; Davies, A.; Saleeb, R.; Duncan, R.; Rickman, C.; Dalgarno, P.; Henderson, R.K. Bit-plane processing techniques for low-light, high-speed imaging with a SPAD-based QIS. In Proceedings of the 2015 International Image Sensor Workshop (IISW), Vaals, The Netherlands, 8–11 June 2015.

29. Dutton, N.A.W.; Parmesan, L.; Gnecchi, S.; Gyongy, I.; Calder, N.; Rae, B.R.; Grant, L.A.; Henderson, R.K. Oversampled ITOF imaging techniques using SPAD-based quanta image sensors. In Proceedings of the 2015 International Image Sensor Workshop (IISW), Vaals, The Netherlands, 8–11 June 2015.

30. Dutton, N.A.; Gyongy, I.; Parmesan, L.; Gnecchi, S.; Calder, N.; Rae, B.; Pellegrini, S.; Grant, L.A.; Henderson, R.K. A SPAD-based QVGA image sensor for single-photon counting and quanta imaging. *IEEE Trans. Electron Devices* **2016**, *63*, 189–196. [CrossRef]

31. Niclass, C.; Rochas, A.; Besse, P.-A.; Popovic, R.S.; Charbon, E. CMOS imager based on single photon avalanche diodes. In Proceedings of the 13th International Conference on Solid-State Sensors, Actuators and Microsystems, Digest of Technical Papers (TRANSDUCERS'05), Seoul, Korea, 5–9 June 2015; Volume 1, pp. 1030–1034.

32. Charbon, E. Will avalanche photodiode arrays ever reach 1 megapixel? In Proceedings of the 2007 International Image Sensor Workshop (IISW), Ogunquit, ME, USA, 7–10 June 2007.

33. Sbaiz, L.; Yang, F.; Charbon, E.; Süsstrunk, S.; Vetterli, M. The gigavision camera. In Proceedings of the 2009 IEEE International Conference on Acoustics, Speech and Signal (ICASSP 2009), Taipei, Taiwan, 19–24 April 2009.

34. Yang, F.; Sbaiz, L.; Charbon, E.; Süsstrunk, S.; Vetterli, M. On pixel detection threshold in the gigavision camera. *Proc. SPIE* **2010**, *7537*. [CrossRef]

35. Yoon, H.J.; Charbon, E. The Gigavision Camera: A 2 Mpixel Image Sensor with 0.56 μm^2 1-T digital pixels. In Proceedings of the 2011 International Image Sensor Workshop (IISW), Hokkaido, Japan, 8–11 June 2011.

36. Yang, F.; Lu, Y.M.; Sbaiz, L.; Vetterli, M. Bits from photons: oversampled image acquisition using binary Poisson statistics. *IEEE Trans. Image Process.* **2012**, *21*, 1421–1436. [CrossRef] [PubMed]

37. Hurter, F.; Driffield, V.C. *Photo-Chemical Investigations and a New Method of Determination of the Sensitiveness of Photographic Plates—Reprinted from The Journal of the Society of Chemical Industry, 31st May 1890. No. 5, Vol IX. The Photographic Researches of Ferdinand Hurter and Vero C*; Driffield, W.B., Ed.; The Royal Photographic Society of Great Britain: London, UK, 1920; pp. 76–122. Available online: https://archive.org/details/memorialvolumeco00hurtiala (accessed on 8 August 2016).

38. Oliver, C.J.; Pike, E.R. Measurement of low light flux by photon counting. *J. Phys. D Appl. Phys.* **1968**, *1*, 1459. [CrossRef]

39. Nieto, J. *New Aspects of Galaxy Photometry*, Proceedings of the Specialized Meeting, Toulouse, France, 17–21 September 1984; Springer-Verlag: Heidlberg, Germany, 1985; Volume 232;

40. Nieto, J.; Thouvenot, E. Recentring and selection of short-exposure images with photon-counting detectors. I-Reliability tests. *Astron. Astrophys.* **1991**, *241*, 663–672.

41. Bondarenko, G.; Dolgoshein, B.; Golovin, V.; Ilyin, A.; Klanner, R.; Popova, E. Limited Geiger-mode silicon photodiode with very high gain. *Nucl. Phys. B Proc. Suppl.* **1998**, *61*, 347–352. [CrossRef]

42. Dierickx, B. Electronic image sensors vs. film: Beyond state of the art. In Proceedings of the OEEPE Workshop on Automation in Digital Photogrammetric Production, Paris, France, 21–24 June 1999.

43. Mackay, C.D.; Tubbs, R.N.; Bell, R.; Burt, D.J.; Jerram, P.; Moody, I. Sub-electron read noise at MHz pixel rates. *Proc. SPIE* **2011**, *4306*, 289–298.

44. Wolfel, S.; Herrmann, S.; Lechner, P.; Lutz, G.; Porro, M.; Richter, R.; Struder, L.; Treis, J. Sub-electron noise measurements on RNDR Devices. In Proceedings of the IEEE Nuclear Science Symposium Conference Record, San Diego, CA, USA, 29 October–1 November 2006; pp. 63–69.

45. Lotto, C.; Seitz, P.; Baechler, T. A sub-electron readout noise CMOS image sensor with pixel-level open-loop voltage amplification. In Proceedings of the IEEE International Solid-State Circuits Conference (ISSCC), San Francisco, CA, USA, 20–24 February 2012.

46. Chen, Y.; Xu, Y.; Chae, Y.; Mierop, A. A 0.7 e− r.m.s.-temporal-readout-noise CMOS image sensor for low-light-level imaging. In Proceedings of the IEEE International Solid-State Circuits Conference (ISSCC), San Francisco, CA, USA, 20–24 February 2012.

47. Yao, Q.; Dierickx, B.; Dupont, B. CMOS image sensor reaching 0.34 e− r.m.s. read noise by inversion-accumulation cycling. In Proceedings of the 2015 International Image Sensor Workshop (IISW), Vaals, The Netherlands, 8–11 June 2015.

48. Boukhayma, A.; Peizerat, A.; Enz, C. A 0.4 e− r.m.s. Temporal Readout Noise, 7.5 µm pitch and a 66% fill factor pixel for low light CMOS image sensors. In Proceedings of the 2015 International Image Sensor Workshop (IISW), Vaals, The Netherlands, 8–11 June 2015.

49. Parks, C.; Kosman, S.; Nelson, E.; Roberts, N.; Yaniga, S. A 30 fps 1920 × 1080 pixel electron multiplying CCD image sensor with per-pixel switchable gain. In Proceedings of the 2015 International Image Sensor Workshop (IISW), Vaals, The Netherlands, 8–11 June 2015.

50. Janesick, J.; Elliott, T.; Andrews, J.; Tower, J. Fundamental performance differences of CMOS and CCD imagers: Part VI. *Proc. SPIE* **2015**, *9591*. [CrossRef]

51. Seo, M.-W.; Kawahito, S.; Kagawa, K.; Yasutomi, K. A 0.27 e- r.m.s. read noise 220-µV/e- conversion gain reset-gate-less CMOS image sensor with 0.11-µm CIS process. *IEEE Electron Device Lett.* **2015**, *36*, 1344–1347.

52. Berezin, V. Active Pixel Sensor with Mixed Analog and Digital Signal integration. U.S. Patent 7,139,025, 21 November 2006.

53. Chan, S.H.; Lu, Y.M. Efficient image reconstruction for gigapixel quantum image sensors. In Proceedings of the 2014 IEEE Global Conference on Signal and Information Processing (GlobalSIP), Atlanta, GA, USA, 3–5 December 2014; pp. 312–316.

54. Misc. Private Communications, 2005–2016.

55. Yadid-Pecht, O.; Staller, C.; Fossum, E.R. Wide intrascene dynamic range CMOS APS using dual sampling. *IEEE Trans. Electron Devices* **1997**, *44*, 1721–1723. [CrossRef]

56. Teranishi, N. Required conditions for photon-counting image sensors. *IEEE Trans. Electron Devices* **2012**, *59*, 2199–2205. [CrossRef]

57. Boukhayma, A.; Peizerat, A.; Enz, C. Noise reduction techniques and scaling effects towards photon counting CMOS image sensors. *Sensors* **2016**, *16*, 514. [CrossRef]

58. Kwon, S.; Kwon, H.; Choi, W.; Song, H.-S.; Lee, H. Effects of shallow trench isolation on low frequency noise characteristics of source-follower transistors in CMOS image sensors. *Solid State Electron.* **2016**, *119*, 29–32. [CrossRef]

59. Black, R.D.; Weissman, M.B.; Restle, P.J. 1/*f* noise in silicon wafers. *J. Appl. Phys.* **1982**, *53*, 6280. [CrossRef]

60. Norton, P.W. On the Origin of 1/*f* Noise in Electronic Devices. Unpublished White Paper. 2011.

61. Suh, S.; Itoh, S.; Aoyama, S.; Kawahito, S. Column-parallel correlated multiple sampling circuits for CMOS image sensors and their noise reduction effects. *Sensors* **2010**, *10*, 9139–9154. [CrossRef] [PubMed]

62. Kusuhara, F.; Wakashima, S.; Nasuno, S.; Kuroda, R.; Sugawa, S. Analysis and reduction of floating diffusion capacitance components of CMOS image sensor for photon-countable sensitivity. In Proceedings of the 2015 International Image Sensor Workshop (IISW), Vaals, The Netherlands, 8–11 June 2015.

63. Guidash, M. Active Pixel Sensor with Punch-Through Reset and Cross-Talk Suppression. U.S. Patent 5,872,371, 16 February 1999.

64. Koskinen, S.; Kalevo, O.; Rissa, R.; Alakarhu, J. Color Filters for Sub-Diffraction Limit-Sized Light Sensors. U.S. Patent 8,134,115, 13 March 2012.

sensors

MDPI

Article

Noise Reduction Effect of Multiple-Sampling-Based Signal-Readout Circuits for Ultra-Low Noise CMOS Image Sensors

Shoji Kawahito * and Min-Woong Seo

Research Institute of Electronics, Shizuoka University, Shizuoka 432-8011, Japan; mwseo@idl.rie.shizuoka.ac.jp
* Correspondence: kawahito@idl.rie.shizuoka.ac.jp; Tel.: +81-53-478-1313

Academic Editor: Eric R. Fossum
Received: 12 August 2016; Accepted: 1 November 2016; Published: 6 November 2016

Abstract: This paper discusses the noise reduction effect of multiple-sampling-based signal readout circuits for implementing ultra-low-noise image sensors. The correlated multiple sampling (CMS) technique has recently become an important technology for high-gain column readout circuits in low-noise CMOS image sensors (CISs). This paper reveals how the column CMS circuits, together with a pixel having a high-conversion-gain charge detector and low-noise transistor, realizes deep sub-electron read noise levels based on the analysis of noise components in the signal readout chain from a pixel to the column analog-to-digital converter (ADC). The noise measurement results of experimental CISs are compared with the noise analysis and the effect of noise reduction to the sampling number is discussed at the deep sub-electron level. Images taken with three CMS gains of two, 16, and 128 show distinct advantage of image contrast for the gain of 128 (noise(median): 0.29 e$^-$$_{rms}$) when compared with the CMS gain of two (2.4 e$^-$$_{rms}$), or 16 (1.1 e$^-$$_{rms}$).

Keywords: ultra low noise; multiple correlated double sampling; correlated multiple sampling; correlated double sampling; differential averager; CMOS image sensor; readout noise; *1/f* noise; RTS noise; noise analysis

---

1. Introduction

Since the introduction of the concept of active-pixel CMOS image sensors (CISs) using in-pixel charge transfer [1,2], CISs have been recognized as image sensors suitable for low-light level imaging, and the introduction of pinned photodiodes in four-transistor (4T) active-pixel CISs has enabled overall image quality control for low-light-level imaging, including those for low dark current, fewer white defects, and no image lag [3–5]. Since the read noise performance of CISs is determined by many factors which are controlled by process, device, and circuit technologies, the read noise of CISs with pinned photodiodes is gradually reduced in the past twenty years as new techniques and technologies are introduced. In the CIS with pinned photodiodes reported in 2001, the read noise was 13.5 e$^-$ [6]. Several CISs with sub-electron [7–9] and deep sub-electron noise [10–12] levels have been reported recently, and the best noise level has reached below 0.3 e$^-$ [13–15]. In an active pixel device called DEPFET with non-destructive multiple readouts of the pixel output, very low noise level of 0.25 e$^-$ [16] and 0.18 e$^-$ [17] have been attained. Roughly speaking, the read noise of CISs is reduced down to one-fiftieth in the past 15 years. High conversion gain is definitely the most important factor for realizing the low read noise. However, a deep sub-electron noise level is not realized without the help of readout-circuit techniques with a high noise reduction capability. For instance, a column high-gain pre-amplifier before an analog serial readout or a column analog-to-digital conversion (ADC) is an effective technique for low-noise CISs [18–20]. A very low noise level of 1.5 e$^-$$_{rms}$ is demonstrated in a pinned-photodiode CIS using a high-gain (gain = 32) column amplifier [18]. For further efficient noise reduction, high-gain pre-amplification using multiple sampling of the pixel output is becoming another important technique for low-noise CISs. A multiple sampling technique known as Fowler sampling is used for reading, non-destructively, the outputs of infrared light image sensors [21], and a technique called multiple correlated double sampling (MCDS) [22], or correlated multiple sampling (CMS), is used for a pixel detector for high-energy particles [22] and column readout circuits for low-noise CISs [23–25]. The authors have recently applied this technique to an experimental image sensor using high-conversion gain pixels and a large sampling number of 128, and deep sub-electron noise level of 0.27 e$^-$$_{rms}$ has been attained [15].

In this paper, to reveal how the column CMS circuits, together with high-conversion-gain pixels and low-noise transistors, realizes deep sub-electron read noise levels in our previous implementation [15], the read noise of signal readout chain from the pixel to column ADC is analyzed and the noise components of the pixel and column amplifiers as a function of the sampling number (=gain) are examined to clarify the dominant noise component at high gain. The noise measurement results of the experimental CIS chip are compared with the noise analysis and the noise reduction effect to the sampling number is discussed. The noise reduction effect as a function of the sampling number is also evaluated by images taken by different CMS gains, and the advantage of image quality with the deep sub-electron noise level is demonstrated.

2. Signal Readout Architecture for Ultra-Low-Noise CISs

2.1. Active Pixel Sensors for High-Conversion Gain

Two types of active pixel sensors (APSs), as shown in Figure 1, are used here for realizing ultra-low-noise CISs together with high-gain column readout circuits. One (Figure 1a) is the well-known APS with four transistors for a source follower (M_1), pixel selection (M_2), charge transfer (M_3), and charge resetting (M_4). The other (Figure 1b) is a special type of APS for higher conversion gain with three transistors and a reset-gateless (RGL) charge resetting technique [15,26]. Both pixels use a pinned photodiode for low dark current and signal readout with perfect charge transfer. In Figure 1a, the size of transistors, wiring, and size of floating diffusion (FD) are carefully designed to minimize the parasitic capacitance of the floating diffusion node and maximize the conversion gain. In Figure 1b, a very high conversion gain is expected because of small parasitic capacitance at the FD node not only by optimizing transistor size and wiring, but also by using a structure to reduce parasitic capacitance due to transistors. To reduce the capacitance from the gate of M_3 to FD, a depleted potential saddle is created between the transfer gate and the FD [25]. To eliminate the capacitance of the reset transistor, the reset transistor is removed and the resetting of charge in the FD is done by pulling the drain junction to a very high level.

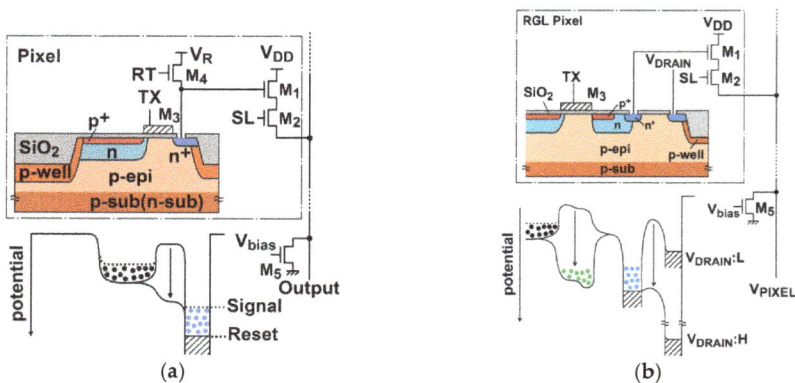

Figure 1. High conversion gain pixels. (**a**) 4T pixel with a pinned photodiode; and (**b**) an RGL high conversion gain pixel.

2.2. Column Readout and ADC Circuits Using Multiple Sampling

A column readout circuit using multiple sampling is shown in Figure 2. The column correlated multiple sampling (CMS) is implemented with a switched-capacitor (SC) integrator. The operation phase diagram and timing diagram of the column CMS circuits are shown in Figures 3 and 4, respectively. At the beginning, the capacitor C_2 of the integrator is reset by turning the on switch controlled by ϕ_R as shown in Figure 3a, while the RT in the pixel in the case of the 4T pixel is set to high for resetting the FD node of the pixel. Then, for multiple sampling of the reset level, the pixel output is sampled by the capacitor C_1 with switches controlled by ϕ_1 and ϕ_{1d} as shown in Figure 3b and the charge in C_1 is transferred to C_2 as shown in Figure 3c by turning switches controlled by ϕ_2 and ϕ_{2d} on. By repeating this operation of Figure 3b,c M times, the M samples of the reset level are integrated over in the integrator. The resulting output of the integrator after M-time sampling is given by $G_I \times M \times \overline{V_{reset}}$, where $\overline{V_{reset}}$ is the average of the reset level of the pixel output and $G_I = C_1/C_2$ is the gain of the integration in one cycle. This integrator output is sampled by a sample-and-hold capacitor and converted to an n-bit digital code by the n-bit column ADC. Similarly, after the charge transfer from the photodiode (PD) to FD by opening the charge transfer (TX) gate, the photo-signal level of the pixel output is sampled M times and the M samples are integrated over in the integrator. The resulting output after M-time sampling is given by $G_I \times M \times \overline{V_{signal}}$, where $\overline{V_{signal}}$ is the average of the photo-signal level of the pixel output. This integrator output is also sampled by a sample-and-hold capacitor and converted to an n-bit digital code by the n-bit column ADC. After the A/D conversion of the integrator output for the reset and photo-signal levels, the difference of those stored in two n-bit memories for reset and signal levels is taken in the digital domain to perform the correlated double sampling (CDS) for cancelling the pixel fixed pattern noise (FPN) and reset noise. This CMS processing, which is a combination of M-time sampling and integration in the analog domain, and the CDS in digital domain, has high suppression effects of thermal and $1/f$ noise and a strong effect of cancelling vertical FPN (VFPN) of CISs, which is caused by the offset deviation of the column readout circuits. The sampling number of the readout circuits based on the CMS technique should be carefully chosen by their applications, e.g., the sensor operations can be determined by following the desired capabilities for applications: (1) high sensitivity with a relatively low frame rate; and (2) high operation speed with an allowable noise level.

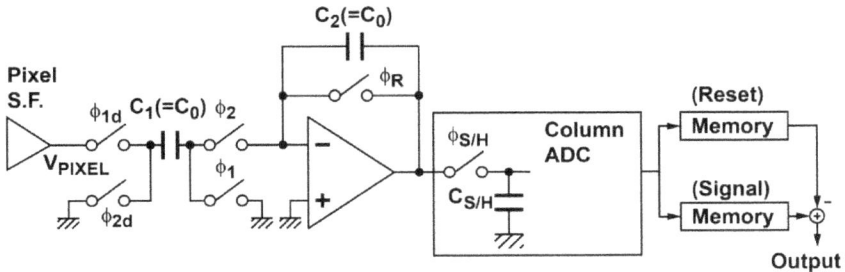

Figure 2. Schematic diagram of the column readout circuits using multiple sampling for low-noise readout.

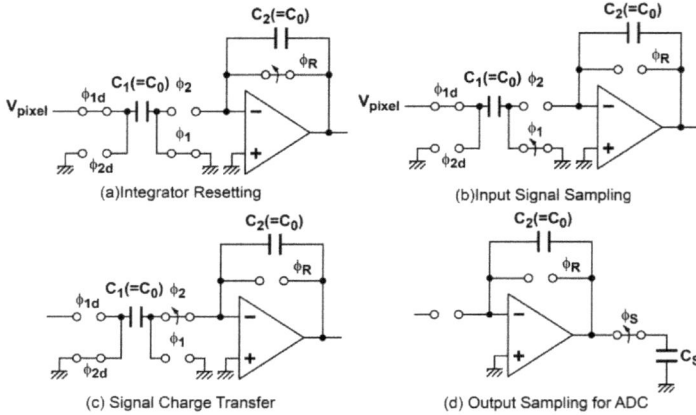

Figure 3. Phase diagram of the column CMS readout circuits.

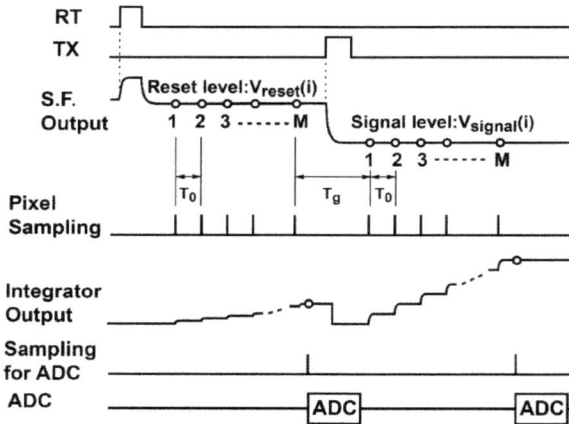

Figure 4. Timing diagram of the CMS.

3. Noise Analysis of Readout Circuits with Multiple Sampling

3.1. Modeling of Noise Sources: Pixel Source Follower and Column Amplifier

An equivalent circuit of the active pixel for the noise modeling is shown in Figure 5. The pixels with high conversion gain shown in Figure 1a,b can use the same equivalent circuit of Figure 5. The conversion gain of the pixel using a source follower amplifier, G_{cSF}, is given by:

$$G_{cSF} = \frac{qG_{SF}}{C_{FD0} + (1 - G_{SF})C_{GS}} \tag{1}$$

where G_{SF} is the source follower gain, C_{GS} is the gate-to-source capacitance of the in-pixel transistor M_1, C_{FD0} is the capacitance at the floating diffusion node other than the term due to C_{GS} and q is the elementary charge. The source follower DC gain G_{SF} is given by:

$$G_{SF} = \frac{g_{mSF}}{g_{oSF} + g_{mSF}} \tag{2}$$

where g_{mSF} is the transconductance of M_1 and g_{oSF} is the output conductance of the source follower, which includes the equivalent conductance component due to the body bias effect of M_1 and the output conductance of M_1 and the current-source load M_4. The gain of the source follower is typically 0.8–0.9. The noise power (squared current) spectrum density S_{InSF} measured at the source follower output [27], including the thermal and $1/f$ (flicker) noise sources, is expressed as:

$$S_{InSF} = 4k_B T \breve{\zeta}_{SF} g_{mSF} + \frac{K_{fSF}}{f} \varsigma_{SF} g_{mSF}^2 \tag{3}$$

where k_B is the Boltzmann constant, T is the absolute temperature, f is the frequency. $\breve{\zeta}_{SF}$ is the excess thermal noise factor of the source follower given by:

$$\breve{\zeta}_{SF} = \breve{\zeta}_P + \frac{g_{mCS}}{g_{mSF}} \breve{\zeta}_{CS} \tag{4}$$

where $\breve{\zeta}_P$ and $\breve{\zeta}_{CS}$ are the excess noise factor of M_1 and M_4, respectively. ς_{SF} is the flicker noise factor to include the influence of the current-source load given by:

$$\varsigma_{SF} = 1 + \frac{K_{fCS}}{K_{fSF}} \left(\frac{g_{mCS}}{g_{mSF}} \right)^2 \tag{5}$$

where K_{fSF} and K_{fCS} are the flicker noise coefficients of M_1 and M_4, respectively.

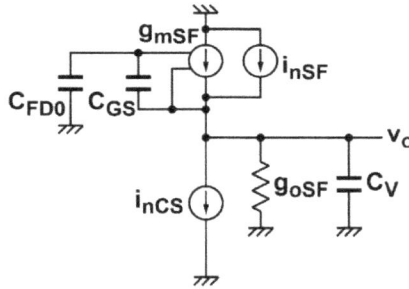

Figure 5. Equivalent circuit of the pixel source follower for noise analysis.

As for an operational amplifier (op-amp) used in the integrator, a high-gain single-pole op-amp using telescopic cascode or folded cascode topology can be used. Figure 6a,b show a telescopic cascode op-amp used in the column readout circuits of this CIS design and its equivalent circuit for noise analysis. In the telescopic cascode op-amp of Figure 6a, the noise of transistors MP5, MP3, MP4, MN4, and MN3 is ignored in the equivalent circuit of Figure 6b. Then the equivalent noise power spectrum S_{InA} measured at the source follower output, including the thermal and $1/f$ (flicker) noise sources, is expressed as:

$$S_{InA} = 4k_B T \breve{\zeta}_A g_{mA} + \varsigma_A \frac{K_{fA}}{f} g_{mA}^2 \tag{6}$$

where $\breve{\zeta}_A$ is the excess thermal noise factor of the op-amp, which includes the influence of all of the transistors given by:

$$\breve{\zeta}_A = 2 \left(\breve{\zeta}_{PA} + \frac{g_{mNA}}{g_{mA}} \breve{\zeta}_{NA} \right) \tag{7}$$

where $\breve{\zeta}_{PA}$ and $\breve{\zeta}_{CS}$ are the excess noise factors of MP_1 (MP_2) and MN_1 (MN_2), respectively. ς_A is the flicker noise factor to include the influence of all the transistors given by:

$$\zeta_A = 2\left(1 + \frac{K_{fNA}}{K_{fPA}}\left(\frac{g_{mNA}}{g_{mA}}\right)^2\right)$$

(8)

where K_{fSF} and K_{fCS} are the flicker noise coefficient of M_1 and M_4, respectively, and the g_{mA} and g_{mNA} are the transconductances of MP_1 (MP_2) and MN_1 (MN_2), respectively.

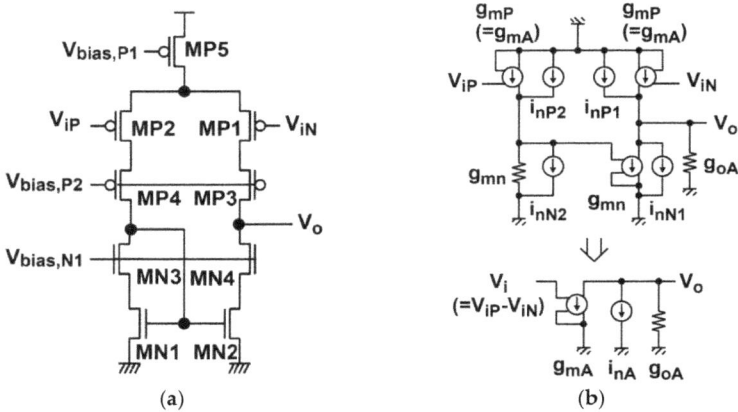

Figure 6. Operational amplifier used in the integrator and its equivalent circuits for noise calculation. (**a**) Circuit schematic; and (**b**) the equivalent circuit for noise analysis.

3.2. Analysis of Noise Components of Readout Circuits

During the signal readout process from the pixel output sampling to A/D conversion, the readout circuits' noise is superimposed on the photo signal at each phase of operation of the CMS readout circuits. The equivalent circuits for noise calculation at each phase of Figure 3 are shown in Figure 7.

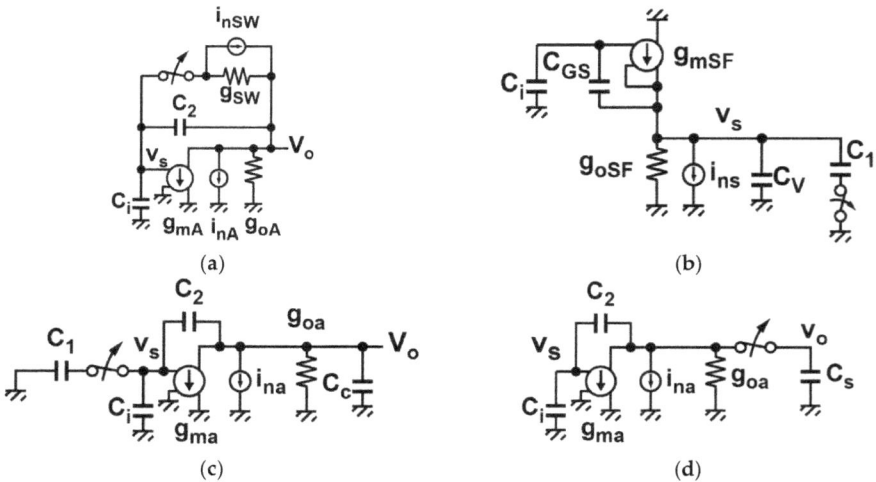

Figure 7. Equivalent circuits for noise calculation at four phases of Figure 3. (**a**) Integrator resetting (Figure 3a); (**b**) input signal sampling (Figure 3b); (**c**) signal charge transfer (Figure 3c); and (**d**) integrator output sampling for ADC (Figure 3d).

3.2.1. Reset Noise of the Integrator

During the resetting phase of the integrator, the thermal noise of the switch by ϕ_R is sampled in the capacitor C_2 and appears at the integrator output. The noise due to the operational amplifier and the influence of input capacitance of the amplifier C_i can be neglected in this phase. Then this noise power component denoted by $P_{nT,rst}$ is approximately given by:

$$P_{nT,rst} = 2\frac{k_B T}{C_2} \tag{9}$$

Due to the digital CDS operation for the output of the integrator, the resetting is done two times for the pixel reset level and signal level, and the reset noise power is increased by a factor of two, as in Equation (9).

3.2.2. Thermal and *1/f* Noise in the Input Signal Sampling Phase

The equivalent circuit in the input sampling phase of the integrator is shown in Figure 7b. The major noise component in this phase is the thermal and *1/f* noise of the pixel source follower and these noises are influenced by the noise-power transfer function of the source follower. Using the equivalent circuits of Figure 5, the noise-power transfer function denoted by $|H_{nSF}(\omega)|^2$ is given by:

$$|H_{nSF}(\omega)|^2 = \frac{G_{nSF}^2}{1 + (\omega/\omega_{cSF})^2} \tag{10}$$

where G_{nSF} is the noise gain factor of the source follower based on the fact that the noise current due to M_1 and M_5 (current source load) is amplified by the positive feedback effect of C_{GS} of the source follower and is expressed as [28]:

$$G_{nSF} = \frac{G_{SF}(C_{FD0} + C_{GS})}{C_{FD0} + (1 - G_{SF})C_{GS}} \tag{11}$$

and ω_{cSF} is the cutoff angular frequency of the source follower with the load capacitance of C_v and sampling capacitance of C_1 which is given by:

$$\omega_{cSF} = \frac{g_{mSF}}{G_{nSF}(C_V + C_1)} \tag{12}$$

Due to the positive feedback effect caused by C_{GS}, the actual transconductance of the source follower is reduced by the same factor of the noise gain G_{nSF}.

In the phase diagram of the CMS readout circuits (Figure 3b), the noise of the pixel source follower is sampled in the capacitor C_1, and then the sampled noise is transferred to C_2. This operation is done M times for both reset and signal levels, and the difference of the integrator output after A/D conversion is taken for the digital CDS. As a result, the noise in this phase, which is finally contained in the digital-domain signal is calculated with the transfer functions of the CMS and the source follower. The noise components in this phase, the thermal ($P_{nT,smpl}$) and the *1/f* ($P_{nF,\,smpl}$) noises, are expressed as:

$$P_{nT,smpl} + P_{nF,smpl} = \int_{-\infty}^{\infty} \frac{S_{InSF}}{g_{mSF}^2}|H_{nSF}(\omega)|^2|H_{CMS}(\omega)|^2 df \tag{13}$$

where $|H_{CMS}(\omega)|^2$ is the power transfer function of the CMS given by [29,30]:

$$|H_{CMS}(\omega)|^2 = \frac{4\sin^2(M\omega T_0/2)\sin^2((M + M_G - 1)\omega T_0/2)}{\sin^2(\omega T_0/2)} \tag{14}$$

For the thermal noise component of Equation (13), a sampled noise of one cycle is calculated by the noise power spectrum and transfer function of the source follower. After the CMS operation, the noise power sampled and accumulated with 2*M* times in the integrator is given by:

$$P_{nT,smpl} = 2G_I{}^2 MG_{nSF}{}^2 \zeta_{SF} \frac{k_B T}{g_{mSF}} w_{cSF} = \frac{2G_I{}^2 MG_{nSF} \zeta_{SF} k_B T}{C_V + C_1} \tag{15}$$

For the *1/f* noise component, Equation (13) can be written as:

$$P_{nF,smpl} = G_I{}^2 M^2 G_{nSF}{}^2 \zeta_{SF} K_{fSF} \int_0^\infty \frac{|H_{nSF}(\omega)|^2 |H_{CMS}(\omega)|^2}{G_{nSF}{}^2 M^2 f} df \tag{16}$$

The integral in Equation (16) is a noise reduction factor of the CMS to *1/f* noise and is defined by:

$$F_{CMS}(M, M_G, x_c) = \int_0^\infty \frac{4\sin^2(Mx/2)\sin^2((M + M_G - 1)x/2)}{M^2 x (1 + (x/x_c)^2)\sin^2(x/2)} dx \tag{17}$$

with the definition of $x = \omega T_0$ and $x_c = \omega_{cSF} T_0$. Then Equation (16) can be expressed as:

$$P_{nF,smpl} = G_I{}^2 M^2 G_{nSF}{}^2 \zeta_{SF} K_{fSF} F_{CMS}(M, M_G, \omega_{cSF} T_0) \tag{18}$$

The factor of the *1/f* noise reduction for the CMS for a large *M* becomes almost the same as that for the case of the noise reduction technique called the differential averager using continuous integration [31]. The ratio of M_G to *M* is denoted by R_G, i.e., $R_G = M_G/M$. Then the noise reduction factor of the CMS can be approximated by a noise reduction factor of the differential averager F_{DA}, which is a function of R_G only and is given by [31]:

$$\frac{F_{DA}(R_G)}{2} = \frac{1}{2} R_G{}^2 \ln R_G + \frac{1}{2}(2 + R_G)^2 \ln(2 + R_G) - (1 + R_G)^2 \ln(1 + R_G) \tag{19}$$

For $R_G << 1$, it is approximated as $F_{DA}(R_G)/2 = 2\ln(2) \cong 1.386$. Equation (19) is a useful equation for calculating the *1/f* noise after the CMS operation without numerical calculation of the integration, as is done in Equation (17). For a large *M*, F_{CMS} can be exactly approximated by F_{DA}. However, for a small *M*, F_{CMS} become larger than F_{DA}. Figure 8 shows the noise reduction factor of the CMS, F_{CMS}, and the differential averager, F_{DA}, as a function of M_G, for the multiple sampling number (*M*) of two, eight, 32, and 128. x_c of 30 is assumed. For efficient noise reduction of the *1/f* noise, the ratio of M_G to *M* or R_G must be kept as small as possible and, from Figure 8, the noise increase is less than 5% if M_G is less than 10% of *M*. In case that M_G is much larger than *M*, it must be noted that the noise reduction effect of the CMS becomes considerably worse than the ideal factor of $2\ln(2) \cong 1.386$.

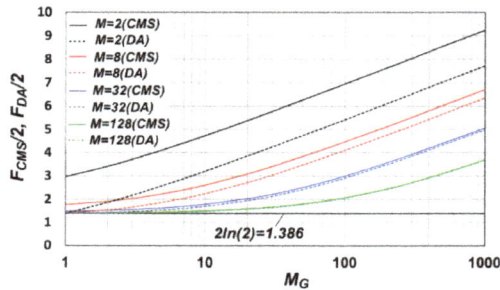

Figure 8. Noise reduction factor of the CMS, F_{CMS}, and differential averager, F_{DA}, as a function of M_G and *M*.

3.2.3. Thermal and *1/f* Noise in the Signal Charge Transfer Phase

In charge transfer phase of Figure 3c, the signal charge sampled in C_1 is transferred to C_2, and then C_1 is disconnected from the input of the op-amp. At this instance, a noise charge caused by the noise of the op-amp used in the SC integrator is sampled in C_1. The sampled noise charge in C_1 is lost in the next input sampling phase. As a result, a noise charge, which is the same amount but opposite polarity as the noise charge in C_1, remains in C_2 of the SC integrator. This noise component is generated in every cycle of the multiple-sampled integration, and the final noise component as a result of the CMS operation is calculated with the noise power transfer function of the SC integrator and CMS using the equivalent circuit of Figure 7c. The power transfer function $|H_{nA}(\omega)|^2$ of the SC integrator to the noise source including the load and sampling capacitances is given by:

$$|H_{nA}(\omega)|^2 = \frac{1}{\beta_A{}^2} \frac{1}{1 + (\omega/\omega_{cA})^2} \tag{20}$$

where β_A is the feedback factor of the SC integrator expressed as:

$$\beta_A = \frac{C_2}{C_2 + C_1 + C_i} \tag{21}$$

and ω_{cA} is the cutoff angular frequency of the SC integrator given by:

$$\omega_{cA} = \frac{g_{mA}\beta_A}{C_{L,trns}} \tag{22}$$

where $C_{L,trns}$ is the load capacitance of the SC integrator in charge transfer phase given by:

$$C_{L,trns} = \frac{C_2(C_1 + C_i)}{C_2 + C_1 + C_i} + C_c \tag{23}$$

In Equation (23), C_c is the additional capacitance at the output for bandwidth limitation of the SC integrator. The noise components in this phase, the thermal ($P_{nT,trns}$) and the *1/f* ($P_{nF, trns}$) noises, are calculated by:

$$P_{nT, trns} + P_{nF, trns} = \beta_S{}^2 \int_{-\infty}^{\infty} \frac{S_{InA}}{g_{mA}{}^2} |H_{nA}(\omega)|^2 |H_{CMS}(\omega)|^2 df \tag{24}$$

where β_S is the noise charge re-sampling factor when the capacitor C_1 is disconnected from the charge summation node of V_s, which is given by:

$$\beta_S = \frac{C_1}{C_1 + C_2 + C_i} \tag{25}$$

The thermal noise component after the CMS operation is calculated as:

$$P_{nT,trns} = 2M\zeta_A \frac{k_B T}{g_{mA}} \frac{\beta_S{}^2}{\beta_A{}^2} \omega_{cA} = 2G_I{}^2 M\zeta_A k_B T \frac{\beta_A}{C_{L,trns}} \tag{26}$$

For the *1/f* noise component, Equation (24) can be written as

$$P_{nF,trns} = G_I{}^2 M^2 \zeta_A K_{fA} F_{CMS}(M, M_G, \omega_{cA} T_0) \tag{27}$$

using Equation (17).

3.2.4. Sampled Noise of the Integrator Output for A/D Conversion

The last component is the sampled noise at the sample-and-hold circuit connected at the integrator. Equivalent circuit in this phase corresponding to the Figure 3d is shown in Figure 7d.

This sample-and-hold circuit is used for column A/D conversion. If the *1/f* noise, due to the amplifier used for the ADC, is ignored because of the low-noise design of the amplifier using relatively large transistor sizes, the thermal noise component ($P_{nT,ADC}$) in the A/D conversion of the integrator output is calculated by:

$$P_{nT,ADC} = \int_{-\infty}^{\infty} \frac{S_{InA}}{g_{mA}^2} |H_{nA2}(\omega)|^2 |H_{CDS}(\omega)|^2 df \tag{28}$$

where $|H_{CDS}(\omega)|^2 = 4\sin^2(\omega T_{CDS}/2)$ is the power transfer function of the CDS operation and $|H_{nA}(\omega)|^2$ is the noise power transfer function of the amplifier given by:

$$|H_{nA}(\omega)|^2 = \frac{1}{\beta_{A1}^2} \frac{1}{1 + (\omega/\omega_{cA2})^2} \tag{29}$$

where β_A is the feedback factor of the SC integrator in the output sampling phase expressed as:

$$\beta_{A1} = \frac{C_2}{C_2 + C_i} \tag{30}$$

and ω_{cA} is the cutoff angular frequency of the SC integrator given by:

$$\omega_{cA} = \frac{g_{mA}\beta_{A1}}{C_{L,ADC}} \tag{31}$$

The thermal noise and *1/f* noise components are calculated as:

$$P_{nT,ADC} = \frac{2\xi_A k_B T}{C_{L,ADC}\beta_{A1}} \tag{32}$$

where $C_{L,ADC}$ is the load capacitance in this phase given by:

$$C_{L,ADC} = \frac{C_2 C_i}{C_2 + C_i} + C_s \tag{33}$$

The factor of two in Equation (32) is based on the fact that the CDS operation doubles the thermal noise power. This noise component generated during the A/D conversion of the integrator output depends on the type of the A/D converter used.

3.2.5. Total Noise

The total noise power referred at the output of the integrator $P_{nCMS,total}$, if all of the noise components are uncorrelated from each other, is given by:

$$P_{nCMS,total} = P_{n,rst} + P_{nT,smpl} + P_{nF,smpl} + P_{nT,trns} + P_{nF,trns} + P_{nT,ADC} \tag{34}$$

Since the gain from the charge to the integrator output is given by $G_I \times M \times G_{cSF}$, the input referred noise is expressed as:

$$N_{nCMS,total} = \frac{\sqrt{P_{nCMS,total}}}{G_I M G_{cSF}} = \frac{\sqrt{P_{n,rst} + P_{nT,smpl} + P_{nF,smpl} + P_{nT,trns} + P_{nF,trns} + P_{nT,ADC}}}{G_I M G_{cSF}} \tag{35}$$

To explicitly show the contribution of the noise components as noise-equivalent charge, the total input referred noise is expressed as:

$$N_{nCMS,total} = \sqrt{N_{n,rst}^2 + N_{nT,smpl}^2 + N_{nF,smpl}^2 + N_{nT,trns}^2 + N_{nF,trns}^2 + N_{nT,ADC}^2} \tag{36}$$

where:

$$N_{n,rst} = \frac{\sqrt{P_{n,rst}}}{G_I M G_{cSF}} = \frac{1}{G_I M G_{cSF}} \sqrt{\frac{2k_B T}{C_2}} \tag{37}$$

$$N_{nT,smpl} = \frac{\sqrt{P_{nT,smpl}}}{G_I M G_{cSF}} = \frac{1}{\sqrt{M} G_{cSF}} \sqrt{\frac{2G_{nSF}\xi_{SF}k_B T}{C_V + C_1}} \tag{38}$$

$$N_{nF,smpl} = \frac{\sqrt{P_{nF,smpl}}}{G_I M G_{cSF}} = \frac{G_{nSF}}{G_{cSF}} \sqrt{2\varsigma_{SF} K_{fSF} F_{CMS}} \tag{39}$$

$$N_{nT,trns} = \frac{\sqrt{P_{nT,trns}}}{G_I M G_{cSF}} = \frac{1}{\sqrt{M} G_{cSF}} \sqrt{\frac{2\xi_A k_B T \beta_A}{C_{L,trns}}} \tag{40}$$

$$N_{nF,trns} = \frac{\sqrt{P_{nF,trns}}}{G_I M G_{cSF}} = \frac{1}{G_{cSF}} \sqrt{2\varsigma_A K_{fA} F_{CMS}} \tag{41}$$

and

$$N_{nT,ADC} = \frac{\sqrt{P_{nT,ADC}}}{G_I M G_{cSF}} = \frac{1}{G_I M G_{cSF}} \sqrt{\frac{2\xi_A k_B T}{\beta_{A1} C_{L,ADC}}} \tag{42}$$

There are three-types of noise components in the CIS with the CMS readout circuits. The first type is the component whose noise amplitude is reduced by a factor of M, as in Equations (37) and (42). These noise components are effectively reduced by increasing the gain M and the total noise is almost unaffected for a large gain. The second type is the component whose noise amplitude is reduced by a factor of $\sqrt{M}$, as in Equations (38) and (40), and dominates the total noise for the middle-gain region. The third type are the components which have a weak dependency on M, as in Equations (39) and (41).

3.3. Noise Calculation for the Designed Ultra-Low-Noise CIS

As described in Section 4, an experimental CIS chip with ultra-low-noise performance is designed and implemented. Using the device parameters used for the design of the CIS chip, the noise components of the readout circuits and the resulting total noise are calculated. Figure 9 shows an example of noise calculation of the CIS using the RGL pixels. The parameters used in this noise calculation are given in Table 1. Table 1 contains parameters for the RGL pixel and the conventional 4T pixel shown in Figure 1. The capacitances are those used for the design of the CIS chip, and the excess noise factors are calculated with the well-known characteristics of the excess noise factor as a function of channel length of nMOS transistor [32]. The *1/f* noise parameters for the amplifier design are calculated with the measured data supplied as the process design kit (PDK) from the CIS foundry. Since no measurement data on the small-size in-pixel transistors are supplied by the PDK, it is estimated by the *1/f* noise measurement data of the 3.3 V medium threshold voltage (V_T) nMOS devices with a size of 10 μm(W)/0.55 μm(L), and the theoretical model of the *1/f* noise parameter (K_f) of the nMOS transistors given by $K_f = k_f / C_{ox}{}^2 WL$, where k_f is a constant which is independent of the dimension of devices, i.e., K_f is inversely proportional to the channel area (W × L). The *1/f* noise also depends on the gate bias condition, and the flicker noise coefficient is increased as the gate bias increases. With the measurement results of the *1/f* noise of the medium V_T device ($K_f = 1.4 \times 10^{-11}$ [V^2] @ Id = 17 μA) and the size dependency of the *1/f* noise, the flicker noise coefficients of the source follower transistors in the RGL pixel and 4T pixel are estimated as 1.0×10^{-9} [V^2] and $K_f = 1.8 \times 10^{-10}$ [V^2], respectively. In this case, the source follower sizes (W/L) for the RGL and 4T pixels are 0.345 μm/0.325 μm and 0.9 μm/0.7 μm, respectively. Sometimes the optimized transistor size can lead to an increased probability that large noise, such as a random telegraph signal (RTS) noise, occurs, but a high conversion gain with the optimized SF size is more beneficial to achieve the low-noise performance. Extremely large noise generated by a smaller transistor size can be overcome by the advanced process technologies and the low-noise transistors [9].

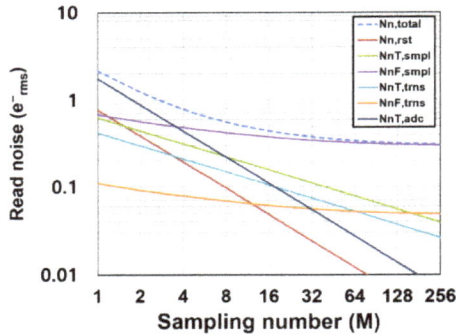

Figure 9. Noise components as a function of the sampling number in the CMS.

Table 1. Device and circuit parameters used for noise calculations.

Parameters	Values (Conventional 4T)	Values (RGL pixel)
Temperature (K)	263	263
G_{cSF} (μV/e$^-$)	135	220
G_{nSF}	2.22	1.21
G_I	0.5	0.5
C_1 (F)	0.5×10^{-12}	0.5×10^{-12}
C_2 (F)	1.0×10^{-12}	1.0×10^{-12}
C_V (F)	0.84×10^{-12}	0.84×10^{-12}
C_i (F)	0.15×10^{-12}	0.15×10^{-12}
C_S (F)	0.5×10^{-12}	0.5×10^{-12}
C_C (F)	0.5×10^{-12}	0.5×10^{-12}
K_{fSF} (V^2)	1.8×10^{-10}	1.0×10^{-9}
K_{fA} (V^2)	0.98×10^{-11}	0.98×10^{-11}
ξ_{SF}	2.15	2.87
ξ_A	2.25	2.25
ζ_{SF}	1.01	1.01
ζ_A	3.94	3.94

Very high conversion gains of 220 μV/e$^-$ and 135 μV/e$^-$ are assumed for the RGL and 4T pixels, respectively, in order to compare with the experimental results described in Section 4. A M_G of 16 is assumed. As shown in Figure 9, for the low-gain region (M: 1–4), the read noise is determined by the ADC noise. This component rapidly decreases by increasing M as a function of $1/M$. In the medium-gain region (M: 4–16), the noise is dominated by the mixture of noise components including the thermal noise components. For the high-gain region (M: larger than 32), the read noise is dominated by the *1/f* noise of the pixel source follower, and because the *1/f* noise component has a slight dependency on M for a large M, the read noise approaches to the lowest limit of noise reduction. The achievable noise level for a large M depends on the *1/f* noise performance of the pixel source follower which is determined by the fabrication process technology and the conversion gain. A deep sub-electron noise level can be realized if a pixel with low *1/f* noise devices and high conversion gain is available. Figure 10 shows the calculated total read noise as a function of M and for different *1/f* noise parameters of the pixel source follower. If the target noise level is 0.2 e$^-$$_{rms}$, a very high CMS gain ($M > 64$) and a low *1/f* noise transistor ($K_f < 0.25 \times 10^{-9}$ [V^2]) is necessary if the conversion gain is unchanged for maintaining the signal dynamic range.

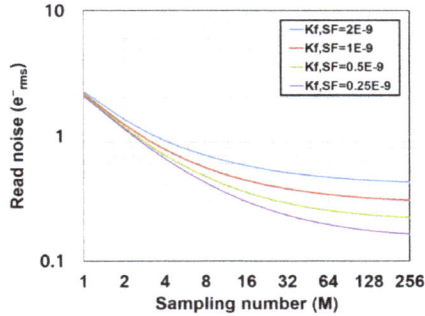

Figure 10. Calculated total read noise as a function of *M* and for different values of $K_{f, SF}$.

4. Implementation and Results

4.1. Implementation

An experimental CMOS image sensor with 32 (V) × 512 (H) RGL active pixels (Figure 1b) and 110 (V) × 512 (H) 4T active pixels (Figure 1a) has been implemented using Dongbu HiTek (Eumseong, Korea) 0.11 μm CIS technology. The block diagram of the CIS chip is shown in Figure 11. In this experimental chip, the CMS circuit is implemented as a column ADC, called the folding-integration ADC [24,25]. This ADC works as a resettable first-order delta-sigma modulator, which is based on the multiple-sampling based integrator shown in Figure 2, but has a negative feedback loop with a one-bit sub-ADC and one-bit DAC for an extended dynamic range. For instance, the output of the conventional multiple-sampling based integrator increases linearly in small input signal region, and then saturates. In the folding integration, however, the analog signal amplitude is kept to a limited range by the folding operation, while applying a high analog gain by the integration. After the folding-integration operation, the integrator output is digitized with another high-resolution ADC, called a cyclic ADC, which is implemented with the same analog circuits as the folding-integration ADC. This column ADC using multiple sampling and the digital CDS has almost the same noise reduction effect as the CMS circuits described in Section 2.

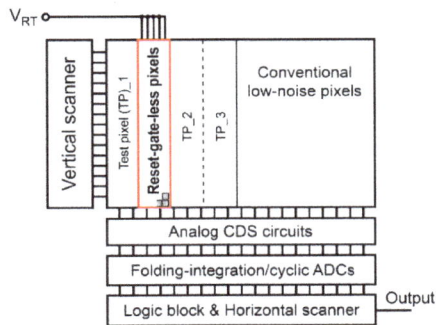

Figure 11. Block diagram of the experimental CIS chip.

The noise analysis given in Section 3 is based on a simplified and more general type of the CMS readout circuits. This simplified analysis is useful for understanding the contribution of noise components at different gain settings of the CMS. To compare the noise measurement results and the noise calculated for the readout circuits actually implemented a few modifications to the noise model are necessary. In the actual implementation, an analog CDS circuit is used in front of the column ADC, as shown in Figure 12. This is for clamping the pedestal level (or reset level) to a fixed voltage level, which is close to the bottom reference level of the ADC to maximize the available voltage range. The reset noise is generated in the analog CDS circuits, but it is cancelled by the final digital CDS operation in the digital domain [33]. The CMS circuits actually used are implemented as a folding integration ADC, of which the analog core is also used for the cascaded A/D conversion using the cyclic ADC, as shown in Figure 12. To include the noise due to the analog CDS circuit, and the influence of the noise increase due to another sampling capacitor C_{1b}, the thermal noises in the input sampling phase and charge transfer phase given by Equations (15) and (26), respectively, are modified as:

$$P_{nT,smpl2} = 2G_I^2 M k_B T \left(\frac{G_{nSF}\tilde{\varsigma}_{SF}}{C_V} + \frac{\tilde{\varsigma}_{CA}+1}{C_1} \right) \tag{43}$$

where $\tilde{\varsigma}_{CA}$ is the excess thermal noise factor of the op-amp for the analog CDS amplifier, and:

$$P_{nT,trns2} = 2M\tilde{\varsigma}_A \frac{k_B T}{g_{mA}} \frac{\beta_{S2}^2}{\beta_{A2}^2} \omega_{cA} = 2M\tilde{\varsigma}_A k_B T \frac{\beta_{S2}^2}{\beta_{A2}^2} \frac{\beta_{A2}}{C_{L,trns2}} \tag{44}$$

where β_{S2} is the noise charge re-sampling factor given by:

$$\beta_{S2} = \frac{2C_1}{2C_1 + C_2 + C_i} \tag{45}$$

β_{A2} is the feedback factor given by:

$$\beta_{A2} = \frac{C_2}{2C_1 + C_2 + C_i} \tag{46}$$

and $C_{L,trns2}$ is the load capacitance:

$$C_{L,trns2} = \frac{(2C_1 + C_i)C_2}{2C_1 + C_2 + C_i} + C_c \tag{47}$$

of the actually implemented CMS circuits as the floding-integration ADC using C_{1a} and C_{b1} whose capacitances are C_1. The input-referred noises of these components are modified from Equations (38) and (40) as:

$$N_{nT,smpl2} = \frac{\sqrt{P_{nT,smpl2}}}{G_I M G_{cSF}} = \frac{\sqrt{2k_B T}}{\sqrt{M}G_{cSF}} \sqrt{\frac{G_{nSF}\tilde{\varsigma}_{SF}}{C_V} + \frac{\tilde{\varsigma}_{CA}+1}{C_1}} \tag{48}$$

and

$$N_{nT,trns2} = \frac{\sqrt{P_{nT,trns}}}{G_I M G_{cSF}} = \frac{2}{\sqrt{M}G_{cSF}} \sqrt{\frac{2\tilde{\varsigma}_A k_B T \beta_{A2}}{C_{L,trns2}}} \tag{49}$$

respectively.

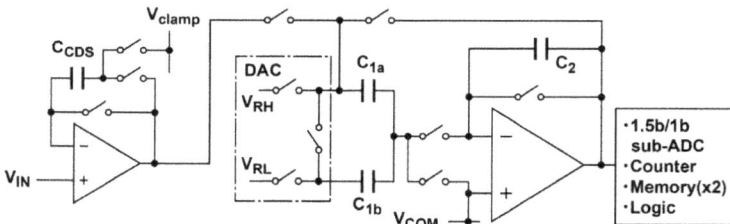

Figure 12. Column analog CDS and ADC circuits.

4.2. Noise Reduction Effect of the CMS

The noise reduction effect of the CMS is experimentally demonstrated in the deep sub-electron noise region. Figure 13 shows the measured and calculated input-referred noise (noise equivalent charge) as a function of the multiple-sampling gain(the sampling number) of the CMS. The noise calculated with the noise model of the CMS circuits is also shown. The timing diagram for reading one horizontal line of the image signal and the value of M, M_G, and the actual readout time of one horizontal line used in this measurement is shown in Figure 14 and Table 2, respectively. In order to reduce the influence of dark current, and to evaluate the noise of readout circuits only, the following data including those of Figure 13 were measured at $-10\ ^\circ$C. Even if the read noise is measured at room temperature, the result is almost the same as the current noise level, but the total noise distribution at room temperature is slightly spread by the influence of dark current, particularly from the FD node.

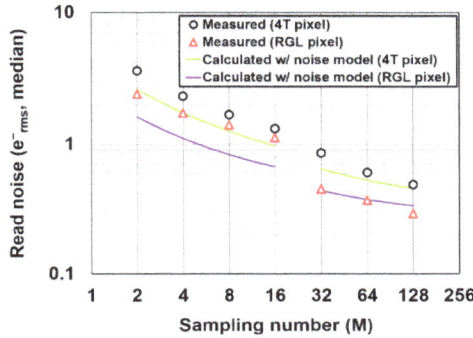

Figure 13. Measured noise as a function of the sampling number and the comparison with the noise calculated with the noise model.

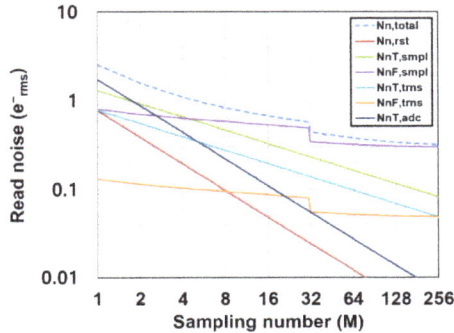

Figure 14. Noise components as a function of the sampling number in the folding-integration ADC.

Table 2. M, M_G and $T_{H\text{-}READ}$ used in the measurements.

M	M$_G$	V$_{H\_READ}$ (µs)
2	268	172
4	264	172
8	256	172
16	240	172
32	16	57.6
64	16	96
128	16	172

As shown in Figure 15 and Table 2, M_G for low gain (M = 2, 4, 8, and 16) is set to large values of more than 200. This causes a lesser *1/f* noise reduction effect, as explained in Equation (17). For high gain (M = 32, 64, and 128), M_G of 16 is used, and a high *1/f* noise reduction effect is expected. The CMS effectively reduces the noise (median) from 3.7 e$^-$ to 0.5 e$^-$ for the 4T pixel, and 2.3 e$^-$ to 0.29 e$^-$ for the RGL pixel, respectively, by increasing the gain from two to 128. The noise calculated with the proposed model does not perfectly explain the experimental results, particularly at the low CMS gain. Since the *1/f* noise suppression capability of the CMS can be degraded by increasing the time from reset to signal samples, and the noise of the small-size transistors in the pixels does not always take the exact *1/f* noise spectrum. These can make the difference between the simulation and measurement. Another possible reason is that the noise of the cyclic ADC is not exactly modeled and other noise components, such as the noise from power supply lines of the substrate, are not included in the noise model. Such noises from power lines of the substrate are often generated due to on-chip digital switching or clocking circuits. Since these noises are not uniform in time, the irregular dependency of the noise reduction to the sampling number of the CMS, or the difference of the calculation and measurement results is likely explained. The measurement results show that the read noise can be further reduced by increasing the CMS gain. This larger dependency of the noise reduction to the CMS gain at high gain (M = 32, 64, and 128) when compared to the theoretical estimation is not clear, but is possibly due to the influence of the additional thermal noise components, which are not modeled in the theory, or RTS (random telegraph signal)-like noise of the in-pixel source follower. The RTS noise or RTS-like noise has a Lorentzian spectrum, or a mixture of Lorentzian spectra and the noise with such a spectrum can be reduced by band-width reduction using a higher CMS gain. The noise of the majority of pixels may take the spectrum of RTS-like noise, not that of the *1/f* noise.

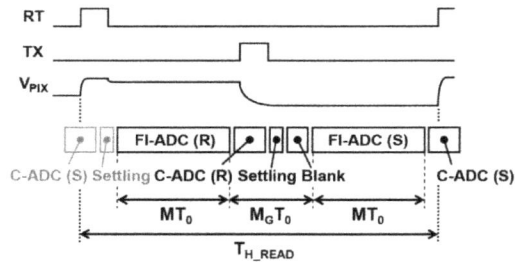

Figure 15. Timing diagram of signal readouts and A/D conversion.

In order to demonstrate the noise reduction effect of CMS in the deep sub-electron region, sample images are taken by three different CMS gains of two, 16, and 128, as shown in Figure 16. With these three gains of two, 16, and 128, the noise levels (median) of 2.4 e$^-$$_{rms}$, 1.1 e$^-$$_{rms}$, and 0.29 e$^-$$_{rms}$, respectively, have been obtained. The character code of "1951" in a part of the USAF (United State Air Force) test chart is used for this imaging test of three different low-noise levels and small signal photoelectron number of less than ten. When compared to the image with the noise level of 1.1 e$^-$$_{rms}$, which is the best noise level of commercially available very-low-noise CISs, the image with the noise level of 0.29 e$^-$$_{rms}$ has advantages in image contrast and recognizability of the character code. In the image with the noise level of 2.4 e$^-$$_{rms}$, it is hard to recognize the character code without prior knowledge that the character code is "1951".

In Figure 17, the cumulative probability plot of noise for the RGL-pixel CIS and 4T-pixel CIS is shown. The CMS gain (M) of 128 is used. The transistor size of the in-pixel source follower of the RGL pixel is 0.325 mm × 0.345 mm, and that of the 4T pixel is 0.7 μm × 0.9 μm. Due to the small gate area of the in-pixel source follower transistor of the RGL pixel, the population of noisy pixels with greater than 1 e$^-$ is higher than that of the 4T-pixel CIS [34].

(a)

(b)

(c)

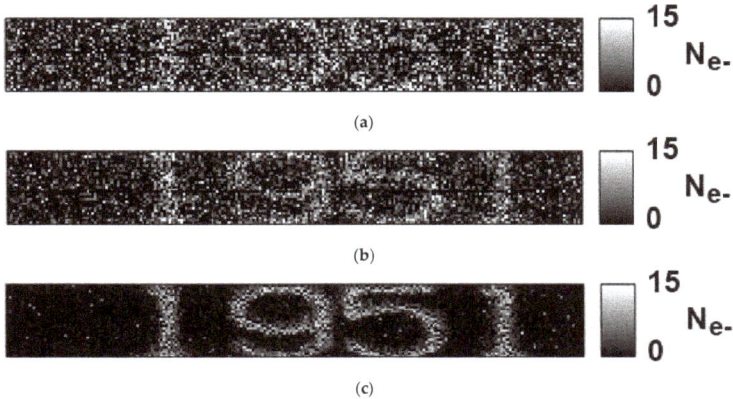

Figure 16. Low-light-level images with three different CMS gains (M = 2, 16, and 128). (**a**) M = 2, noise (median): 2.4 e$^-$ $_{rms}$; (**b**) M = 16, noise (median): 1.1 e$^-$ $_{rms}$; and (**c**) M = 128, noise (median): 0.29 e$^-$ $_{rms}$.

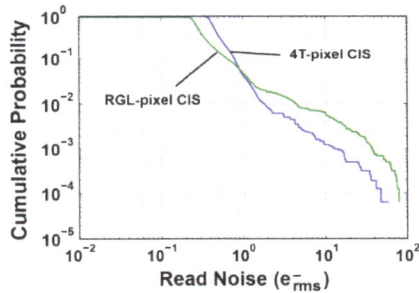

Figure 17. Cumulative probability plots of noise in the RGL-pixel CIS and 4T-pixel CIS.

5. Conclusions

This paper describes a noise model for explaining the ultra-low noise level of CMOS image sensors, and the noise reduction effect of the multiple-sampling-based readout circuits used. The use of very high multiple-sampling gain of correlated multiple sampling (CMS) circuits for signal readout sufficiently reduces the noise components of readout circuits, other than the $1/f$ noise of the in-pixel source follower, and the resulting noise level of CMOS image sensors can be smaller than 0.3 e$^-$ using a high conversion gain pixel, high CMS gain (> 100), and a low-noise in-pixel transistor. Though the noise model does not perfectly explain the noise reduction effect of the CMS circuits, it can be used for theoretically predicting the deep sub-electron noise level in the design of CMOS image sensors by knowing the circuit and device parameters. A comparison of images taken with read noise levels of 1.1 e$^-$ and 0.29 e$^-$ have shown distinct merit in image contrast by reducing the read noise of the deep sub-electron noise level.

Acknowledgments: This work was supported in part by the Japan Society for the Promotion of Science (JSPS) KAKENHI, the Grant-in-Aid for Scientific Research (S) under Grant 25220905, Grant-in-Aid for Scientific Research on Innovative Areas 25109003, the JST COI-STREAM program and the JST A-STEP. Authors wish to thank Nobukazu Teranishi, Keita Yasutomi and Keiichiro Kagawa of Shizuoka University, Satoshi Aoyama and Takashi Watanabe of Brookman Technology Inc. for helpful discussion. Authors appreciate Dongbu HiTek for CIS chip fabrication.

Author Contributions: Shoji Kawahito wrote the paper, proposed the noise model, analyzed the noise of image sensors, Min-Woong Seo designed the CIS chip, did experiments and measured the data.

Conflicts of Interest: The authors declare no conflict of interest.

Abbreviations

The following abbreviations are used in this manuscript:

CIS	CMOS image sensor
CDS	correlated double sampling
CMS	correlated double sampling,
ADC	analog to digital converter
MCDS	multiple correlated double sampling

References

1. Fossum, E.R. Active pixel sensors: Are CCDs dinosaurs? *Proc. IEEE* **1993**, *1900*, 2–14.
2. Mendis, S.; Kemeny, S.E.; Fossum, E.R. CMOS active pixel image sensor. *IEEE Trans. Electron Devices* **1994**, *41*, 452–453. [CrossRef]
3. Lee, P.R.K.; Gee, R.C.; Guidash, R.M.; Lee, T.-H.; Fossum, E.R. An active pixel sensor fabricated using CMOS/CCD process technology. In Proceedings of the IEEE Workshop CCD and Advanced Image Sensors, Dana Point, CA, USA, 20–22 April 1995; pp. 115–119.
4. Teranishi, N.; Kohno, A.; Ishihara, Y.; Oda, E.; Arai, K. No image lag photodiode structure in the interline CCD image sensor. In Proceedings of the IEDM '98 Technical Digest International Electron Devices Meeting, San Francisco, CA, USA, 6–9 December 1998; pp. 324–327.
5. Fossum, E.R.; Hondongwa, D.B. A review of the pinned photodiode for CCD and CMOS image sensors. *IEEE J. Electron Devices Soc.* **2014**, *2*, 33–43. [CrossRef]
6. Inoue, S.; Sakurai, K.; Ueno, I.; Koizumi, T.; Hiyama, H.; Asaba, T.; Sugawa, S.; Maeda, A.; Higashitani, K.; Kato, H.; et al. A 3.25-Mpixel APS-C size CMOS image sensor. In Proceedings of the IEEE Workshop on Charge-Coupled Devices and Advanced Image Sensors, Lake Tahoe, NV, USA, 7–9 June 2001.
7. Fowler, B.; Liu, C.; Mims, S.; Balicki, J.; Li, W.; Do, H.; Vu, P. Wide dynamic range low-light-level CMOS image sensor. In Proceedings of the 2009 International Image Sensor Workshop, Bergen, Norway, 26–28 June 2009; pp. 340–343.
8. Lotto, C.; Seitz, P.; Baechler, T. A sub-electron readout noise CMOS image sensor with pixel-level open-loop voltage amplification. In Proceedings of the 2011 IEEE International Solid-State Circuits Conference, San Francisco, CA, USA, 20–24 February 2011; pp. 402–403.
9. Chen, Y.; Xu, Y.; Mierop, A.; Wang, X.; Theuwissen, A. A 0.7 e$^-$ temporal readout noise CMOS image sensor for low-light-level imaging. In Proceedings of the 2012 IEEE International Solid-State Circuits Conference, San Francisco, CA, USA, 19–23 February 2012; pp. 384–385.
10. Boukhayma, A.; Peizeat, A.; Enz, C. A 0.4 e$^-$rms temporal readout noise 7.5 μm pitch and a 66% fill factor pixel for low light CMOS image sensors. In Proceedings of the 2015 International Image Sensor Workshop, Vaals, the Netherlands, 8–11 June 2015; pp. 365–368.
11. Yao, Q.; Dierickx, B.; Dupont, B.; Rutterns, G. CMOS image sensor reaching 0.34 e$^-$rms read noise by inversion-accumulation cycling. In Proceedings of the 2015 International Image Sensor Workshop, Vaals, The Netherlands, 8–11 June 2015; pp. 369–372.
12. Wakabayashi, S.; Kusuhara, F.; Kuroda, R.; Sugawa, S. A linear response single exposure CMOS image sensor with 0.5 e$^-$ readout noise and 76 ke$^-$ full well capacity. In Proceedings of the 2015 Symposium on VLSI Circuits (VLSI Circuits), Kyoto, Japan, 17–19 June 2015; pp. 88–89.
13. Ma, J.; Fossum, E. Quanta image sensor jot with sub 0.3 e$^-$rms read noise and photon counting capability. *IEEE Electron Device Lett.* **2015**, *36*, 926–928. [CrossRef]
14. Ma, J.; Starkey, D.; Rao, A.; Odame, K.; Fossum, E.R. Characterization of quanta image sensor pump-gate jots with deep sub-electron read noise. *J. Electron Devices Soc.* **2015**, *3*, 472–480. [CrossRef]
15. Seo, M.W.; Kawahito, S.; Kagawa, K.; Yasutomi, K. A 0.27 e$^-$rms read noise 220 μV/e$^-$ conversion gain reset-gate-less CMOS image sensor with 0.11 μm CIS process. *IEEE Electron Device Lett.* **2015**, *36*, 1344–1347.
16. Wolfel, S.; Herrmann, S.; Lechner, P.; Lutz, G.; Porro, M.; Richter, R.H.; Struder, L.; Treis, J. A novel way of single optical photon detection: Beating *1/f* noise limit with ultra-high resolution DEPFET-RNDR devices. *IEEE Trans. Nucl. Sci.* **2007**, *54*, 1311–1318. [CrossRef]

17. Lutz, G.; Porro, M.; Aschauer, S.; Wolfel, S.; Struder, L. The DEPFET sensor-amplifier structure: A method to beat *1/f* noise and reach sub-electron noise in pixel detectors. *Sensors* **2016**, *16*, 608. [CrossRef]

18. Krymski, A.; Khaliullin, N.; Rhodes, H. A 2 e$^-$ noise 1.3-megapixel CMOS sensor. In Proceedings of the IEEE Workshop on CCD and Advanced Image Sensors, Elmau, Germany, 15–17 May 2003; pp. 1–6.

19. Kawahito, S.; Sakakibara, M.; Handoko, D.; Nakmura, N.; Satoh, H.; Higashi, M.; Mabuchi, K.; Sumi, H. A column-based pixel-gain-adaptive CMOS image sensor for low-light-level imaging. In Proceedings of the 2003 IEEE International Solid-State Circuits Conference, San Francisco, CA, USA, 13 February 2003; pp. 224–225.

20. Sakakibara, M.; Kawahito, S.; Handoko, D.; Nakmura, N.; Satoh, H.; Higashi, M.; Mabuchi, K.; Sumi, H. A high-sensitivity CMOS image sensor with gain-adaptive column amplifiers. *IEEE J. Solid State Circuits* **2005**, *40*, 1147–1156. [CrossRef]

21. Fowler, A.M.; Gatley, I. Noise reduction strategy for hybrid IR focal plane arrays. *Proc. SPIE* **1991**, *1541*, 127–133.

22. Porro, M.; Fiorini, C.; Studer, L. Theoretical comparison between two different filtering techniques suitable for VLSI spectroscopic amplifier ROTOR. *Nucl. Instrum. Methods Phys. Res. A* **2003**, *512*, 179–190. [CrossRef]

23. Kawahito, S.; Kawai, N. Column parallel signal processing techniques for reducing thermal and RTS noises in CMOS image sensors. In Proceedings of the IEEE International Image Sensor Workshop, Ogunquit, ME, USA, 7–10 June 2007; pp. 226–229.

24. Seo, M.W.; Suh, S.H.; Iida, T.; Takasawa, T.; Isobe, K.; Watanabe, T.; Itoh, S.; Yasutomi, K.; Kawahito, S. A low-noise high intrascene dynamic range CMOS image sensor with a 13 to 19b variable-resolution column-parallel folding-integration/cyclic ADC. *IEEE J. Solid State Circuits* **2012**, *47*, 272–283. [CrossRef]

25. Seo, M.-W.; Sawamoto, T.; Akahori, T.; Iida, T.; Takasawa, T.; Yasutomi, K.; Kawahito, S. A low noise wide dynamic range CMOS image sensor with low-noise transistors and 17b column-parallel ADCs. *IEEE Sens. J.* **2013**, *13*, 2922–2929. [CrossRef]

26. Guidash, M. Active Pixel Sensor with Punch-through Reset and Cross-Talk Suppression. U.S. Patent 5,872,371, 16 February 1999.

27. Seitz, P.; Theuwissen, A.J.P. *Single-Photon Imaging*; Springer: Berlin, Germany, 2011; pp. 197–217.

28. Yadid-Pecht, O.; Fossum, E.R.; Pain, B. Optimization of noise and responsivity in CMOS active pixel sensors for detection of ultra low-light level. *Proc. SPIE* **1997**, *3019*, 123–136.

29. Suh, S.G.; Itoh, S.; Aoyama, S.; Kawahito, S. Column parallel correlated multiple sampling circuits for CMOS image sensors and their noise reduction effect. *Sensors* **2010**, *10*, 9139–9154. [CrossRef]

30. Kawai, N.; Kawahito, S. Effectiveness of a correlated multiple sampling differential averager for *1/f* noise. *IEICE Express Lett.* **2005**, *2*, 379–383. [CrossRef]

31. Hopkinson, G.R.; Lumb, D.H. Noise reduction techniques for CCD image sensors. *J. Phys. E Sci. Instrum.* **1982**, *15*, 1214–1222. [CrossRef]

32. Goo, J.-S.; Choi, C.-H.; Abramo, A.; Ahn, J.-G.; Yu, Z.; Lee, T.-H.; Dutton, R.W. Physical origin of the excess thermal noise in short channel MOSFETs. *IEEE Electron Device Lett.* **2001**, *22*, 101–103.

33. Kawai, N.; Kawahito, S. Noise analysis of high-gain low-noise column readout circuits for CMOS image sensors. *IEEE Trans. Electron Devices* **2004**, *51*, 185–194. [CrossRef]

34. Findlater, K.M.; Vaillant, J.M.; Baxter, D.J.; Augier, C.; Herault, D.; Henderson, R.K.; Hurwitz, J.E.D.; Grant, L.A.; Volle, J.M. Source follower noise limitations in CMOS active pixel sensors. *Proc. SPIE* **2004**, *5251*, 187–195.

Chapter 2:
Photon Counting with
Avalanche-Based Devices

sensors

MDPI

Article

Single-Photon Avalanche Diode with Enhanced NIR-Sensitivity for Automotive LIDAR Systems

Isamu Takai *, Hiroyuki Matsubara, Mineki Soga, Mitsuhiko Ohta, Masaru Ogawa and Tatsuya Yamashita

Toyota Central R&D Labs., Inc., 41-1, Yokomichi, Nagakute, Aichi 480-1192, Japan; hmatsu@mosk.tytlabs.co.jp (H.M.); msoga@mosk.tytlabs.co.jp (M.S.); ohtam@mosk.tytlabs.co.jp (M.O.); ogawa@mosk.tytlabs.co.jp (M.O.); tatsu-y@mosk.tytlabs.co.jp (T.Y.)
* Correspondence: takai@mosk.tytlabs.co.jp; Tel.: +81-561-71-7795

Academic Editor: Nobukazu Teranishi
Received: 20 January 2016; Accepted: 25 March 2016; Published: 30 March 2016

Abstract: A single-photon avalanche diode (SPAD) with enhanced near-infrared (NIR) sensitivity has been developed, based on 0.18 μm CMOS technology, for use in future automotive light detection and ranging (LIDAR) systems. The newly proposed SPAD operating in Geiger mode achieves a high NIR photon detection efficiency (PDE) without compromising the fill factor (FF) and a low breakdown voltage of approximately 20.5 V. These properties are obtained by employing two custom layers that are designed to provide a full-depletion layer with a high electric field profile. Experimental evaluation of the proposed SPAD reveals an FF of 33.1% and a PDE of 19.4% at 870 nm, which is the laser wavelength of our LIDAR system. The dark count rate (DCR) measurements shows that DCR levels of the proposed SPAD have a small effect on the ranging performance, even if the worst DCR (12.7 kcps) SPAD among the test samples is used. Furthermore, with an eye toward vehicle installations, the DCR is measured over a wide temperature range of 25–132 °C. The ranging experiment demonstrates that target distances are successfully measured in the distance range of 50–180 cm.

Keywords: avalanche photodiodes; light detection and ranging (LIDAR); single-photon avalanche diode (SPAD); time-of-flight (TOF); depth sensor; rangefinder; 3-D imaging; single-photon detector; advanced driver assistance system (ADAS)

---

1. Introduction

Recently, advanced driver assistance systems (ADASs) have been designed and developed not only to make automobiles more comfortable, but also to reduce the number of traffic accidents [1]. Forward collision warning (FCW), autonomous emergency breaking (AEB), and pedestrian detection, to cite a few examples, rely on various technologically advanced sensors. High-accuracy ranging sensors are particularly important components of ADASs. Among ranging sensor technologies, millimeter-wave RADARs and stereo-vision cameras remain the key sensors of choice. However, these sensors do have limitations, so a number of ADAS implementations rely on the data fusion of two or more sensors.

In this context, we have been developing an optical long-range sensor technology based on the time-of-flight (TOF) principle for next-generation ADASs [2–4]. This light detection and ranging (LIDAR) sensor offers a very good balance of overall performance in terms of spatial (image) resolution, field-of-view, precision, and depth range. In our LIDAR sensor, avalanche photodiodes operating in so-called Geiger mode have been employed as photodetectors (*i.e.*, optical receivers). The device, referred to as a single-photon avalanche diode (SPAD), is a highly sensitive photodetector capable of outputting a precise and digital trigger signal upon the detection of ultralow-power signals, down to the single photon level. Moreover, since the SPAD can be fabricated using general complementary metal-oxide-semiconductor (CMOS) technology, SPADs, their front-end circuits, and a complete digital signal processor (DSP) have been successfully and integrally designed as a low-cost system-on-a-chip implementation. Our latest LIDAR system [3] using the CMOS SPAD technology has achieved 100-m ranging even under the strong outdoor lighting environment of 70 klux.

Although the SPAD is sensitive to a single photon, not all photons are detected. For example, photons that penetrate the SPAD are not detected. Also, photo-generated electrons do not always trigger an avalanche breakdown because it is a probabilistic process. Thus, a higher photon detection efficiency (PDE) is crucial for any sensing application. Until recently, CMOS SPAD technologies have shown a high PDE only in the visible light band. However, in the near-infrared (NIR) band between 800 nm and 1 μm, CMOS SPAD devices have exhibited rather low PDEs. For example, at the wavelength of interest in our application (870 nm), the PDEs have typically been below 5%. In order to increase the measurement range and/or rate of automotive LIDAR systems that generally employ NIR light sources, a much higher PDE in the NIR band is highly desirable. More generally, however, the ranging performance depends on the overall sensitivity, which, in turn, is characterized by the product of the PDE and the fill factor (FF) of the SPAD. Therefore, key to realizing a high-performance automotive LIDAR system is having both a high PDE in the NIR band and a high FF.

Significant improvements have recently been achieved in the NIR PDE [5,6]. For example, the current state-of-the-art PDE of 20% at 870 nm has been obtained with an excess bias (V_E) of 12 V [6]. Moreover, a fill factor of 21.6% has been reported [7]. This paper presents a new SPAD structure that achieves a similar PDE performance to that reported in [6], while also improving the sensitivity by optimizing the FF. Furthermore, the performance of this SPAD is described on the basis of various experimental results.

2. Design and Chip Implementation of NIR-Sensitivity-Enhanced SPADs

Figure 1 shows the cross-sectional structure of the newly designed SPAD, which achieves high sensitivity in the NIR band without compromising the FF. The SPAD lies in the p-epitaxial layer of a CMOS wafer and comprises two SPAD-specific custom layers, namely n-SPAD and p-SPAD. Isolation between adjacent SPADs (*i.e.*, a guard ring) is achieved by the superposition of existing p-well and deep p-well layers, which are available in most modern CMOS processes.

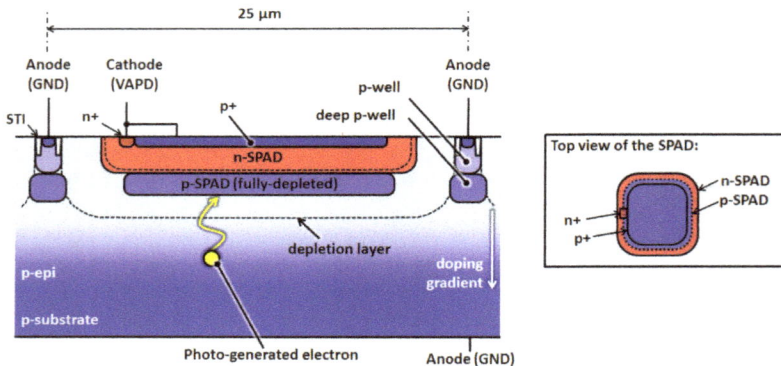

Figure 1. Simplified cross-sectional structure of the proposed SPAD. The top view of the SPAD is shown in the inset to illustrate how the SPAD forms an electrical contact with the n+ layer. Two custom layers, n-SPAD and p-SPAD, are employed for this SPAD, and the p-SPAD is fully depleted when the SPAD is biased at or above its breakdown voltage. Moreover, the p-epitaxial layer has a gradient doping profile. The thick depletion layer and doping profile efficiently collect electrons generated by the NIR light, which results in a high NIR PDE.

Since the absorption coefficient for NIR light is low compared to that for visible light, NIR light can penetrate deeper before it is completely absorbed in the silicon substrate. By increasing the vertical thickness of a depletion layer, the detection probability of incident NIR light is enhanced. The depletion layer thickness can be increased by increasing the SPAD bias voltage. However, a high bias voltage potentially leads to edge breakdown, which would interfere with the photon counting operation. To avoid edge breakdown and achieve high PDE in the NIR band, the distance between the SPAD and guard ring was increased, which reduces the FF. Conversely, a high-FF SPAD, achievable by thinning the depletion layer, has been proposed, which results in a lower PDE in the NIR band. Thus, previous SPAD designs have involved a tradeoff between the PDE and FF for sensing NIR lights, which represents an obstacle to higher sensitivity.

The proposed structure simultaneously achieves high PDE and FF. The doping concentrations of the two custom layers are optimized to obtain a high electric field for the avalanche multiplication of photo-generated electrons. Importantly, they are carefully designed so that the p-SPAD layer becomes fully depleted when the SPAD is biased at or above its breakdown voltage. Both a high electric field and a thick depletion layer are achieved under a low bias voltage, and thus, edge breakdown is avoided. As a result, the NIR PDE is enhanced without compromising the FF.

Furthermore, this structure also exploits a p-epitaxial layer that features a gradient doping profile; *i.e.*, a profile where the doping concentration increases with depth. Such a gradient doping profile further improves the PDE by promoting upward migration and efficient collection of photo-generated electrons toward the avalanche multiplication region. This technique has been used in image sensors to improve the quantum efficiency. It has also been reported in CMOS SPADs [5].

Additionally, the proposed SPAD also has a shallow p+ layer at the surface of the SPAD in order to collect surface-generated carriers, which reduces its intrinsic dark noise. Thus, the proposed structure prevents dark noise from triggering false avalanche events.

Figure 2a shows a photograph of a test chip that was fabricated to characterize the proposed SPAD using a 0.18-μm CMOS process. The test chip contains an array of 5 × 3 SPADs and three quenching/readout circuits. Only the innermost three SPADs are connected to these circuits, as shown in the figure, while the remaining 12 SPADs are used as dummies because the characteristics of the outermost devices in a device array are generally unreliable. The distance between adjacent SPADs is 25 μm. Figure 2b shows schematics of quenching and readout circuits containing the SPADs.

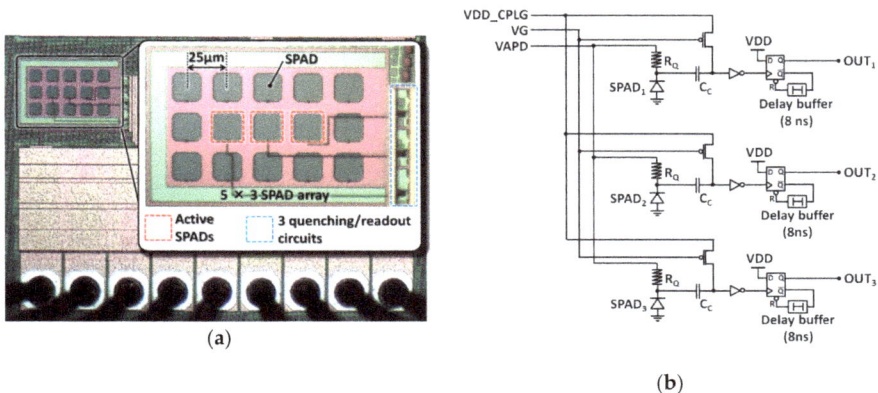

Figure 2. (a) Photomicrograph of a sample chip containing an array of 5 × 3 SPADs and three quenching/readout circuits. Only the innermost three SPADs are connected to the circuits. The remaining 12 SPADs are used as "dummy" devices; (b) a schematic of three quenching and readout circuits containing SPADs.

SPAD quenching and recharge is typically performed passively by high-resistivity polysilicon resistors of $R_Q = 300$ kΩ. When a photon enters the SPAD, negative voltage pulses with an amplitude of approximately $V_E = VAPD - V_{BD}$ are generated at the SPAD cathode, where V_{BD} is the SPAD breakdown voltage, $VAPD$ is the positive bias voltage, and V_E is the SPAD excess bias. The generated pulse is then capacitively coupled via a 5 fF metal fringe capacitor (C_C) to the input of a CMOS inverter that serves as a comparator. The inverter input is biased to VDD_CPLG by a thick-oxide PMOS transistor with a constant gate bias VG. The inverter is designed with thick-oxide transistors and is powered by the core 1.8 V supply voltage in this process. Positive pulses at the inverter output are transformed into rectangular pulses of approximately 8.5 ns by a D-flip-flop-based monostable circuit. This readout circuit imposes a minimum dead time of approximately 17 ns, regardless of the actual SPAD recharge time.

3. Experimental Results

Figure 3 shows static SPAD current characteristics as a function of the reverse bias voltage supplied to the SPAD cathode. The current measurements are conducted under dark and light conditions with an integration time of 320 ms using a semiconductor parameter analyzer. From this result, the V_{BD} of the proposed SPAD is approximately 20.5 V.

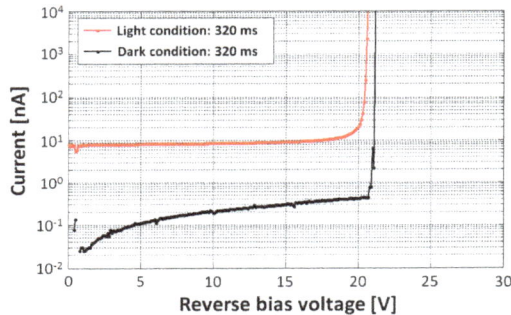

Figure 3. Static SPAD current characteristics as a function of the reverse bias voltage supplied to the SPAD cathode.

Figure 4a shows the microscope-based measurement system used to accurately measure the FF of the proposed SPAD.

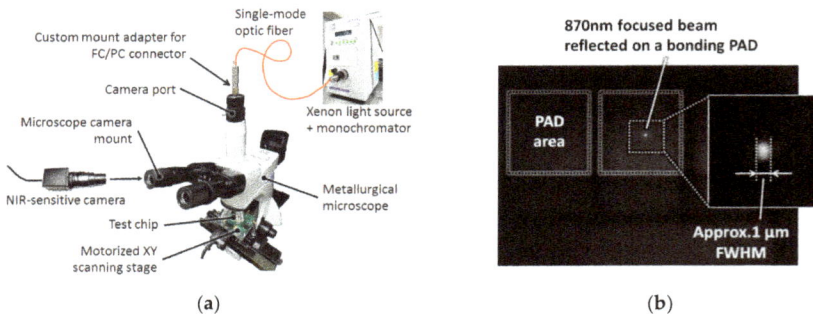

Figure 4. (**a**) Measurement system for high-resolution XY mapping of the photon response of a SPAD at any desired optical wavelength. An objective lens with an NA of 0.6 and ×40 is used in the microscope; (**b**) Image captured using an NIR-sensitive camera mounted on one of the microscope eyepieces.

This system relies on the fact that all of the light coming out of a point source placed in the focal plane of the camera port is focused onto a point on the sample surface. Therefore, it provides an XY mapping of the photon counting rate of a SPAD at any desired optical wavelength with a high spatial resolution. In the microscope, an objective lens with an NA of 0.6 is used. Figure 4b shows the experimentally confirmed resolution of the proposed system and demonstrates that the spot size of an 870-nm focused beam is approximately 1 μm FWHM (full width at half maximum).

Figure 5 shows a normalized map of the photon response of a SPAD with a resolution of 0.5×0.5 μm, using the system shown in Figure 4, measured with a V_E of 5 V and 870-nm light. The measurement area is 30×30 μm, fully encompassing the 25×25 μm SPAD area. This result experimentally demonstrates that the proposed SPAD achieves a high FF of 33.1% FWHM.

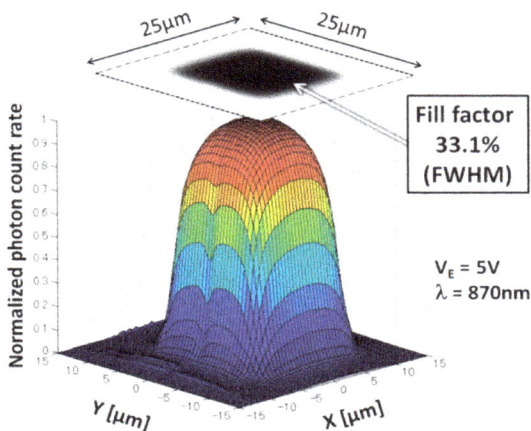

Figure 5. Normalized photon counting rate map of the proposed 25×25 μm^2 SPAD. V_E is 5 V and the incident cone of 870 nm light is approximately 74°.

Figure 6a shows a light source, an integrating sphere, and a reference photodiode based measurement system to measure the PDE of the proposed SPAD. A light source radiates light with wavelengths ranging from 350 to 1100 nm into the integrating sphere.

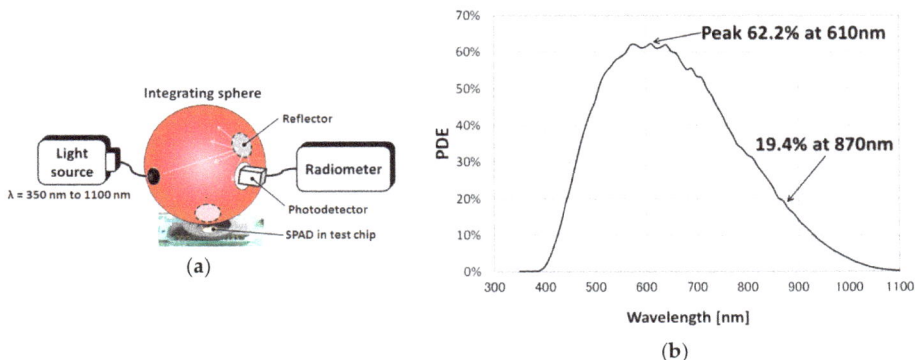

Figure 6. Measurement system (**a**) used to determine the PDE of the proposed SPAD. This system consists mainly of a light source, an integrating sphere, and a photodetector connected to a radiometer; (**b**) PDE as a function of light wavelength at an excess bias of 5 V.

Figure 6b shows the measured PDE of the proposed SPAD as a function of light wavelength with a V_E of 5 V. The proposed SPAD has a peak PDE of 62.2% at 610 nm and achieves a high PDE of 19.4% at 870 nm, which is the laser light wavelength of our LIDAR system. To the best of our knowledge, this is the highest PDE yet achieved for a CMOS SPAD under similar V_E bias conditions.

Figure 7 shows dark count rate (DCR) measurement results. The dark counts are caused by false output pulses, which are mainly induced by intrinsic dark noise in the SPAD, despite the absence of light incidence. A high DCR worsens the ranging performance, since false pulses obscure true pulses generated by the desired laser light. Figure 7a shows DCRs as a function of V_E for 18 SPAD samples in six test chips at room temperature. As seen in this figure, the DCR values vary greatly among samples. Such large variances in DCR are common, as evidenced by [5,6]. At a V_E of 5 V, DCRs of half of SPAD samples are less than 100 counts per second (cps), and the maximum DCR is 12.672 kcps, *i.e.*, false pulses are generated in cycles of 78.914 μs. However, these DCR values have little impact on the ranging performance. For example, since a 150 m ranging operation requires only 1 μs measurement time, dark counts (*i.e.*, false pulses) only affect one operation in approximately every 78 operations, even if the worst SPAD sample is used. Figure 7b shows the DCR as a function of chip temperature over the range of 25–132 °C. As the figure shows, the DCR grows exponentially with chip temperature, as dark noise is directly correlated with temperature.

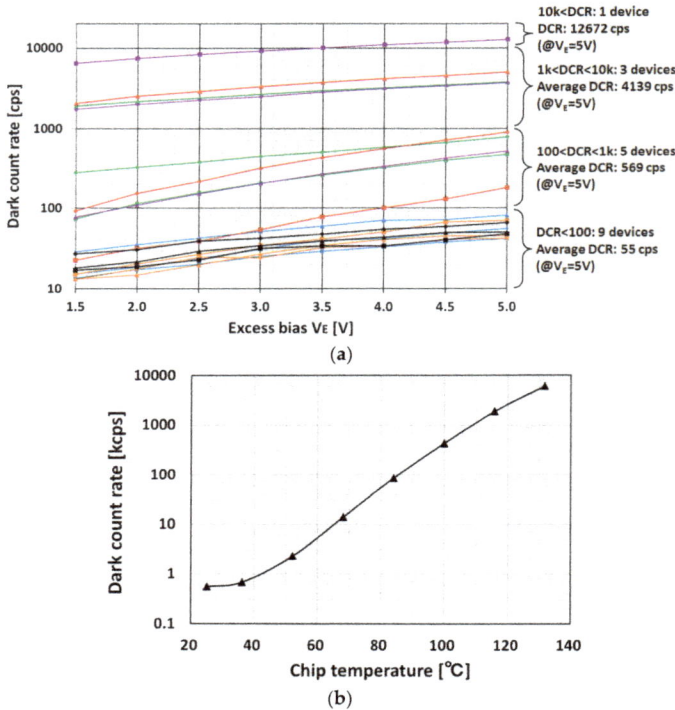

Figure 7. (**a**) Measurement results of DCR as a function of excess bias V_E, using 18 SPAD samples in 6 test chips at room temperature; (**b**) Measurement results of DCR as a function of chip temperature, using one SPAD sample.

The room-temperature DCR is less than 1 kcps. When the chip temperature increases, however, the DCR reaches 10 kcps at 65 °C, 100 kcps at 85 °C, 1 Mcps at 110 °C, and 6 Mcps at 132 °C. As noted previously, a 150 m ranging operation is completed in 1 μs, so the false pulse affects every ranging operation above 110 °C. However, our final chip will implement spatiotemporal correlation technology based on a macro-pixel structure [3], which will effectively suppress the impact of DCR at high temperatures. The present measurement results offer important insight for the application of the proposed SPAD to actual automotive systems that must operate at high temperatures.

Figure 8a shows the elementary system used for measurements of the TOF and target distances using the proposed SPAD. Mainly, this experiment ensures that the proposed SPAD can accurately respond to narrow and high-speed laser pulses. This is a first step in the development of a LIDAR system. The present elementary system consists of a picosecond laser source (λ = 635 nm, FWHM $\approx$ 100 ps), a test chip containing SPADs to which a VAPD of 25.5 V is supplied, an oscilloscope (25 GSPS) to record trigger signals from the laser source and output signals of the SPAD with a period of 25 ns, a reflector as a measurement target, and an offline PC. When the pulsed laser light reflected by the reflector enters the SPAD in the test chip, a signal formed to an 8.5 ns PW is output from the chip. Finally, the PC superimposes approximately 5000 frames of the output signal data recorded by the oscilloscope and produces a TOF image similar to an eye diagram. In this measurement, the target distance ranges from 50 to 180 cm. Figure 8b shows representative TOF images at 50, 90, 130, and 170 cm. As seen in the figure, the pulse delay time (t_d) based on the TOF at 50 cm increases linearly with increasing target distance.

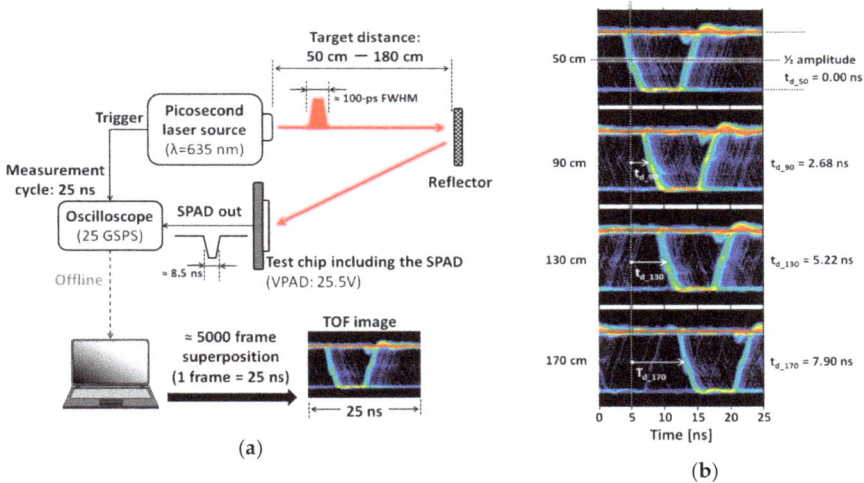

Figure 8. (**a**) Measurement system for the TOF and target distances; (**b**) TOF images superimposing 5000-frame SPAD out signals at 50, 90, 130, and 170 cm. From the TOF images, each delay time (t_d) over various distances is measured, based on the result at 50 cm.

Figure 9 shows measured distances to the target reflector at room temperature, based on the delay times (t_{d_50}–t_{d_180}) measured by the system in Figure 8. Figure 9a shows distances calculated from each t_d as a function of target distance (the actual distance to the target reflector). As shown in the figure, the measured distances are comparable to the actual target distances. Figure 9b shows the error as a function of target distance. As seen in Figure 9b, the errors are all less than $\pm$2%. These results demonstrate that the proposed SPAD can respond to narrow and high-speed laser pulses and can thus be used for ranging systems.

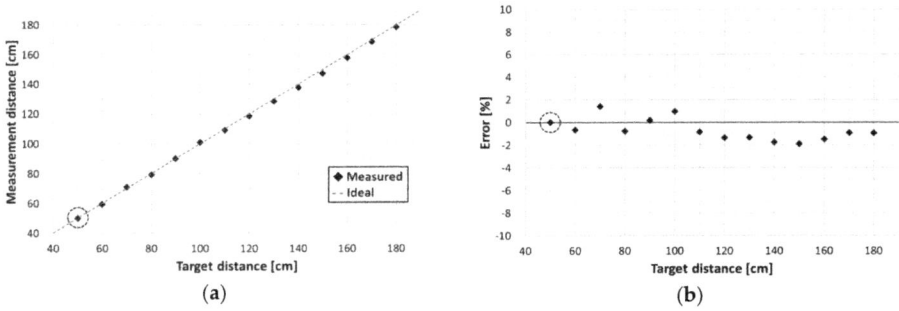

Figure 9. Measurement results of target distances at room temperature. (**a**) Measured distances to the target reflector as a function of actual target distance, based on the result obtained at 50 cm; (**b**) Measurement errors as a function of target distance, based on the result obtained at 50 cm.

4. Conclusions

This paper presents a new SPAD structure to achieve a higher-performance, e.g., longer-range, higher-precision, and higher-measurement-rate, LIDAR system for future ADASs. In the proposed SPAD, two custom layers, p-SPAD and n-SPAD, are employed using the existing 0.18-μm CMOS technology. The doping concentrations of the customized layers are optimized to achieve a high electric field, while the p-type layer becomes fully depleted when the device is biased to operate in Geiger mode. This full-depletion layer enables the efficient collection of electrons generated by NIR light, without compromising the FF. Moreover, the p-epitaxial layer, which has a gradient doping profile, further enhances the NIR PDE.

In this study, various experiments are conducted to characterize the proposed SPAD. In SPAD PDE measurements, PDEs of 62.2% at 610 nm and 19.4% at 870 nm are attained. To the best of our knowledge, this is the highest PDE achieved for a CMOS SPAD under similar excess bias conditions, *i.e.*, 5 V. Moreover, the actual FF is measured by the special measurement system, and a high FF of 33.1% is obtained at 870 nm. Since SPAD sensitivity is characterized by the product of the PDE and FF, the proposed structure imparts the SPAD with higher NIR sensitivity, eliminating the PDE-FF tradeoff. The DCR levels of the proposed SPAD have only a small effect on the ranging performance, even when the worst SPAD among the test samples is used. The DCR measurement results at high temperature reveal new insight for actual vehicle installations. The ranging experiment demonstrates that target distances are successfully measured within an error of ±2%. We believe that the proposed SPAD will pave the way for higher-performance LIDAR and ADAS systems.

Acknowledgments: This paper is based on results obtained from a project commissioned by the New Energy and Industrial Technology Development Organization (NEDO). We would like to express our deepest gratitude to Cristiano Niclass. Advice and comments by Kota Ito (Toyota Central R&D Labs, Inc.) have been a great help.

Author Contributions: M. Soga and T. Yamashita contributed to the realization of the SPAD with enhanced NIR-sensitivity; I. Takai, H. Matsubara, M. Soga, and M. Ogawa conceived and designed the experiments; I. Takai and H. Matsubara performed the experiments; I. Takai, H. Matsubara, M. Ohta, and M. Ogawa contributed to the development of the systems for the experiments; All authors contributed to analyses and considerations of the experimental results; I. Takai wrote the paper.

Conflicts of Interest: The authors declare no conflict of interest.

References

1. Ninomiya, Y. Special feature: Active safety. *R&D Rev. Toyota CRDL* **2012**, *43*, 1–6.
2. Matsubara, H.; Soga, M.; Niclass, C.; Ito, K.; Aoyagi, I.; Kagami, M. Development of next generation LIDAR. *R&D Rev. Toyota CRDL* **2012**, *43*, 7–12.

3.   Niclass, C.; Soga, M.; Matsubara, H.; Ogawa, M.; Kagami, M. A 0.18-μm CMOS SoC for a 100-m-range 10-frame/s 200 × 96-pixel time-of-flight depth sensor. *IEEE J. Solid-State Circuits* **2014**, *49*, 315–330. [CrossRef]
4.   Niclass, C.; Matsubara, H.; Soga, M.; Ohta, M.; Ogawa, M.; Yamashita, T. A NIR-sensitivity-enhanced single-photon avalanche diode in 0.18 μm CMOS. In Proceedings of the International Image Sensor Workshop, Vaals, The Netherlands, 8–11 June 2015; pp. 340–343.
5.   Webster, E.A.G.; Richardson, J.A.; Grant, L.A.; Renshaw, D.; Henderson, R.K. A Single-Photon Avalanche Diode in 90-nm CMOS Imaging Technology With 44% Photon Detection Efficiency at 690 nm. *IEEE Electron Device Lett.* **2012**, *33*, 694–696. [CrossRef]
6.   Webster, E.A.G.; Grant, L.A.; Henderson, R.K. A High-Performance Single-Photon Avalanche Diode in 130-nm CMOS Imaging Technology. *IEEE Electron Device Lett.* **2012**, *33*, 1589–1591. [CrossRef]
7.   Webster, E.A.G.; Walker, R.J.; Henderson, R.K.; Grant, L.A. A silicon photomultiplier with >30% detection efficiency from 450 to 750 nm and 11.6 μm pitch NMOS-only pixel with 21.6% fill factor in 130 nm CMOS. In Proceedings of the European Solid-State Device Research Conference (ESSDERC), Bordeaux, France, 17–21 September 2012; pp. 238–241.

sensors

Review

Geiger-Mode Avalanche Photodiode Arrays Integrated to All-Digital CMOS Circuits

Brian Aull

Massachusetts Institute of Technology Lincoln Laboratory, 244 Wood St, Lexington, MA 02420, USA; aull@ll.mit.edu; Tel.: +1-781-981-4676

Academic Editor: Edoardo Charbon
Received: 21 January 2016; Accepted: 1 April 2016; Published: 8 April 2016

Abstract: This article reviews MIT Lincoln Laboratory's work over the past 20 years to develop photon-sensitive image sensors based on arrays of silicon Geiger-mode avalanche photodiodes. Integration of these detectors to all-digital CMOS readout circuits enable exquisitely sensitive solid-state imagers for lidar, wavefront sensing, and passive imaging.

Keywords: imagers; imaging; avalanche photodiodes; Geiger-mode avalanche photodiodes; lidar; ladar; wavefront sensing; photon counting

1. Introduction

The MIT Lincoln Laboratory (Lexington, MA, USA) has been a world leader in the development of specialized high-performance charge-coupled-device imagers (CCDs) [1]. CCDs have outstanding performance and support advanced functions such as charge-domain image stabilization, electronic shuttering, and blooming control. The ultimate sensitivity limitation of a CCD is set by the readout noise of the output amplifier that senses the charge packets and converts them to analog voltage levels. The faster the readout rate, the more severe the readout noise penalty. This sensitivity limitation becomes important in photon-starved applications, such as night vision or high-temporal-resolution imaging.

Interest in such scenarios lead to Lincoln's development of photon-counting image sensors, primarily based on arrays of custom-fabricated Geiger-mode avalanche photodiodes (GMAPDs) integrated with all-digital CMOS readout circuits. The term "photon counting" is used broadly here to mean that each photon arrival is digitally recorded by the pixel circuit. The pixel could be designed to either time stamp photons or count them. While the primary focus of this article is silicon GMAPDs [2], Lincoln Laboratory has developed GMAPD arrays based on compound semiconductors sensitive at wavelengths ranging from 1.06 μm [3] to the mid-wave infrared [4].

The principal advantage of a photon counting image sensor is that it eliminates readout noise. Because digitization occurs within the pixel, there is no analog circuitry in the readout path, and therefore no analog circuit noise. This means that there is no readout noise penalty for operating at high frame rates, using short integration times, or dividing the incoming light into multiple spectral or spatial channels. In-pixel digitization therefore enables high-frame-rate imaging [5,6], multispectral imaging [7], and spatial oversampling [8].

A second advantage of a photon-counting image sensor is that it facilitates in-pixel time-to-digital conversion. The ability to digitally time stamp individual photon arrivals is an enabler for exquisitely sensitive lidar imaging systems. The GMAPD arrays can also be used as laser communication receivers in systems with challenging link budgets. Use of pulse-position-modulated formats allows for multiple bits of information to be encoded in a single transmitted photon [8].

The integration of GMAPDs with digital CMOS can take advantage of Moore's Law scaling and 3D integration to incorporate on-focal-plane processing functions such as imaging stabilization, spatial filtering, change detection, and tracking of objects of interest without mechanical scanning.

Potential disadvantages of GMAPDs include afterpulsing and crosstalk. Afterpulsing denotes false detection events triggered by carriers generated in previous events. To avoid afterpulsing the avalanche photodiode (APD) must be debiased for a sufficient time after an event to give such residual carriers time to recombine or be collected. Since the APD is unresponsive during this time, this leads to blocking loss and causes the count rate to saturate at high optical fluxes. Crosstalk is mediated by hot-carrier light emission by the avalanche process and is discussed at the end of Section 4.

2. Geiger-mode APDs and Photon-to-Digital Conversion

There are many types of detectors that have single-photon sensitivity, including photomultiplier tubes, image intensifiers, superconducting nanowires, and CCDs with avalanche gain registers [9]. GMAPDs, also known as single-photon avalanche detectors or SPADs [10], have a combination of advantages over these other photon-counting technologies. It is an all-solid-state device technology, scalable to many pixels, capable of room-temperature or thermoelectrically-cooled operation, and as noted already, amenable to on-focal-plane digitization and processing.

An avalanche photodiode is a *p-n* junction diode whose doping profile supports high electric fields near the junction at operational bias. A photoelectron or photohole created by the absorption of a photon is accelerated to sufficient energy to initiate a chain of impact ionization events, creating offspring electron-hole pairs and leading to internal gain. Once this avalanche process is underway, there is a competition between carrier generation and carrier extraction. Below the avalanche breakdown voltage, carriers are being extracted faster than they are being generated. The current flow decays, leading to self-termination of the avalanche process. This gives finite gain and therefore produces a photocurrent that is proportional to the intensity of the incident light. This is the traditional mode of operation used in optical communication receivers, known as linear mode. The gain is an increasing function of bias, and there is a random variation of gain that degrades signal-to-noise ratio.

The APD can also be operated above the avalanche breakdown voltage. When an avalanche starts, carrier generation predominates over extraction leading to exponential growth of the current. This growth is arrested by series resistance (in most cases mediated by space charge buildup). This mode of operation is known as Geiger mode, in which a single photon can initiate an avalanche that is self-sustaining. Of course, the APD must then be reset by reducing the bias to below breakdown long enough to terminate the avalanche, a process known as quenching.

The physics of Geiger-mode operation lends itself to a startlingly simple way of interfacing the GMAPD to a digital CMOS circuit: a direct connection to the input of a logic element, illustrated in Figure 1. The *p*-side of the APD is tied to a negative bias voltage slightly less in magnitude than the breakdown. The *n*-side is connected to a logic circuit that performs simple voltage sensing. Initially, the n-side of the APD is set to a logic-high voltage by briefly turning on a pull-up transistor. Once armed in this fashion, the APD has a reverse bias several volts above breakdown. When an avalanche is initiated, the APD turns on and then discharges its own parasitic capacitance. Once at or below breakdown, the avalanche terminates, leaving the *n*-side at a voltage close to a logic 0. The APD is effectively a CMOS-compatible digital element. Also, like a CMOS logic element, the APD draws current only when it is switching. This "photon-to-digital conversion" scheme has been the basis of all of Lincoln's GMAPD imaging devices, with minor modifications. One can, for example, add a pull-down transistor to speed up the APD discharge, size the transistors in the sensing gate to adjust the triggering threshold, or cascode transistors to augment the voltage swing. Voltage sensing gives more timing latency and timing jitter than current sensing circuits [11]. This drawback has been more than offset by the advantages of the photon-to-digital conversion scheme: simplicity, robustness, and freedom from static power dissipation.

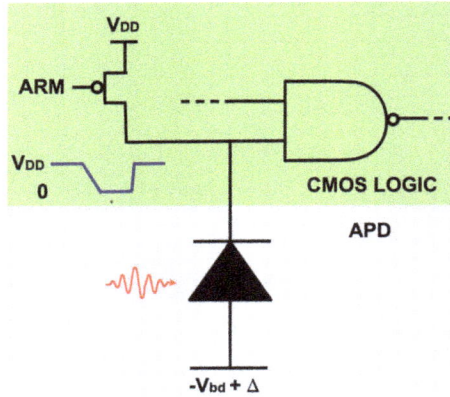

Figure 1. Photon-to-digital conversion. The GMAPD is biased so that it produces a CMOS-logic-compatible voltage pulse, enabling a direct connection to pixel logic.

3. Fabrication and Integration with CMOS

Lincoln Laboratory's silicon APD arrays are fabricated in house in its Microelectronics Laboratory, currently on 200-mm silicon substrates. Typically, the substrate is heavily p doped, with a lightly p-doped epitaxial layer on which the APD structure is fabricated. The doping profile of the APD is defined by a series of patterned ion implantation steps. After back-end processing to make contact pads, the APD array is integrated with a CMOS readout circuit, using either bump bonding or a 3D-integration technique. During this process the opaque silicon substrate must be removed to enable back-side illumination. The back side is passivated with a p-doped contact layer. This layer is typically implanted and then activated using laser annealing, although molecular beam epitaxy has also been used to grow the contact layer. In many cases an antireflection coating and a sparse contact metal pattern is also added. A simplified cross section of the pixel structure is shown in Figure 2. The APD is a separate-absorber-multiplier structure, and the figure also shows the electric field profile at operational bias. A photon incident from the back side creates an electron-hole pair in the absorber. The hole is collected at the back side, and the electron drifts to the multiplier region, where the avalanche is initiated.

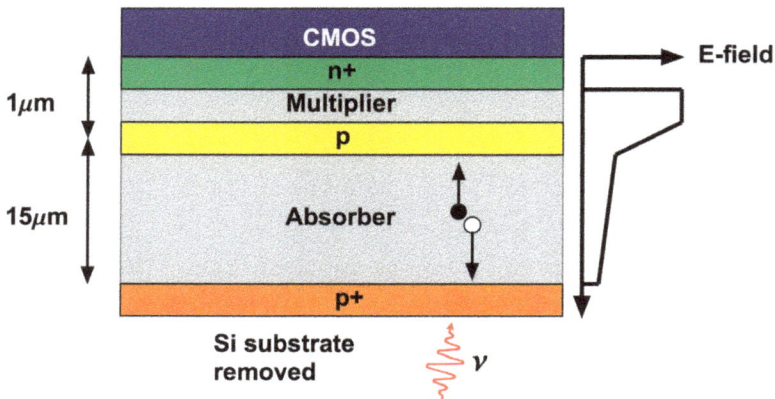

Figure 2. Simplified cross section showing pixel structure, field profile, and operation.

A number of investigators have demonstrated front-illuminated silicon GMAPD arrays using a monolithic device structure in which the photodiode is incorporated into the CMOS readout circuit using a standard CMOS foundry process [10]. This approach facilitates rapid and low-cost prototyping. Because of the thin device structure, monolithic CMOS APDs can time stamp photon detections with very low (tens of ps) timing jitter. However, they share pixel area with the readout circuitry and this limits fill factor. Moreover, fabrication using a standard CMOS process flow precludes customization of photodiode doping profiles and layer thicknesses. Photon detection efficiency is typically poor in the red and near-IR spectral regions. This is the reason why Lincoln has pursued back-illuminated APD arrays hybridized to foundry-fabricated CMOS.

Hybridization of compound semiconductor detector arrays (InGaAs or HgCdTe) is common. These detectors, however, are fabricated in layers heteroepitaxially grown on transparent substrates that serve as a mechanical support during bump bonding and device operation. Silicon APDs are fabricated on homoepitaxially grown layers on a substrate that is opaque at wavelengths of operation. The substrate must be entirely removed either before or after bump bonding, entailing difficulties associated with handling of a thin (15 µm) detector layer or with transfer of this layer to a transparent substrate.

The first GMAPD arrays made at Lincoln were hybridized by a technique known as bridge bonding. Individual CMOS die were epoxied to an APD wafer, providing immediate mechanical support, but with no electrical connection. After removal of the APD wafer substrate, large vias were etched through the detector layer between pixels to expose connection pads, and a metal strap patterned to complete the electrical connection. This technique worked but its use is limited to devices with coarse (>50 µm) pixel pitch and low fill factor [2].

Another technique pursued is transfer and bump bonding [12], shown in Figure 3. Once the APD arrays are fabricated, the wafer is bonded to a temporary silicon handle wafer. The APD substrate is removed and then the device bonded to a transparent fused silica substrate. The temporary handle is removed, and then bump bonding to CMOS readout chips can be carried out in the same manner as for the heteroepitaxial detector materials. Both the temporary handle and the transparent substrate are bonded using oxide-oxide bonding with no adhesive. The transfer process provides APD arrays that can then be quickly bumped to any CMOS readout chip that matches the format and pixel pitch. It is, however, a relatively complex process because of the two-step transfer involved.

Figure 3. Transfer and bump bonding.

Lincoln Laboratory pioneered a 3D-integration process using its in-house fully depleted silicon-on-insulator (SOI) CMOS process. One of the first imager demonstrations was a 64 × 64-pixel GMAPD array for lidar [13]. The APD was integrated with two tiers of SOI CMOS readout circuitry. Figure 4 illustrates the process. After APD and SOI CMOS fabrication is complete, the APD wafer (tier 1) and the first CMOS wafer (tier 2) are precision aligned and oxide bonded face to face. Then the SOI handle wafer is removed, using the SOI buried oxide as an etch stop. Concentric vias are then etched through the oxide layers and filled with tungsten to interconnect the last metal of the SOI wafer with the top metal of the APD wafer. These tungsten plugs are micron-scale in diameter and only a few microns in height, much smaller than bump bonds or through-silicon vias used in some wafer stacking processes. The process can be repeated, adding the tier-3 CMOS wafer to the stack. Because the SOI handle wafer is removed, the process is amenable to multiple-tier structures. Figure 5 shows a cross-sectional micrograph of a pixel of the 64 × 64-pixel GMAPD lidar imager.

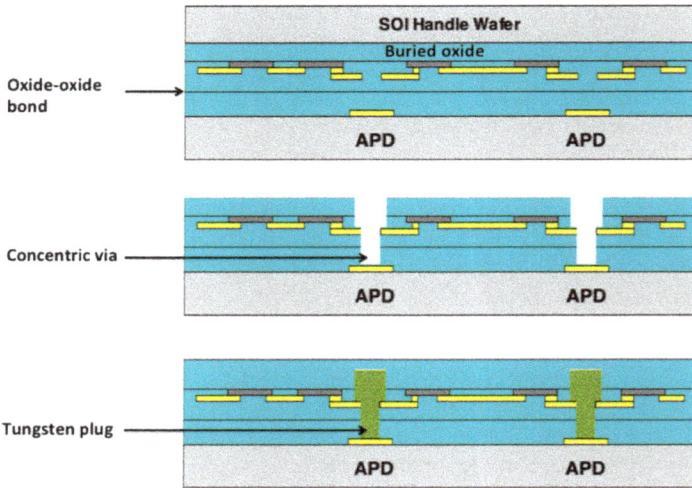

Figure 4. Steps in the Lincoln 3D-integration process to integrate the APD wafer (tier 1) with the first SOI CMOS wafer (tier 2).

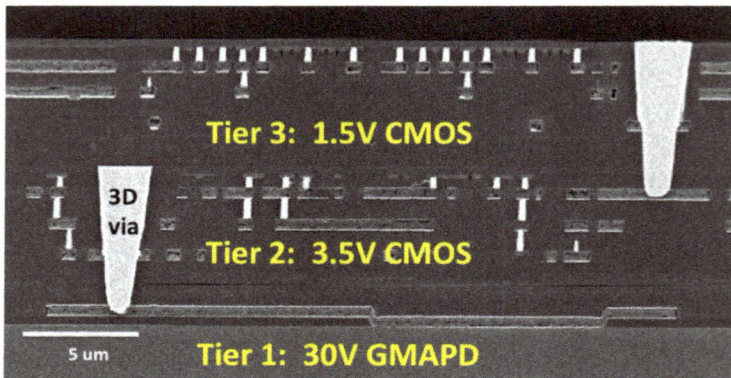

Figure 5. Cross-sectional SEM through a pixel of the 64×64 GMAPD array.

3D integration is an important technology direction for advanced image sensors. In the device structure shown in Figure 5, tier 1 contains the photodetectors, tier 2 has first-level signal conditioning circuits, and tier 3 has high-speed logic for photon time stamping. The ability to divide the pixel circuitry among multiple tiers enables a sophisticated pixel circuit to fit into a smaller area than would be possible for a conventional 2D-integrated circuit. A tier could be dedicated to image processing functions to extract information at the pixel level and reduce readout bandwidth. Having multiple tiers also enables the designer to combine multiple circuit or device technologies into a single integrated device, as in Figure 5.

The Lincoln 3D process is an example of a *via-last* process. The connection via from each tier of SOI CMOS down to the previous tier is made after wafer bonding and SOI handle removal. This approach is amenable to multi-tier structures, but requires dedicated full-wafer SOI CMOS for each tier, with the exception of tier 1. One can also use a *via-first* approach [14], illustrated in Figure 6. Metal posts (which perform the same function as bump bonds, but are much smaller) are patterned on each wafer and planarized along with the bonding oxide. When the wafers are bonded and then heated to strengthen the oxide-oxide bond, the posts expand and fuse. In a single step, therefore, mechanical bonding and electrical connection are achieved simultaneously. The CMOS wafer also functions as a permanent mechanical support for APD substrate removal, backside processing, and subsequent imager operation. The *via-first* approach allows for the use of any CMOS process, although it is not easily extendable to multi-tier structures.

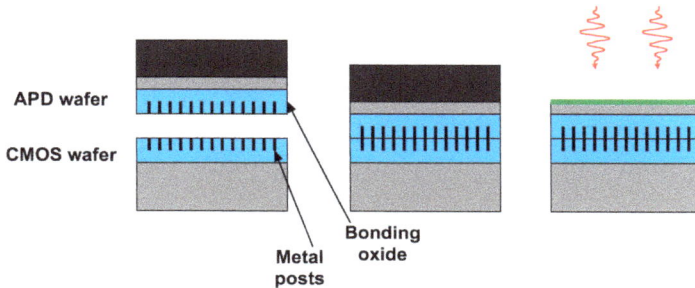

Figure 6. *Via-first* 3D integration.

4. Application to Lidar Imaging

The first application of Lincoln's GMAPD arrays was optical flash radar for three-dimensional imaging. This is known as ladar or lidar, and we will use the second term. In a flash lidar system the scene is illuminated by a short laser pulse, and imaged onto an array of detectors, each of which measures photon arrival time, and therefore depth to the corresponding point in the scene. By using an array of detectors, one can avoid the mechanical scanning needed in single-detector systems. Because the available light is divided among many pixels, however, the average return signal can be weak, sometimes less than a single photon per pixel. In addition, the pixel circuit is called upon to perform high-temporal-bandwidth time stamping. For linear-mode detectors and analog preamplifiers, sensitivity and speed are conflicting requirements. High noise bandwidth comes along with high signal bandwidth. The user of the system is forced to transmit enough photons so that the amplified return in each pixel can be discriminated from the noise floor. For a given amount of average laser power, this mandates low pulse repetition frequency with high energy per laser pulse.

The use of a Geiger-mode detector array offers a solution. The GMAPD is triggered by a single photon to produce a digital pulse, which is then digitally time stamped by the pixel circuit. There is no tradeoff between sensitivity and timing resolution. The lidar system is operated to take advantage of single-photon time stamping. The laser is operated at high pulse repetition frequency and low energy per laser pulse, so that each pixel gets average returns of a fraction of a photon. The image is built up by combining multiple frames. This mode of operation has two advantages. First, no photons are "wasted" overcoming a noise floor; timing information is obtained from each detected photon. Second, if a pixel sees returns from multiple depths, a histogram of timing values can be built up that reproduces the temporal profile that would be seen by a linear-mode detector operating with a much stronger signal.

The first proof of concept of a GMAPD array for lidar was a front-illuminated 4 × 4-pixel APD die piggy backed onto a 16-channel timing chip and the connections made with wire bonds. This device is shown in Figure 7. The CMOS circuit was fabricated through MOSIS using an HP 0.5-μm foundry process. The timing was done by broadcasting a clock to all the pixels, each of which had a pseudorandom counter based on a linear feedback shift register.

Figure 7. 4 × 4 GMAPD array wire bonded to CMOS timing circuits.

The APD doping profile used in the lidar devices gives a low fill factor. A cross section through two adjacent pixels is shown in Figure 8. The center of the APD has the separate-absorber-multiplier structure depicted in Figure 2, but the n^+-doped region extends out beyond the multiplier, creating a peripheral portion of the junction that collects electrons thermally generated in the region between pixels; these electrons do not trigger Geiger-mode events. A microlens array is integrated on the back side to concentrate incident light onto the responsive portions of the APDs. Alternatively, the lidar transmitter can be designed to project an array of spots onto the scene, which are then imaged on the APDs.

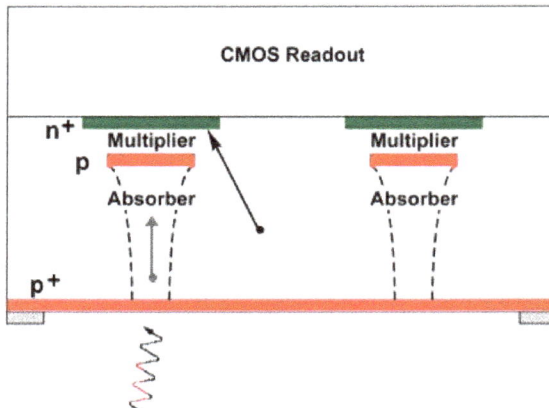

Figure 8. Low-fill-factor APD design used in lidar sensors.

Following this proof of concept, the Laboratory developed a series of 32 × 32 arrays bridge bonded to MOSIS-fabricated CMOS circuits, and used them to build systems to perform foliage penetration and terrain mapping. Figure 9 shows the results of a foliage penetration experiment [15]. Lidar data was collected through a forest canopy from different heights on a tower. These multiple views enable returns to be collected from different angles through gaps in the foliage, thereby filling in the details about objects obscured by the foliage in conventional imagery. Figure 9b is a composite 3D image with the early returns from foliage filtered out, revealing vehicles, picnic tables, and a gazebo.

(a)   (b)

Figure 9. (**a**) Forest canopy as viewed from a nearby tower (**b**) Composite lidar image obtained by combining data collects from four different heights and then filtering out the early returns.

Lincoln developed the Airborne Ladar Imaging Research Testbed (ALIRT) system, an airborne lidar for terrain mapping. Initially, the system operated at 780 nm and used a 32×32 silicon GMAPD array. Because of the availability of efficient lasers at 1060 nm, however, the Laboratory developed short-wave-IR-sensitive GMAPDs based on InGaAsP detectors grown on InP substrates. These APDs also have the separate-absorber-multiplier structure, but the doping profile is created in the epitaxial growth and the pixels are isolated by mesa etch. 128×32 lidar image sensors were built by bump bonding the APD arrays to a CMOS timing circuit. Because of the exquisite sensitivity of the GMAPDs, the ALIRT system could collect wide-area terrain maps fifteen times faster than commercially available mapping lidars.

One example of the many missions flown by ALIRT was in support of humanitarian efforts in Haiti after the 2010 earthquake [16]. Figure 10 shows a lidar image of a bridge in Port-au-Prince, Haiti. ALIRT collected terrain maps of the city over time, which enabled relief workers to know where tents were being erected or taken down, what roads were blocked, and where it was safe to land helicopters.

Figure 10. Lidar image of a bridge in Port-au-Prince, Haiti.

A successor to the ALIRT system has an area coverage rate of 400 km^2/h at 25-cm ground sampling distance. It can rapidly map a broad region and supply detailed three-dimensional images of every terrain feature or manmade structure over which it flies. It sees through foliage or dense dust clouds [17].

Lincoln demonstrated another lidar focal plane that is highly significant from a technology evolution standpoint. Figure 11a shows the 3D-integrated 64 × 64-pixel lidar focal plane whose three-tier pixel cross section is shown in Figure 5. Figure 11b is a lidar image of a cone obtained using this device. To our knowledge, this was the first demonstration of a three-tier integrated circuit of any kind, based on a process that supports dense, arbitrarily placed micron-scale inter-tier connection vias (as opposed to chip stacking with peripheral wire bonds or ball grid arrays).

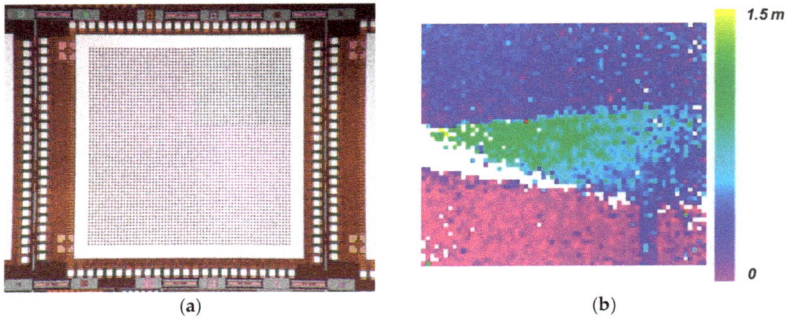

(a)   (b)

Figure 11. (**a**) 64 × 64-pixel lidar focal plane fabricated by 3D integration of a GMAPD array with two tiers of SOI CMOS circuitry (**b**) Lidar image of a cone obtained by illumination from a doubled Nd:YAG microchip laser.

3D integration, as already pointed out, allows one to mix different technologies and put more circuitry within the area of a pixel. More importantly, however, it enables new imager architectures. Raw pixel data flows in parallel up through tiers of on-focal-plane processing circuits that extract information and reduce readout bandwidth.

5. A Photon-Counting Wavefront Sensor

Adaptive optics systems for ground-based astronomy and space surveillance require sensors to measure the distortion of a wavefront (from either a bright star or an artificially created beacon) due to atmospheric turbulence. The Shack-Hartmann technique [18] uses arrays of lenslets that focus the light on quad-cell detectors; the displacement of a light spot from the center of a quad cell determines the partition of intensity among the four pixels. This in turn indicates the local wavefront tilt. Lincoln Laboratory has used its CCD imagers to build Shack-Hartmann wavefront sensors. However, in scenarios with weak beacon signals and fast wavefront update rates, the performance of CCDs is limited by readout noise. To address this problem, Lincoln developed Geiger-mode quad-cell arrays, to measure the number of photons from a beacon. Since the quad cell must be responsive to photons incident in between pixels, the low-fill-factor design shown in Figure 8 is not suitable. A high-fill-factor design was devised [19] and its cross section is shown in Figure 12.

The upper p-type layer is implanted at high energy through an oxide mesa so that the doping profile peaks at a relatively shallow depth (1 µm) in the center of each diode and deeper (2 µm) around the periphery and between pixels. The shallow portion of this stepped implant separates the absorber and multiplier portions of each detector. The step lowers the electric field at the periphery of the diode, preventing edge breakdown. The peripheral part of the diode functions as a guard ring, collecting surface-generated dark current without triggering Geiger-mode events. The deep portion of the implant, which is partially undepleted, prevents the guard ring from collecting photoelectrons generated in the absorber; as indicated in the figure, these photoelectrons reach a nearby multiplier region by a combination of diffusion and drift.

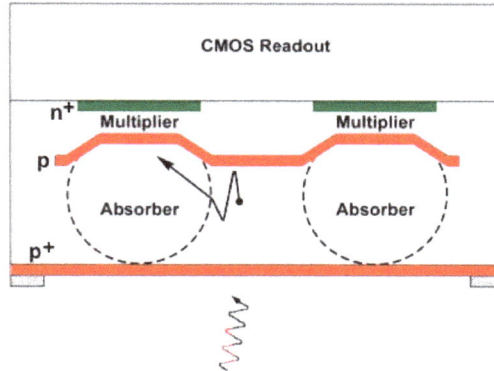

Figure 12. High-fill-factor APD design used in quad cells and passive imaging devices.

16 × 16 and 32 × 32-subaperture quad-cell arrays were fabricated and hybridized to readout chips that count the number of detection events with a 10-bit counter in each pixel. The device could be operated with 20-μs wavefront update latency while introducing no readout noise. Hybridization was accomplished first by bridge bonding and then later on by transfer and bump bonding. Pixel pitch within each quad cell was 50 μm, with a 200-μm spacing between subapertures [20].

To verify the functionality and contiguous spatial response, a focused 5-μm-diameter spot from a blue (450-nm) LED was raster scanned over the area of a quad cell and, at each 5-μm step, recording the photon counts from each of the four pixels. Figure 13a is a contour plot of the count rate of the lower right pixel as a function of the position of the light spot. Figure 13b shows the aggregate count rate from all four pixels. This data shows a monotonic transition of detection activity from one pixel to its neighbor as the light spot is being moved across the midpoint. It also shows no droop in the aggregate response in the central region.

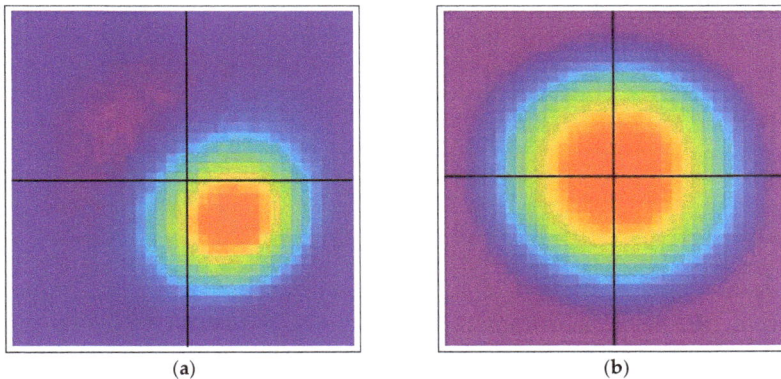

(a)   (b)

Figure 13. (**a**) Color contour plot of the count rate from the lower right detector as a function of the position of a small light spot raster scanned over the area of the quad cell; (**b**) Contour plot of the sum of the count rates from all four pixels as a function of light spot position.

These quad-cell arrays report out raw pixel intensity values, and the wavefront tilt calculation is done off chip. However, one can envision incorporating computational functions such as centroid and tracking into the pixel circuitry.

6. Passive Imaging

In 2011, the Laboratory demonstrated a 256 × 256 passive photon counting imager with 25-μm pixel pitch [21]. The GMAPD array was based on the high-fill-factor design (see Figure 12) and was integrated to a foundry-fabricated CMOS readout using the transfer and bump bonding technique. To our knowledge, this was the first ever back-illuminated passive image sensor with this large a format based on hybridization of a GMAPD array to a CMOS readout. Figure 14 shows one of the first images taken with the device, a church steeple located about 3.5 km from the camera system. The black specks in the image are bump bond defects. Use of a 3D-integration technique results in far better image cosmetics.

Figure 14. Image of church steeple.

The GMAPDs are armed, queried, and reset under supervision of external polling clocks. The CMOS pixel circuit has two readout modes. On the one hand, one can read out binary images in which each pixel reports whether or not it had a detection event since the last readout. This readout mode is called binary readout mode. On the other hand, one can read out the overflow bit of a 7-bit pseudorandom counter in each pixel; if set, this bit represents 127 cumulative detection events and it is reset by the readout operation. This readout mode is called overflow readout mode. At the conclusion of multiple overflow-bit readouts, the entire 7-bit counter can be read out to get the remainder. Binary readout mode and overflow readout mode are implemented by separate clocking systems that address rows or columns, respectively, in a rolling readout that does not blind the detectors. The overflow-bit readout mode provides for readout bandwidth reduction and dynamic range extension without having to put a large counter in the pixel. The overflow bits can be streamed to an Field-Programmable Gate Array (FPGA) that effectively provides the most significant bits of a longer counter. The image in Figure 14 was obtained by operating a 8 kiloframes/s in binary readout mode, digitally adding many frames off chip and performing a flat field correction.

Figure 15 shows images resulting from post-readout digital summation of short-time binary frames. Since there is no readout noise in a photon counting device, there is no noise penalty for such post-readout summation. One can imagine processing the binary frames before summing them to correct for scene motion or platform vibration. Even if each binary frame has a signal level less than one photoelectron per pixel, enough pixels have events to create discernable spatial structure, making such "smart integration" possible. This would not be feasible with a conventional analog imager, as even 1 electron rms readout noise would be too much.

A technological challenge with dense-pitch Geiger-mode APD arrays is optical crosstalk. During a detection event, the carriers traversing the avalanche region lose some energy by optical emission, and the near-infrared photons given off can spuriously trigger neighboring pixels. The quad-cell arrays [22] and the passive imager yielded valuable data on this phenomenon. Above a certain bias voltage, crosstalk-induced events dominate the dark count rate. Optical-crosstalk-based analytical models and Monte Carlo simulations match the spatial, temporal, and statistical characteristics of the dark count activity observed in the image data. The APDs in the passive imager did not have aggressive crosstalk reduction features, because that would have added yet another element of yield risk in the effort to prove out a new hybridization technique. The devices are operated at sufficiently low bias to avoid the crosstalk-dominated dark-count-rate regime. This limits the photon detection efficiency to the 10%–20% range. With the maturation of 3D integration techniques, current efforts are focused on crosstalk reduction.

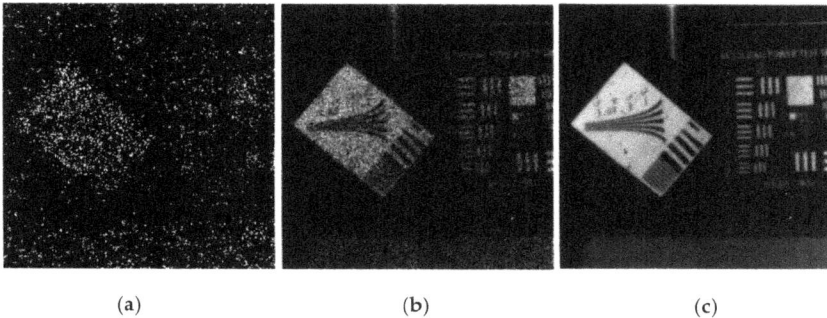

| (a) | (b) | (c) |

Figure 15. (a) One binary frame; (b) the digital sum of 64 binary frames; and (c) the digital sum of 32,768 binary frames.

7. The Future of GMAPD Imager Technology

The past 20 years have seen great progress in solid-state image sensors based on custom Geiger-mode avalanche photodiode arrays hybridized or 3D-integrated to all-digital CMOS readouts. Application to lidar represented the "low-hanging fruit," as compelling system functionality could be realized with small imager format, coarse pixel pitch, low fill factor, and dark count rates in the tens of kcounts/s. Passive imaging is far more demanding of APD performance and, for back-illuminated silicon arrays, much more dependent on the maturation of hybridization techniques. Now that 3D integration methods have become available, optical crosstalk reduction is the next task. The successful demonstration of a 256 × 256 passive photon counting imager without crosstalk reduction is an encouraging result. A number of measures, such as capacitance scaling and low-reflectivity contact metal, can now be implemented to improve sensitivity.

All-digital CMOS tiers can exploit Moore's Law scaling to realize increasingly sophisticated on-focal-plane processing functions. For lidar, these include in-pixel histogramming, tracking of objects of interest, multi-frame coincidence to reject background, and data thinning. For passive imaging, these include image stabilization, temporal change detection, and spatial filtering. Ultimately, one could implement a deeply scaled CMOS tier that could be programmed like an FPGA and support multiple firmware-defined functions with no hardware redesign.

While the focus of this review is imaging, GMAPD arrays sensitive in the short-wave infrared have potential as agile laser communications receivers. By spreading the received laser flux over multiple pixels, the effective APD reset time can be much shorter than the single-detector reset time, enabling high-data-rate free space optical communications links. In a remarkable technology demonstration, researchers at the Laboratory demonstrated the feasibility of an optical data link from a science satellite orbiting Mars. The laboratory demonstration used a pulse-position modulated format and error correction coding to achieve 0.5 photons/bit [23].

An imaging system traditionally consists of three distinct subsystems: (1) A bulky optical train that merely carries out an isomorphic transformation from object space to image space; (2) an image sensor that produces a high-aggregate-bandwidth stream of raw image data; and (3) a post-processing system to extract information of interest. Lincoln Laboratory's long-term vision is to merge these functions, so that the work of information extraction is carried out by co-designed computational optics and smart focal planes. The use of coded apertures or light-field camera architectures can be combined with digital time stamping of photon arrivals to extract information about a scene, including regions that are obscured from the view of a conventional camera. The shift of computation burden to the optics and imager could also reduce readout bandwidth and enable low size, weight, and power.

Conflicts of Interest: The authors declare no conflict of interest.

References

1. Burke, B.E.; Gregory, J.A.; Cooper, M.; Loomis, A.H.; Young, D.J.; Lind, T.A.; Doherty, P.; Daniels, P.; Landers, D.J.; Ciampi, J.; *et al.* CCD Imager Development for Astronomy. *Linc. Lab. J.* **2007**, *16*, 393–412.
2. Aull, B.F.; Loomis, A.H.; Young, D.J.; Heinrichs, R.M.; Felton, B.J.; Daniels, P.J.; Landers, D.J. Geiger-Mode Avalanche Photodiodes for Three-Dimensional Imaging. *Linc. Lab. J.* **2002**, *13*, 335–350.
3. McIntosh, K.A.; Donnelly, J.P.; Oakley, D.C.; Napoleone, A.; Calawa, S.D.; Mahoney, L.J.; Molvar, K.M.; Duerr, E.K.; Shaver, D.C. Development of Geiger-mode APD arrays for 1.06 μm. In Proceedings of the 15th Annual Meeting of the IEEE Lasers and Electro-Optics Society, Glasgow, Scotland, 10–14 November 2002; pp. 760–761.
4. Duerr, E.K.; Manfra, M.J.; Diagne, M.A.; Bailey, R.J.; Zayhowski, J.J.; Donnelly, J.P.; Connors, M.K.; Grzesik, M.J.; Turner, G.W. Antimonide-based Geiger-mode avalanche photodiodes for SWIR and MWIR photon-counting. *Proc. SPIE* **2010**, *7681*. [CrossRef]
5. Frechette, J.P.; Grossmann, P.J.; Busacker, D.E.; Jordy, G.J.; Duerr, E.K.; McIntosh, K.A.; Oakley, D.C.; Bailey, R.J.; Ruff, A.C.; Brattain, M.A.; *et al.* Readout circuitry for continuous high-rate photon detection with arrays of InP Geiger-mode avalanche photodiodes. *Proc. SPIE* **2012**, *8375*. [CrossRef]
6. Gersbach, M.; Trimananda, R.; Maruyama, Y.; Fishburn, M.; Stoppa, D.; Richardson, J.; Walker, R.; Henderson, R.K.; Charbon, E. High Frame-rate TCSPC-FLIM Using a Novel SPAD-based Image Sensor. *Proc. SPIE* **2010**, *7780*. [CrossRef]
7. Gordon, K.J.; Hiskett, P.A.; Lamb, R.A. Advanced 3D Imaging Lidar Concepts for Long Range Sensing. *Proc. SPIE* **2014**, *9114*. [CrossRef]
8. Grein, M.E.; Elgin, L.E.; Robinson, B.S.; Caplan, D.O.; Stevens, M.L.; Hamilton, S.A.; Boroson, D.M.; Langrock, C.; Fejer, M.M. Efficient communication at telecom wavelengths using wavelength conversion and silicon photon-counting detectors. *Proc. SPIE* **2007**, *6709*. [CrossRef]
9. Hadfield, R. Single-photon detectors for optical quantum information applications. *Nat. Photonics* **2009**, *3*, 696–705. [CrossRef]
10. Charbon, E. Towards large scale CMOS single-photon detector arrays for lab-on-chip application. *J. Phys. D Appl. Phys.* **2008**, *41*. [CrossRef]
11. Cova, S.; Ghioni, M.; Lacaita, A.; Samori, C.; Zappa, F. Avalanche photodiodes and quenching circuits for single-photon detection. *Appl. Opt.* **1996**, *35*, 1956–1976. [CrossRef] [PubMed]
12. Schuette, D.R.; Westhoff, R.C.; Loomis, A.H.; Young, D.J.; Ciampi, J.S.; Aull, B.F.; Reich, R.K. Hybridization process for back-illuminated silicon Geiger-mode avalanche photodiode arrays. *Proc. SPIE* **2010**, *7681*. [CrossRef]
13. Aull, B.F.; Burns, J.; Chen, C.; Felton, B.; Hanson, H.; Keast, C.; Knecht, J.; Loomis, A.; Renzi, M.; Soares, A.; *et al.* Laser radar imager based on three-dimensional integration of Geiger-mode avalanche photodiodes with two SOI timing-circuit layers. In Proceedings of the 2006 International Solid State Circuits Conference Digest of Technical Papers, San Francisco, CA, USA, 6–9 February 2006; pp. 304–305.
14. Enquist, P.; Fountain, G.; Petteway, C.; Hollingsworth, A.; Grady, H. Low Cost of Ownership Scalable Copper Direct Bond Interconnect 3D IC Technology for Three Dimensional Integrated Circuit Applications. In Proceedings of the IEEE International Conference on 3D System Integration, San Francisco, CA, USA, 28–30 September 2009.
15. Aull, B.F. 3D Imaging with Geiger-mode Avalanche Photodiodes. *Opt. Photonics News* **2005**, *16*, 42–46. [CrossRef]

16. Neuenschwander, A.L.; Crawford, M.M.; Magruder, L.A.; Weed, C.A.; Cannat, R.; Fried, D.; Knowlton, R.; Heinrichs, R. Terrain classification of ladar data over Haitian urban environments using a lower envelope follower and adaptive gradient operator. *Proc. SPIE* **2010**, *7684*. [CrossRef]

17. The Discussion of this in the Context of Lincoln Laboratories Effort in Humanitarian Assistance and Disaster Relief. Available online: http://phys.org/news/2015–10-technology-disasters.html (accessed on 5 April 2016).

18. Platt, B.C.; Shack, R. History and principles of Shack-Hartmann wavefront sensing. *J. Refract. Surg.* **2001**, *17*, S573–S577. [PubMed]

19. Renzi, M.J.; Aull, B.F.; Reich, R.K.; Kosicki, B.B. High-fill-factor avalanche photodiode. U.S. Patent No. 8,093,624, 10 January 2012.

20. Aull, B.F.; Renzi, M.J.; Loomis, A.H.; Young, D.J.; Felton, B.J.; Lind, T.A.; Craig, D.M.; Johnson, R.L. Geiger-Mode Quad-Cell Array for Adaptive Optics. In Proceedings of the Conference on Lasers and Electrooptics, San Jose, CA, USA, 4–9 May 2008.

21. Aull, B.F.; Schuette, D.R.; Young, D.J.; Craig, D.M.; Felton, B.J.; Warner, K. A Study of Crosstalk in a 256 × 256 Photon Counting Imager Based on Silicon Geiger-Mode Avalanche Photodiodes. *IEEE Sens. J.* **2015**, *15*, 2123–2132. [CrossRef]

22. Aull, B.F.; Reich, R.K.; Ward, C.M.; Craig, D.M.; Young, D.J.; Johnson, R.L. Detection statistics in Geiger-mode avalanche photodiode quad-cell arrays with crosstalk and dead time. *IEEE Sens. J.* **2015**, *15*, 2133–2143. [CrossRef]

23. Mendenhall, J.A.; Candell, L.M.; Hopman, P.I.; Zogbi, G.; Boroson, D.M.; Caplan, D.O.; Digenis, G.J.; Hearn, D.R.; Shoup, R.C. Design of an Optical Photon Counting Array Receiver System for Deep-Space Communications. *IEEE Proc.* **2007**, *95*, 2059–2069. [CrossRef]

sensors

MDPI

Review

Compact SPAD-Based Pixel Architectures for Time-Resolved Image Sensors

Matteo Perenzoni [1],*, Lucio Pancheri [2] and David Stoppa [1]

[1]   Integrated Radiation and Image Sensors, Fondazione Bruno Kessler, Trento 38123, Italy; stoppa@fbk.eu
[2]   Department of Industrial Engineering, University of Trento, Trento 38123, Italy; lucio.pancheri@unitn.it
*   Correspondence: perenzoni@fbk.eu; Tel.: +39-0461-314-533

Academic Editor: Edoardo Charbon
Received: 7 March 2016; Accepted: 16 May 2016; Published: 23 May 2016

Abstract: This paper reviews the state of the art of single-photon avalanche diode (SPAD) image sensors for time-resolved imaging. The focus of the paper is on pixel architectures featuring small pixel size (<25 µm) and high fill factor (>20%) as a key enabling technology for the successful implementation of high spatial resolution SPAD-based image sensors. A summary of the main CMOS SPAD implementations, their characteristics and integration challenges, is provided from the perspective of targeting large pixel arrays, where one of the key drivers is the spatial uniformity. The main analog techniques aimed at time-gated photon counting and photon timestamping suitable for compact and low-power pixels are critically discussed. The main features of these solutions are the adoption of analog counting techniques and time-to-analog conversion, in NMOS-only pixels. Reliable quantum-limited single-photon counting, self-referenced analog-to-digital conversion, time gating down to 0.75 ns and timestamping with 368 ps jitter are achieved.

Keywords: single-photon avalanche diode; SPAD; time-resolved imaging; time-gating photon counting

1. Introduction

Solid-state image sensors with nanosecond and sub-nanosecond timing resolution are needed in applications such as optical ranging, fluorescence microscopy and Raman spectroscopy [1–3]. Research is moving in two main directions: on the one hand lock-in pixels with high shutter efficiencies and high frequency operation are already used in 3D Time-of-Flight cameras [4]. On the other hand, time-resolved pixels based on single-photon avalanche diodes (SPADs) appear more and more a feasible and competitive perspective.

The last years have seen CMOS SPAD-based sensors enter the consumer market with products based on single-pixel detectors or small arrays. For imaging applications, however, the requirements are much more demanding and several technological challenges still need to be tackled. In addition to device optimization, the use of SPADs in image sensors imposes additional requirements such as yield, uniformity of breakdown voltage and photon detection efficiency (PDE), reduction of optical cross-talk and minimization of guard rings.

Several challenges also need to be solved with respect to the readout electronics. Storing the timing information at the pixel level requires either fast time-gating or in-pixel time-tagging. In addition, especially for large arrays, the pixels should store timing information on multiple photons, to reduce the bandwidth needed for the array readout. These operations should be performed with a minimum area overhead in order to maintain a small pixel pitch and a good fill factor.

As happened in CMOS image sensor development, the first proof-of-concept designs have exploited standard CMOS processes [5–8]. Although not optimized in many respects, CMOS proved to be a good platform to test the feasibility of time-resolved image sensors at the architectural level.

As the focus passes from proof-of-concept to applications, the optimization of fabrication technologies is unavoidable. Currently, the possibilities for commercial exploitation of CMOS SPAD arrays are increasing and new companies are entering the field, proposing solutions based on customized technologies [9]. Optimized single-photon image sensors will not only be appealing for time-resolved imaging, but also for low-light level imaging, for example in security and scientific applications.

This paper reviews the work on SPAD pixel arrays in the recent years, focusing on pixel architectures suitable for the realization of large pixel arrays. Section 2 presents an overview of the main SPAD characteristics, with particular attention to array integration challenges. In Section 3, several pixel architectures are analyzed and discussed. Finally, a roadmap for future developments is traced.

2. SPADs in CMOS Technologies

2.1. SPAD Structure

Since their first proof-of-concept in the early 2000s, CMOS SPAD detectors have been integrated in many technology nodes, using both standard, High Voltage or CIS processes. Most of the devices demonstrated so far are based on a p+/nwell junction, which is intrinsically isolated from the p-type substrate [10–12]. While the guard ring can be implemented in different ways, the most common embodiments use a deep nwell to separate the guard ring from the substrate. Different guard ring solutions are reviewed in [13].

The immediate advantage of this configuration is the possibility of accessing both the anode and cathode, and thus the direct coupling to the readout electronics operating at low voltage. To obtain the structure in Figure 1a, only an additional deep-nwell implantation is needed in addition to the layers typically available in a standard CMOS. A higher red and NIR sensitivity can be obtained with a deeper junction, for example a pwell/deep-nwell, as shown in Figure 1b [12,14].

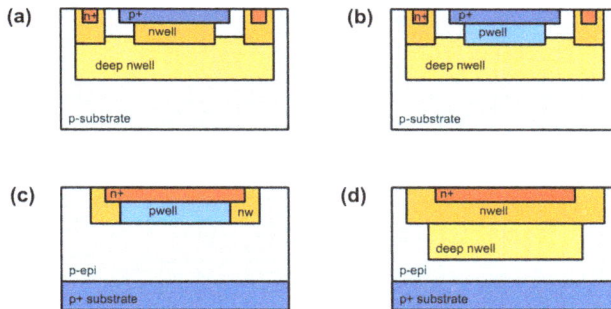

Figure 1. Cross sections of different CMOS SPAD devices (**a**) p+/nwell; (**b**) pwell/deep-nwell; (**c**) n+/pwell; (**d**) deep-nwell/p-epi/p+ sub.

N-in-p device structures have been proposed, with the active junction based on a or n+/pwell junction or a deep-nwell/p-epi/p+sub (Figure 1c,d) [15,16]. The last options potentially offer an increased photon detection efficiency (PDE) in the red and NIR spectral region, but require a quenching resistor at the high voltage node and a capacitive decoupling of the readout circuit. Their use in a pixel is thus not straightforward.

2.2. Figures of Merit

The excellent timing resolution, in the picosecond range, is one of the main advantages of SPAD devices. It has been demonstrated that even unoptimized processes can lead to devices having a jitter of few tens of ps FWHM [17]. One of the key factors to minimize the device jitter is the reduction of photo-carrier diffusion towards the high-field region. To this extent, p+/nwell structures typically feature a very low jitter, as the diffusion tail is often reduced to a minimum, while the jitter is larger in pwell/deep-nwell devices [14,18].

The dark count rate (DCR) is usually the most important source of noise in SPADs, since its fluctuations set the minimum detectable light signal. In a SPAD array, even though all the devices are nominally equal, DCR can have an enormous variability, spanning several orders of magnitude (Figure 2). What is really important in arrays is therefore not the DCR of a single device, but the whole distribution. If small SPADs are considered, two common types of distribution are typically found. A first case where most of the devices have a similar DCR, while a small percentage have large value, as shown in Figure 2a. It is not infrequent to find distributions where a plateau does not exist, as in Figure 2b. The different behaviors can be explained by considering the dominant sources of DCR for different SPAD structures.

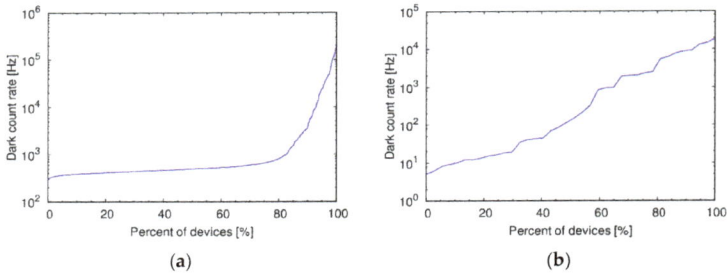

(a)   (b)

Figure 2. Dark Count Rate distribution of SPADs fabricated in different CMOS process technologies (a) 0.35 μm High Voltage with 130-μm^2 active area [19]; (b) 0.7 μm High Voltage with 100-μm^2 active area [11].

Some insight on the origin of DCR can be obtained by analyzing its temperature dependence. Figure 3 shows the DCR temperature dependence of four devices with the same area, representative of the behavior of a whole distribution. The three SPADs with high DCR have an activation energy lower than $E_G/2$, indicating the presence of trap-assisted tunneling. For the SPAD with low DCR, the activation energy has a transition between E_G and 0.2 V as the temperature decreases. In this device, DCR is dominated by injection of minority carriers from the neutral regions at high temperatures, and from tunneling at low temperatures. The relative weight between the different components determine the shape of the DCR distribution. If the devices are small and the amount of contaminants is low, DCR will mostly be dominated by minority carrier injection or band-to-band tunneling, and will be very uniform. If, on the contrary, the amount of contaminants is relatively high, the distribution will not show a plateau.

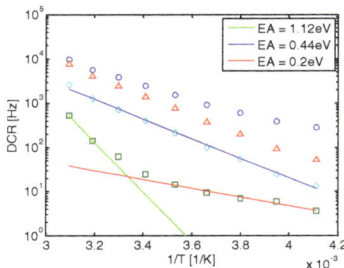

Figure 3. DCR temperature dependence for 10-μm diameter SPADs in 150 nm standard CMOS.

The integration of SPADs in deep submicron technology nodes has several obvious advantages. On one hand, being an intrinsically digital device, the possibility of introducing in-pixel dense digital readout electronics such as counters or TDCs is appealing. On the other hand, it is increasingly clear that a doping profile customization is absolutely needed to obtain detectors with acceptable performance in advanced nodes. The first attempts to fabricate SPADs in unmodified 90 and 65 nm processes have led to devices with very high DCR, mainly due to tunneling [15]. In fact, the well doping concentration steadily increases with device scaling. As a general consideration, SPADs integrated in standard processes with no process modifications have hardly an optimized DCR.

A customization of the doping profiles is required to optimize the SPAD characteristics. On the one hand, DCR can be reduced by carefully tuning the electric field profile in the avalanche region [20], so as to minimize the contribution of tunneling. This is also an advantage if the device needs to be cooled, as the absence of tunneling provides a more efficient DCR reduction with decreasing temperature. In fact, once the doping profiles are optimized for tunneling reduction, the device DCR is mainly due to the contamination by heavy metals. Using both excellent starting material and a clean production line, as in most CIS processes, is fundamental for the production of low-noise devices. The use of advanced imaging processes, in fact, can lead to devices having good characteristics both regarding PDE and DCR, as was demonstrated in a few cases [14,21].

The electric field profile affects the PDE through the optimization of avalanche triggering probability. In a graded profile the breakdown probability reaches very large values at smaller voltages than in step profile junction with similar breakdown voltages [18]. A good PDE at low excess bias can thus be obtained by combining a graded profile with an optimized optical stack, which is readily available in a CIS process [14].

Afterpulsing is a source of correlated noise, and in time resolved applications it can lead to measurement distortions if it is not minimized. Since its origin lies in the presence of deep trapping centers, the quality of the process can also contribute to its reduction. In addition, a careful design of the quenching circuit, ensuring a minimum stray capacitance and fast avalanche extinction, helps keeping afterpulsing under control. In general, in small CMOS devices, the total afterpulsing rate can be maintained at acceptable levels with a few tens ns hold-off time [11,14,22]. One notable exception to this general observation is a 0.35 HV processes, which has a long tails and needs a hold-off time of 100s ns [23]. This peculiar behavior, limited to a single 0.35 HV process has been reported by several authors, but the reason of the long afterpulsing tail has not yet been completely understood.

Optical cross-talk is a critical parameter in SPAD arrays, arising from the emission of optical photons during an avalanche event [24]. Measurements on dense p+/nwell SPAD arrays have shown that cross-talk rates can be as high as a few percent for nearest neighbor devices if the fill factor is in the order of 70% [25]. For optimal performance, therefore, cross-talk should be minimized by introducing deep trench isolation, as currently done in silicon photomultiplier by most manufacturers [26].

2.3. Uniformity

The uniformity of breakdown voltage is of paramount importance for the integration of SPAD image sensors, since all the SPADs are biased at the same voltage. Non-uniformities in the order of 100 mV can be tolerated if SPADs are operated at excess bias voltages of a few volts. A good uniformity of PDE along the pixel array is also required in image sensors. Experimental investigations on 150 nm CMOS SPADs have shown that breakdown voltage non-uniformity is larger in small devices, reaching values in the order of 1 V for 5-μm diameter SPADs, while 10-μm devices have a peak-to-peak non-uniformity lower than 0.5 V and in larger ones it is in the order of 100 mV (Figure 4). The effect of device size is also visible in the average breakdown voltage. The trend is toward a decreasing average breakdown voltage with device area. Surprisingly, however, once the avalanche has been triggered, the PDE results very uniform in all the tested detectors, with non-uniformities lower than 1% [27].

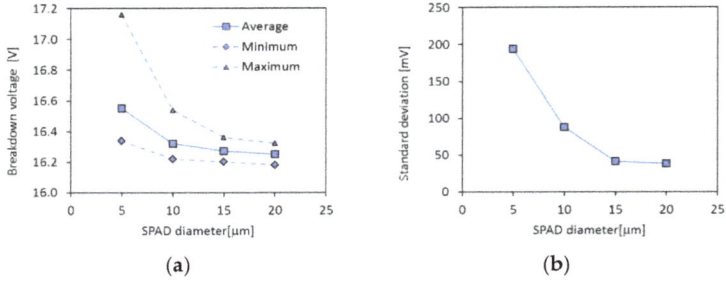

Figure 4. Measured breakdown voltage as a function of SPAD diameter. Measurements were done on 150-nm CMOS SPADs [27]. (**a**) Breakdown voltage distribution; (**b**) breakdown voltage non-uniformity (standard deviation).

2.4. Layout

High fill factor pixel arrays can be obtained only with a careful minimization of guard ring and of the area occupied by in-pixel readout electronics. Deep-nwell sharing between different SPADs and between SPADs and electronic readout circuits can be used to effectively increase the device packing and therefore to help shrinking pixel size [28]. If every SPAD has a separate deep-nwell, as in Figure 5a, the bias can be applied either at the anode or at the cathode, and quenching circuitry can be connected to the other terminal. In the second case, the total device capacitance will be lower, but the distance between different devices should be maintained large enough to avoid punch through between the deep nwells. This layout solution was used in the early proof-of-concept arrays with large pitch and small fill factor [5,29].

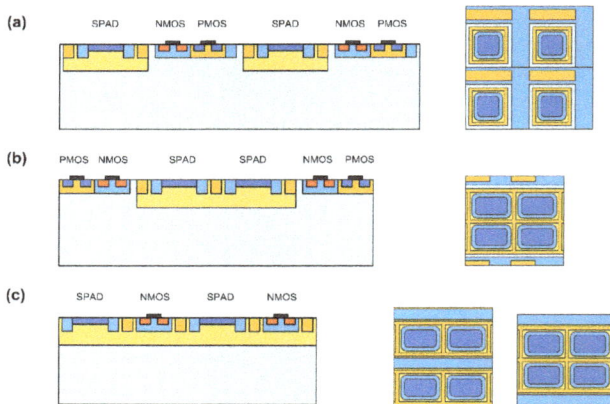

Figure 5. Cross section and 4 × 4 pixel layout of (**a**) SPADs integrated in separate deep nwell; (**b**) SPADs sharing the same deep nwell; (**c**) SPADs and NMOS transistors sharing the same deep nwell.

Deep-nwell sharing can be exploited to reduce the dead area at the borders of the deep nwell, as shown in Figure 5b. With deep-nwell sharing, first proposed in [11] the cathode of all SPADs are in common and the quenching should be performed at the anode side. Densely-packed SPAD arrays can be easily obtained using currently available technologies if digital readout circuits are placed outside the deep nwell. High fill factors, limited only by the device guard ring, have been demonstrated, although the detectors were confined in arrays including only a few lines of SPADs [30–33].

The packing desity can be further increased if the readout n-type MOSFETs are included in the same deep nwell with the SPADs, as in Figure 5c. In this case, the pwell with the MOSFETs must be biased at a voltage close to the one applied to SPAD anode, and thus the breakdown voltage of the pwell/deep-nwell junction should be high enough to avoid early brekdown problems. The main disadvantage of this solution, which has been adopted in [19,34–37], is that p-type transistors cannot be used inside the pixel electronics.

3. Compact Pixels for Time-Resolved Imaging

The realization of photon-counting or time-stamping function within a given time window is straightforward with digital logic when using SPADs, since they provide a digital output. Although the fully digital solution meets the requirements of robustness and ease of implementation, the area occupation of the circuitry becomes extremely large. This is in contrast with the application field of this class of imagers, which is typically fluorescence lifetime imaging microscopy (FLIM), where also good spatial resolution and high efficiency are needed, driving towards small pixels with high fill factor.

3.1. Pixel Architectures

In order to keep the area small, analog solutions have been employed [19,35–38] by making use of analog counters and time-to-analog converters. The main techniques enabling a small pixel pitch are the implementation of the pixel circuitry using NMOS only so as to avoid nwells, the reuse of transistors for different purposes, a simple active-pixel readout with source follower and selection switch, and almost minimum-sized NMOS and capacitors, often making use of parasitic capacitances as storage nodes.

In [19] a pixel pitch of 25 µm with a 20.8% fill-factor was demonstrated in a 0.35 µm CMOS technology thanks to extensive use of analog techniques. The pixel schematic is depicted in Figure 6: the front-end is composed by a quenching transistor M1 and a clamp M2 which limits the voltage swing, and a disabling transistor M3. The gating circuit implements a pulse shortener by performing a logic AND between the SPAD pulse and its delayed and inverted companion, transmitting the WINn pulse to the analog counter. The latter signal defines the gating window, so when disabled ($\overline{WIN} = 1$) no pulse is generated. The analog counter then discharges the integration capacitance by an amount proportional to the pulse width. This architecture reaches a gating window width down to 1 ns, but can also accommodate larger integration time, operating as a global shutter photon-counting pixel. Moreover, it can operate multiple gating cycles for a relatively long time, limited only by the charge leakage of the analog memory (hundreds of milliseconds): indeed, thanks to the decoupling effect of M7, the repeated activation of the $\overline{WIN}$ signal does not cause an output swing reduction due to charge injection.

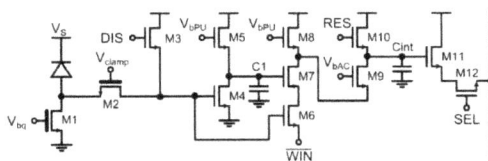

Figure 6. Time-gated pixel with analog counter, enabling gating windows from ≈1 ns up to several hundreds of milliseconds [19].

The main issue of this pixel structure is the non-uniformity of the counting step in the array, which is driven by many mismatch constraints: the resulting pulse width of the gating circuit, mismatch of integration capacitances, distribution of the gating window signal, mismatch of the current limiter M9.

In order to limit it to the measured value of 11.9%, non-minimum sized transistors have been used in the critical signal path although an unprecedented fill-factor was obtained.

Concerning the minimum time-gating, the precision for very small windows is limited by the pulse shortener: indeed, when a SPAD event occurs during the opening or closing of the gating window, the pulse may be sliced by the WINn signal and therefore a reduced step is recorded in the integration capacitance.

A strong reduction of the number of transistors per pixel has been achieved in [35] by replacing the gating circuitry with a single NMOS performing both clamping and time-gating. As shown in Figure 7, the SPAD is passively quenched and the pulse transmission can be inhibited by M2. The analog counter performs a controlled discharge of the integration capacitor through M5 and M6. The time gating performed by this circuit can only be coarse, meaning that it has to be larger than the dead time of the SPAD (typ. 10–100 ns) in order to have predictable voltage steps at the analog counter. Indeed, differently from the previous architecture, there is no pulse shortener and the full SPAD pulse has to be transmitted to the analog counter. Some small injection contribution is expected to impact on the output signal swing, due to WIN coupling through the gate-drain capacitance of M5 to the integration capacitor. Non-uniformities in this scheme are determined by mismatch between integration capacitors, current limiting transistor M6 but also differences in SPAD breakdown voltage and dead-time which change the shape of the pulse. Remarkable fill-factor of 26.8% in a 8-μm pitch is achieved in a 130 nm CIS technology. An on-chip 1-bit digital conversion has been implemented so as to achieve a fast single photon oversampled imager.

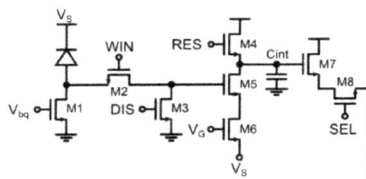

Figure 7. Analog counting pixel with coarse time gating for windows width larger than the SPAD dead time (*i.e.*, >10 ns) [35].

Extensive reuse of transistors and a deferred counting technique has been implemented in [36] in order to further reduce the transistor count and at the same time allow for precise and short gating windows. This has been achieved at the expense of generality: in this implementation, no continuous integration is possible, but only periodic excitation/gating schemes are allowed. As shown in Figure 8 the frontend is composed by a switch M1 acting both as precharge and disable with independent gate and source voltages, while M2 clamps and samples the SPAD pulse. The main difference with respect to the previous schemes is that, due to the repetitive nature of the measurement, the counting operation is performed after the sampling of the SPAD state in the observation window. Again, counting is performed by subtracting a controlled amount of charge from an integration capacitor.

The deferred counting scheme makes the counting step independent from the SPAD pulse nature, width, and position within the gating time window, achieving sharp window edges down to 200 ps rise/fall time and 750 ns minimum gating window width. This advantage was exploited using almost minimum-sized transistors in order to obtain a pixel pitch of 15 μm and 21% fill-factor. At the same time, a self-referenced column-wise analog to digital conversion cancels the residual non-uniformity of 15.7% by using the very same pixel as a ramp generator for a single-slope ADC.

As a drawback, this scheme does not allow for more than one pulse to be counted for a single excitation/gating cycle, which is actually not a limitation when the gating window has to be smaller than the dead-time of a SPAD. For example, typical FLIM decay times fall within the range 1–10 ns, typically smaller than the SPAD recharge time. Another limitation is given by the fact that even if no events are recorded, the deferred counter is stimulated with the digital signals, and therefore charge is injected and accumulated, reducing the available output voltage swing thus limiting the dynamic range to few hundreds of counted photons between each readout.

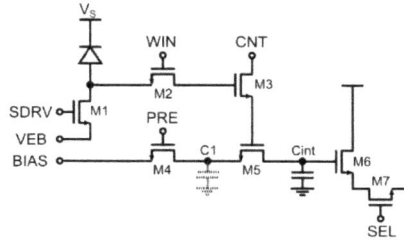

Figure 8. Analog time-gated pixel with transistor reuse and deferred analog counting [36].

Another function suitable for compact SPAD imaging arrays is the photon timestamping by using time-to-analog converters (TAC). In [38] a fairly complex TAC with in pixel digital conversion of the analog value was presented, but finally achieving a low fill-factor of 1% in a 50-μm pitch. Optimization of the area occupation can surely be improved by moving the conversion off-pixel, as described in [37]. With an approach similar to the analog photon counting described in the previous paragraphs, TAC circuitry can also be implemented using only NMOS transistors.

In Figure 9 the schematic shows a frontend with a standard passive quenching with disabling circuitry, driving a dynamic memory composed by M4 and M5. Whenever a SPAD pulse occurs, the memory is fully discharged, sampling onto the storage capacitor the present reference voltage fed to the whole array, representing time in the analog domain. This solution virtually eliminates many of the non-uniformity sources affecting analog counters: among remaining issues impacting uniformity and linearity, there are leakage from the storage capacitor and proper distribution of the fast-varying reference signal to the whole array. A remarkable 20% fill-factor in a 8 μm pitch pixel is achieved, with an overall jitter of 368 ps rms.

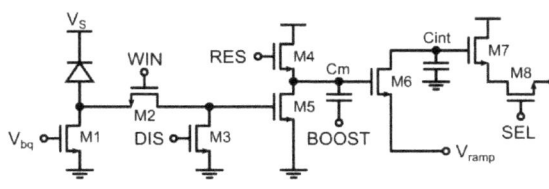

Figure 9. Compact NMOS-only TAC-based pixel [38].

3.2. Analysis and Comparison

The time-gated analog counters presented in [19] and [36] can be directly compared as they perform a very similar operation, with the main difference that the deferred counting scheme of [36] does not allow long integration time with global shutter.

As a first comparison, the analog photon counting may be addressed: by selecting a single pixel and recording the output analog values, it is possible to reconstruct an histogram showing the characteristic peaks corresponding to the detection of 0, 1, 2, . . . , photons. As shown in Figure 10, both histograms show clearly the single photon detection resolution, but Figure 10a highlights a better noise performance with almost isolated peaks. The higher noise visible in Figure 10b is due to the smaller in-pixel capacitance values used for the analog counting, leading to larger kTC noise contribution.

In both cases, the kTC noise is repeatedly accumulated during the charge transfer, and progressively confuses individual photon peaks; anyway, this occurs when the shot noise is already the dominant noise source [36].

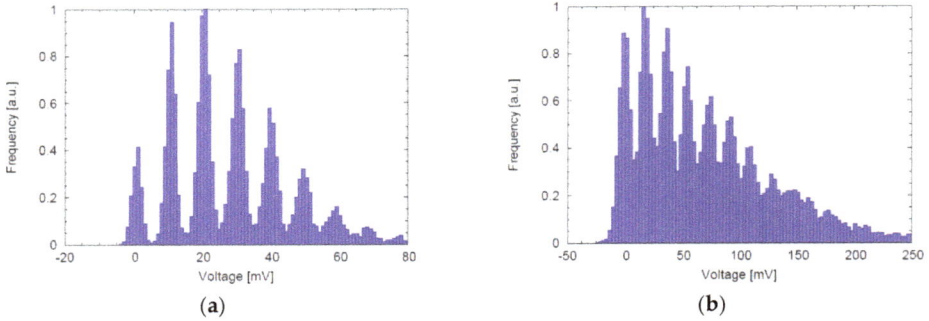

Figure 10. Pixel analog output histograms for the analog counting pixels of [19] and [36], depicted in (a) and (b), respectively.

A second comparison can be made by exciting the sensor with a short laser pulse, in this case a 70-ps FWHM laser, progressively increasing the relative delay. The analog output then follows the shape of the gating window performed by the pixel. In Figure 11 this measurement has been repeated for windows width of approximately 1, 2, 3, 4, 5 ns; in both cases, the different window widths have been programmed to the internal logic which is stabilized against variations using a locked control loop with a reference clock. In this case it can be observed that edges of gating windows are better defined in Figure 11b: the motivation lays in the technique used to start and close the windows, avoiding the use of a monostable.

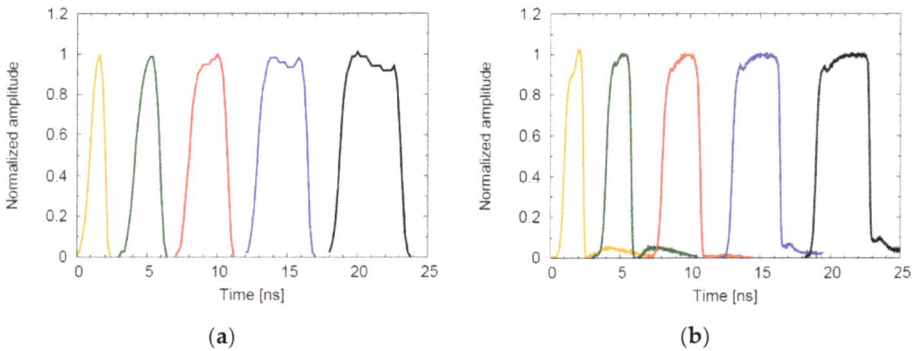

Figure 11. Time-gating windows for different widths for the analog counting pixels of [19] and [36], depicted in (a) and (b), respectively.

Table 1 summarizes the main parameters of all the analyzed SPAD image sensors, in particular for what concerns their compactness, size, and timing performance.

Table 1. Comparison between recent compact analog-based SPAD sensors for time-resolved imaging.

	[19]	[35]	[36]	[37]
Process	0.35 μm HV	0.13 μm CIS	0.35 μm HV	0.13 μm CIS
Supply	3.3 V	1.2 V	3.3 V	1.2 V
Array size	32 × 32	320 × 240	160 × 120	256 × 256
Pixel pitch	25 μm	8 μm	15 μm	8 μm
Fill-factor	20.8%	26.8%	21%	20%
NMOS per pixel	12T + 2C	8T + 1C	7T+1C	8T + 2C
Timing	Gating >1.1 ns	Gating > 10 ns	Gating > 0.75 ns	TAC > 0.37 ns
Interface	Analog	Analog/Digital 1 b	Analog/Digital 8 b	Analog/Digital 2 b
Consumption	33 mW	69.5 mW	20.6 mW [a] 157 mW [b]	n.a.

[a] Analog readout mode; [b] Digital readout mode.

4. Discussion and Conclusions

The way towards high-resolution single-photon image sensors with timing capabilities (either implementing time-gating or time-stamping pixels) has to be pursued through two parallel approaches: on one side the optimization of the device, and on the other side with area-efficient circuit topologies.

The device optimization will inevitably go through the use of specialized process options, as already happened to conventional CIS processes. Indeed, SPAD devices will surely benefit from CIS (clean) processes with custom profile, which will enable reduction of tunneling and optimization of photon detection probability [39]. Additional modules, such as deep trench isolation in order to reduce cross talk, especially for deep-junction SPADs, will also improve the performance in densely packed arrays.

As far as pixel circuits is concerned, they must employ area-efficient analog and all-NMOS topologies in order to exploit shared nwell SPAD layout, passive quenching, transistors reuse, and possibly self-referenced conversion for increased reliability and readout speed in perspective of high-resolution imagers.

Finally, in the near future SPAD-based imagers could definetely take advantage of the advent of 3D-stacked fabrication technologies exploiting the co-integration of specialized CIS back-side illuminated sensing layer with deep-submicron digital CMOS. BSI-compatible SPADs offer higher sensitivity in the NIR region that combined with the stacking on advanced digital CMOS technologies will enable the realisation of pixel pitches in the order of a few micrometers and high fill factor [40,41]. At the same time efficient processing of the generated data flow will still be possible. These features, not only will drastically improve the performance of SPAD-based range cameras, but will open the way to new applications that are now dominated by other detector technologies, such as high dynamic range, high sensitivity and high speed intensity cameras.

Author Contributions: The authors equally contributed in the preparation of the manuscript text while L. Pancheri and M. Perenzoni drew most of the figures (equally).

Conflicts of Interest: The authors declare no conflict of interest.

References

1. Aull, B.; Loomis, A.; Young, D. Geiger-mode avalanche photodiodes for three-dimensional imaging. *Linc. Lab. J.* **2002**, *13*, 335–350.
2. Lakowicz, J.R. *Principles of Fluorescence Spectroscopy*, 3rd ed.; Springer US: Boston, MA, USA, 2006.
3. Kostamovaara, J.; Tenhunen, J.; Kögler, M.; Nissinen, I.; Nissinen, J.; Keränen, P. Fluorescence suppression in Raman spectroscopy using a time-gated CMOS SPAD. *Opt. Express* **2013**, *21*, 31632–31645. [CrossRef] [PubMed]
4. Remondino, F.; Stoppa, D. *TOF Range-Imaging Cameras*; Springer: Heidelberg, Germany, 2013; Volume 68121.

5. Niclass, C.; Rochas, A.; Besse, P.-A.; Charbon, E. Design and characterization of a CMOS 3-D image sensor based on single photon avalanche diodes. *IEEE J. Solid State Circuits* **2005**, *40*, 1847–1854. [CrossRef]

6. Stoppa, D.; Pancheri, L.; Scandiuzzo, M.; Gonzo, L.; Betta, G.-F.D.; Simoni, A. A CMOS 3-D Imager Based on Single Photon Avalanche Diode. *IEEE Trans. Circuits Syst. I Regul. Pap.* **2007**, *54*, 4–12. [CrossRef]

7. Niclass, C.; Favi, C.; Kluter, T.; Gersbach, M.; Charbon, E. A 128 × 128 single-photon image sensor with column-level 10-bit time-to-digital converter array. *IEEE J. Solid State Circuits* **2008**, *43*, 2977–2989. [CrossRef]

8. Guerrieri, F.; Tisa, S.; Tosi, A.; Zappa, F. Two-Dimensional SPAD Imaging Camera for Photon Counting. *IEEE Photonics J.* **2010**, *2*, 759–774. [CrossRef]

9. Mori, M.; Sakata, Y.; Usuda, M.; Yamahira, S.; Kasuga, S.; Hirose, Y.; Kato, Y.; Tanaka, T. A 1280 × 720 single-photon-detecting image sensor with 100 dB dynamic range using a sensitivity-boosting technique. In Proceedings of the IEEE International Solid-State Circuits Conference (ISSCC), San Francisco, CA, USA, 31 January–4 February 2016; pp. 120–121.

10. Rochas, A.; Gani, M.; Furrer, B.; Besse, P.A.; Popovic, R.S.; Ribordy, G.; Gisin, N. Single photon detector fabricated in a complementary metal-oxide-semiconductor high-voltage technology. *Rev. Sci. Instrum.* **2003**, *74*, 3263–3270. [CrossRef]

11. Pancheri, L.; Stoppa, D. Low-Noise CMOS single-photon avalanche diodes with 32 ns dead time. In Proceedings of the 37th European Solid State Device Research Conference, Munich, Germany, 11–13 September 2007; pp. 362–365.

12. Pancheri, L.; Stoppa, D. Low-noise single photon avalanche diodes in 0.15 μm CMOS technology. In Proceedings of the 41th European Solid State Device Research Conference, Helsinki, Swedish, 12–16 September 2011; pp. 179–182.

13. Dalla Betta, G.-F.; Pancheri, L.; Stoppa, D.; Henderson, R.; Richardson, J. Avalanche Photodiodes in Submicron CMOS Technologies for High-Sensitivity Imaging. In *Advances in Photodiodes*; Dalla Betta, G.-F., Ed.; InTech: Vienna, Austria, 2011; pp. 225–248.

14. Richardson, J.A.; Webster, E.A.G.; Grant, L.A.; Henderson, R.K. Scaleable Single-Photon Avalanche Diode Structures in Nanometer CMOS Technology. *IEEE Trans. Electron. Devices* **2011**, *58*, 2028–2035. [CrossRef]

15. Karami, M.A.; Yoon, H.J.; Charbon, E. Single-photon avalanche diodes in sub-100 nm standard CMOS technologies. In Proceedings of the International Image Sensor Workshop, Hokkaido, Japan, 8–11 June 2011.

16. Webster, E.A.G.; Grant, L.A.; Henderson, R.K. A High-Performance Single-Photon Avalanche Diode in 130-nm CMOS Imaging Technology. *IEEE Electron. Device Lett.* **2012**, *33*, 1589–1591. [CrossRef]

17. Tisa, S.; Guerrieri, F.; Zappa, F. Variable-load quenching circuit for single-photon avalanche diodes. *Opt. Express* **2008**, *16*, 2232–2244. [CrossRef] [PubMed]

18. Pancheri, L.; Stoppa, D.; Dalla Betta, G.-F. Characterization and Modeling of Breakdown Probability in Sub-Micrometer CMOS SPADs. *IEEE J. Sel. Top. Quantum Electron.* **2014**, *20*, 328–335. [CrossRef]

19. Pancheri, L.; Massari, N.; Stoppa, D. SPAD image sensor with analog counting pixel for time-resolved fluorescence detection. *IEEE Trans. Electron. Devices* **2013**, *60*, 3442–3449. [CrossRef]

20. Ghioni, M.; Gulinatti, A.; Rech, I.; Zappa, F.; Cova, S. Progress in Silicon Single-Photon Avalanche Diodes. *IEEE J. Sel. Top. Quantum Electron.* **2007**, *13*, 852–862. [CrossRef]

21. Villa, F.; Bronzi, D.; Zou, Y.; Scarcella, C.; Boso, G.; Tisa, S.; Tosi, A.; Zappa, F.; Durini, D.; Weyers, S.; et al. CMOS SPADs with up to 500 μm diameter and 55% detection efficiency at 420 nm. *J. Mod. Opt.* **2014**, *61*, 102–115. [CrossRef]

22. Niclass, C.; Soga, M. A miniature actively recharged single-photon detector free of afterpulsing effects with 6 ns dead time in a 0.18 μm CMOS technology. In Proceedings of the 2010 International Electron Devices Meeting, San Francisco, CA, USA, 6–8 December 2010; pp. 14.3.1–14.3.4.

23. Stoppa, D.; Mosconi, D.; Pancheri, L.; Gonzo, L. Single-Photon Avalanche Diode CMOS Sensor for Time-Resolved Fluorescence Measurements. *IEEE Sens. J.* **2009**, *9*, 1084–1090. [CrossRef]

24. Rech, I.; Ingargiola, A.; Spinelli, R.; Labanca, I.; Marangoni, S.; Ghioni, M.; Cova, S. Optical crosstalk in single photon avalanche diode arrays: A new complete model. *Opt. Express* **2008**, *16*, 8381–8394. [CrossRef] [PubMed]

25. Xu, H.; Pancheri, L.; Braga, L.H.C.; Dalla Betta, G.-F.; Stoppa, D. Crosstalk Characterization of Single-photon Avalanche Diode (SPAD) Arrays in CMOS 150 nm Technology. *Procedia Eng.* **2014**, *87*, 1270–1273. [CrossRef]

26. Piemonte, C.; Acerbi, F.; Ferri, A.; Gola, A.; Paternoster, G.; Regazzoni, V.; Zappala, G.; Zorzi, N. Performance of NUV-HD Silicon Photomultiplier Technology. *IEEE Trans. Electron. Devices* **2016**, *63*, 1–6. [CrossRef]

27. Pancheri, L.; Dalla Betta, G.-F.; Campos Braga, L.H.; Xu, H.; Stoppa, D. A single-photon avalanche diode test chip in 150 nm CMOS technology. In Proceedings of the 2014 International Conference on Microelectronic Test Structures (ICMTS), Udine, Italy, 24–27 March 2014; pp. 161–164.

28. Henderson, R.K.; Webster, E.A.G.; Walker, R.; Richardson, J.A.; Grant, L.A. A 3 × 3, 5 µm pitch, 3-transistor single photon avalanche diode array with integrated 11 V bias generation in 90 nm CMOS technology. In Proceedings of the 2010 International Electron Devices Meeting, San Francisco, CA, USA, 6–8 December 2010; pp. 14.2.1–14.2.4.

29. Veerappan, C.; Richardson, J.; Walker, R.; Li, D.-U.; Fishburn, M.W.; Maruyama, Y.; Stoppa, D.; Borghetti, F.; Gersbach, M.; Henderson, R.K.; *et al.* A 16 0 × 128 single-photon image sensor with on-pixel 55 ps 10 b time-to-digital converter. In Proceedings of the 2011 IEEE International Solid-State Circuits Conference, San Francisco, CA, USA, 20–24 February 2011; pp. 312–314.

30. Pancheri, L.; Stoppa, D. A SPAD-based pixel linear array for high-speed time-gated Fluorescence Lifetime Imaging. In Proceedings of the ESSCIRC 2009—35th European Solid-State Circuits Conference, Athens, Greece, 14–18 September 2009; pp. 428–431.

31. Niclass, C.; Ito, K.; Soga, M.; Matsubara, H.; Aoyagi, I.; Kato, S.; Kagami, M. Design and characterization of a 256 × 64-pixel single-photon imager in CMOS for a MEMS-based laser scanning time-of-flight sensor. *Opt. Express* **2012**, *20*, 11863–11881. [CrossRef] [PubMed]

32. Braga, L.H.C.; Gasparini, L.; Grant, L.; Henderson, R.K.; Massari, N.; Perenzoni, M.; Stoppa, D.; Walker, R. A Fully Digital 8 × 16 SiPM Array for PET Applications With Per-Pixel TDCs and Real-Time Energy Output. *IEEE J. Solid State Circuits* **2014**, *49*, 301–314. [CrossRef]

33. Perenzoni, M.; Perenzoni, D.; Stoppa, D. A 64 × 64-pixel digital silicon photomultiplier direct ToF sensor with 100 M photons/s/pixel background rejection and imaging/altimeter mode with 0.14% precision up to 6 km for spacecraft navigation and landing. In Proceedings of the 2016 IEEE International Solid-State Circuits Conference (ISSCC), San Francisco, CA, USA, 31 January–4 February 2016; pp. 118–119.

34. Burri, S.; Maruyama, Y.; Michalet, X.; Regazzoni, F.; Bruschini, C.; Charbon, E. Architecture and applications of a high resolution gated SPAD image sensor. *Opt. Express* **2014**, *22*, 17573–17589. [CrossRef] [PubMed]

35. Dutton, N.A.W.; Gyongy, I.; Parmesan, L.; Gnecchi, S.; Calder, N.; Rae, B.R.; Pellegrini, S.; Grant, L.A.; Henderson, R.K. A SPAD-Based QVGA Image Sensor for Single-Photon Counting and Quanta Imaging. *IEEE Trans. Electron. Devices* **2015**, *63*, 189–196. [CrossRef]

36. Perenzoni, M.; Massari, N.; Perenzoni, D.; Gasparini, L.; Stoppa, D. A 160 × 120 Pixel Analog-Counting Single-Photon Imager with Time-Gating and Self-Referenced Column-Parallel A/D Conversion for Fluorescence Lifetime Imaging. *IEEE J. Solid State Circuits* **2016**, *51*, 155–167.

37. Parmesan, L.; Dutton, N.A.W.; Caldery, N.J.; Krstaji, N.; Holmes, A.J.; Grant, L.A.; Henderson, R.K. A 256 × 256 SPAD array with in-pixel Time to Amplitude Conversion for Fluorescence Lifetime Imaging Microscopy. In Proceedings of the International Image Sensors Workshop 2015, Vaals, The Netherlands, 8–11 June 2015.

38. Stoppa, D.; Borghetti, F.; Richardson, J.; Walker, R.; Grant, L.; Henderson, R.K.; Gersbach, M.; Charbon, E. A 32 × 32-pixel array with in-pixel photon counting and arrival time measurement in the analog domain. In Proceedings of the ESSCIRC 2009, Athens, Greece, 14–18 September 2009; pp. 204–207.

39. Bronzi, D.; Villa, F.; Tisa, S.; Tosi, A.; Zappa, F. SPAD Figures of Merit for Photon-Counting, Photon-Timing, and Imaging Applications: A Review. *IEEE Sens. J.* **2016**, *16*, 3–12. [CrossRef]

40. Kondo, T.; Takemoto, Y.; Kobayashi, K.; Tsukimura, M.; Takazawa, N.; Kato, H.; Suzuki, S.; Aoki, J.; Saito, H.; Gomi, Y.; *et al.* A 3D stacked CMOS image sensor with 16Mpixel global-shutter mode and 2 M pixel 10000 fps mode using 4 million interconnections. In Proceedings of the 2015 Symposium on VLSI Circuits (VLSI Circuits), Kyoto, Japan, 17–19 June 2015; pp. C90–C91.

41. Aull, B.F.; Schuette, D.R.; Young, D.J.; Craig, D.M.; Felton, B.J.; Warner, K. A Study of Crosstalk in a 256 × 256 Photon Counting Imager Based on Silicon Geiger-Mode Avalanche Photodiodes. *IEEE Sens. J.* **2015**, *15*, 2123–2132. [CrossRef]

sensors

MDPI

Article

Photon-Counting Arrays for Time-Resolved Imaging

I. Michel Antolovic [1,†], Samuel Burri [2,†], Ron A. Hoebe [3], Yuki Maruyama [4], Claudio Bruschini [2] and Edoardo Charbon [1,2,*]

[1] Applied Quantum Architecture Lab (AQUA), Quantum Engineering Department, Delft University of Technology, Delft 2628CD, The Netherlands; i.m.antolovic@tudelft.nl
[2] Advanced Quantum Architecture lab (AQUA), Microengineering Department, Ecole Polytechnique Fédérale de Lausanne (EPFL), Lausanne 1015, Switzerland; samuel.burri@epfl.ch (S.B.); claudio.bruschini@epfl.ch (C.B.)
[3] Academic Medical Centre, University of Amsterdam, Amsterdam 110DD, The Netherlands; r.a.hoebe@amc.uva.nl
[4] Jet Propulsion Lab, Pasadena, CA 91109, USA; y.maruyama@tudelft.nl
* Correspondence: edoardo.charbon@epfl.ch; Tel.: +41-021-693-6487
† These authors contributed equally to this work.

Academic Editor: David Stoppa
Received: 18 February 2016; Accepted: 16 June 2016; Published: 29 June 2016

Abstract: The paper presents a camera comprising 512×128 pixels capable of single-photon detection and gating with a maximum frame rate of 156 kfps. The photon capture is performed through a gated single-photon avalanche diode that generates a digital pulse upon photon detection and through a digital one-bit counter. Gray levels are obtained through multiple counting and accumulation, while time-resolved imaging is achieved through a 4-ns gating window controlled with subnanosecond accuracy by a field-programmable gate array. The sensor, which is equipped with microlenses to enhance its effective fill factor, was electro-optically characterized in terms of sensitivity and uniformity. Several examples of capture of fast events are shown to demonstrate the suitability of the approach.

Keywords: single-photon avalanche diode; SPAD; fluorescence; fluorescence lifetime imaging microscopy; FLIM; fluorescence correlation spectroscopy; FCS

1. Introduction

Photon counting and single-photon detection have been available at least since the 1930s, with the phtomultiplier tube (PMT) first and microchannel plate (MCP) later. These devices enable relatively high sensitivity, known in this context as photon detection efficiency (PDE), and low dark counts, quantified in terms of dark count rate (DCR), however they are generally bulky and they require high voltages to operate, typically hundreds to thousands of volts. In the 1940s, researchers started working with solid-state diodes operating in avalanche mode, known as avalanche photodiodes (APDs); these devices were refined through the 1950s and 1960s to be then implemented in planar processes. The devices required p-n junctions with guard rings and enhancement regions to prevent premature breakdown at the edge of the junction. With the improvement of semiconductor processes and the availability of more options, Cova and others began experimenting with Geiger-mode APDs or single-photon avalanche diodes (SPADs) in the 1970s and 1980s, recognizing the potential of these devices in capturing fast processes [1]. In the early 2000s, SPADs could be implemented in high-voltage processes first [2] and in standard CMOS image sensor processes later [3–6].

With the availability of SPADs in deep-submicron CMOS processes, it became conceivable to implement useful functionality in situ, possibly in pixel, so as to count photons and to time stamp them upon detection. The major consequence of this trend was massively parallel timestamping with picosecond resolution (LSB), with a possible explosion of data generated on chip [7]. Timestamping impinging photons individually has several uses. For instance, through time-correlated single-photon counting (TCSPC) [8], it becomes possible to accurately characterize photo responses of fluorophores when excited by fast light pulses. Fluorophores exhibit a time-dependent behavior, known as lifetime, that is specific to the fluorophore and/or the environment it is in [9]. Fluorescence lifetime imaging microscopy (FLIM) [10] is a technique used to characterize lifetime in fluorophores using multiple excitations and histogramming; in general, FLIM may be used in a confocal microscope with a single pixel, but it can also be used in widefield microscopy with a large number of pixels capable of performing TCSPC independently, thus speeding up lifetime capture by several orders of magnitude [11,12].

The first pixel with embedded time-to-digital converter (TDC) for in situ TCSPC was introduced in the project MEGAFRAME [13,14]. The drawback of this approach was the reduced fill factor that in turn required microlenses to recover, at least in part, lost sensitivity. As an alternative, researchers proposed to use simpler pixels, with a single digital counter [15] or with dual digital counters [16]. The use of analog counters was also proposed to ensure large resolution at low cost in terms of fill factor [17–19]. Due to the lack of a TDC though, these methods require a precise gate and significant algorithmic complexity [20,21].

In this paper, we describe a photon counting imager comprising a programmable global shutter with sub-150 ps skew and a minimum width of 4ns for time-resolved imaging applications. The image sensor comprises an array of 512×128 SPAD pixels that are read out in rolling mode, while the shutter itself is global. Since each pixel has a one-bit counter embedded in it, a frame is read as a binary matrix and can be converted to a multi-bit matrix externally by adding up subsequent frames, as first shown in [22]. The chip was demonstrated for fast fluorescence imaging and could be used for FLIM in [23]. Throughout the paper, significant attention was given to circuit details that led to the exceptional skew and to tradeoffs used during the design to achieve the target readout speed. A complete dynamic and static characterization of the chip was also provided with images exemplifying the suitability of the approach.

The paper is organized as follows: Section 2 describes the architecture of the sensor and its components. Section 3 analyzes the implications of using binary pixels towards image quality and Sections 4 and 5 report the optical and electrical characterization of the sensor in the context of the target applications. Section 6 concludes the paper.

2. Sensor Architecture

A single-photon avalanche diode (SPAD) is a p-n junction biased above breakdown, so as to operate in Geiger mode. In this design, SPADs are similar to those in [24], comprising a circular p+ active region over n-well, whereas premature edge breakdown is prevented by means of p-well guard rings. Figure 1 shows the cross-section of a planar implementation of a SPAD consistent with CMOS processes. The SPADs are passively quenched, while an active recharge technique is provided [1,2].

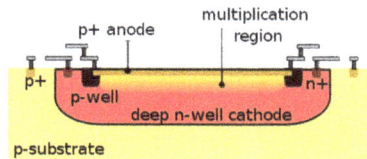

Figure 1. Planar single-photon avalanche diode (SPAD): the p+-n-well junction is biased above breakdown so as to achieve infinite optical gain. A guard ring prevents premature edge breakdown by reducing the electric field in areas at risk, such as the corners of the junction. The deep n-well acts as an insulation mechanism to minimize electrical crosstalk, while optical crosstalk is generally reduced using deep trench isolation (not available in this CMOS technology).

The pixel can perform photon counting by means of a one-bit counter, implemented as a static latch. The pixel achieves gated operation by way of three transistors acting as switches. The pixel counter content is transferred to the exterior of the sensor via a fast digital readout channel capable of transferring a complete frame in 6.4 µs. The sensor has a global shutter that gates all the pixels simultaneously for a time as short as 3.8 ns.

The pixel conceptual diagram is shown in Figure 2a. The MOS switch "SPADOFF" is activated to bring the SPAD below breakdown, thereby quenching any ongoing avalanche and preventing any future avalanche in the same frame (see Figure 2b). The second MOS switch, "RECHARGE", is designed to bring the bias voltage close to ground, thereby rapidly recharging the SPAD to its idle bias. This action reactivates the SPAD and, to avoid direct conduction from VDD to ground, it should never be performed simultaneously to "SPADOFF". The last MOS switch, "GATE", is used to prevent the one-bit counter from being accidentally set during the gating operations.

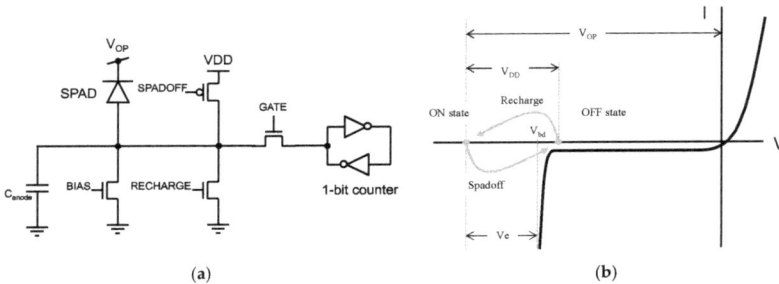

Figure 2. Pixel architecture. (**a**) SPAD configuration with parasitic load (C_{anode}); the MOS transistors controlled by signals "SPADOFF", "GATE", and "RECHARGE" act as switches that implement gating. The quenching transistor controlled by "BIAS" acts as a non-linear resistor used for quenching. The one-bit counter (represented here as a simplified latch) is used to record photon detection; (**b**) I-V characteristics of the SPAD pixel. In the ON state, the SPAD is biased above breakdown (V_{bd}) by a voltage known as excess bias (V_e). When "SPADOFF" is activated, the SPAD bias is pushed below breakdown; "GATE" is deactivated to isolate the counter; this is the OFF state. To bring the SPAD back to the ON state, "GATE" is activated, "SPADOFF" is deactivated, and "RECHARGE" is activated for a short time, typically nanoseconds, thereby bringing the bias to the initial state above breakdown. Photon detection triggers a similar cycle with a resulting change of state of the one-bit counter from "L" to "H".

The actual implementation of the pixel is shown in Figure 3. The recharging transistor is controlled by a global "RECHARGE" signal. The switches are implemented as NMOS transistors, while the latch is implemented by way of four NMOS transistors, connected as back-to-back NMOS inverters to eliminate the need for PMOS transistors. The pull-up transistors are critical to control power consumption in the latch during idle phases and settling time during set/reset phases. These transistors can be controlled using an external voltage, "TOPGATE". The column pull-up transistor is biased so as to minimize the power required to bring the column to 'L' while ensuring a readout cycle of 6.4 µs/128 = 50 ns.

The pixel content is stored in a latch at the bottom of the column (not shown in the figure) that stores its value for 50 ns while the other three columns are multiplexed out to the external PAD. A 4:1 multiplexer serializes the output of the latches of four columns to the PADs; it is operated at four times that speed, i.e., 4/50 ns = 80 MHz, which is the maximal operating speed of the PADs in this technology and represents a good speed-power tradeoff.

The block diagram of the sensor, known as SwissSPAD, is shown in Figure 4a. The timing diagram for the pixel is shown in Figure 4b. The chip features a balanced network that distributes a low-skew version of signals "SPADOFF", "RECHARGE", and "GATE" (Figure 4c). Due to their nanosecond length, "RECHARGE", and "GATE" are distributed as three precisely timed rising-edge signals that are recomposed in situ by means of a pulse generator (PG) shown in Figure 4d.

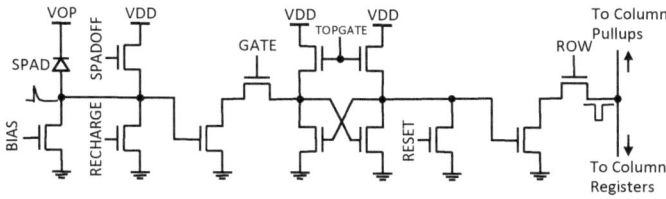

Figure 3. Pixel schematic. The counter is implemented as a latch, which is reset by "RESET" and biased by "TOPGATE". The content of the counter is read out using a rolling shutter mechanism by setting "ROW" to "H"; when the counter has recorded a photon, a pull-down transistor sets the column to "L" and this state is transferred to a latch at the bottom of the column (not shown in the figure) and then to a PAD via a multiplexer for external processing. A pull-up ensures return of the column to "H" state when no photons are detected. "RECHARGE" and "SPADOFF" are used to turn on and off the SPAD, while "GATE" is used as a pass gate and "BIAS" determines the equivalent resistance used as ballast.

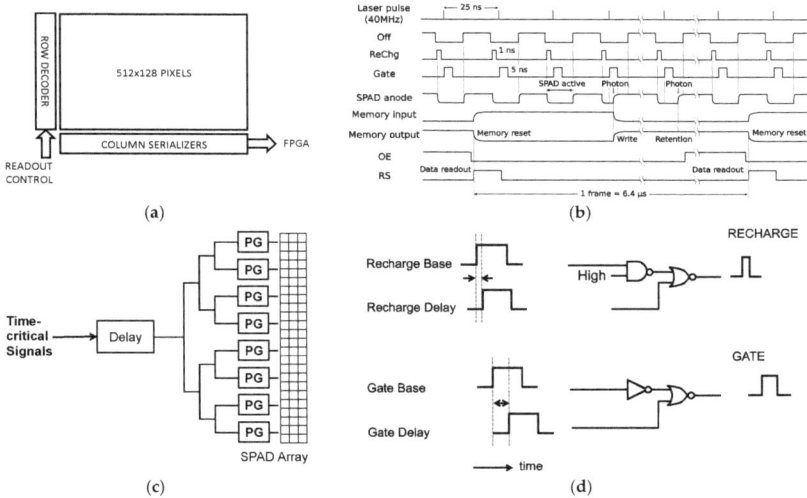

Figure 4. Block diagram of the sensor: (**a**) overall block diagram; (**b**) timing diagram; (**c**) balanced tree network for time-critical signal distribution; (**d**) pulse generation mechanism for in situ generation of accurate pulses insensitive to rise/fall time asymmetries.

The timing diagram shows a typical readout cycle, wherein a memory reset is performed at the beginning of the cycle and a series of gating operations follows, until the next readout is performed. The gates are spaced an arbitrary time period (25 ns in this example) and are generally synchronized with a light source. This is done to maximize the effective spatio-temporal fill factor when a fast but dim response is expected from a pulsed light source, as, for example in FLIM. Fewer gates or even a single gate is possible, however, it results in an effective spatio-temporal FF computed as

$$FF = FF_G \cdot DC = FF_G \frac{N \cdot t_{GATE}}{T_{FRAME}} = FF_G \cdot f_{GATE} \cdot t_{GATE} \qquad (1)$$

where FF_G is the geometric fill factor, DC the duty cycle of gating, N the number of gates in a frame, t_{GATE} the time length of the gate, T_{FRAME} the period of the frame, and f_{GATE} the frequency of the gate. Note that even though we reduce the temporal FF, light will usually impinge within the gate synchronized with a laser, while the DCR will be reduced. The duty cycle is generally selected to be a fraction of the lifetime of a fluorophore. The delay between gate and excitation light is varied a minimum of 20 ps and a maximum of T_{FRAME}/N, so as to scan the entire laser period $1/f_{GATE}$. N is chosen to minimize the pile-up effect. Since the counter only counts, at most, one event, no accumulation is possible during a frame but only digitally after multiple frames. This enables us to construct gray levels in images at the expense of a reduced frame rate [22].

3. Binary Pixels

SwissSPAD is an all-digital, clock-driven sensor comprising pixels that can only detect one photon in a frame: we call these pixels "binary pixels". Photons impinge a binary pixel with an expected arrival rate χ (photons per second) and are distributed in time following a Poisson distribution (the probability of k counts per second is $p\,(cps = k) = \frac{\chi^k e^{-\chi}}{k!}$). Thus, the probability of detecting one or more photons per second is $p\,(cps > 0) = 1 - e^{-\chi}$. For a non-unity photon detection probability (PDP) and non-zero FF, the probability of photon detection per frame $p\,(cpf > 0) = 1 - e^{-\chi \cdot PDP \cdot FF \cdot T_{FRAME}}$. The expected photon counts per second measured in the pixel will thus become [25]

$$\mathrm{E}\,(C_M) = \frac{1 - e^{-\chi \cdot PDP \cdot FF \cdot T_{FRAME}}}{T_{FRAME}} \tag{2}$$

where C_M is the measured SPAD count rate. Thus, even if one photon per frame is expected to impinge on the pixel, the pixel will detect it, on average, a fraction of the time, i.e., it will detect a fraction of a count, on average, per frame. Dark noise in a SPAD is dominated by three sources: thermal (trap-assisted and tunneling), noise, and afterpulsing. Assuming a large dead time, afterpulsing can be ignored and thus, with the exception of hot pixels, most exhibit a noise approaching Poisson statistics. The rate of occurrence of this noise is quantified by dark count rate (DCR). Thanks to its Poissonian nature, DCR is added to the equation as follows

$$\mathrm{E}\,(C_M) = \frac{1 - e^{-(\chi \cdot PDP \cdot FF + DCR) \cdot T_{FRAME}}}{T_{FRAME}} \tag{3}$$

From this equation, one can derive the correction factor for the expected detected SPAD count rate $\mathrm{E}\,(C_D) = \chi \cdot PDP \cdot FF + DCR$, by simply solving the equation w.r.t $\mathrm{E}(C_D)$, as follows

$$C_D \approx \frac{-\ln\,(1 - C_M \cdot T_{FRAME})}{T_{FRAME}} \tag{4}$$

Note that $\mathrm{E}(C_M)$ and $\mathrm{E}(C_D)$ were replaced by C_M, and C_D, respectively, since it is assumed that the correction is applied to a single sample generated by the detector and not the expected value achieved over a very large number of measurements. As can be seen from the equation, this correction is only needed for high values of C_M, above 15 kcps.

However, in this condition, the asymptotic behavior can be used to extend the dynamic range of the pixel, as has been known in the silicon photomultiplier community for several years [26] and in the radiation community from the 1970s [27]. This can be done both in time and in space, whenever multiple pixels are added to make a larger one [28,29]. Figure 5 shows the theoretical and measured response of a binary pixel in clock-driven and in event-driven modes, as compared to the linear response of a non-binary pixel. In clock-driven mode, SPAD recharge or memory reset is applied periodically, asynchronously with respect to SPAD activity, while in event-driven mode recharge is done T_{dead} after a SPAD avalanche, thus synchronously with SPAD activity. While clock-driven resets at high frequency are not used in single SPAD devices because of possible afterpulsing, arrays with long T_{dead} do not show increased afterpulsing. Recent work is indicating a trend towards higher pixel resolution and advanced processing [30,31].

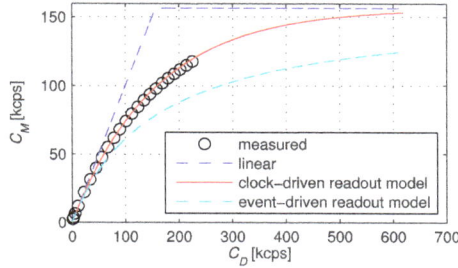

Figure 5. Theoretically predicted and actually recorded (measured) photons in all-digital image sensors as a function of impinging photons, as they are detected and multiplied in a binary pixel. The plot shows clock-driven and event-driven modes in comparison with a non-binary pixel.

4. Sensor Fabrication

The sensor microphotograph is shown in Figure 6; the inset shows a detail of the pixels. An array of microlenses (CSEM, Basel, Switzerland) was deposited on the chip matching the pixel pitch to improve light collection through light concentration [32,33]. The microlens array, shown in an artist's rendering in Figure 7a [34], was measured and simulated as a function of the f-number of the main objective lens, yielding the plot of Figure 7b.

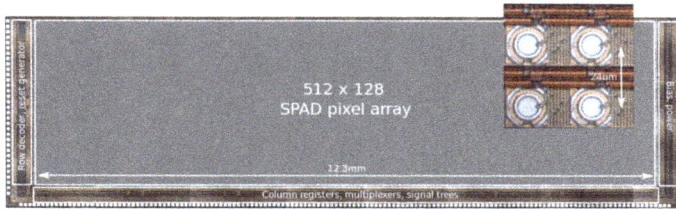

Figure 6. Photomicrograph of the sensor; it was fabricated in a 0.35 µm CMOS process with a pitch of 24 µm. The inset shows a detail of four pixels achieving a fill factor of 5%.

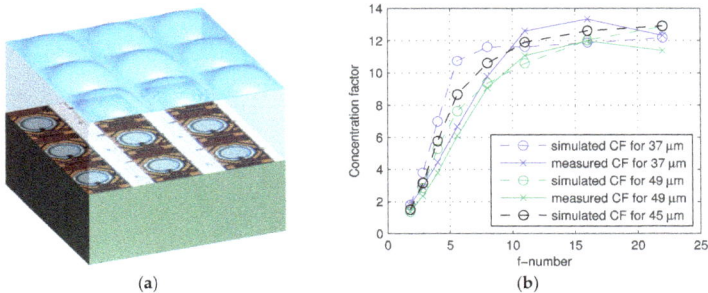

Figure 7. (**a**) Artist's rendering of the microlens array deposited on the sensor (Courtesy: Juan Mata Pavia); (**b**) measured and simulated concentration factors as a function of the f-number in the main lens [25]. The graph shows simulated and measured data for chips with 37, 45, and 49 µm microlens height.

Thus, the effective fill factor achieved with a lens of f/10 was 60% with high reliability and reproducibility over the entire array. The pixel PDP and DCR are plotted in Figure 8 at room temperature.

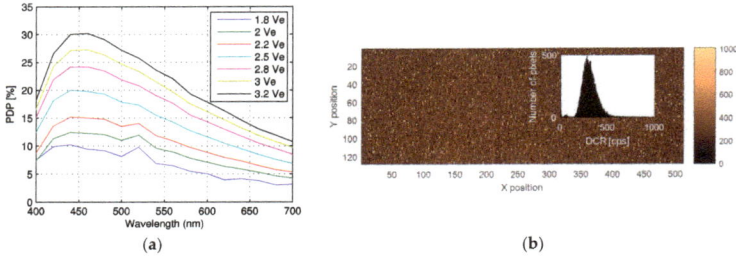

(a)   (b)

Figure 8. Pixel characterization: (**a**) PDP vs. wavelength at given excess bias voltages; (**b**) DCR distribution at 40 °C for V_e = 4.5 V, with an average of 1169 cps and a median of 302 cps. In the inset the DCR distribution is shown.

The sensor was also characterized in terms of afterpulsing. The measurement was achieved by means of the inter-arrival response method introduced in [35]. The pixels were exposed to a uniform wide-spectrum light source and the inter-arrival time of the response was stored in the FPGA for an integration time of 80 s. A histogram was then constructed confirming the exponential behavior of the response due to the Poissonian nature impinging photons. Afterpulsing probability $APP(t)$ is approximated as

$$APP\left(t\right) \approx \frac{\int_t^{\infty}\left[h_M\left(\tau\right) - h_F\left(\tau\right)\right]d\tau}{\int_0^{\infty}h_M\left(\tau\right)d\tau}, \quad t > t_{DT} \tag{5}$$

In the equation, $h_M\left(\tau\right)$ is the measured histogram, $h_F\left(\tau\right)$ the exponential fit, and t_{DT} the dead time of the pixel, in this case 6.4 μs. Crosstalk probability, or simply crosstalk, $XT\left(\tau\right)$ is computed in a similar way, wherein the inter-arrival time is measured between two adjacent pixels, as

$$XT\left(t\right) \approx \frac{\int_t^{\infty}\left[h_{ij}\left(\tau\right) - h_F\left(\tau\right)\right]d\tau}{\int_0^{\infty}h_{ij}\left(\tau\right)d\tau} \tag{6}$$

where $h_{ij}\left(\tau\right)$ is the inter-arrival time histogram measured between pixels i and j. Afterpulsing and crosstalk are reported in Figure 9a,b, respectively.

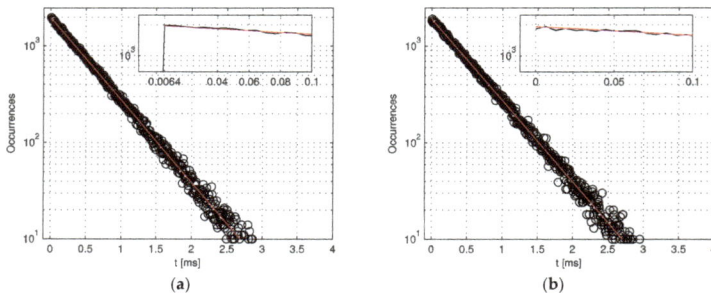

(a)   (b)

Figure 9. Afterpulsing and crosstalk: (**a**) inter-arrival histogram showing no measurable afterpulsing above the dead time of 6.4 μs, the inset shows a zoom of the plot; (**b**) crosstalk measured using the same method. Note that in this case the dead time is zero, given by the fact that two pixels are independently firing upon detection of a photon [25].

5. Results

The sensor was used to image a large number of biological samples using fluorescence intensity and fluorescence lifetime imaging microscopy. Fluorescence intensity was achieved using a setup based on a dual port Leica SR GSD super resolution microscope (Leica Microsystems, Wetzlar, Germany) where SwissSPAD and an Andor iXon3 897 BV EMCCD were placed on the two ports of the microscope using the same illumination conditions for comparison purposes (Figure 10). As an illustration, several biological samples are shown hereafter. First, let us consider BPAE cells labeled with MitoTracker Red CMX Ros, Alexa Fluor 488, and DAPI dyes. SwissSPAD was used at V_e = 4.5 V. The EMCCD raw intensity image was converted to a photon count image using counts $D = (d - b) \times g_{amp}/g_{EM}$, where d is the digital intensity value, b the bias offset, g_{amp} the preamp gain value and g_{EM} the EM gain. Due to pixel size differences, 2×2 SPAD pixels and 3×3 EMCCD pixels were binned to obtain counts for the same area. MATLAB software was used to find the overlapping area of the two images and compare the intensities.

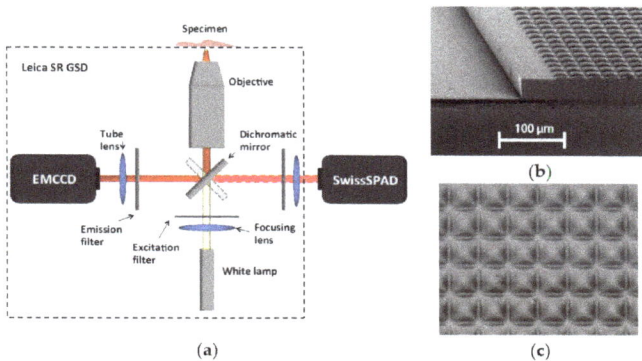

Figure 10. (**a**) Schematic microscope setup used in the imaging experiments based on the Leica SR GSD microscope (Leica Microsystems Wetzlar, Germany); (**b**) SEM scan of the microlens array deposited on the sensor; (**c**) Optical micrograph of the microlens array.

Figure 11 shows the images obtained with the EMCCD (a) and SwissSPAD (b); the exposure times were 10 ms and 73.4 ms, respectively, to match the number of collected photons. The scale shows the number of collected photons per exposure. Figure 12 shows a widefield image of a cellular cluster magnified $10\times$ using the same microscope setup.

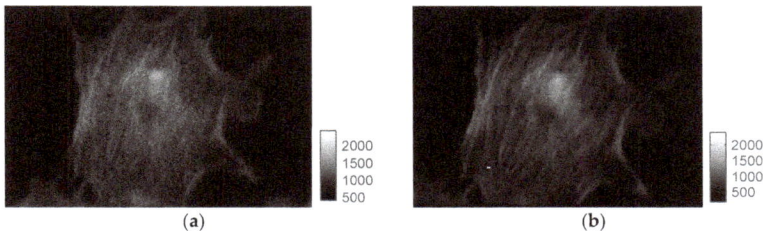

Figure 11. BPAE cell imaging. The cells were labeled with MitoTracker Red CMX Ros, Alexa Fluor 488, and DAPI dyes and imaged in a microscope using two ports matching the photon counts per pixel in each sensor. (**a**) shows the EMCCD and (**b**) the SwissSPAD images obtained with 10 ms and 73.4 ms exposure, respectively. The scales indicate photon counts per frame.

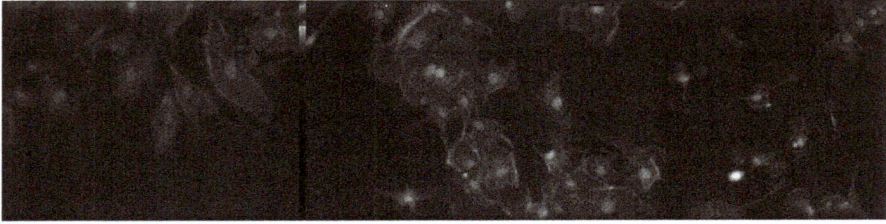

Figure 12. Widefield fluorescence imaging of BPEA QGS cellular clusters labeled with MitoTracker Red CMX Ros, Alexa Fluor 488, and DAPI dyes (magnification = 10×). The dead column in the image is due to a false connection between the imager and the FPGA.

Fluorescence lifetime images could be obtained by sliding the gate start time from 0 to 10 ns with a step of 20 ps and integrating 255 frames per step, whereas the gate timing performance is essential for high quality FLIM images. The timing performance of the gate is summarized in Figure 13: the response of the sensor is shown in Figure 13a for a random pixel with minimum gate width and the uniformity of its position and length is shown in Figure 13b. Figure 13 shows steep edges of the counting response of the sensor when gating is used, and should not be mistaken with the signal shape of the "Gate" signal. The photosensitive window is defined by the falling edge of "Recharge" and falling edge of "Gate" (if "Gate" occurs after "Recharge"). A large vertical dimension with 128 pixels resulting in 3 mm long metal wires introduces an undesired RC component, limiting the minimal gate width. Although the metal wires of "Gate" and "Recharge" are equal, "Recharge" was designed larger to enable a shorter signal. This introduces a mismatch in "Gate" and "Recharge" RC, and nonuniformity of photosensitive window widths. Smaller technology nodes will decrease transistor gate capacitances, and wider metal lines through the column can reduce the resistance while keeping the parasitic capacitance dominated by the lateral component constant. The use of repeaters is also an option. Both the metal widening and repeater though can reduce fill factor. A smaller pixel pitch will also reduce the RC component of the line. The right edge of Figure 13a corresponds to the falling edge of the "Recharge" signal and it represents the critical edge for FLIM. A theoretical approach of a FLIM measurement with gating is a convolution between an exponential distribution signal and a rectangular gate signal. The fall time of the falling edge should be small in comparison to the lifetime of the exponential distribution to assure high precision measurements. This fall time is a similar measure as the instrumentation response function (IRF) in TCSPC.

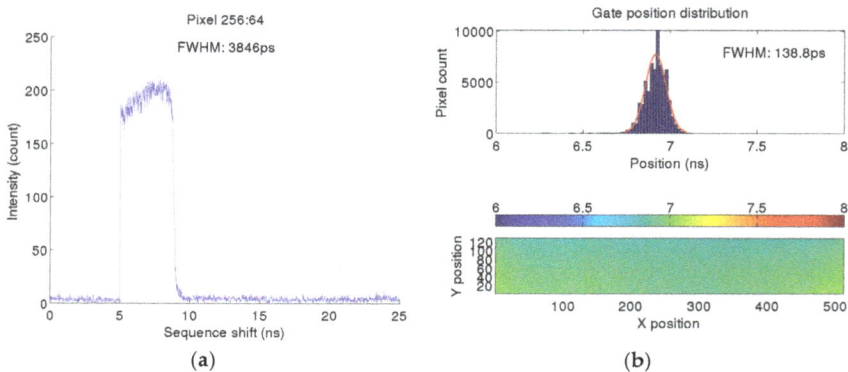

Figure 13. Timing characterization of the gating mechanism: (**a**) characteristic sensitivity at the minimum gate width of 3.8 ns; (**b**) gate position (in time) statistics as a function of pixel position [23].

Figure 14 reports fluorescence intensity images of samples stained with Safranin and Fast Green, having peak excitation wavelengths of 530 nm and 620 nm, respectively. Filtering was used in two subsequent exposures of the same sample and software based recomposition was then applied. Pictures on Figures 11, 12, and 14 were corrected for DCR and possible count compression using Equation (4).

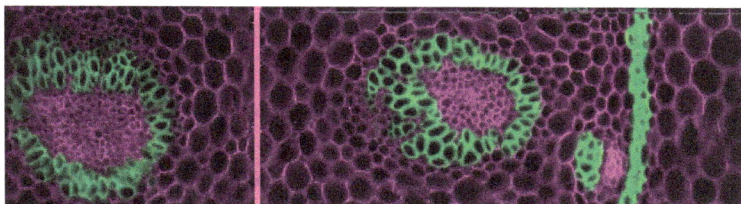

Figure 14. Composite fluorescence image of a thin slice of a plant root stained with a mixture of fluorescent dyes. The picture was obtained with two 20 nm interference filters centered at 530 and 640 nm, respectively. The dead column in the image is due to a false connection between the imager and the FPGA.

Thanks to the frame readout period of 6.4 μs, one can achieve a maximum frame rate of 156 kfps, whereas a Virtex™ IV or Spartan™ 6 FPGA (Xilinx, San Jose, CA, USA.) is used to acquire and store the one-bit frames. Figure 15 shows the physical appearance of the system, whereas a daughterboard hosting SwissSPAD is electrically connected to a motherboard hosting two Xilinx-IV FPGAs for acquisition and formatting of the data that are then sent to a Mac/PC through USB2 link.

Figure 15. Physical appearance of the camera based on SwissSPAD; a daughterboard hosting the chip is electrically connected to a motherboard hosting two Xilinx-IV that process the raw frames generated by the chip and format them to send them through a USB2 link.

To construct gray level images, the one-bit frames may be accumulated in the FPGA and transferred to the Mac/PC through a USB-2 or USB-3 link; by doubling the number of frames, the pixel intensity resolution increases by one bit [22]. The tradeoff between pixel effective number of bit (ENOB) and effective frames-per-second (EFPS) is shown in Figure 16. While the speed of a DDR memory is high enough to keep up with the data rate generated by the FPGA at any ENOB, USB-2, and USB-3 links to the PC do not allow continuous recording at all ENOBs, making an intermediate memory like DDR2 necessary. An USB-2 link can transfer eight-bit frames continuously.

The sequence of Figure 17 shows images of an analog oscilloscope obtained without accumulation (156 kfps, 1 bit) and with several levels of accumulation from 4× to 65,536×, resulting in a pixel ENOB of 2 and 16 bits, respectively, during the cumulative frame. Unlike conventional cameras, the pixel ENOB is derived by the simple expression

$$ENOB = \log_2 \left(N_{pixel} \right)$$

where N_{pixel} is the maximum possible number of accumulated counts in the pixel during a cumulative frame. The SNR per pixel, and over the entire sensor, is approximated by

$$SNR_{pixel} = \sqrt{N_{pixel}}$$

This approximation assumes no readout noise and lower count rates with linear response, thus a Poisson limited noise at any frame rates and no saturation. The images in Figure 17 show Poisson limited noise in the images at five different exposure times.

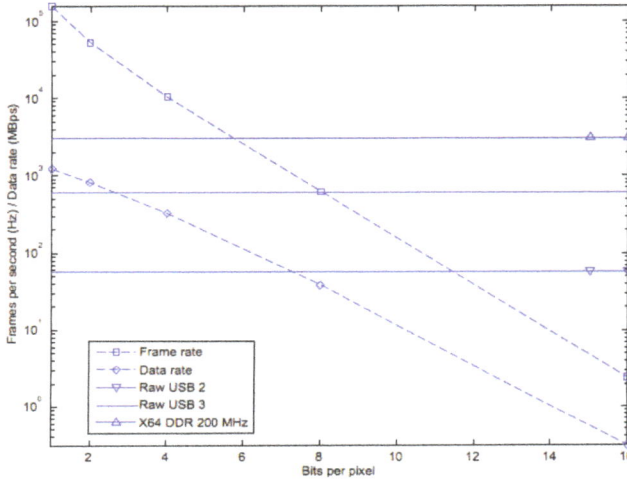

Figure 16. Tradeoff between pixel ENOB and EFPS in SwissSPAD. X64 DDR, USB-2, and USB-3 are shown as horizontal lines to represent maximal supported data rates.

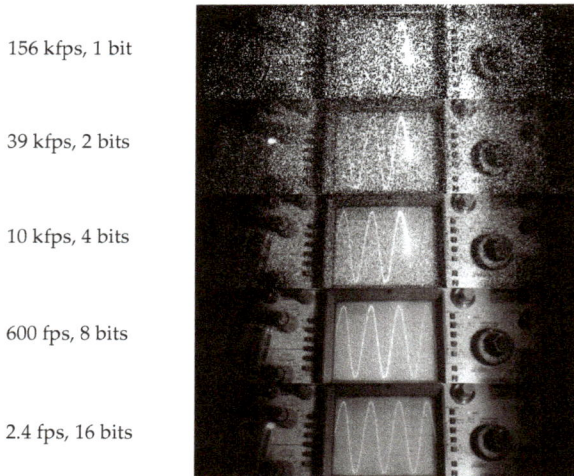

Figure 17. Tradeoff between effective number of bits (ENOB) and frame rate, from 156 kfps to 2.4 fps [23].

In Figure 18a sequence of a fast event is shown; the frame rate was fix at 1200 frames-per-second, of which one frame every 100 ms is depicted. A global shutter was used achieving deep-subnanosecond uncertainty of the gate width and position.

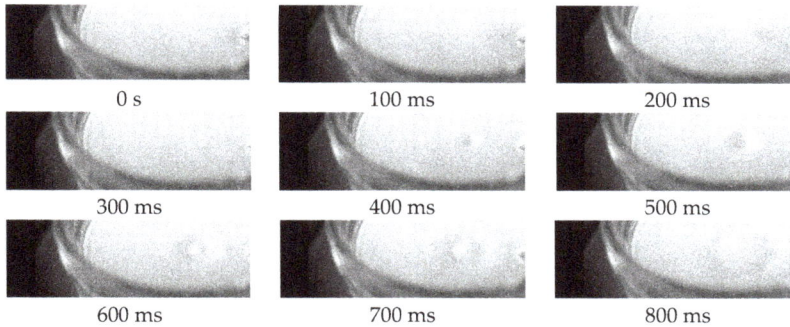

0 s	100 ms	200 ms
300 ms	400 ms	500 ms
600 ms	700 ms	800 ms

Figure 18. Sequence of a drop fall recorded at 1200 fps; the frame-to-frame time interval shown in the sequence was 100 ms.

The ability of SwissSPAD to acquire lifetime images was demonstrated in a lab setup using point detection. Indocyanine green (ICG) in milk with a concentration of 40 μM was excited using a 790 nm laser with 55 ps pulse width and 100MHz repetition rate synchronized with the SwissSPAD gating. Fluorescence intensity from the excited spot was measured for 512 gate windows offset by a fraction of the repetition period (25 ps). From the response similar to the IRF shown in Figure 13 convolved with an exponential decay the lifetime is extracted by fitting against a set of models constructed from the IRF used in these measurements. Figure 19 shows the per pixel extracted lifetime and normalized intensity over the excited spot. The extracted lifetimes with μ = 636 ps and σ = 56 ps overestimate the 580 ps reference lifetime given in literature [36]. Homulle et al. showed in [37] how the accuracy of lifetime extraction from gated measurements can be improved through refinement of the modeling and simulation.

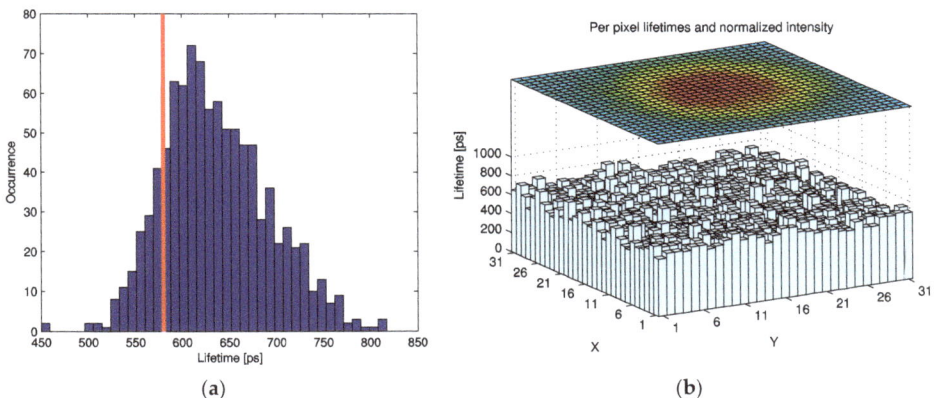

(a)   (b)

Figure 19. (a) FLIM results show extracted lifetimes distribution of 31 × 31 pixels compared to reference lifetime of 40 μM ICG in milk (red). (b) shows the comparison of intensity and lifetime per pixel.

The sensor specifications are listed in Table 1. All the measurements were performed at room temperature, unless otherwise indicated.

Table 1. Specifications of the sensor SwissSPAD. All the values are reported at room temperature.

Parameter	Value	Condition
Peak PDE	20%	450 nm
Max. frame rate	156 kfps	1 bit ENOB [1]
Readout noise	0 cps	
Dark counts	200 cps	25 °C
Pixel pitch	24 μm	
Active area	28.3 μm^2	Drawn area
Imager size	12.3 × 3.1 mm^2	
Operating temperature	25 °C	
Afterpulsing probability	0.3%	1 μs dead time
Crosstalk	<0.3%	
PRNU	<1.8% [2]	
Min. gate width	3.8 ns	
Gate skew	<150 ps	Sigma

[1] ENOB denotes effective number of bits for the pixel resolution. [2] PRNU denotes photo-response non-uniformity without compensation.

The sensor is currently used by a number of researchers in different institutions under use warranty based on GNU policies.

6. Conclusions

We reported on a 512 × 128 SPAD image sensor operating at a maximum speed of 156 kfps with a low-skew global shutter. The sensor has been successfully used in a variety of applications, involving time-resolved capture of fast events. It can be operated in TCSPC mode to achieve, for instance, FLIM images. The sensor is highly versatile and allows one to achieve tradeoffs between frame speed and resolution. Thanks to its zero readout noise, the sensor noise is always Poisson limited, thus enabling investigations in photon starved regimes. Future work includes extensive modeling for super-resolution microscopy and the design of a larger array with reduced DCR and increased PDE.

Supplementary Materials: The full videos are available online at http://aqua.epfl.ch and http://www.aqua.ewi.tudelft.nl/movies/.

Acknowledgments: This research was funded in part by the Dutch Technology Foundation (STW), by the Swiss National Science Foundation (SNF) under Grant SNF 51NF40-144633, by the NCCR MICS, as well as by CCES through the SwissEx project. The microlenses on the SwissSPAD chip were fabricated by CSEM Muttenz, Switzerland. The authors thank Xilinx Inc. (Santa Jose, CA, USA) for the generous donation of Virtex™ 4 FPGAs used in this work.

Author Contributions: I.M.A. measured the SwissSPAD sensor and fluorescence intensity, and co-developed the theory used in the analysis; S.B. designed the SwissSPAD sensor and the firmware for the FPGAs, and measured the fluorescence lifetime; R.A.H. developed the experiments in super-resolution fluorescence microscopy and provided samples and microscope systems; Y.M. co-designed the pixel; C.B. co-designed the gating mechanism and the microlens array, C.B. co-directed the work; E.C. co-designed the pixel and the sensor, and co-developed the theory; E.C. co-directed the work and wrote the paper.

Conflicts of Interest: The authors declare no conflict of interest.

Abbreviations

The following abbreviations are used in this manuscript:

SPAD	Single-photon avalanche diode
FLIM	Fluorescence lifetime imaging microscopy
EMCCD	Electromultiplied charge-coupled device
SNR	Signal-to-noise ratio
PRNU	Photo-response non-uniformity
DCR	Dark count rate
PDP	Photon detection probability

PDE	Photon detection efficiency
BPAE	Bovine pulmonary artery endothelial cells
DAPI	4′,6-Diamidino-2-phenylindole
IRF	Instrumentation response function
TCSPC	Time-correlated single photon counting
ENOB	Effective number of bits
ICG	Indocyanine green

References

1. Ghioni, M.; Gulinatti, A.; Rech, I.; Zappa, F.; Cova, S. Progress in silicon single-photon avalanche diodes. *IEEE J. Sel. Top. Quantum Electron.* **2007**, *13*, 852–862. [CrossRef]
2. Rochas, A.; Gosch, M.; Serov, A.; Besse, P.A.; Popovic, R.S.; Lasser, T.; Rigler, R. First fully integrated 2-D array of single-photon detectors in standard CMOS technology. *IEEE Photon. Technol. Lett.* **2003**, *15*, 963–965. [CrossRef]
3. Niclass, C.; Sergio, M.; Charbon, E. A single photon avalanche diode array fabricated in 0.35-μm CMOS and based on an event-driven readout for TCSPC experiments. *Proc. SPIE* **2006**, *6372*, 63720S.
4. Webster, E.A.G.; Richardson, J.A.; Grant, L.A.; Renshaw, D.; Henderson, R.K. A single-photon avalanche diode in 90-nm CMOS Imaging technology with 44% photon detection efficiency at 690 nm. *IEEE Electron Device Lett.* **2012**, *33*, 694–696. [CrossRef]
5. Webster, E.A.G.; Grant, L.A.; Henderson, R.K. A high-performance single-photon avalanche diode in 130-nm CMOS imaging technology. *IEEE Electron Device Lett.* **2012**, *33*, 1589–1591. [CrossRef]
6. Veerappan, C.; Charbon, E. A substrate isolated CMOS SPAD enabling wide spectral response and low electrical crosstalk. *IEEE J. Sel. Top. Quantum Electron.* **2014**, *20*, 3801507. [CrossRef]
7. Charbon, E. *Single Photon Imagers*; Proc. CLEO, Sci. Innov.: San Jose, CA, USA, 2014.
8. Becker, W. *The bh TCSPC Handbook*; Becker & Hickl: Berlin, Germany, 2014.
9. Gersbach, M.; Boiko, D.L.; Niclass, C.; Petersen, C.C.H.; Charbon, E. Fast-fluorescence dynamics in nonratiometric calcium indicators. *Opt. Lett.* **2009**, *34*, 362–364. [CrossRef] [PubMed]
10. Gadella, T.W.J.; Jovin, T.M.; Clegg, R.M. Fluorescence lifetime imaging microscopy (FLIM): Spatial resolution of microstructures on the nanosecond time scale. *Biophys. Chem.* **1993**, *48*, 221–239. [CrossRef]
11. Gersbach, M.; Maruyama, Y.; Trimananda, R.; Fishburn, M.W.; Stoppa, D.; Richardson, J.A.; Walker, R.; Henderson, R.K.; Charbon, E. A time-resolved, low-noise single-photon image sensor fabricated in deep-submicron CMOS technology. *IEEE J. Solid-State Circuits* **2012**, *47*, 1394–1407. [CrossRef]
12. Tosi, A.; Villa, F.; Bronzi, D. Low-noise CMOS SPAD arrays with in-pixel time-to-digital converters. *Proc. SPIE* **2014**, *9114*, 91140C.
13. Stoppa, D.; Borghetti, F.; Richardson, J.A.; Walker, R.; Grant, L.; Henderson, R.K.; Gersbach, M.; Charbon, E. A 32 × 32-pixel array with in-pixel photon counting and arrival time measurement in the analog domain. In Proceedings of the European Solid-State Circuits Conference (ESSCIRC'09), Athens, Greece, 14–18 September 2009; pp. 204–207.
14. Veerappan, C.; Richardson, J.A.; Walker, R.; Li, D.-U.; Fishburn, M.W.; Maruyama, Y.; Stoppa, D.; Borghetti, F.; Gersbach, M.; Henderson, R.K.; et al. A 160 × 128 single-photon image sensor with on-pixel 55ps 10b time-to-digital converter. In Proceedings of the 2011 IEEE International Solid-State Circuits Conference, San Francisco, CA, USA, 20–24 February 2011; pp. 312–314.
15. Villa, F.; Lussana, R.; Bronzi, D.; Tisa, S.; Tosi, A.; Zappa, F.; Mora, A.D.; Contini, D.; Durini, D.; Weyers, S.; et al. CMOS imager with 1024 SPADs and TDCs for single-photon timing and 3-D time-of-flight. *IEEE J. Sel. Top. Quantum Electron.* **2014**, *20*, 364–373. [CrossRef]
16. Niclass, C.; Favi, C.; Kluter, T.; Monnier, F. Single-photon synchronous detection. *IEEE J. Solid State Circuits* **2009**, *44*, 1977–1989. [CrossRef]
17. Pancheri, L.; Massari, N.; Borghetti, F.; Stoppa, D. A 32 × 32 SPAD pixel array with nanosecond gating and analog readout. In Proceedings of the International Image Sensor Workshop (IISW), Hokkaido, Japan, 8–11 June 2011.
18. Pancheri, L.; Massari, N.; Stoppa, D. SPAD image sensor with analog counting pixel for time-resolved fluorescence detection. *IEEE Trans. Electron Devices* **2013**, *60*, 3442–3449. [CrossRef]

19. Dutton, N.A.W.; Gyongy, I.; Parmesan, L.; Gnecchi, S.; Calder, N.; Rae, B.R.; Pellegrini, S.; Grant, L.A.; Henderson, R.K. A SPAD-based QVGA image sensor for single-photon counting and quanta imaging. *IEEE Trans. Electron Devices* **2016**, *63*, 189–196. [CrossRef]
20. Li, D.-U.; Arlt, J.; Richardson, J.A.; Walker, R.; Buts, A.; Stoppa, D.; Charbon, E.; Henderson, R.K. Real-time fluorescence lifetime imaging system with a 32 × 32 0.13 microm CMOS low dark-count single-photon avalanche diode array. *Opt. Exp.* **2010**, *18*, 10257–10269. [CrossRef] [PubMed]
21. Stegehuis, P.L.; Boonstra, M.C.; de Rooij, K.E.; Powolny, F.E.; Sinisi, R.; Homulle, H.; Bruschini, C.; Charbon, E.; van de Velde, C.J.H.; Lelieveldt, B.P.E.; et al. Fluorescence lifetime imaging to differentiate bound from unbound ICG-cRGD both in vitro and in vivo. *Proc. SPIE* **2015**, *9313*. [CrossRef]
22. Niclass, C.; Rochas, A.; Besse, P.-A.; Popovic, R.S.; Charbon, E. CMOS imager based on single photon avalanche diodes. In Proceedings of the 13th International Conference on Solid-State Sensors, Actuators and Microsystems, 2005. Digest of Technical Papers, TRANSDUCERS'05, Seoul, Korea, 5–9 June 2005; Volume 1, pp. 1030–1034.
23. Burri, S.; Maruyama, Y.; Michalet, X.; Regazzoni, F.; Bruschini, C.; Charbon, E. Architecture and applications of a high resolution gated SPAD image sensor. *Opt. Exp.* **2014**, *22*, 17573–17589.
24. Maruyama, Y.; Blacksberg, J.; Charbon, E. A 1024 × 8, 700-ps Time-Gated SPAD Line Sensor for Planetary Surface Exploration with Laser Raman Spectroscopy and LIBS. *IEEE J. Solid-State Circuits* **2014**, *49*, 179–189. [CrossRef]
25. Antolovic, I.M.; Burri, S.; Bruschini, C.; Hoebe, R.; Charbon, E. Nonuniformity analysis of a 65-kpixel CMOS SPAD imager. *IEEE Trans. Electron Devices* **2016**, *63*, 57–64. [CrossRef]
26. Bondarenko, G.; Dolgoshein, B.; Golovin, V.; Ilyin, A.; Klanner, R.; Popova, E. Limited Geiger-mode silicon photodiode with very high gain. *Nucl. Phys. B Proc. Suppl.* **1988**, *61*, 347–352. [CrossRef]
27. Knoll, G.F. *Radiation Detection and Measurement*; Wiley: Hoboken, NJ, USA, 2000; pp. 119–127.
28. Sbaiz, L.; Yang, F.; Charbon, E.; Susstrunk, S.; Vetterli, M. The gigavision camera. In Proceedings of the 2009 IEEE International Conference on Acoustics, Speech and Signal Processing, Taipei, Taiwan, 19–24 April 2009; pp. 1093–1096.
29. Fossum, E.R. Modeling the performance of single-bit and multi-bit quanta image sensors. *IEEE J. Electron Devices Soc.* **2013**, *1*, 166–174. [CrossRef]
30. Villa, F.; Bronzi, D.; Bellisai, S.; Boso, G.; Shehata, A.B.; Scarcella, C.; Tosi, A.; Zappa, F.; Tisa, S.; Durini, D.; et al. SPAD imagers for remote sensing at the single-photon level. *Proc. SPIE* **2012**, *8542*, 85420G.
31. Dutton, N.A.W.; Parmesan, L.; Holmes, A.J.; Grant, L.A.; Henderson, R.K. 320 × 240 oversampled digital single photon counting image sensor. In Proceedings of the 2014 Symposium on VLSI Circuits Digest of Technical Papers, Honolulu, HI, USA, 10–13 June 2014; pp. 1–2.
32. Donati, S.; Martini, G.; Norgia, M. Microconcentrators to recover fill-factor in image photodetectors with pixel on-board processing circuits. *Opt. Exp.* **2007**, *15*, 18066–18075. [CrossRef]
33. Donati, S.; Martini, G.; Randone, E. Improving photodetector per- formance by means of microoptics concentrators. *J. Lightw. Technol.* **2011**, *29*, 661–665. [CrossRef]
34. Pavia, J.M.; Wolf, M.; Charbon, E. Measurement and modeling of microlenses fabricated on single-photon avalanche diode arrays for fill factor recovery. *Opt. Exp.* **2014**, *22*, 4202–4213. [CrossRef] [PubMed]
35. Fishburn, M.W. *Fundamentals of CMOS Single-Photon Avalanche Diodes*; Delft University of Technology: Delft, The Netherlands, 2012.
36. Gerega, A.; Zolek, N.; Soltysinski, T.; Milej, D.; Sawosz, P.; Toczylowska, B.; Liebert, A. Wavelength-resolved measurements of fluorescence lifetime of indocyanine green. *J. Biomed. Opt.* **2011**, *16*, 067010. [CrossRef] [PubMed]
37. Homulle, H.A.R.; Powolny, F.; Stegehuis, P.L.; Dijkstra, J.; Li, D.-U.; Homicsko, K.; Rimoldi, D.; Muehlethaler, K.; Prior, J.O.; Sinisi, R.; et al. Compact solid-state CMOS single-photon detector array for in vivo NIR fluorescence lifetime oncology measurements. *Biomed. Opt. Express* **2016**, *7*, 1797–1814. [CrossRef] [PubMed]

sensors

MDPI

Article

Single Photon Counting Performance and Noise Analysis of CMOS SPAD-Based Image Sensors

Neale A. W. Dutton [1,*], Istvan Gyongy [2], Luca Parmesan [2] and Robert K. Henderson [2]

[1] STMicroelectronics Imaging Division, Pinkhill, Edinburgh EH12 7BF, UK
[2] CMOS Sensors and Systems Group, School of Engineering, The University of Edinburgh,
    Edinburgh EH9 3JL, UK; Istvan.Gyongy@ed.ac.uk (I.G.); l.parmesan@ed.ac.uk (L.P.);
    robert.henderson@ed.ac.uk (R.K.H.)
* Correspondence: neale.dutton@st.com; Tel.: +44-131-336-6000

Academic Editor: Eric R. Fossum
Received: 21 January 2016; Accepted: 12 July 2016; Published: 20 July 2016

Abstract: SPAD-based solid state CMOS image sensors utilising analogue integrators have attained deep sub-electron read noise (DSERN) permitting single photon counting (SPC) imaging. A new method is proposed to determine the read noise in DSERN image sensors by evaluating the peak separation and width (PSW) of single photon peaks in a photon counting histogram (PCH). The technique is used to identify and analyse cumulative noise in analogue integrating SPC SPAD-based pixels. The DSERN of our SPAD image sensor is exploited to confirm recent multi-photon threshold quanta image sensor (QIS) theory. Finally, various single and multiple photon spatio-temporal oversampling techniques are reviewed.

Keywords: single photon avalanche diode; SPAD; CMOS image sensor; CIS; single photon counting; SPC; quanta image sensor; QIS; spatio-temporal oversampling

1. Introduction

Imaging a few photons per pixel, per frame, demands pixels operating in the single photon counting regime. This challenge is encountered in either low-light or high-speed imaging; at long (ms to s) integration times and low photon flux, or short (µs or less) integration times and high photon flux, respectively. Examples are high-speed cameras for engine and exhaust combustion analysis, low-light or night-vision cameras for defence [1], staring applications in astronomy and many scientific applications such as, spectroscopy, fluorescence lifetime imaging microscopy (FLIM) [2,3], positron emission tomography (PET) [4], fluorescence correlation spectroscopy (FCS) [5], Förster Resonance Emission Tomography (FRET) [6], and in automotive applications for LIDAR [7].

For true photoelectron (or photon) counting to be reached, the ratio of the input sensitivity or signal to the noise of the imaging system must be sufficiently high to allow discrete and resolvable signal levels for each photoelectron to be discriminated. Referring the readout noise to the input sensitivity in photoelectrons, the single photon counting regime is theoretically entered below 0.5 e$^-$ input referred read noise (RN) [8], but practically there is a 90% accuracy of determining the number of photoelectrons at 0.3 e$^-$ RN, and approaching 100% accuracy at 0.15 e$^-$ RN [9]. These probability figures, assume RN is Gaussian distributed and the discrimination thresholds between one photoelectron signal, to the next, are set precisely mid-way and do not take into account fixed pattern noise (FPN) or gain variations in photo-response non-uniformity (PRNU). Such sensors in this photon-counting regime with approximately <0.3 e$^-$ RN may be referred to as deep sub-electron read noise (DSERN) image sensors [10].

With high charge to voltage factor (CVF) sensitivity (or conversion gain (CG)), DSERN pixels have limited photoelectron or photon counting capability (full well capacity), and therefore restricted dynamic range (DR). DR may be extended by a range of techniques: exposure control with the capture of multiple sequential images [11], pixel design with dual integrations (e.g., lateral overflow integration capacitors (LOFIC) [12]), or by combining multiple pixel samples through spatio-temporal oversampling [13,14]. In the latter the number of oversampled frames is traded off against the frame rate.

This paper evaluates the single photon counting and noise characteristics of our recent work on SPAD-based image sensors [15–17] and analyses the benefits, tradeoffs and noise performance of various spatio-temporal oversampling techniques [18,19]. A new method of determining RN, CVF and other imaging measurements of DSERN image sensors is described.

2. Solid-State Single Photon Counting Imaging Background

Since the late 1980s, single photon counting (SPC) and time-gated imaging have been dominated by photo-cathode based intensifier techniques achieving high signal amplification through the "photo-intensification" of the generated electron cascade through the photo-electric effect using existing charge-coupled device (CCD) and CMOS image sensors (CIS) [1]. However, there are a number of drawbacks which limit their usage dependent on the application. Namely, the wavelength (colour) and spin properties of the photons are lost. Systems have high cost and are physically bulky due to the requirement of operation in a vacuum. Furthermore, photo-cathodes are sensitive to magnetic fields, they have high (kV) operating voltage and also cannot be used in vivo. Solid-state photon counting image sensor technologies, developed over the last 16 years, address some of these issues.

The electron-multiplying CCD (EMCCD) was first demonstrated in 2001 [20], and has recently achieved 0.45 e$^-$ RN [21]. However, dark current is amplified through the electron multiplication process, and therefore external cooling is employed [22]. The first solid-state CIS pixel array with DSERN appeared in 2015, achieving best-case 0.22 e$^-$ RN in a remarkable 1.4 μm pixel pitch (PP) with 403 μV/e$^-$ CVF [10]. Later, the first photon-counting CMOS imager achieved 0.27 e$^-$ RN, by external cooling and a high CVF of 220 μV/e$^-$ was realised by removing the reset transistor [23]. Oversampling ADCs have been employed in CIS to reduce all sources of readout noise (1/f, systematic temporal, source follower thermal, etc.) by correlated multiple sampling (CMS). The lowest published CIS RN in voltage (estimated by the author as CVF multiplied by RN) through four sample CMS is 31.7 μV RMS [24]. Therefore, with CVF surpassing 400 μV/e$^-$ and RN as low as 31.7 μV RMS, CIS with sub 0.15 e$^-$ RN appears not an unreasonable assumption in the near future.

Single photon avalanche diode (SPAD) image sensors emerged in 2002 with bump-bonded SPADs [25] onto a digital counter or time-to-digital converter (TDC) per SPAD device recording the time of arrival of single photons. High temporal resolution ($\approx$50 ps [26]) permits time resolved imaging such as capturing light-in-flight [27], and seeing round corners [28]. These time correlated single photon counting (TCSPC) sensors have favoured the temporal precision of the photon's arrival over spatial resolution (>44 μm) and fill-factor (<4%) which has, so far, restricted the wider adoption of these sensors. The digital circuit providing photon counting or timing occupies the majority of the pixel area to the detriment of photon detection. Chip stacking technology and the use of advanced digital CMOS process technologies are two methods that pitch reduction and fill factor increase will be achieved for SPAD-based image sensors in the future. Regardless of the technology, to realise high fill factor SPAD pixels, a trade-off is made between optical efficiency versus in-pixel functionality or the number of in-pixel transistors; low-transistor count analogue circuits will always be more compact than digital circuits. Our recent research has focused on time resolved photon counting applications using alternative analogue pixel designs that achieve higher fill factor and smaller pixel pitch, namely analogue counters [15], time-to-amplitude converters (TAC) [29,30] and single bit binary memories [17].

Binary SPAD-based imagers, with the capability of recording one SPAD avalanche within an integration time, were first published in 2011 [31] and have recently been published at 65 k binary pixels [32] and in our work at 77 k binary pixels [17]. Binary black and white imaging is not inherently practical for many imaging applications, therefore spatio-temporal oversampling is employed to create gray levels [14,19]. SPAD-based image sensors based on analogue counting techniques first appeared in [33] and have recently been demonstrated with 8 to 15 μm PP commensurate with CCD, EMCCD and sCMOS image sensors, and fill factor (FF) as high as 26.8% [15,16,34]. These sensors achieve time-gating comparable to gated photo-cathodes in the nanosecond [18] and sub nanosecond range [34]. Analogue-based SPAD imagers employ conventional CIS readout techniques and so, to aid comparison with CCD and CIS, equivalent metrics may be applied such as:

- Sensitivity, of the counter circuit to one SPAD avalanche event in mV/SPAD event, equivalent to CVF (or CG).
- Maximum number of SPAD events equivalent to full well.
- Input referred RN normalising voltage RMS RN to one SPAD event instead of one photoelectron.

These equivalencies are used throughout this paper. SPAD-based image sensors are the first solid-state imaging technology to have demonstrated sub 0.15 e⁻ RN, and as such provide a look-ahead to the signal and noise characteristics of DSERN image sensors in CMOS and other technologies.

3. Single Photon Counting Noise Modelling and Analysis

The first part of this section details a model of read noise and sensitivity (or CVF) developed to characterise our recent work in SPAD-based imaging. The second part discusses three noise measurement methods for DSERN image sensors based on the photon counting histogram (PCH). The use of single photon counting histograms are not new to the imaging community but the analysis presented here seeks to model and quantify the noise measurements that may be obtained from the PCH. A discrete Poisson probability density function (PDF) may represent photoelectrons (or photons) either from multiple reads of a single pixel or a single read of multiple pixels. For a single pixel "i", the PDF for the captured photoelectrons k may be represented as:

$$P(i,k) = \frac{\lambda^k exp(-\lambda)}{k!} : k \in \mathbb{Z} \tag{1}$$

where λ = mean number of photoelectrons in the integration period. PRNU may be modelled to first order as a normal distribution with mean CVF μ_{CVF} and variance σ_{CVF}^2. For each electron k, the ideal voltage domain input signal S_{IN} is created with the signal from each electrons at a separation $v_{(i,k)}$ equal to the CVF for that pixel "i":

$$v(i,k) = k \cdot CVF(i) \tag{2}$$

$$S_{IN}(v_{k,i}) = P(i,k) \tag{3}$$

For each electron k, assuming the read noise is dominated by thermal noise it follows a Gaussian distribution. Read noise σ_{RN} is applied on each electron's output signal S_k for the range $v = 0$ to $(n.CVF)$: where n is the maximum number of electrons in the Poissonian PDF in Equation (1):

$$S_k(v) = \frac{1}{\sigma_{RN}\sqrt{2\pi}} \cdot exp\left(-\frac{(v - S_{IN}(v_k))^2}{2\sigma_{RN}^2}\right) \tag{4}$$

The voltage domain output signal is then represented as the summation of each of the constituent signals for each electron within the PDF:

$$S_{OUT}(v) = \sum_{k=0}^{n} S_k(v) \tag{5}$$

Figure 1 provides a photon counting histogram (PCH) example of the output of the model given by Equation (5) with 10 mV/SPAD event (or 10 mV/e⁻ equivalent) and 0.1 e⁻ equivalent RN. As seen in the figure, discrete peaks are visible in the PCH. The RN distribution around each photon counting peak can be determined using three recent methods:

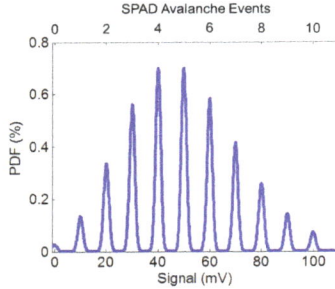

Figure 1. Photon counting histogram (PCH) generated by the read noise model with CVF equivalent of 10 mV/e, mean $\lambda = 5$ e$^-$ exposure and 0.1 e$^-$ equivalent RN.

3.1. Valley to Peak Ratio Method

Fossum et al. proposed the Valley to Peak ratio Method (VPM) detailed in [10,35]. This measures the peak height and the neighbouring valley height (or dip between photon peaks) in the PCH. The VPM has an upper and lower RN measurement limit. Although theoretically possible, it is difficult in practice to obtain peaks and valleys in PCHs in the region of 0.5 e$^-$ to $\approx$0.45 e$^-$ RN giving an upper limit to VPM. At the lower limit, below 0.15 e$^-$ RN, the VPM is inherently restricted as the valley has reached the "floor" of the PCH (zero counts in more than one adjacent bin), and a companion method is needed.

3.2. Peak Separation and Width Method

The Peak Separation and Width (PSW) method is proposed in this paper, and has been used in this paper to measure the SPAD-based image sensors in our recent work [15,16]. The previous VPM measurement evaluates vertically in the PCH, whereas this PSW method operates in the voltage domain or horizontally in the PCH. By determining, the centroid of each single photon counting peak (whether by taking the peak position, or using a centroid weight algorithm, or similar), the peak separation data may provide a number of measurements:

- The sensitivity or CVF per pixel ("i") is established by mean peak separation in a per-pixel PCH.
- The PRNU and the average CVF of the sensor are evaluated through a histogram of the compiled peak separation data from step 1 above, taking RMS and mean respectively.
- Vertical, horizontal and pixel to pixel FPN (VPFN, HFPN, PPFPN) are exhibited as horizontal offsets to the peaks, in the set of per pixel PCHs.

The width of each peak is measured to deduce the noise characteristics of the sensor. The full width half maximum (FWHM) of each peak is captured (preferably using interpolative fitting between PCH bins to lessen errors from quantisation and non-linearity in calculations). Assuming the noise around each peak is normally distributed, the FWHM may be converted to standard deviation using the conventional expression:

$$\sigma = \frac{FWHM}{2\sqrt{2\ln 2}} \rightarrow \sigma \approx \frac{FWHM}{2.3548} \tag{6}$$

The interested reader may create a more complete noise model by expanding Equations (4) and (6) to take into account other read noise sources (reset, flicker, etc.). Ideally the peak width remains constant across the full signal range, and RN is determined by the mean of the peak width data. However, if a signal dependent noise source is present then the peak widths will increase (and peak heights decrease) for increasing signal. There is no lower limit to the PSW method. However, the upper limit is set by the height of the valley between two peaks: by definition this valley must be lower than half of the two adjacent peak heights which evaluates at <0.3 e$^-$ RN approximately.

3.3. Regressive Modelling and Fitting Method

The third method fits and scales the noise model described above, against a PCH (whether a single exposures of a full sensor or multiple exposures per pixel). This method has been used in [23] to graphically confirm the correct evaluation of RN and mean exposure. This method is expanded here to encompass the previous two methods. First the VPM and PSW are used (as appropriate given their respective limits) to obtain an estimate of RN and CVF to restrict the scaling and fitting "search" domain. Next the iterative process begins, recording the goodness of fit of the recorded PCH to the modelled PCH and continuing the regression analysis (by whichever chosen fitting method).

Like the PSW method, this regression analysis should be performed per pixel to obtain the CVF, PRNU and FPN distributions of the image sensor. Furthermore, as in PSW, ADC non-linearity will affect the regression analysis so some method of interpolation between PCH bins may be necessary. The downside to this method, is its computationally intensive nature and the requirement to have a consistent mean number of photons for exact fitting. The Poisson distribution in Equation (1) assumes a constant mean number of photoelectrons (i.e., constant light level) through successive reads of a single pixel, and a constant light level across the array with equal sensitivity (0% PRNU). The advantage of the method is that the model can be expanded to account for known converter non-linearity or other noise sources, such as described in the following sections.

4. Analogue Counter and Photon Counting Performance

Single photon counting is achieved in the analogue domain with a SPAD avalanche pulse triggering an integrator circuit based on the principle of the charge transfer amplifier (CTA) whose operation is briefly described here, and in further detail in [15,16].

In reference to Figure 2, the SPAD is connected to a passive quench and recharge transistor with static DC bias voltage "V_Q" controlling the recharge or "dead" time of the diode. This is connected to the analogue counting CTA circuit via a two transistor global shutter time gate. The CTA is reset by pulling the main capacitor "C" to the high reset voltage V_{RT}. The CTA operates by the input gate voltage (in this case the SPAD anode voltage) increasing above the threshold voltage of the input source follower. Charge flows from the main capacitor "C" to the parasitic capacitor "C_P" and the voltage rises on the parasitic node. The rising voltage pushes the source follower into the cut-off region and the charge flow halts, causing a discrete charge packet to be transferred from the main capacitor for each input pulse. The SPAD anode begins recharging and the lower transistor in the CTA discharges the parasitic capacitance which is achieved with a static bias voltage "V_{DC}" applied keeping this transistor, below threshold, in weak inversion.

The voltage step sensitivity (CVF equivalent) of CTA pixels is determined by the fixed capacitor ratio (parasitic capacitance "C_P" divided by integration capacitor "C") scaling down the input voltage spike. The CTA voltage step ("ΔV_{CTA}") is bias controllable by "V_{SOURCE}" and given to a first order by the equation:

$$\Delta V_{CTA} = \left(\frac{C_P}{C}\right) \cdot (V_{EB} - V_{SOURCE} - V_{TH}) \tag{7}$$

where V_{EB} is the excess bias of the SPAD above the breakdown voltage V_{BD}, V_{SOURCE} is the global CTA source bias voltage, and V_{TH} is the threshold voltage of the CTA input transistor.

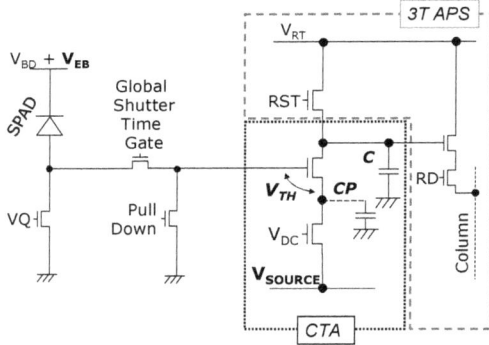

Figure 2. Charge transfer amplifier (CTA) analogue integrator pixel with active pixel sensor (APS) readout for global shutter or time-gated SPAD-based photon counting imaging.

Figure 3a illustrates an example of the output of one test structure pixel recorded with 1000 repetitions of 30 μs integration time and ADC conversion from [15]. 1000 repetitions were chosen to give an adequate number of data samples versus experimental time. The SPAD is biased at 2.7 V V_{EB} above breakdown voltage $V_{BD} \approx 13.4$ V. The discrete peaks under a classical Poisson distribution are clearly evident indicating the photon counting in this example is shot noise limited. Figure 3b is the side-by-side modelled PCH from a manual regressive modelling and fitting method analysis. The parameters were chosen for the closest found fit, although an offset in the *x*-axis is still present. In Figure 3a, there is a slight "in-filling" of some data values between the peaks. This is attributed to a distortion mechanism in the passively operated CTA circuit due to the imperfect reset, or incomplete discharge, of the parasitic capacitance C_P for short inter-arrival times of two SPAD avalanche events less than 100 ns apart.

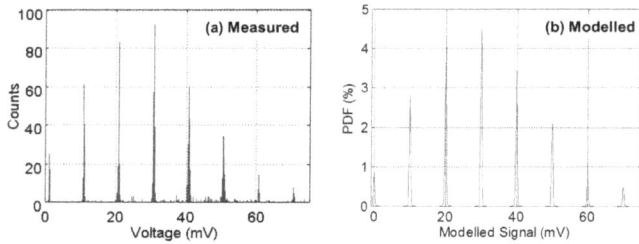

Figure 3. (**a**) Measured PCH of the analogue counting pixel test structure in [15]; (**b**) Modelled PCH with mean λ = 3 SPAD events, CVF equivalent of 10 mV/SPAD event and equivalent 0.02 e$^-$ RN.

The PSW method is performed for the image sensor in [16] to determine the response of the analogue counter to the SPAD excess bias and the source bias voltage. Figure 4 illustrates the relationship of the mean peak separation or image sensor sensitivity to both the SPAD excess bias and the CTA source voltage. The absolute value of the linear gradient fitting parameter indicates the capacitor ratio whilst the offset parameter indicates the other terms in the CTA equation.

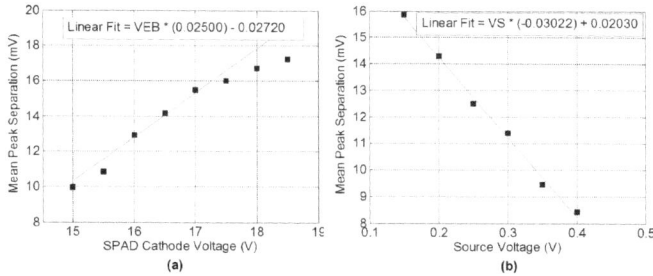

Figure 4. Measured mean peak separation from a set of PCHs, (**a**) The relationship of counter sensitivity to SPAD operating voltage; (**b**) The relationship to CTA V_{SOURCE} voltage.

The linear full well (defined as a deviation of 3% in sensor output from an ideal linear response) is measured against the CTA V_{SOURCE} bias, and the data are presented in Table 1. This demonstrates the trade-off of increasing full well against lower sensitivity and increasing RN.

Table 1. Photon counting performance of 320 × 240 SPAD-based image sensor [16].

V_{SOURCE} Bias Voltage (mV)	Linear Full Well Voltage (mV)	Sensitivity from Linear Fit (mV/SPAD Event)	Input Referred Read Noise (SPAD Events)	Equivalent Linear Full Well (SPAD Events)
200	802.8	14.26	0.064	56
300	722.1	11.23	0.082	64
400	651.4	8.21	0.113	79
500	648.3	5.19	0.178	125

5. Analogue Counter Cumulative Noise

Through noise measurement and iterative modelling, it is established that the analogue integrator circuits employed in SPAD-based counting pixels suffer from cumulative noise. For each SPAD event, noise affecting the counter circuit modulates the circuit sensitivity, and as the pixel integrates, the noise cumulates. Although the passive CTA pixel suffers from the "in-filling" distortion mechanism described in the previous section, all analogue integrator structures such as CTAs or switched current sources (SCS) [36,37] circuits will suffer from cumulative noise to a certain degree. The two main sources of cumulative noise are thermal noise through the switched path (which exhibits as a kT/C noise on the in-pixel capacitor, with the SPAD dead time, or counter switch time, controlling the thermal noise bandwidth) and systematic temporal noise on the common supplies. Of course, for long integration times, $1/f$ noise in the counter circuit and low frequency temporal noise on the common supplies will also modulate the integrator sensitivity and contribute cumulative noise.

The PSW method is employed on one pixel in the test array in [15] to evaluate for this cumulative and signal dependent noise source. Multiple experiments were captured (each with an individual PCH as seen in Figure 3), and for each experiment the integration time (from 1 μs to 100 μs) was increased to obtain greater number of SPAD events. An example of the combined PCH is modelled in Figure 5a. Figure 5b extracts the increasing peak width indicating the presence of a cumulative noise source (σ_C) from measured data. A linear fit (solid black line) identifies an σ_C = 86.9 μV RMS noise increase per SPAD event. The model shown in Figure 5a is matched with 86.9 μV RMS noise per counter step and the modelled FWHM response is shown alongside (dashed red line) in Figure 5b.

The cumulative noise modelled response S_N after N steps can be modelled to first order by expanding Equation (4) into an iterative expression assuming the cumulative noise is Gaussian. The initial reset level S_0 (N = 0) is assumed constant with no FPN and no noise terms (a Dirac function). The first modelled counter step S_1 has σ_C cumulative noise applied. The second step S_2 is the convolved response of the first counter step with the same Gaussian cumulative noise, and so on, as an iterative convolution for subsequent counter steps as shown in Equation (8):

$$S_N(v_N) = \frac{1}{\sigma_C\sqrt{2\pi}} \cdot exp\left(-\frac{(v_N - S_{N-1}(v_{N-1}))^2}{2\sigma_C{}^2}\right) \tag{8}$$

where the v_N represents the voltage range of interest.

The same PSW procedure is performed for the full 320 × 240 image sensor in [16]. The imager has 700 µV RMS noise per SPAD event, an increase of approximately 8 times. This is attributed to an increase of kT/C noise due to both the main and parasitic capacitors decreasing in size between the sensors, the capacitance ratio increasing from approximately doubling from 0.013 to 0.03, and an increase in temporal noise due to many more pixels active on the same supplies. Although it is noted, that some fraction of the increase may also be attributed to ≈1% PRNU which would manifest similarly with a ≈100 µV RMS broadening of the peaks per counted photon.

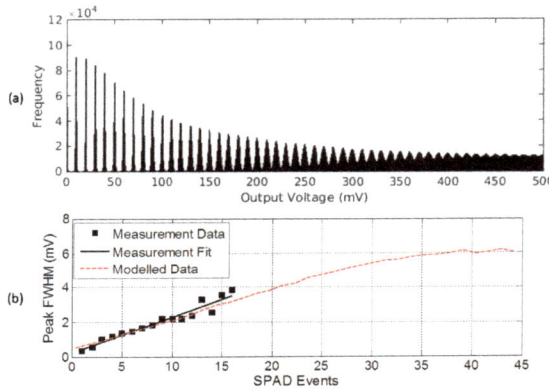

Figure 5. (**a**) Modelled multiple exposure PCH of a signal dependent cumulative noise source in the SPAD-based analogue counter structure [15]; (**b**) Measured and modelled peak FHWM, the first order linear fit has parameters: offset 204.7 µV with cumulative noise FWHM 225.9 µV/SPAD event = 86.9 µV/SPAD event RMS. The modelled data has cumulative noise 86.9 µV/SPAD event applied.

Figure 6a gives an example PCH from the imager. Figure 6b is the PCHs of the noise model applying 700 µV RMS cumulative noise and 0.06 e⁻ RN, and Figure 6c applying only RN. Figure 6b has a much closer fit to the captured PCH, whereas Figure 6c indicates the shape of a PCH that a CIS DSERN sensor with 0.06 e⁻ RN should achieve. With such a cumulative noise source, the equivalent input referred read noise increases depending on exposure. Table 2 presents the signal against the equivalent input referred noise figures for both the imager and test structure.

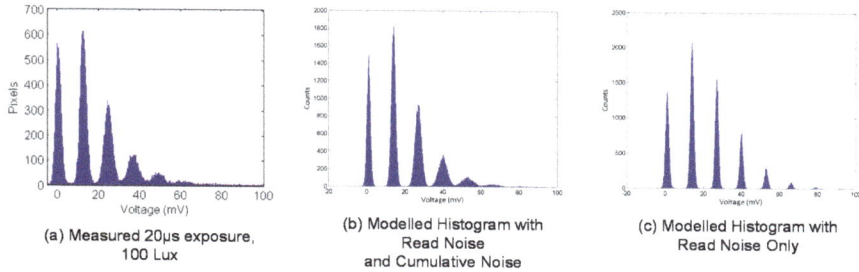

(a) Measured 20µs exposure, 100 Lux

(b) Modelled Histogram with Read Noise and Cumulative Noise

(c) Modelled Histogram with Read Noise Only

Figure 6. (a) Measured PCH for all pixels in the 320 × 240 image sensor in [16]; (b) Modelled PCH (mean $\lambda = 1.5\,e^-$) accounting for both cumulative noise and read noise showing close fit to the measured PCH; (c) Modelled PCH with read noise only showing a different response.

Table 2. Equivalent noise at a range of SPAD events.

Equivalent Input Referred Total Noise	No. of SPAD Avalanche Events	
	Image Sensor [16]	Test Structure [15]
$0.15\,e^-$	2	19
$0.3\,e^-$	5	45
$1\,e^-$	19	160

6. Spatio-Temporal Oversampling of Photon Counting Pixels

As analogue SPAD pixels suffer from increasing cumulative noise at higher photon counts and the effective full well is restricted, oversampling individual frames at low photon counts provides a means to create an image of high dynamic range with low overall noise. This section addresses trade-offs, and details different methods, of spatio-temporal oversampling of photon counting pixels. The Quanta Image Sensor (QIS) framework proposed by Fossum [38], extrapolates the imaging trends of pixel shrink, increasing CVF, decreasing RN, decreasing full well and spatio-temporal oversampling to a concept of a SPC image sensor where a "pixel" is the spatio-temporal sum of multiple integrations of multiple sub-pixels ("jots").

The small full well of photon counting pixels, in the order of magnitude of 100's of photoelectrons or photons, limits a sensor's dynamic range. Spatio-temporal oversampling of multiple pixels may be performed to increase the full well past a single pixel's limit. Furthermore, for DSERN photon counting pixels with cumulative noise such as the SPAD-based analogue pixels described in this paper, the level of the photon counting oversampling threshold (i.e., if pixel output >1 photon or if >2 photons, etc.) sets the noise of the oversampled output image; a higher oversampling threshold induces greater noise in the output frame image. However, this threshold is traded off against the frame rate and the oversampled full well. A signal level of N photoelectrons can be reached with less oversampled frames (and greater output frame rate) with a higher oversampling threshold of the pixel signal. By setting the threshold above the thermal and $1/f$ noise floor, the oversampled is truly shot noise limited as little or no thermal and $1/f$ noise accumulates.

6.1. Single Photon Binary Quanta Imaging

Using SPAD-based single photon image sensors with binary response, a variety of oversampling techniques have been evaluated in our recent work [16–19] and in the work of others [14,39]. "Field" images are individual reads from the image sensor and the oversampled frame is a summation of fields. The simplest technique in order to oversample a set of binary single photon field images, is to temporally or spatially sum a set of input binary pixel (or "jot") values, to create an output "macro" pixel with grey levels. This is the equivalent operation of a first-order low-pass infinite impulse response (IIR) filter with a periodic filter reset operation as shown in Figure 7a. Considerin temporal oversampling only (as demonstrated in [17]), to achieve a certain output frame rate in FPS, with oversampled ratio OSR and input binary field rate f, the output rate is: FPS = f/OSR and inversely the IIR reset period = OSR/f, thus attaining an output bit depth of B = Log2(OSR), increasing the image bit depth by a factor of OSR or 2^B. It is clear that to attain frame rates >30 FPS, at bit depths B > 5 bit, a high field rate f > 1 k fields/s is required from the sensor. In [17], we demonstrated 7b bit depth at 40 FPS, and 8b at 20 FPS with 5.12 k global shutter fields per second.

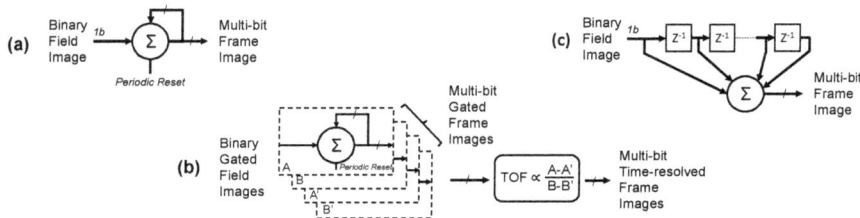

Figure 7. Per-pixel spatio-temporal oversampling techniques. (**a**) Intensity image using IIR with periodic reset [17]; (**b**) Time-resolved image: four IIR per pixel [18]; (**c**) High frame rate intensity image using first-order FIR per pixel [19].

SPADs with picosecond temporal precision enable Indirect Time of Flight (ITOF) imaging to be performed. Previous examples are pulsed ITOF using analogue pixels [33] and continuous wave ITOF using digital pixels [40]. However, both approaches had very large pixel pitch and low fill factor. A similar oversampling technique was applied in [18] with compact binary SPAD pixels, to investigate time-gated binary image oversampling to produce a high resolution QVGA Indirect Time of Flight (ITOF) output image as shown in Figure 7b. Two primary gated field images (A & B) are sequentially captured in interleaved fashion synchronous to a pulsed laser. Two secondary gated images (A' & B') are set with the same time-gate without the laser for background removal. With four field images, the output time-resolved frame rate is therefore a quarter of the previous intensity-only technique (assuming a pipelined division operation).

A third technique in [19], addresses the low frame rate, and evaluates a continuous-time moving average operation by applying a first-order low-pass finite impulse response (FIR) filter. As shown in Figure 7c, the FIR is implemented as a shift-register of length equal to the over-sampling ratio (i.e., a FIR with number of taps = OSR) and a tracking counter. The benefit of this technique is the output frame rate has no relationship with the OSR and is equal to the input field rate of the sensor. The frame rate increase over the IIR technique is at the cost of the shift register per pixel. Longer integration time increases temporal blur, therefore, higher OSR increases image lag of fast moving scene elements. On the other hand, an increased bit depth (from greater OSR) decreases quantisation noise in areas of slow movement in an imaged scene.

We compare our recent work in this area, to two others demonstrating high binary field rates with column parallel single bit flash ADCs for single bit QIS in Table 3. In a 3T CIS implementation [41], amplification and CDS is employed and suitable for pixels with low signal swing (i.e., CVF ⩽ input-referred offset and read noise). In our work [16] and another SPAD-based example [42], no CDS or column amplifier circuits are required as the pixel sensitivity is >1 V/SPAD event which is much greater than offsets and RN. The RN and non-linear exposure characteristics of such oversampled binary imagers are theoretically described in [9] and experimentally confirmed in our work in [16,18]. The measured bit error rate is 0.0017 providing an equivalent DSERN of 0.168 e⁻ Without CDS timing and increased column current, the field rate more than doubles [16].

Table 3. Binary capture, oversampled output, quanta image sensor comparison table. FOM† = Sensor power/(No. of Pixel × FPS × N), where N = ADC resolution = 1b for these sensors.

Reference	[41]	[42]	[16,17]
Sensor Name	QIS Pathfinder	SwissSPAD	SPC Imager
Process Technology	180 nm CMOS	0.35 μm HV CMOS	130 nm Imaging CMOS
Array Size	1376 × 768	512 × 128	320 × 240
Photo-detector	"Pump-gate Jot" PD	SPAD	SPAD
NMOS Pixel Transistors	3	11	9
Fill Factor (%)	45	5	26.8
Pixel Pitch (μm)	3.6	24	8
Microlensing	N	Y (12× concentration factor)	N
Shuttering	Rolling	Global	Global
CDS	True CDS	None	None
Parallel Data Channels	32	128	16
Max. Field Rate (FPS)	1000	150,000	20,000
Sensor Data Rate	1 Gbps	10.24 Gbps	1.54 Gbps
Pixel CVF or Equivalent	120 μV/e^-	>1 V per SPAD Event	>1 V per SPAD Event
Bit Error Rate	Not Reported	Not Reported	1.7×10^{-3} BER
Read Noise (e^-) or Equivalent	Not Reported	Not Reported	0.168 e^-
Power During Operation	20 mW	1650 mW	40.8 mW
Power FOM†	2.5 pJ/b (ADC only) 19 pJ/b (Full Sensor)	168 pJ/b (Full Sensor + SPADs)	104 pJ/b (Full Sensor + SPADs)

6.2. Multi-Photon Binary Quanta Imaging

As previously discussed, setting the oversampling threshold greater than a single counted photon provides a benefit to output frame rate assuming the sensor output data rate remains the same. By setting the oversampling threshold at two photons rather than one, half number of field readouts are required to reach a certain oversampled signal level as each binary bit now represents more than one photon. However, for the SPAD-based analogue counting pixel this is at the cost of oversampling greater cumulative noise, FPN or PRNU with each successive field image.

An experiment is performed on the image sensor [16] recording the "bit density" (the number of pixels outputting a logical high indicating the multi-photon counting threshold is reached) against increasing integration time for a fixed light level. The pixel array in configured in analogue counting CTA mode with V_{SOURCE} = 0.15 V. Figure 8 highlights the normalized bit density (D) to normalized exposure (H), where 1.0 H = 5 μs integration time, for an incrementing comparator threshold capturing two to eight photons. The theoretical curves from [9] are plotted alongside for comparison. As no CDS is implemented, the high FPN due to column comparator mismatch and source follower threshold variation will effectively induce a PRNU in the measured data for all pixels which is seen as the discrepancy between ideal and measured data particularly in the plotted line for the four photon threshold. The closest fit in terms of photon number (2 to 8) is listed in the legend alongside.

Figure 9 is the measured normalized RMS noise which has the characteristic shape from Fossum's theoretical Quanta Image Sensor paper in [9]. A few remarkable characteristics of multi-photon threshold binary imaging that are experimentally verified in this noise plot. The rising slope of each of the noise plots indicates the shot-noise dominant region. The 2-photon line demonstrates the "soft-knee" shot noise compression with a smooth roll-off after the peak after H = 1.0 as expected in 1-photon or 2-photon threshold QIS. The subsequent increasing thresholds show a horizontal shift in the exposure x-axis as a higher number of photons (or equivalent SPAD events) are required to trigger the binary output. This can also be observed in the horizontal shift in the D-LogH plot in

Figure 8. The maximum noise in the 8-photon threshold is measured as 1.52 times higher than the 2-photon threshold where the theory [9] suggests it should be no more than square root of two higher (1.412 times).

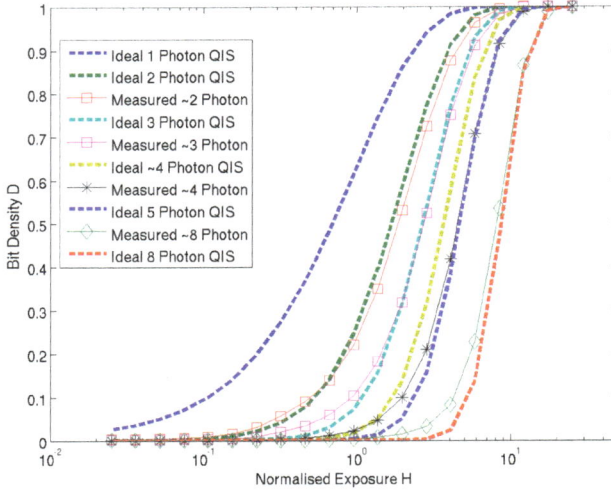

Figure 8. Multi-photon threshold oversampled binary imaging normalised bit density to exposure. Ideal curves from [9] are presented alongside measured results.

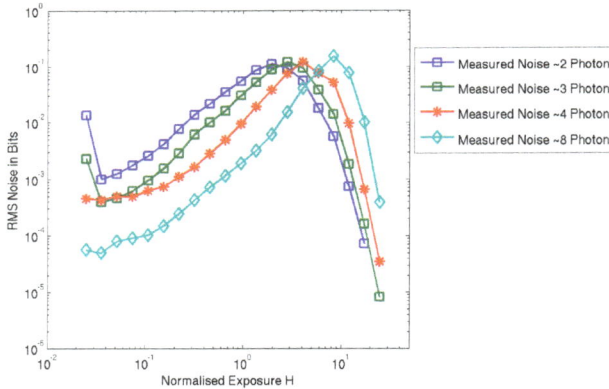

Figure 9. Measured RMS noise in multi-photon threshold oversampled binary imaging.

7. Discussion

Table 4 provides a comparison table highlighting a selection of state of the art solid-state photon counting image sensors in the three different technologies (CIS, EMCCD and SPAD). This section discusses and compares the performance of SPAD-based image sensors based on analogue integration. SPAD based image sensors have the highest CVF of solid-state SPC image sensors. Moreover, the pixel size of the SPAD analogue-based imagers is commensurate with EMCCD and sCMOS scientific imagers, although FF is lower. With the exception of the LOFIC pixel which has dual CVF's, like the recent CIS DSERN pixels, the increase in CVF of SPAD pixels yields a reduced full well in the order of 100's photo-electrons or integrated SPAD events.

Table 4. Solid state single photon counting image sensor comparison table.

Reference	[12]	[10]	[23]	[21]	[37]	[15]	[16]	[26]
Photodetector	PIN PD + LOFIC	"Pump-gate Jot" PIN PD	PIN PD	EMCCD	SPAD	SPAD	SPAD	SPAD
Pixel Circuit	5T + LOFIC	4T	4T	CCD	Active CTA 8T	Passive CTA 11T	Passive CTA 9T	7b Counter >100T
Array Size	1280 × 960	1	35 × 512	1920 × 1080	160 × 120	3 × 3	320 × 240	32 × 32
Pixel Size (μm)	5.6	1.4	11.2 × 5.6	5.5	15	9.8	8	50
Fill Factor (%)	30.4	-	-	50	21	3.12	26.8	1
Pixel CVF or Equivalent	240 μV/e$^-$	403 μV/e$^-$	220	Gain dependent from 44 μV/e$^-$	16.5 mV/SPAD event	13.1 to 2 mV/SPAD Event	17.4 to 8.4 mV/SPAD event	1 DN/SPAD Event
Full Well or Equivalent	200 ke$^-$	210 e$^-$	1500 e$^-$	20 ke$^-$ to 160 e$^-$	41	80 to 360	56 to 125	127
Read Noise (or Equivalent)	0.41 e$^-$	0.22 e$^-$	0.27 e$^-$	0.45 e$^-$	0.08 e$^-$	<0.01 e$^-$ to 0.22 e$^-$	0.06 e$^-$ to 0.18 e$^-$	0
Excess Noise	-	-	-	Y	-	Y	-	-
Cumulative Noise	-	-	-	-	Y*	Y	Y	-
Measured Cumulative Noise	-	-	-	-	Not Measured	86.9 μV RMS/SPAD Event	700 μV RMS/SPAD Event	-
Time Gating Width or Temporal Resolution	-	-	-	-	0.75 ns	100 ns	30 ns	52 ps

* As based on a CTA analogue integrator structure, the presence of cumulative noise is assumed by the author.

SPADs have the advantage of picosecond temporal resolution. Analogue pixels with low transistor counts permit nanosecond and sub-nanosecond time-gated SPC imaging to be realized where digital pixels further permit TCSPC imaging with 10's ps time resolution at the cost of low spatial resolution.

In terms of RN, SPAD analogue integrators share a similar noise characteristic with 3 transistor (3T) CIS pixels in that the integration node is not fully depleted and so suffers from kT/C noise. Our test structure [15] cancels the kT/C noise by implementing 3T-pixel true CDS timing and furthermore implemented 4096 sample CMS to yield <0.01 e^- equivalent RN in the best case. However, both 3T timing and >1 k sample CMS is very restrictive in an image sensor design preventing, for example, the global shutter or global time-gated operation that our recent work and [37] implements. Therefore delta-reset sampling CDS [43] is implemented in our SPAD analogue counter image sensor which adds a noise component of 100's μV RMS kT/C to the RN. However, the equivalent CVF of the SPAD-based analogue pixels in the 10 mV range is high enough to compensate, as demonstrated by the 0.06 e^- RN figure which is the lowest in the published SPC image sensor literature.

In comparison to other works, analogue integrators suffer from cumulative noise limiting the photon number resolution. Spatio-temporal oversampling, at a few photons per pixel level, mitigates the noise integration whilst extending the photon number resolution although high frame rates are required.

8. Conclusions

Our recent work on SPAD-based photon counting image sensors is analysed for photon counting performance and deep sub electron equivalent noise characteristics. A noise model is developed to include both CIS RN and the cumulative noise specific to analogue integrator circuits. When combined, the three new methods (VPM, PSW and regressive analysis) of determining RN form a new powerful set of tools for the measurement of most SPC and DSERN image sensor characteristics alongside the existing techniques such as photon transfer curve analysis.

These single-photon and multi-photon methods of binary image capture have the attractive quality of similar noise and signal characteristics of photographic film. Future development of these binary photon-counting image sensors is an interesting and new avenue of research. The tradeoff between in-pixel cumulative and spatio-temporal oversampling is examined. Analogue SPC pixels have DSERN but exhibit cumulative noise limiting photon number resolution. As a result they are best operated at low photon number in combination with digital oversampling. A very large dynamic range is conceivably possible, combining the multi-photon counting with an oversampled frame store, which would extend the limited dynamic range of the analogue counter. Furthermore, the frame rate penalty of oversampling is addressed by a continuous-time moving average technique.

The capability of an image sensor to capture the arrival of a single photon, is the fundamental limit to the detection of quantised electromagnetic radiation. Each of the three solid-state SPC image sensor technologies, CMOS SPAD, EMCCD and DSERN CIS have specific advantages that will individually serve a variety of photon counting applications.

Acknowledgments: The following people are gratefully acknowledged for their support in this work: Lindsay Grant, Bruce Rae, Sara Pellegrini, Tarek Al Abbas, Graeme Storm, Kevin Moore, Pascal Mellot, Salvatore Gnecchi, and all our co-authors in our recent works. Thanks to ST Crolles for silicon fabrication. This work is primarily supported by STMicroelectronics Imaging Division and the research leading to these results has received funding from the European Research Council under the EU's Seventh Framework Programme (FP/2007-2013)/ERC Grant Agreement n.339747.

Author Contributions: N.D., I.G. and R.H. conceived and designed the experiments; N.D. and I.G. performed the experiments; N.D. modelled and analyzed the data; N.D. and R.H. conceived and designed the pixel test structure, N.D., L.P. and R.H. conceived and designed the image sensor. N.D. wrote the paper; N.D., L.P., and R.H. contributed edits to the paper.

Conflicts of Interest: The authors declare no conflict of interest. The founding sponsors had no role in the design of the study; in the collection, analyses, or interpretation of data; in the writing of the manuscript, and in the decision to publish the results.

Abbreviations

The following abbreviations are used in this manuscript:

CG	Conversion Gain
CIS	CMOS Image Sensor
CTA	Charge Transfer Amplifier
CVF	Charge to Voltage Conversion Factor
DSERN	Deep Sub Electron Read Noise
EMCCD	Electron Multiplied Charge Coupled Device
FIR	Finite Impulse Response Filter
IIR	Infinite Impulse Response Filter
PCH	Photon Counting Histogram
PSW	Peak Seperation and Width method
QIS	Quanta Image Sensor
RN	Read Noise
SPAD	Single Photon Avalanche Diode
SPC	Single Photon Counting
TCSPC	Time Correlated Single Photon Counting
VPM	Valley to Peak method

References

1. Seitz, P.; Theuwissen, A. *Single Photon Imaging*, 1st ed.; Springer: Heidelberg, Germany, 2011.
2. Dutton, N.A.W.; Gnecchi, S.; Parmesan, L.; Holmes, A.J.; Rae, B.; Grant, L.A.; Henderson, R.K. A Time Correlated Single Photon Counting Sensor with 14 GS/s Histogramming Time to Digital Converter. In Proceedings of the IEEE International Solid-State Circuits Conference—ISSCC Digest of Technical Papers, San Francisco, CA, USA, 22–26 February 2015.
3. Li, D.-U.; Walker, R.; Richardson, J.; Rae, B.; Buts, A.; Renshaw, D.; Henderson, R. FPGA implementation of a video-rate fluorescence lifetime imaging system with a 32 × 32 CMOS single-photon avalanche diode array. In Proceedings of the 2009 IEEE International Symposium on Circuits and Systems, Taipei, Taiwan, 24–27 May 2009; pp. 3082–3085.
4. Braga, L.H.C.; Gasparini, L.; Grant, L.; Henderson, R.K.; Massari, N.; Perenzoni, M.; Stoppa, D.; Walker, R. A Fully Digital 8 × 16 SiPM Array for PET Applications With Per-Pixel TDCs and Real-Time Energy Output. *IEEE J. Solid-State Circuits* **2014**, *49*, 301–314. [CrossRef]
5. Burri, S.; Powolny, F.; Bruschini, C.E.; Michalet, X.; Regazzoni, F.; Charbon, E. 65 k pixel, 150 k frames-per-second camera with global gating and micro-lenses suitable for life-time imaging. In Proceedings of the SPIE Photonics Europe, Brussels, Belgium, 14–17 April 2014.
6. Poland, S.P.; Krstajić, N.; Monypenny, J.; Coelho, S.; Tyndall, D.; Walker, R.J.; Devauges, V.; Richardson, J.; Dutton, N.; Barber, P.; et al. A high speed multifocal multiphoton fluorescence lifetime imaging microscope for live-cell FRET imaging. *Biomed. Opt. Express* **2015**, *6*, 277–296. [CrossRef] [PubMed]
7. Niclass, C.; Ito, K.; Soga, M.; Matsubara, H.; Aoyagi, I.; Kato, S.; Kagami, M. Design and characterization of a 256 × 64-pixel single-photon imager in CMOS for a MEMS-based laser scanning time-of-flight sensor. *Opt. Express* **2012**, *20*, 11863–11881. [CrossRef] [PubMed]
8. Teranishi, N. Required Conditions for Photon-Counting Image Sensors. *IEEE Trans. Electron Devices* **2012**, *59*, 2199–2205. [CrossRef]
9. Fossum, E.R. Modeling the Performance of Single-Bit and Multi-Bit Quanta Image Sensors. *IEEE J. Electron Devices Soc.* **2013**, *1*, 166–174. [CrossRef]
10. Ma, J.; Fossum, E. Quanta Image Sensor Jot with Sub 0.3 e-rms Read Noise and Photon Counting Capability. *IEEE Electron Device Lett.* **2015**, *36*, 926–928. [CrossRef]
11. Bamji, C.S.; O'Connor, P.; Elkhatib, T.; Mehta, S.; Thompson, B.; Prather, L.A.; Snow, D.; Akkaya, O.C.; Daniel, A.; Payne, A.D.; et al. A 0.13 μm CMOS System-on-Chip for a 512 × 424 Time-of-Flight Image Sensor with Multi-Frequency Photo-Demodulation up to 130 MHz and 2 GS/s ADC. *IEEE J. Solid State Circuits* **2015**, *50*, 303–319. [CrossRef]

12. Nasuno, S.; Wakashima, S.; Kusuhara, F.; Kuroda, R.; Sugawa, S. A CMOS Image Sensor with 240 µV/e$^-$ Conversion Gain, 200 ke$^-$ Full Well Capacity and 190–1000 nm Spectral Response. In Proceedings of the International Image Sensor Workshop, Vaals, The Netherlands, 8–11 June 2015.

13. Vogelsang, T.; Guidash, M.; Xue, S. Overcoming the Full Well Capacity Limit: High Dynamic Range Imaging Using Multi-Bit Temporal Oversampling and Conditional Reset. In Proceedings of the International Image Sensor Workshop, Snowbird, UT, USA, 12–16 June 2013.

14. Yang, F. Bits from Photons: Oversampled Binary Image Acquisition. Ph.D. Thesis, EPFL, Lausanne, Switzerland, 20 February 2012.

15. Dutton, N.A.W.; Grant, L.A.; Henderson, R.K. 9.8 µm SPAD-based Analogue Single Photon Counting Pixel with Bias Controlled Sensitivity. In Proceedings of the International Image Sensors Workshop, Snowbird, UT, USA, 12–16 June 2013.

16. Dutton, N.A.W.; Gyongy, I.; Parmesan, L.; Gnecchi, S.; Calder, N.; Rae, B.R.; Pellegrini, S.; Grant, L.A.; Henderson, R.K. A SPAD-Based QVGA Image Sensor for Single-Photon Counting and Quanta Imaging. *IEEE Trans. Electron Devices* **2016**, *63*, 189–196. [CrossRef]

17. Dutton, N.A.W.; Parmesan, L.; Holmes, A.J.; Grant, L.A.; Henderson, R.K. 320 × 240 Oversampled Digital Single Photon Counting Image Sensor. In Proceedings of the 2014 Symposium on VLSI Circuits Digest of Technical Papers, Honolulu, HI, USA, 10–13 June 2014.

18. Dutton, N.A.W.; Parmesan, L.; Gnecchi, S.; Gyongy, I.; Calder, N.J.; Rae, B.R.; Grant, L.A.; Henderson, R.K. Oversampled ITOF Imaging Techniques using SPAD-based Quanta Image Sensors. In Proceedings of the International Image Sensor Workshop, Vaals, The Netherlands, 8–11 June 2015.

19. Gyongy, I.; Dutton, N.A.W.; Parmesan, L.; Davies, A.; Saleeb, R.; Duncan, R.; Rickman, C.; Dalgarno, P.; Henderson, R.K. Bit-plane Processing Techniques for Low-Light, High Speed Imaging with a SPAD-based QIS. In Proceedings of the International Image Sensor Workshop, Vaals, The Netherlands, 8–11 June 2015.

20. Hynecek, J. Impactron-a new solid state image intensifier. *IEEE Trans. Electron Devices* **2001**, *48*, 2238–2241. [CrossRef]

21. Parks, C.; Kosman, S.; Nelson, E.; Roberts, N.; Yaniga, S. A 30 fps 1920 × 1080 pixel Electron Multiplying CCD Image Sensor with Per-Pixel Switchable Gain. In Proceedings of the International Image Sensor Workshop, Vaals, The Netherlands, 8–11 June 2015.

22. Robbins, M.S.; Hadwen, B.J. The noise performance of electron multiplying charge-coupled devices. *IEEE Trans. Electron Devices* **2003**, *50*, 1227–1232. [CrossRef]

23. Seo, M.-W.; Kawahito, S.; Kagawa, K.; Yasutomi, K. A 0.27 e$^-$ Read Noise 220-µV Conversion Gain Reset-Gate-Less CMOS Image Sensor. *IEEE Electron Device Lett.* **2015**, *36*, 1344–1347.

24. Chen, Y.; Xu, Y.; Chae, Y.; Mierop, A.; Wang, X.; Theuwissen, A. A 0.7 e$^-$ rms-temporal-readout-noise CMOS image sensor for low-light-level imaging. In Proceedings of the 2012 IEEE International Solid-State Circuits Conference, San Francisco, CA, USA, 19–23 February 2012; pp. 384–386.

25. Aull, B.F.; Loomis, A.H.; Young, D.J.; Heinrichs, R.M.; Felton, B.J.; Daniels, P.J.; Landers, D.J. Geiger-Mode Avalanche Photodiodes for Three Dimensional Imaging. *Linc. Lab. J.* **2002**, *13*, 335–350.

26. Richardson, J.; Walker, R.; Grant, L.; Stoppa, D.; Borghetti, F.; Charbon, E.; Gersbach, M.; Henderson, R.K. A 32 × 32 50 ps Resolution 10 bit Time to Digital Converter Array in 130 nm CMOS for Time Correlated Imaging. In Proceedings of the IEEE Custom Integrated Circuits Conference, San Jose, CA, USA, 13–16 September 2009; pp. 77–80.

27. Gariepy, G.; Krstajić, N.; Henderson, R.; Li, C.; Thomson, R.R.; Buller, G.S.; Heshmat, B.; Raskar, R.; Leach, J.; Faccio, D. Single-photon sensitive light-in-fight imaging. *Nat. Commun.* **2015**, *6*, 6021. [CrossRef] [PubMed]

28. Gariepy, G.; Tonolini, F.; Henderson, R.; Leach, J.; Faccio, D. Detection and tracking of moving objects hidden from view. *Nat. Photonics* **2015**, *10*, 23–26. [CrossRef]

29. Parmesan, L.; Dutton, N.A.W.; Calder, N.J.; Holmes, A.J.; Grant, L.A.; Henderson, R.K. A 9.8 µm Sample and Hold Time to Amplitude Converter CMOS SPAD Pixel. In Proceedings of the 44th European Solid State Device Research Conference (ESSDERC), Venice, Italy, 22–26 September 2014.

30. Parmesan, L.; Dutton, N.A.W.; Calder, N.J.; Grant, L.A.; Henderson, R.K. A 256 × 256 SPAD array with in-pixel Time to Amplitude Conversion for Fluorescence Lifetime Imaging Microscopy. In Proceedings of the International Image Sensor Workshop, Vaals, The Netherlands, 8–11 June 2015.

31. Maruyama, Y.; Charbon, E. A Time-Gated 128 × 128 CMOS SPAD Array for On-Chip Fluorescence Detection. In Proceedings of the 2011 International Image Sensors Workshop, Hokkaido, Japan, 8–11 June 2011.

32. Pavia, J.M.; Wolf, M.; Charbon, E. Measurement and modeling of microlenses fabricated on single-photon avalanche diode arrays for fill factor recovery. *Opt. Express* **2014**, *22*, 4202–4213. [CrossRef] [PubMed]

33. Stoppa, D.; Pancheri, L.; Scandiuzzo, M.; Gonzo, L.; Dalla Betta, G.-F.; Simoni, A. A CMOS 3-D Imager Based on Single Photon Avalanche Diode. *IEEE Trans. Circuits Syst. I Regul. Pap.* **2007**, *54*, 4–12. [CrossRef]

34. Perenzoni, M.; Massari, N.; Perenzoni, D.; Gasparini, L.; Stoppa, D. A 160 × 120 Pixel Analog-Counting Single-Photon Imager with Time-Gating and Self-Referenced Column-Parallel A/D Conversion for Fluorescence Lifetime Imaging. *IEEE J. Solid-State Circuits* **2016**, *51*, 155–167.

35. Ma, J.; Starkey, D.; Rao, A.; Odame, K.; Fossum, E.R. Characterization of Quanta Image Sensor Pump-Gate Jots With Deep Sub-Electron Read Noise. *IEEE J. Electron Devices Soc.* **2015**, *3*, 472–480. [CrossRef]

36. Pancheri, L.; Massari, N.; Stoppa, D. SPAD Image Sensor with Analog Counting Pixel for Time-Resolved Fluorescence Detection. *IEEE Trans. Electron Devices* **2013**, *60*, 3442–3449. [CrossRef]

37. Perenzoni, M.; Massari, N.; Perenzoni, D.; Gasparini, L.; Stoppa, D. A 160 × 120-pixel analog-counting single-photon imager with Sub-ns time-gating and self-referenced column-parallel A/D conversion for fluorescence lifetime imaging. In Proceedings of the IEEE International Solid-State Circuits Conference—ISSCC Digest of Technical Papers, San Francisco, CA, USA, 22–26 February. 2015; pp. 1–3.

38. Fossum, E.R. Gigapixel Digital Film Sensor (DFS) Proposal. In Nanospace Manipulation of Photons and Electrons for Nanovision Systems, Proceedings of The 7th Takayanagi Kenjiro Memorial Symposium and the 2nd International Symposium on Nanovision Science, Hamamatsu, Japan, 25–26 October 2005.

39. Yang, F.; Sbaiz, L.; Charbon, E.; Susstrunk, S.; Vetterli, M. Image reconstruction in the gigavision camera. In Proceedings of the 2009 IEEE 12th International Conference on Computer Vision Workshops (ICCV Workshops), Kyoto, Japan, 27 September–4 October 2009; pp. 2212–2219.

40. Niclass, C.; Favi, C.; Kluter, T.; Monnier, F.; Charbon, E. Single-Photon Synchronous Detection. *IEEE J. Solid-State Circuits* **2009**, *44*, 1977–1989. [CrossRef]

41. Masoodian, S.; Rao, A.; Ma, J.; Odame, K.; Fossum, E.R. A 2.5 pJ/b Binary Image Sensor as a Pathfinder for Quanta Image Sensors. *IEEE Trans. Electron Devices* **2016**, *63*, 100–105. [CrossRef]

42. Burri, S.; Maruyama, Y.; Michalet, X.; Regazzoni, F.; Bruschini, C.; Charbon, E. Architecture and applications of a high resolution gated SPAD image sensor. *Opt. Express* **2014**, *22*, 17573–17589. [CrossRef] [PubMed]

43. Nakamura, J. *Image Sensors and Signal Processing for Digital Still Cameras*, 1st ed.; CRC Press: Boca Raton, FL, USA, 2006.

sensors

MDPI

Article

Development of Gated Pinned Avalanche Photodiode Pixels for High-Speed Low-Light Imaging

Tomislav Resetar [1,2,*], Koen De Munck [2], Luc Haspeslagh [2], Maarten Rosmeulen [2], Andreas Süss [2], Robert Puers [1,2] and Chris Van Hoof [1,2]

[1] KU Leuven, ESAT, Kasteelpark Arenberg 10, B-3001 Leuven, Belgium; puers@esat.kuleuven.be (R.P.); chris.vanhoof@imec.be (C.V.H.)

[2] Imec, Kapeldreef 75, B-3001 Leuven, Belgium; koen.demunck@imec.be (K.D.M.); luc.haspeslagh@imec.be (L.H.); Maarten.Rosmeulen@imec.be (M.R.); andreas.suess@imec.be (A.S.)

*   Correspondence: tomislav.resetar@imec.be; Tel.: +32-16-28-16-52

Academic Editor: Albert Theuwissen
Received: 29 February 2016; Accepted: 10 August 2016; Published: 15 August 2016

Abstract: This work explores the benefits of linear-mode avalanche photodiodes (APDs) in high-speed CMOS imaging as compared to different approaches present in literature. Analysis of APDs biased below their breakdown voltage employed in single-photon counting mode is also discussed, showing a potentially interesting alternative to existing Geiger-mode APDs. An overview of the recently presented gated pinned avalanche photodiode pixel concept is provided, as well as the first experimental results on a 8×16 pixel test array. Full feasibility of the proposed pixel concept is not demonstrated; however, informative data is obtained from the sensor operating under -32 V substrate bias and clearly exhibiting wavelength-dependent gain in frontside illumination. The readout of the chip designed in standard 130 nm CMOS technology shows no dependence on the high-voltage bias. Readout noise level of 15 e^- rms, full well capacity of 8000 e^-, and the conversion gain of 75 $\mu V/e^-$ are extracted from the photon-transfer measurements. The gain characteristics of the avalanche junction are characterized on separate test diodes showing a multiplication factor of 1.6 for red light in frontside illumination.

Keywords: APD; avalanche photodiode; CIS; CMOS; high-speed; image sensor; PAPD; pinned; pixel

1. Introduction

An image sensor that is suitable for low-light imaging on one hand needs to have good optical properties to collect as many incident photons as possible, and, on the other hand, requires low readout noise levels to bring the signal-to-noise ratio (SNR) close to the Poissonian limit. Suppressing the impact of readout noise can be achieved either by lowering the noise level itself, or by employing a multiplication mechanism in the charge domain. Recent progress in lowering the readout noise of CMOS image sensors (CIS) [1–4] leaves less and less room for the benefits of electron-multiplying (EM) devices such as EMCCDs, EMCMOS, intensified CCDs (ICCDs) or avalanche photodiodes (APDs), due to the inherent noisiness of the impact-ionization mechanism. Nevertheless, a more thorough comparison of sensors can be made that takes into account their optical properties as well as their speed properties. To compare the low-light performance of different image sensors, one can define a figure of merit (FOM) as the minimum incident photon count I that is required to achieve signal-to-noise ratio (SNR) equal to one. This approach is similar to luminance-SNR of 10 (YSNR10) FOM first proposed for mobile-phone cameras [5]. A sensor with lower FOM is therefore desirable for good low-light performance.

$$FOM = I|_{SNR=1} \tag{1}$$

In order to evaluate the FOM as a function of frame rate (fr), the pixel SNR can be expressed as follows:

$$SNR(fr) = \frac{QE \cdot ff \cdot M \cdot I}{\sqrt{M^2 \cdot F \cdot (QE \cdot ff \cdot I + a \cdot D/fr) + (1-a) \cdot D/fr + \sigma_R^2}} \tag{2}$$

taking into account the following parameters: quantum efficiency QE, fill factor ff, multiplication factor M, excess noise factor F, readout noise σ_R in electrons rms, and the number of dark electrons per second D, where the factor a describes the effective fraction of amplified dark current carriers. In this model, the FOM depends on the frame rate only through the dark current. This is especially important in the electron-multiplying devices where—dependent on the position of the EM stage—dark signal carriers can be multiplied together with the photogenerated signal carriers.

Figure 1 shows FOM plotted with respect to frame rate for several sensors taken from literature, representing the different imaging techniques with their properties listed in Table 1. It can be observed that the lowest FOMs are obtained by sensors that are used in relatively low-speed applications, namely EMCCDs [6], scientific CMOS (sCMOS) [7], and ICCDs [8]. Detailed surveys are available in the literature that cover the trade-offs between those techniques [9]. CCD-based approaches are usually limited in speed by the bandwidth of the output amplifier, or in the case of the EMCCD, the speed of the electron-multiplying stage. Standard four-transistor CIS are not optimized for high-speed operation, and are limited either by the readout speed, by rolling-shutter artifacts, or by image-lag constraints [10]. For these reasons, techniques based on high-speed CIS (HS-CIS) [11], avalanche photodiodes (APDs) [12], or single-photon APDs (SPADs) [13] are used in high-speed imaging. As can be seen from Figure 1, those techniques typically have substantially larger FOMs, leaving room for improvement [14].

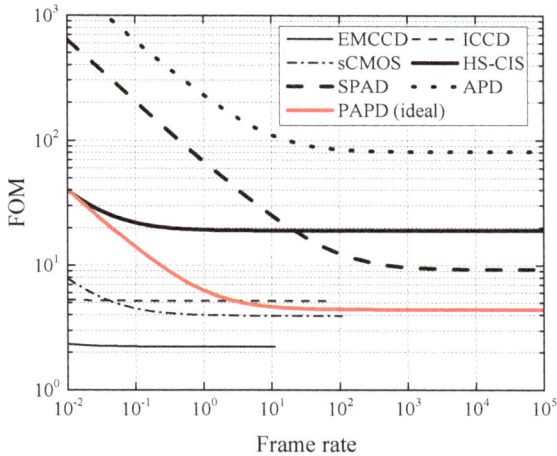

Figure 1. Minimum number of photons incident on a pixel that is required for $SNR = 1$ as a function of frame rate for several typical sensors from literature. Lower figure of merit (FOM) indicates better low-light performance. The red line represents the ideal performance of the gated pinned avalanche photodiode (PAPD) pixel described in this work. APD: avalanche photodiode; EM: electron-multiplying; HS-CIS: high-speed CMOS image sensor; SPAD: single-photon APD; sCMOS: scientific CMOS; ICCD: intensified CCD.

Table 1. Sensors from Figure 1 and their properties related to Equation (2).

Sensor	QE_{max}	ff	σ_R	D	a	M	F	fr_{max}	Ref.	Comment
EMCCD	0.9	1.0	6.0	1.0×10^{-3} *	1.0	1000	2	10	[6]	* deep cooling
ICCD	0.5	1.0	5.4	300	0.0	500	2.6	60	[8]	/
sCMOS	0.6	1.0 *	1.8	0.14	0.0	1.0	1.0	100	[7]	* included in QE
HS-CIS	0.8 *	0.37	5.1	1.0 *	0.0	1.0	1.0	(15×10^3)	[11]	* assumed values
SPAD	0.4	0.27	0.17	47	1.0	1.0 *	1.0 *	(16×10^3)	[13]	* not applicable
APD	0.23	0.26	26	5.0×10^4	0.0	20	4.5	(800 *)	[12]	* time-of-flight
PAPD *	0.8	1.0	10	10	0.5	3.0	2.1	/	[14]	* ideal values

High-speed operation of CMOS-based sensors is typically achieved at the cost of both optical properties and readout noise. The thickness of the epitaxial layer needs to be limited in order to avoid slow moving charges generated outside of the depletion region, causing image lag and cross-talk. This effect is especially present in backside-illuminated (BSI) sensors [15]. Limiting the epitaxial layer thickness results in QE reduction for longer wavelengths. HS-CIS are operated in global shutter mode and therefore require more complex pixel architectures than the standard four-transistor pinned photodiode (4-T PPD) pixels. Additional in-pixel storage nodes are employed if correlated double sampling (CDS) capability is desired, giving rise to signal-fidelity problems and sacrificing the pixel photoactive area [11,16]. Due to short frame-times, multiple sampling by the analog-to-digital converters (ADCs) may not be possible, further increasing the total readout noise. These limitations make the charge-multiplying approach using wide-depletion region APDs an interesting area of research, owing to the signal amplification and high speed capability of these devices.

2. Prospects of Linear-Mode APDs in High-Speed Low-Light Imaging

Generally, linear-mode APDs and SPADs suffer from poor fill factors when placed in focal-plane arrays. This is due to the presence of guard rings that are needed to prevent premature edge breakdown and additional circuitry that is needed for their operation. Limited literature is available on linear-mode APD CMOS image sensors in contrast to the SPAD-based ones [13,17]. A first attempt has been made in 2001 [18] with poor avalanche noise performance due to the hole-initiated avalanche. More recent work has been presented driven from time-of-flight applications with better noise performance due to the dead space effect [12]. There seems to be two opposite directions in which the problem of the excessive noise factor can be addressed: either very high or very low electric field magnitudes should be targeted. On the one hand, very high electric fields can be formed by increasing the doping concentration of the avalanche junction, thereby forming a very narrow depletion region in which an electron has only a few highly probable opportunities to avalanche. Problems related to this approach stem from the high tunneling currents and poor optical properties due to the very narrow depletion region [19]. On the other hand, lowering the noise factor can be achieved by aiming for lowly-doped junctions that result in low peak electric fields at which the hole-ionization contribution is less pronounced. This is typically expressed by the ratio between hole and electron impact ionization coefficients $k = \beta/\alpha$ [20]. This approach is successfully applied in EMCCDs, where the charge packet passes through several hundred low-probability ionization opportunities, resulting in a minimum possible noise factor $F = 2$ that is independent of the multiplication factor M [21]. In linear mode APDs, such low k values are hardly achievable in silicon, since this would require lowly doped junctions that would have to be biased at several hundred volts to achieve the desired gain. If one wants to limit the bias voltage below 100 V, values of k below 0.1 seem to be out of reach, resulting in a gain-dependent noise factor [18,22]. This means that—unlike the EMCCD case—the excessive noise factor F is an increasing function of gain M, and therefore an optimal gain exists for SNR improvement for a certain illumination level. In this analysis, the noise factor dependence on gain is described by the following expression [23]:

$$F(M) = Mk + (2 - 1/M)(1 - k). \tag{3}$$

Relative SNR improvement can be estimated as the ratio between the SNR of a device with avalanche multiplication SNR_{av} and the SNR of the same device without avalanche multiplication SNR_{nav}:

$$\frac{SNR_{av}}{SNR_{nav}} = \sqrt{\frac{I + \sigma_R^2}{FI + \sigma_R^2/M^2}}. \qquad (4)$$

The SNR expression is taken from Equation (2) by omitting the optical factors QE and ff and assuming negligible impact of the dark signal. Relative SNR improvements that can be expected from an APD with $k = 0.1$ and readout noise $\sigma_R = 10$ e⁻ rms from Equation (4) for different electron counts N are presented in Figure 2. It can be seen that SNR improvements can be expected for low photon counts, and that, for the presented case, the optimum gain is lower than 5. Increasing the multiplication factor to higher values eventually results in SNR deterioration. It should also be noted that the absolute SNR values at those light levels are already very low, and that single-photon resolution cannot be expected from this approach. The following section therefore discusses further possibilities of employing APDs in single-photon counting.

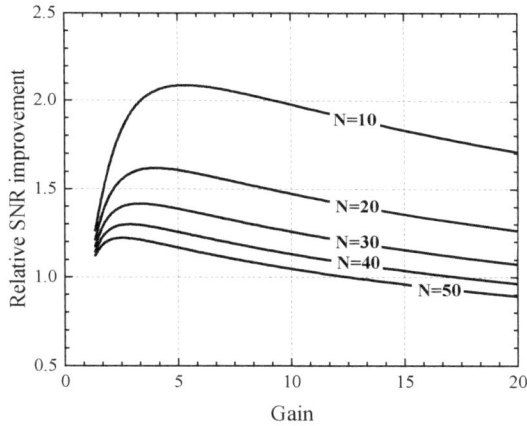

Figure 2. Relative signal-to-noise ratio (SNR) improvement with respect to applied gain for an APD with $k = 0.1$ and readout noise $\sigma_R = 10$ e⁻ rms for different electron counts N.

3. Photon Counting with APDs below the Breakdown Voltage

Apart from SPADs that are biased beyond their breakdown voltage (V_{BD}), single-photon detection is also possible with APDs operating below V_{BD}. Very few reports exist in the literature exploring this approach, since the photon detection efficiency is normally assumed to be low [24], and challenges related to bias voltage stability could be present due to the sharp gain characteristics of APDs. Nevertheless, work evaluating this technique for light detection and ranging (LIDAR) applications with APD arrays fabricated in III-V materials report encouraging results [25,26]. More recently, a similar thresholding technique was employed to boost the dynamic range of a standard CMOS image sensor [27]. A short evaluation of the single photon detection probability (PDP) of CMOS APDs with low-noise readout is provided here.

The probability density function (PDF) of APD output carrier count m, dependent on the multiplication factor M, ratio of hole and electron ionization probability k, and input carrier count N is derived by McIntyre and Conradi in [28,29]. An approximation of the analytical expression of the PDF that is suitable for numerical calculation is given by:

$$PDF \approx \frac{N}{\sqrt{2\pi m(m-N)[N+k(m-N)]}} \times \left[1 - \frac{m/M-N}{m-N}\right]^{(m-N)} \times \left[1 + \frac{(1-k)(m-NM)/M}{N+k(m-N)}\right]^{\frac{N+k(m-N)}{1-k}}. \qquad (5)$$

Assuming a single input carrier $N = 1$ amplified with an average gain of $M = 100$, the single photon detection probability (PDP) can be expressed as:

$$PDP = \int_{N_{th}}^{\infty} PDF dm \qquad (6)$$

where the comparator threshold carrier count is set to two times the the readout noise value $N_{th} = 2\sigma_R$ for a 2.3% probability of false detection. Figure 3 represents the PDP with respect to k for different readout noise values. It can be seen that an APD with $k = 0.3$, biased below breakdown having a readout noise level of $\sigma_R = 5\,e^-$ can achieve a PDP as high as 35%, which is comparable to the performance of the state-of-the art CMOS SPADs [17]. This approach might therefore be attractive since there is no need for quenching circuitry, and the trade-off between PDP on one hand and dark count rate (DCR) and after pulsing on the other could be mitigated. Moreover, due to lower carrier densities, cross-talk could be improved in shared-well APDs, giving rise to higher fill-factors and smaller pixels [30].

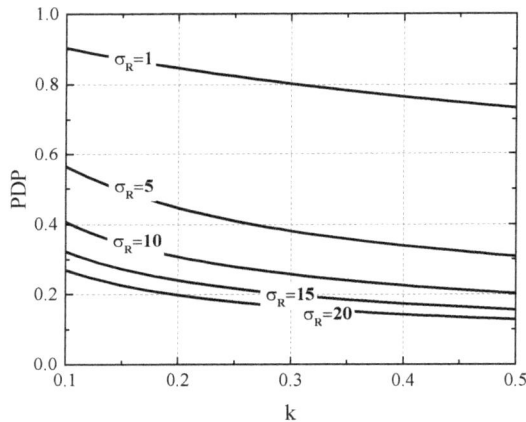

Figure 3. Photon detection probability (PDP) of an APD with $M = 100$ with respect to k for different readout noise levels in electrons rms. The detector threshold is set to two times the readout noise rms value.

4. Backside-Illuminated Gated PAPD Pixel Concept

Recently, a gated pinned APD (PAPD) pixel concept was proposed to explore the possibility of combining standard CIS technology with avalanche signal multiplication, and at the same time providing good optical and speed properties due to full-depletion of the epitaxial layer [14,31]. Figure 4 shows the main principle of operation, with four stages of PAPD operation: 1—BSI light absorption; 2—multiplication; 3—collection; and 4—transfer. As illustrated in Figure 5, besides the high negative voltage V_{BCK}, the pixel can be operated like a standard PPD, since all transistors are isolated from the high voltage. The fill factor problem of conventional APDs and SPADs is mitigated in this approach, since all the pixels in the array share the same avalanche junction and the guard ring preventing premature edge breakdown is implemented only at the edges of the pixel array. The multiplication junction is formed by high-energy boron and phosphorus ion implantations that are added to the standard CIS flow. The doses and energies of the implants were optimized to satisfy two main criteria: firstly, the creation of the high field necessary for impact-ionization at the desired V_{BCK}; and secondly, that both the absorption and the collection regions are fully depleted to avoid shorts and cross-talk between consecutive pixels. The pixel therefore operates at the border of the punch-through breakdown, and special care needs to be taken to maintain a sufficient potential barrier at full depletion. Figure 6 shows a 2D electric field and electrostatic field profile from technology computer-aided design (TCAD) when the pixel is biased at $V_{BCK} = -36$ V.

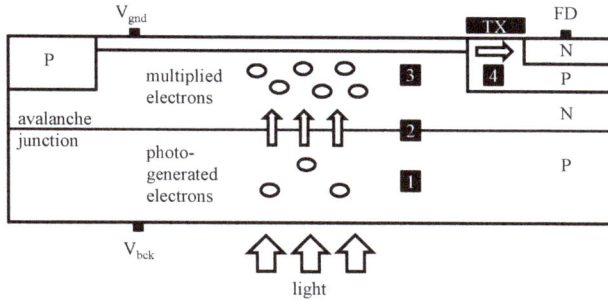

Figure 4. Backside-illuminated (BSI)-gated PAPD pixel concept description in four main phases: 1—light absorption; 2—electron amplification; 3—collection; and 4—transfer. TX: transfer-gate, FD: floating diffusion, P: p-type doped region, N: n-type doped region.

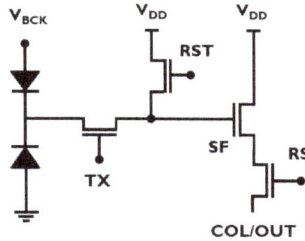

Figure 5. Circuit representation of the gated PAPD pixel.

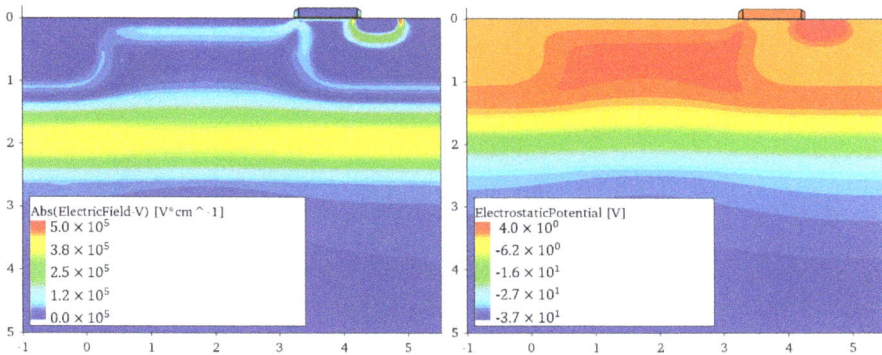

Figure 6. PAPD 2D electric field profile (**left**) and electrostatic potential profile (**right**) from technology computer-aided design (TCAD) at the backside bias voltage $V_{BCK} = -36$ V.

In order to demonstrate the proof of concept and characterize the basic pixel metrics, a 8×32 test pixel array was designed. In addition to the pixel array, a basic readout circuitry was integrated on chip in order to enable row and column addressing and standard four-transistor pixel operation. The readout was designed with 3.3 V CMOS transistors placed in isolated deep wells, as illustrated in Figure 7. It is important that no premature avalanche breakdown happens in the readout part of the chip. From the design perspective, all sharp corners on the outer well edges were avoided in layout, and electrostatic discharge protection structures were modified to withstand the high negative bias. From the technology perspective, with the chosen p-type epitaxial layer with 5.5 µm thickness, the readout avalanche breakdown voltage is -80 V, which imposes the upper limit for the choice of the V_{BCK} for the pixel operation. Choosing a thicker and more highly resistive epitaxial layer would in principle increase the readout breakdown and enable higher V_{BCK} values, which would result in increased QE as well as lower k values of the avalanche junction, and thus better SNR performance. In this work, a more conservative approach was taken by choosing V_{BCK} values between -30 and -40 V.

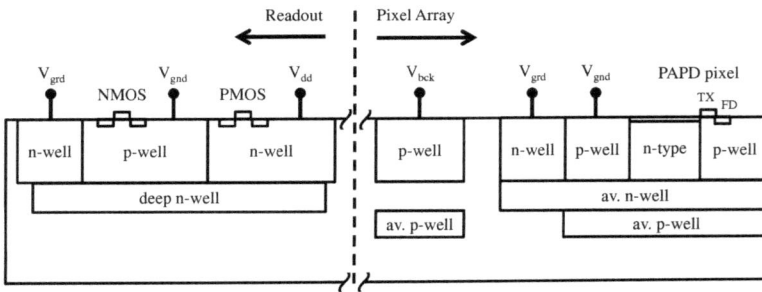

Figure 7. Readout isolation and the pixel array guard-ring implementation.

5. Experimental Results and Discussion

In this section, preliminary experimental results of the pixel performance are presented. The pixel array consists of 32 rows of floating-diffusion (FD)-shared pixels with 10 µm pitch. It should be pointed out that full characterization was not possible at this time due to an error in the processing of the pixel p-well. As a consequence, the deepest boron implant was not present in the device. For that reason, optimal doping conditions determined in the TCAD analysis were not met, causing shorting of the FD-shared pixels. However, due to wider barriers between pixels that do not share the FD node, sufficient isolation was achieved in binning mode of operation. Therefore, the FD-shared pixels were binned so that the effective pixel size was 10×20 µm and the array size was 8×16 pixels. Figure 8 shows the first image of a half-covered PAPD test array taken at half of the full-well capacity. No indications of pixel shorts or blooming were observed at operating voltage $V_{BCK} = -32$ V at these signal levels. The sharpness of the edge should be attributed to the difficulties of projecting a sharp edge onto a small array in our setup. The modulation transfer function (MTF) of PAPD pixels is expected to be high, due to the full depletion of the epitaxial layer. The non-uniformity observed in this image is a combined effect of dark signal non-uniformity (DSNU) and gain non-uniformity.

The PAPD pixel is designed to operate at a fixed backside bias voltage. Increase of the backside voltage results in a sharp increase of the I_{BCK} current and, consequently, in pixel failure. In agreement with the simulations, only a narrow range of 3 V below the punch-through breakdown exists in which the array can be characterized. At lower biases, the pixels are naturally shorted due to the non-depleted n-type layer. The gain properties of the junction therefore cannot be straightforwardly evaluated by sweeping the V_{BCK}. The pixels presented in this work are biased at $V_{BCK} = -32$ V, which is several volts below the expected value from TCAD that is required for observation of the predicted optimal gains from Figure 2. The shift in the operating V_{BCK} is caused by the non-optimal doping conditions due to the mentioned process error. A typical I_{BCK} current of around 1 nA is measured on an area of 0.15 mm^2 at the bulk contact of the chip. The pixel dark current of 1×10^5 e$^-$ per second was measured at room temperature, which is two orders of magnitude higher than typically expected in this technology [32]. Due to the reduced boron dose under the shallow trench isolation (STI) region, it is very likely that the depletion region extends to the STI walls. Nevertheless, the dark current is not considered a crucial performance parameter, since this pixel is intended for high-speed operation.

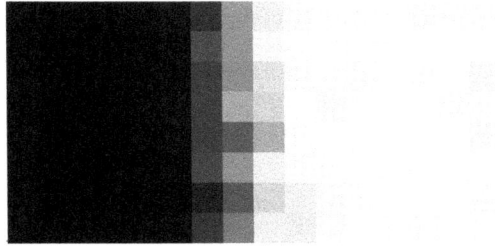

Figure 8. Image taken with a PAPD test array biased at −32 V back bias with left part of the array covered at roughly half full-well capacity.

The pixel is primarily being developed for backside illumination where, in principle, the peak backside QE as high as 90% can be achieved with application of proper anti-reflection coatings [33]. However, for proof of concept purposes, no backside processing was applied, and, therefore, only frontside illumination (FSI) characterization was performed. Measured RGB photon-transfer curves (PTCs) of PAPD pixels are presented in Figure 9. The PTCs were acquired by changing the integration time of a pixel under constant illumination. It can be seen that, due to the avalanche multiplication, the PTCs are shifting upwards when illuminated by light of longer wavelength. It should be noted that the observed shift in the PTC is a combined effect of the multiplication factor M and the excessive noise factor F. For that reason, it is impossible to separate the gain from the noise contribution in the PTC curve. The excessive noise factor can in principle be evaluated with the aid of independent multiplication factor measurements on separate test structures, as discussed at the end of this section. In the present case, due to the limited multiplication factor at this V_{BCK}, the excess noise factor is difficult to estimate with desirable accuracy. Conversion gain of 75 µV/e$^-$ was extracted from the blue curve data, which is assumed to have a negligible multiplication factor in FSI. Readout noise of $\sigma_R = 15$ e$^-$ rms is observed, which, together with large image lag, indicates that the diode is operating in partially-pinned regime and thus suffering from kTC noise. Full-well capacity of 8000 e$^-$ was measured, and is currently limited by the FD node capacitance, which is expected due to the pixel binning. The basic pixel characteristics are summarized in Table 2.

Table 2. Summary of pixel characteristics. FSI: frontside illumination.

Design and Technology		Measured Values	
Pixel count	8 × 16	Backside bias voltage	−32 V
Pixel dimensions	10 µm × 20 µm (binned)	Dark current @ 25 °C	1×10^5 e$^-$/s/pixel
Fill factor	40% FSI	Multiplication factor @ 635 nm FSI	1.6
Transistors per pixel	4	Readout noise	15 e$^-$ rms
Technology	130 nm 3.3 V CIS	Full-well capacity	8000 e$^-$
		Conversion gain @ 470 nm	75 µV/e$^-$

Separate test diodes were designed to characterize the multiplication factor of the avalanche junction for a full V_{BCK} range. A simple model was developed to fit the wavelength dependence of the multiplication factor, which is explained in Appendix A. This model enables the estimation of the pure electron injection multiplication factor which would be observed for blue and green light in BSI. The multiplication factor characteristics are presented in Figure 10 for blue ($\lambda_b = 470$ nm), green ($\lambda_g = 525$ nm), and red ($\lambda_r = 635$ nm) light. It can be seen that biasing the pixel at $V_{BCK} = -32$ V results in a multiplication factor $M = 1.6$ for red light, corresponding to $M = 2.1$ for pure electron injection. This multiplication factor is below the optimal values between 3 and 5, as suggested in Figure 2. According to the presented model, FSI multiplication factors between 1.9 and 2.7 would be desired for red light illumination. It should be noted that in BSI, avalanche gain should depend significantly less on light wavelength than in FSI. For the chosen epitaxial layer thickness of 5.5 μm, carriers that are generated in the 2.5 μm thick region below the avalanche junction all experience the same multiplication factor. Significant wavelength-dependence of the multiplication factor in BSI is therefore expected for wavelengths longer than approximately 580 nm. This effect can be further decreased by placing the avalanche junction closer to the surface, or choosing a thicker epitaxial layer at the expense of higher backside bias needed for the same avalanche gain.

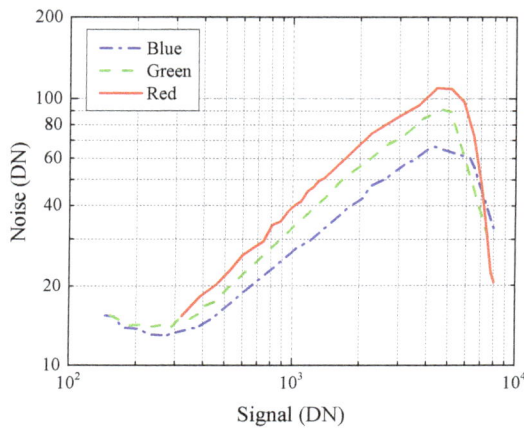

Figure 9. Photon-transfer curves (PTC) for red, blue, and green light, showing the combined impact of avalanche multiplication and noise.

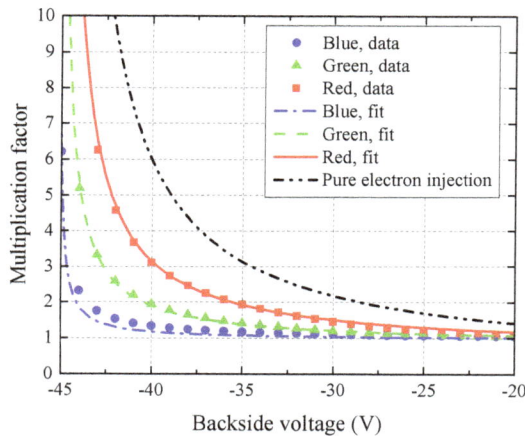

Figure 10. Measurements and analytical fit of avalanche multiplication factor for red, green, and blue light.

6. Conclusions and Outlook

This work shows that, in theory, it is possible to obtain SNR improvements by employing avalanche multiplication in low-light high-speed CMOS imaging. A pixel concept employing avalanche multiplication is proposed which mitigates the need for guardrings and complex circuitry in the pixel array, and is expected to offer good optical and speed properties due to the full depletion of the epitaxial layer. First experimental results of FSI characterization of the proposed pixel are presented. Wavelength-dependent photon-transfer curves due to avalanche multiplication of signal is reported. Nevertheless, because of the unexpected absence of the deepest p-well implant caused by a process error, optimal biasing conditions could not be reached and full depletion of the collection area was not achieved. Further work is therefore needed to obtain devices with higher multiplication factors so that the excess noise factor can be reliably measured in order to evaluate the theoretical predictions of SNR improvement. Aside from proportional signal multiplication, APDs can also be used as photon-counters when biased below their breakdown voltage. Theoretical analysis provided in this paper shows the feasibility of this approach without sacrificing the photon-detection efficiency compared to SPADs. Pixels with substantially higher gains than that presented in this work should therefore be developed by adjusting the implantation doses in order to experimentally demonstrate this approach.

Author Contributions: Tomislav Resetar wrote the article, performed the analysis, calculations and measurements. Koen De Munck is the main inventor of the gated PAPD concept and supervised the pixel development and characterization. Luc Haspeslagh contributed to the pixel fabrication and process development. Maarten Rosmeulen and Andreas Suss contributed to the paper rationale as well as pixel characterization and results interpretation. Robert Puers and Chris Van Hoof contributed in academic supervision and results interpretation.

Conflicts of Interest: The authors declare no conflict of interest.

Appendix Wavelength-Dependent Multiplication Factor Model

The multiplication factor dependence on the applied reverse bias voltage V_{BCK} of an APD is conventionally expressed as:

$$M(V_{BCK}) = \left[1 - \left(\frac{V_{BCK}}{V_{BD}}\right)^n\right]^{-1} \tag{A1}$$

where V_{BD} is the junction breakdown voltage and n is a fitting parameter. Even though this parameter can be used to fit the multiplication data obtained at different wavelengths, here a more physical approach is taken in which a common n is found for all wavelengths; thereby, M becomes only a property of the multiplication junction. If we consider an APD illuminated from the front side and we assume electrons to be the predominant carrier type in the multiplication process, three distinct regions can be defined. The first region is between the silicon surface and the multiplication junction at depth x_2; the second region is between the multiplication junction and the end of the epitaxial layer x_3; and the third region is the highly doped substrate. According to the Beer-Lambert law, the wavelength-dependent portion of the carriers generated in the first region is expressed as:

$$C_1(\lambda) = \frac{1 - \exp(-\alpha(\lambda)x_2)}{1 - \exp(-\alpha(\lambda)x_3)} \tag{A2}$$

where $\alpha(\lambda)$ is a wavelength-dependent absorption coefficient for silicon. Similarly, the electrons generated in the second region are multiplied with the multiplication factor M from Equation (A1):

$$C_2(\lambda, V_{BCK}) = M(V_{BCK})\frac{\exp(-\alpha(\lambda)x_2) - \exp(-\alpha(\lambda)x_3)}{1 - \exp(-\alpha(\lambda)x_3)} \tag{A3}$$

The electrons generated in the substrate are considered lost due to the rapid recombination process. The final wavelength-dependent multiplication factor can therefore be expressed as a sum of the above contributions:

$$M(\lambda, V_{\text{BCK}}) = C_1(\lambda) + C_2(\lambda, V_{\text{BCK}}) \tag{A4}$$

References

1. Ma, J.; Fossum, E.R. Quanta Image Sensor Jot with Sub 0.3 e$^-$ rms Read Noise and Photon Counting Capability. *IEEE Electron Device Lett.* **2015**, *36*, 926–928.
2. Seo, M.W.; Kawahito, S.; Kagawa, K.; Yasutomi, K. A 0.27 e$^-$ rms Read Noise 220-μV/e$^-$ Conversion Gain Reset-Gate-Less CMOS Image Sensor with 0.11-μm CIS Process. *IEEE Electron Device Lett.* **2015**, *36*, 1344–1347.
3. Yeh, S.F.; Chou, K.Y.; Tu, H.Y.; Chao, C.P.; Hsueh, F.L. A 0.66 e$^-$ rms temporal-readout-noise 3D-stacked CMOS image sensor with conditional correlated multiple sampling (CCMS) technique. In Proceedings of the 2015 Symposium on VLSI Circuits (VLSI Circuits), Kyoto, Japan, 17–19 June 2015; pp. C84–C85.
4. Boukhayma, A.; Peizerat, A.; Enz, C. A 0.4e$^-$ rms Temporal Readout Noise 7.5 μm Pitch and a 66% Fill Factor Pixel for Low Light CMOS Image Sensors. In Proceedings of the 2015 International Image Sensor Workshop (IISW), Vaals, The Netherlands, 8–11 June 2015.
5. Alakarhu, J. Image sensors and image quality in mobile phones. In Proceedings of the 2007 International Image Sensor Workshop (IISW), Ogunquit, ME, USA, 7–10 June 2007.
6. e2v. *CCD201-20 Back-Illuminated 2-Phase IMO Series Electron-Multiplying CCD Sensor*; A1A-100013 Version 4; e2v: Chelmsford, UK, 2015.
7. Andor. *Zyla 5.5 sCMOS*; LZYLASS 1115 R2; Andor: Belfast, UK, 2015.
8. Dussault, D.; Hoess, P. Noise performance comparison of ICCD with CCD and EMCCD cameras. In Proceedings of the International Society for Optics and Photonics, SPIE 49th Annual Meeting, Optical Science and Technology, Denver, CO, USA, 2–6 August 2004; pp. 195–204.
9. Moomaw, B. Camera technologies for low light imaging: overview and relative advantages. *Methods Cell Biol.* **2013**, *114*, 243–283.
10. Fossum, E.R.; Hondongwa, D.B. A Review of the Pinned Photodiode for CCD and CMOS Image Sensors. *IEEE J. Electron Dev. Soc.* **2014**, *2*, 33.
11. Tochigi, Y.; Hanzawa, K.; Kato, Y.; Kuroda, R.; Mutoh, H.; Hirose, R.; Tominaga, H.; Takubo, K.; Kondo, Y.; Sugawa, S. A global-shutter CMOS image sensor with readout speed of 1-Tpixel/s burst and 780-Mpixel/s continuous. *IEEE J. Solid State Circuits* **2013**, *48*, 329–338.
12. Shcherbakova, O. 3D Camera Based on Gain-Modulated CMOS Avalanche Photodiodes. Ph.D. Thesis, University of Trento, Trento, Italy, 2013.
13. Dutton, N.; Gyongy, I.; Parmesan, L.; Gnecchi, S.; Calder, N.; Rae, B.; Pellegrini, S.; Grant, L.; Henderson, R. A SPAD-Based QVGA Image Sensor for Single-Photon Counting and Quanta Imaging. *IEEE Trans. Electron Devices* **2016**, *63*, 189–196.
14. Resetar, T.; de Munck, K.; Haspeslagh, L.; de Moor, P.; Goetschalckx, P.; Puers, R.; van Hoof, C. Backside-Illuminated 4-T Pinned Avalanche Photodiode Pixel for Readout Noise-Limited Applications. In Proceedings of the 2015 International Image Sensor Workshop (IISW), Vaals, The Netherlands, 8–11 June 2015.
15. Turchetta, R. Notes about the limits of ultra-high speed solid-state imagers. In Proceedings of the 2015 International Image Sensor Workshop (IISW), Vaals, The Netherlands, 8–11 June 2015.
16. Xu, R.; Liu, B.; Yuan, J. A 1500 fps Highly Sensitive 256 × 256 CMOS Imaging Sensor with In-Pixel Calibration. *IEEE J. Solid State Circuits* **2012**, *47*, 1408–1418.
17. Bronzi, D.; Villa, F.; Tisa, S.; Tosi, A.; Zappa, F. SPAD Figures of Merit for Photon-Counting, Photon-Timing, and Imaging Applications: A Review. *IEEE Sens. J.* **2016**, *16*, 3–12.
18. Biber, A.I. Avalanche Photodiode Image Sensing in Standard Silicon BiCMOS Technology. Ph.D. Thesis, Eidgenössische Technische Hochschule (ETH), Zürich, Switzerland, 2000.
19. Dalla Betta, G.; Pancheri, L.; Stoppa, D.; Henderson, R.; Richardson, J. Avalanche Photodiodes in Submicron CMOS Technologies for High-Sensitivity Imaging. In *Advances in Photodiodes*; Dalla Betta, G., Ed.; InTech Open Access Publisher: Rijeka, Croatia, 2011; pp. 226–248.
20. McIntyre, R. Multiplication noise in uniform avalanche diodes. *IEEE Trans. Electron Devices* **1966**, *13*, 164–168.

21. Robbins, M. Electron-Multiplying Charge Coupled Devices–EMCCDs. In *Single-Photon Imaging*; Seitz, P., Theuwissen, A.J.P., Eds.; Springer: Berlin, Germany, 2011; pp. 103–121.

22. Sze, S.M.; Ng, K.K. *Physics of Semiconductor Devices*; John Wiley & Sons: New York, NY, USA, 2006.

23. Webb, P.; McIntyre, R.; Conradi, J. Properties of avalanche photodiodes. *RCA Rev.* **1974**, *35*, 234–278.

24. Zappa, F.; Tisa, S.; Tosi, A.; Cova, S. Principles and features of single-photon avalanche diode arrays. *Sens. Actuators A Phys.* **2007**, *140*, 103–112.

25. Williams, G.M.; Huntington, A.S. Probabilistic analysis of linear mode vs. Geiger mode APD FPAs for advanced LADAR enabled interceptors. *Proc. SPIE* **2006**, *6220*, doi:10.1117/12.668668.

26. Huntington, A.S.; Compton, M.A.; Williams, G.M. Linear-mode single-photon APD detectors. *Proc. SPIE* **2007**, *6771*, doi:10.1117/12.751925.

27. Mori, M.; Sakata, Y.; Usuda, M.; Yamahira, S.; Kasuga, S.; Hirose, Y.; Kato, Y.; Tanaka, T. A 1280 × 720 single-photon-detecting image sensor with 100 dB dynamic range using a sensitivity-boosting technique. In Proceedings of the 2016 IEEE International Solid-State Circuits Conference (ISSCC), San Francisco, CA, USA, 31 January–4 Febuary 2016; pp. 120–121.

28. McIntyre, R.J. The distribution of gains in uniformly multiplying avalanche photodiodes: Theory. *IEEE Trans. Electron Devices* **1972**, *19*, 703–713.

29. Conradi, J. The distribution of gains in uniformly multiplying avalanche photodiodes: Experimental. *IEEE Trans. Electron Devices* **1972**, *19*, 713–718.

30. Vila, A.; Vilella, E.; Alonso, O.; Dieguez, A. Crosstalk-Free Single Photon Avalanche Photodiodes Located in a Shared Well. *IEEE Electron Device Lett.* **2014**, *35*, 99–101.

31. De Munck, K.; Resetar, T. Pinned Photodiode (PPD) Pixel Architecture with Separate Avalanche Region. U.S. Patent 20,140,374,867 A1, 25 December 2014.

32. De Moor, P. IMEC's Offering for Space Imagers. In *CNES Workshop on CMOS Image Sensors for High Performance Applications*; CNES: Touluse, France, 2016.

33. De Moor, P.; Robbelein, J.; Haspeslagh, L.; Boulenc, P.; Ercan, A.; Minoglou, K.; Lauwers, A.; Munck, K.D.; Rosmeulen, M. Enhanced time delay integration imaging using embedded CCD in CMOS technology. In Proceedings of the 2014 IEEE International Electron Devices Meeting, San Francisco, CA, USA, 15–17 December 2014; doi:10.1109/IEDM.2014.7046984.

Chapter 3:
Other Devices, Materials and Applications for Photon Counting

sensors

MDPI

Article

The DEPFET Sensor-Amplifier Structure: A Method to Beat 1/f Noise and Reach Sub-Electron Noise in Pixel Detectors

Gerhard Lutz [1], Matteo Porro [2], Stefan Aschauer [1], Stefan Wölfel [3] and Lothar Strüder [1,4,*]

[1] PNSENSOR GmbH, München D-81739, Germany; gerhard.lutz@pnsensor.de (G.L.); stefan.aschauer@pnsensor.de (S.A.)
[2] European X-ray Free-Electron Laser Facility GmbH, Hamburg D-22761, Germany; matteo.porro@xfel.eu
[3] Gründeläckerstr. 28, Dormitz D-91077, Germany; stefan.woelfel@steppke.de
[4] Experimental Physics, University of Siegen, Walter-Flex-Str. 3, Siegen D-87068, Germany
*   Correspondence: lothar.strueder@pnsensor.com; Tel.: +49-89-309-087-241

Academic Editor: Albert Theuwissen
Received: 7 February 2016; Accepted: 18 April 2016; Published: 28 April 2016

Abstract: Depleted field effect transistors (DEPFET) are used to achieve very low noise signal charge readout with sub-electron measurement precision. This is accomplished by repeatedly reading an identical charge, thereby suppressing not only the white serial noise but also the usually constant 1/f noise. The repetitive non-destructive readout (RNDR) DEPFET is an ideal central element for an active pixel sensor (APS) pixel. The theory has been derived thoroughly and results have been verified on RNDR-DEPFET prototypes. A charge measurement precision of 0.18 electrons has been achieved. The device is well-suited for spectroscopic X-ray imaging and for optical photon counting in pixel sensors, even at high photon numbers in the same cell.

Keywords: DEPFET; photon detection; sub-electron precision; charge measurement; pixel detector; X-ray spectroscopy

---

1. Introduction

The most prominent request for deep sub-electron noise performance arises from imaging, spectroscopy and photon counting in the visible domain. Either low light level applications or the constant shrinking of pixel sizes reduces the number of signal charges to be detected, and a read noise reduction is therefore required to obtain a decent signal-to-noise ratio.

Single photon detection in the wavelength range from 300 nm (E = 4 eV) to 1.100 nm (E = 1.1 eV) is of interest for many applications in science and industry. In the wavelength range under discussion (300 nm to 1.100 nm), every photon penetrating and converting in the silicon produces one electron-hole pair [1]. For photon energies above 4.3 eV the creation of two electron-hole pairs starts to emerge. In the case of near-infrared photons at 1100 nm (1.1 eV) the detector has to be sensitive over a deep sensor volume, e.g., 500 µm to achieve a quantum efficiency above 35% as the attenuation length for this wavelength is approximately 400 µm. At wavelengths in the ultra-violet region of 300 nm, the absorption length in silicon is only 8 nm, *i.e.*, the absorption of the photon happens very close to the radiation entrance window. Detectors covering this full bandwidth have to cope with a difference of the absorption depth of the photon in the sensor of a factor of 5000 while the quantum efficiency ideally has to be high and constant.

The DEPFET detectors (depleted field effect transistor) perform close to the above experimental requirements. DEPFETs represent a sensor and amplifier structure simultaneously. The generated signal electrons are collected and confined in a potential minimum underneath the transistor gate, where the stored signal charges modulate the DEPFET current. This allows for fast gating, analog storage, repetitive non-destructive readout and the absence of reset noise. For the sake of simplicity we confine our study to silicon as a detector and electronics material.

To get a truly linear amplifier system for single photon counting, the readout noise of the sensor has to be significantly lower than 1 electron (rms), e.g., 0.2 electrons (rms). Floating gate amplifier systems have been studied in the past with some success (see e.g., [2]). The repetitive reading of the same charge package allows for reducing the 1/f noise contribution with approximately the square root of the number of readings for a given fixed signal processing time.

We will first describe the DEPFET concept and basic operation including its use in X-ray sensor systems in Section 2. Section 3 will present the mathematical treatment of 1/f noise and briefly discuss the influence and limitations of other noise components. In Section 4 we describe the experimental verification with the help of test devices to check the models and to demonstrate the single photon resolution up to several hundred optical photons. The best noise performance achieved was 0.18 electrons (rms). A final outlook will be provided regarding how a DEPFET floating gate sensor-amplifier can be converted into an optical sensor system.

2. The Depleted p-Channel MOSFET (DEPFET)—A Detector-Amplifier Structure

The DEPFET structure, invented in 1985 by Kemmer and Lutz [3], possesses unique properties that make it extremely useful as a radiation sensor, in particular as a basic cell of a pixel detector. The DEPFET combines the properties of sensors, amplifiers and signal charge storage and allows for non-destructive reading. A variety of DEPFET structures have been invented, with properties such as gateability and signal compression [4], macro-pixel DEPFETs (a combination of DEPFET and silicon drift detector [5,6]) and DEPFETs with intermediate signal charge storage [7]. DEPFETs with repetitive non-destructive readout (RNDR) that allow sub-electron charge measurement precision [8] are the focus of our paper.

Basically, the DEPFET (Figure 1) is a field effect transistor with the source (S), drain (D) and (external) gate (G) located on the front side of a wafer and a large area diode on the back used to fully deplete the bulk. With suitable doping, a potential minimum below the transistor channel is created. Signal charge created anywhere within the depleted bulk assembles in the potential minimum thereby creating mirror charges in the channel and increasing the transistor current. Due to the current steering function, the potential minimum is called the internal gate (IG). Furthermore, a DEPFET contains a device for removing all charge from the IG. With an empty internal gate, a base current I_0 is flowing through the transistor. At the time a radiation signal is created, the conversion electrons will rapidly move towards the IG and increase the transistor current to I_1. Emptying the IG restores the current to its original value I_0. Either one of these current steps (rise or fall) is a measure for the signal charge.

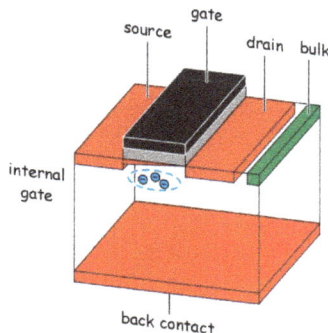

Figure 1. The concept of a DEPFET: The signal electrons are collected in a potential minimum (internal gate) located below the channel of a FET located on top of the fully depleted bulk.

The following physical properties contribute to the noise: thermal fluctuations of the distribution of charge carriers in the transistor channel are the source of white serial noise; thermal fluctuations of the leakage current in the detector volume that lead to parallel white noise; and capture and reemission of charge carriers within the transistor channel in defect locations close to the silicon-SiO_2 interface. Trapping and de-trapping times are different for each individual trapping center. The superposition of many such locations results in a noise spectrum that is approximated by a $1/f$ (serial) spectrum.

As mentioned before the charge is by design not destroyed during the readout process. To preserve the charge it is sufficient to shift the signal charge from the IG to a storage position and use the resulting current step caused by this operation as a measure of the signal charge. Shifting the charge back to the IG and out again results in a second measurement of the charge. Repeating this cycle n times results in a measurement improvement of approximately a factor sqrt(n). This holds not only for serial white noise but also for $1/f$ noise. The proposal for repetitive non-destructive readout (RNDR) was already contained in the original publication [3]. There, charge shifting was done between the Internal Gates of two closely spaced DEPFETs connected by a CCD-like transfer structure (Figure 2).

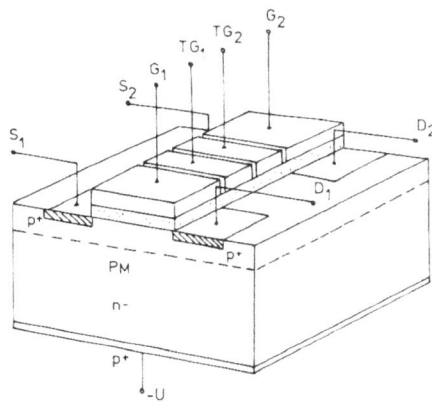

Figure 2. Ping-pong arrangement of DEPFETs in the original publication.

A mathematical correct treatment of noise and measurement precision will be given in the following section and the experimental results will be presented thereafter.

3. DEPFET Readout

In order to measure the signal charge collected in the internal gate of the DEPFET, it is necessary to perform two evaluations. We can assume that before the signal charge arrival, the internal gate is completely empty, since all the charge (both the signal charge of the previous measurement and the leakage current charge) has been removed by a clear pulse. The output of the DEPFET, *i.e.*, the drain current or the source voltage, is measured. This corresponds to evaluation of the baseline of the system, the output corresponding to the empty internal gate. Then, the signal charge is collected into the internal gate of the DEPFET and a second measurement of the device output is performed: the baseline + the signal are evaluated. The difference of the two evaluations (baseline only and baseline + signal) gives the information about the amount of the signal charge. Therefore, every complete measurement is always composed of two evaluations:

- Baseline
- Baseline + Signal

Since the signal arrival time is known, a time variant filter is used for the system readout. One common readout method in APS systems is correlated double sampling (CDS). In this scheme, two correlated samples of the voltage output of the device are taken and subtracted from each other, one sample corresponding to the baseline and one sample corresponding to the baseline + signal. Instead of taking only one sample for each single evaluation, an average of several samples can be taken, performing a multi-correlated double sampling (MCDS) [9]. In most cases this would improve the noise performance of the system. Another readout possibility is to integrate the output current of the device for a certain amount of time instead of sampling its voltage output. This would correspond to an ideal MCDS with an infinite number of samples. In this work we refer to such a current integrating filter that provides a triangular weighting function, *i.e.*, the optimum time-limited filter for white voltage noise. This is the dominant noise source at high speed. In the real case, a trapezoidal weighting function must be used. In fact, a flat-top is necessary to let the output of the DEPFET settle after transferring the charge and to reset some stages of the readout electronics. If the flat-top is relatively short with respect to the total length of the weighting function, the results reported in this paper are not considerably affected. Such a filter has already been successfully implemented in multi-channel readout ASICs for DEPFETs and pnCCDs [10,11].

The operation of subtracting the baseline from the baseline + signal evaluation, independently from the acquisition method (voltage sampling or current integration), results in a high-pass filter. The low-pass limitation is given by the bandwidth of the sampling circuit in the case of CDS or MCDS, and by the integrating process itself in the case of the device current evaluation.

4. RNDR-DEPFET

The RNDR-DEPFET device [12] is composed of two adjacent DEPFET structures, with two individual and insulated internal gates (Figure 3).

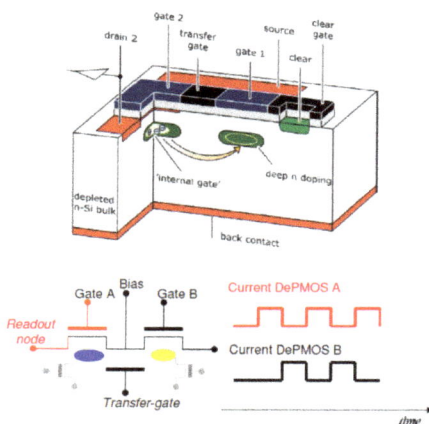

Figure 3. Simplified drawing of a section of a RNDR-DEPFET. The device is composed of two adjacent DEPFET structures with two independent internal gates. Thanks to a transfer gate, the signal charge can be transferred from the internal gate of one DEPFET to the internal gate of the other one. Moving the charge back and forth, it is possible to reproduce and read out the output signal arbitrary often. During operation, when the internal gate of one device is full, the internal gate of the other device is empty and vice versa. This is schematically represented by the plot of the output current of the two transistors. When one transistor has the maximum output current (the internal gate is full), the other transistor has zero signal output current. When the charge is then transferred, the situation is the opposite. For multiple readout it is possible to read out both the transistors or only one. If only one transistor is read out, the other one acts just as a storage device for the signal charge.

The charge in the internal gate of one device can be transferred back and forth to the internal gate of the other device, thanks to one (or more) transfer gate(s). For a proper operation, when the internal gate of one device is full, the internal gate of the other one must be empty and vice versa. Moving the signal charge from one device to the other allows one to reproduce the DEPFET output signal arbitrarily often. In this way the signal charge can be read out non-destructively many times. The main limitation is given by the leakage current that fills the internal gate and spoils the original signal [12]. It is possible to read out the signal from both devices or to evaluate the output of only one device. In the last case the other device is used just as storage for the signal charge. For the sake of simplicity we studied a case in which we read out only one device. It is worth pointing out that, in order to read-read out the signal many times, it is necessary to transfer it from one DEPFET to the other. In fact, as already stated, every measurement is composed of two evaluations: baseline and signal + baseline. It follows that, if we want to read out n times the signal from one of the two devices of the RNDR-DEPFET, e.g., DEPFET (A) in Figure 3, we have to reproduce n times both the baseline, *i.e.*, the output corresponding to empty internal gate, and the signal+ baseline, *i.e.*, the output corresponding to the internal gate filled by signal charge. If we want to reproduce the baseline of transistor (A), without destroying the signal information, we have to move the signal charge into transistor (B), which acts, as already stated, as a charge storage device. Moving back the charge to transistor (A) the output corresponding to baseline + signal is then reproduced. This procedure can be repeated arbitrarily often. The RNDR procedure must not be confused with the MCDS. The MCDS, in fact, refers to the case in which a single baseline or signal + baseline evaluation is obtained by averaging a certain number of samples. The RNDR refers to the case in which the complete measurement procedure (subtraction of the baseline from the signal + baseline) is repeated several times. As already stated, this requires the reproduction both of the signal and of the baseline.

5. DEPFET Noise Analysis

5.1. Noise in Single Readout

In order to better appreciate the benefits of the repetitive non-destructive readout technique, it is useful to discuss the achievable noise performance of a traditional spectroscopic system performing a single readout.

The spectroscopic chain can be represented as in Figure 4 and its associated equivalent noise charge can be expressed with the well-known formula [13]:

$$ENC^2 = \frac{a}{\tau}C_{TOT}^2 A_1 + 2\pi a_f C_{TOT}^2 A_2 + b\tau A_3 \tag{1}$$

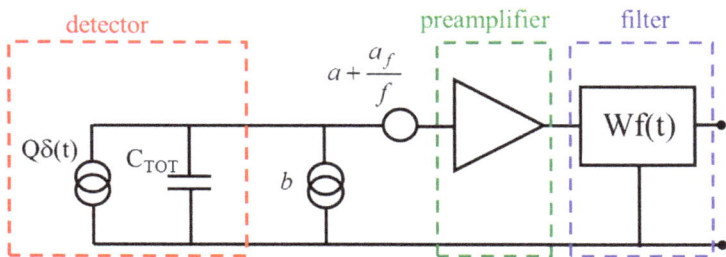

Figure 4. Schematic of a typical spectroscopic chain. The detector is modeled as a delta-like current source in parallel to a capacitance. The input referred noise sources are represented. In this work we consider only the series noise sources (white and 1/f) and a filter triangular weighting function Wf(t).

The three terms of the ENC formula represent the main three noise contributions, *i.e.*, the series white, the series $1/f$ and the parallel white noise contributions.

- C_{TOT} is the equivalent input capacitance of the system.
- a, a_f and b are the series white, the series $1/f$ and the parallel white physical noise sources, referred to the input.
- A_1, A_2 and A_3 are the filter parameters. They depend on the shape of the weighting function $Wf(t)$ implemented by the readout electronics.
- τ is the shaping time of the readout filter and is an expression of the time needed to perform one measurement.

In this work we focus on the series noise contributions (white and $1/f$) and we neglect the effect of the leakage current, which can be minimized by cooling. For this reason the A_3 coefficient is not considered in the following text. From Equation (1) it is evident the $1/f$ component of the ENC is independent from the shaping time of the system, but depends only on the physical $1/f$ noise source and on the type of filter, which determines the coefficient A_2. This means that the term of the ENC related to the $1/f$ noise is independent from the time used to process one signal. In contrast, the ENC term due to the white voltage noise decreases as the shaping time increases. So, once the type of signal processing has been defined, e.g., a triangular weighting function, in order to increase the signal-to-noise ratio, it is possible to increase the shaping time of the filter. This is true as long as the measurement time is so large that the system becomes dominated by $1/f$ noise. At this point a further increase of the shaping time would not bring any benefit, since the $1/f$ noise contribution does not scale with τ. Figure 5 shows the same signal processed by three triangular weighting functions having three different time lengths (τ, $\frac{1}{2}\tau$ and $\frac{1}{4}\tau$). The change of the shaping time modifies only the white noise component of the ENC.

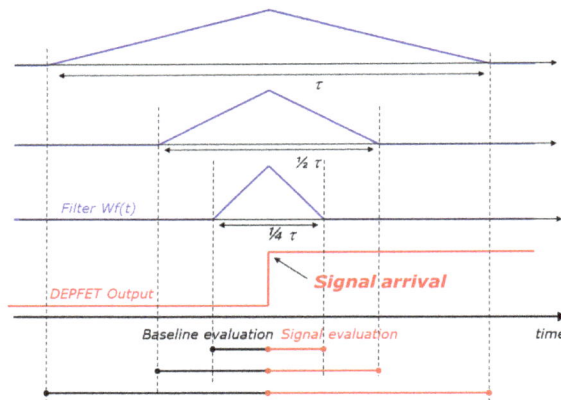

Figure 5. The same signal (red line) is processed by three triangular weighing functions with three different shaping times. The ENC component due to the white voltage noise goes down as the shaping time increases, while the ENC component related to the $1/f$ noise stays constant. Increasing the shaping time the ENC goes down up to the point in which it is dominated by the $1/f$ noise. A further increase of the shaping time would not provide any benefit.

5.2. Noise in Repetitive Non-Destructive Readout

We can assume to fix the total measurement time τ_{TOT}, e.g., we choose a τ_{TOT} for which the $1/f$ noise contribution is dominant. This means that the ENC would not significantly decrease for any $\tau > \tau_{TOT}$. If we make one measurement exploiting the whole time interval τ_{TOT}, thanks to Equation (1) and remembering that the ENC is obtained equating the Noise-to-Signal ratio (N/S) to one (see [13–15]), it is possible to write:

$$\left(\frac{N}{S}\right)^2_{white,\ single} \propto A_1 \frac{1}{\tau_{TOT}} \qquad (2)$$

and

$$\left(\frac{N}{S}\right)^2_{1/f,\ single} \propto A_2 \qquad (3)$$

We can then reproduce, within the time interval τ_{TOT}, the signal n times. This can be done by moving the charge back and forth to the internal gate of one device of the RNDR-DEPFET structure and using the other device as charge storage. In this way it is possible to measure the signal n times and make an average of the measurements. We hypothesize that the signal we reproduce is always the same, *i.e.*, that the signal charge is not spoiled by leakage current electrons that can cumulate in the internal gate. Since τ_{TOT} is fixed, the time available for each single measurement is τ_{TOT}/n, as shown in Figure 6.

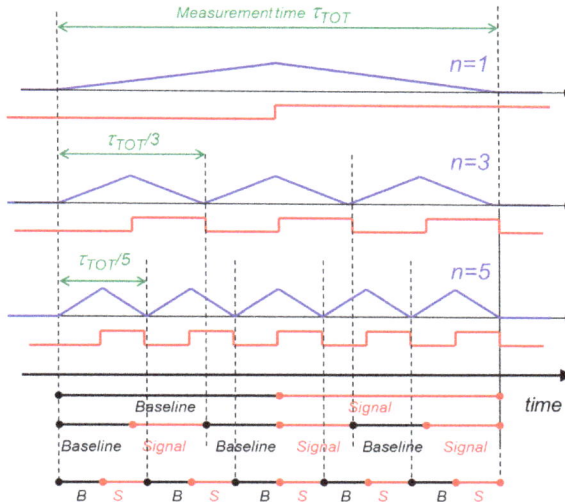

Figure 6. Schematic representation of the multiple non-destructive readout with fixed total measurement time τ_{TOT}. In the upper drawing, only one readout of the signal (red line) is performed, exploiting the whole τ_{TOT}. In the other two cases, the signal is reproduced and read out 3 and 5 times respectively. Since τ_{TOT} is fixed, the time available for each single measurement is $\tau_{TOT}/3$ and $\tau_{TOT}/5$. The reduction of the shaping time for each single measurement (look at the blue weighting functions) turns out in an increased r.m.s. value of the white noise of the individual measurements. The averaging effect of the n readouts (3 and 5 respectively) compensates this noise increment. Therefore the ENC component related to the white voltage noise does not change with the number of measurements in a fixed time interval. For the $1/f$ noise the situation is different. The r.m.s. value of noise of one measurement is independent from the measurement time. This means a single measurement of time length τ_{TOT}, $\tau_{TOT}/3$ or $\tau_{TOT}/5$ would result in the same r.m.s. noise. In this case, the averaging effect of n measurements makes the ENC go down with approximately $\sqrt{n}$. In the figure only the positive step of the signal is measured, *i.e.*, the signal is measured only when it is injected into the internal gate. In theory it would be possible to measure the signal charge also when it is removed from the internal gate, evaluating the falling edge of the DEPFET output. This would increase of a factor two the number of measurements for a certain number of charge transfers.

Since the n measurements are almost independent, making an average the noise sums up quadratically, while the signal sums up linearly. Now let us consider the case of the white voltage noise. If we measure the signal n times within the same total time, the r.m.s. value of the noise of each single measurement is increased because the measurement time has been reduced by a factor of 1/n (see Figure 6). If we want to express the noise-to-signal ratio of the average of n measurements, we have to quadratically sum up the noise of the single measurements and linearly sum up the signal. If we assume that the amplitude of a single signal is normalized to one, then the amplitude of the sum of the n signals is just n. Therefore it is possible to write:

$$\left(\frac{N}{S}\right)^2_{white,\ multiple} \propto \frac{n\left(A_1\frac{n}{\tau_{TOT}}\right)}{n^2} = A_1\frac{1}{\tau_{TOT}} \tag{4}$$

Comparing Equations (2) and (4) it is evident that the multiple readout has no effect on the ENC component due to the white voltage noise, when the measurement time is fixed. The benefit of averaging many readouts is compensated by the increase of the r.m.s. noise of the single measurements, due to the shorter shaping time. For the 1/f noise the situation is different, since in this case the r.m.s. value of the noise is independent from the measurement time. Therefore, a measurement performed in a time τ_{TOT} and a measurement performed in a time τ_{TOT}/n lead to the same 1/f component of the ENC. This means that averaging n measurements with a shaping time τ_{TOT}/n is better than performing only one measurement with the longest possible shaping time τ_{TOT}. In fact the noise-to-signal for the 1/f noise, in the case of n measurements, can be expressed as:

$$\left(\frac{N}{S}\right)^2_{1/f,\ multiple} \propto \frac{nA_2}{n^2} = \frac{A_2}{n}$$

This means that the ENC component related to the 1/f noise scales approximately with $\sqrt{n}$. Actually, to be precise, the (N/S) scales as:

$$\left(\frac{N}{S}\right)^2_{1/f,\ multiple} \propto \frac{A_2}{n^\alpha}$$

where α is a coefficient very close to one. This is due to the fact that the noise of the different measurements is not completely uncorrelated. To make a rigorous calculation the reader can refer to [14–16].

In summary, when the total measurement time is fixed:

- $\left(\frac{N}{S}\right)^2_{white}$ is independent from the number of measurements n
- $\left(\frac{N}{S}\right)^2_{1/f}$ scales approximately as 1/n

For the sake of completeness, even if not analyzed in this work, we can also mention that the noise due to the leakage current in the internal gate does not scale and increases with the total measurement time (see [12]). If the time of a single measurement is fixed, *i.e.*, if the total measurement time increases with number of measurements n:

- the white series noise and the 1/f series noise contributions scale as 1/n
- the noise due to the leakage current that fills the internal gate increases with n, *i.e.*, with the total measurement time.

From the above considerations, it follows that the repetitive non-destructive readout technique should be used when the 1/f noise is dominant. Given a system with defined input noise sources, it is in general possible to increase the readout time in order to make the white series noise negligible.

Only at this point it is convenient to apply the RNDR processing scheme. Of course, in a real case, the total measurement time duration is limited by experimental constraints and by the leakage current that fills the internal gate of the devices, degrading the noise properties of the RNDR-DEPFET (see [12]).

On the other hand, the minimum time length of a single measurement is also limited in reality by technical constraints. In this case, in order to achieve a number of measurements n, sufficient to decrease the 1/f component to the desired value, it can be necessary to increase the total measurement time of the system, operating at a lower rate. In practice it is convenient to operate with flexible readout electronics [10,11] which allows one to change the time duration of the weighting function and the number of possible readout cycles. With this kind of tunability it is possible to trade speed for resolution with respect to the different experimental requirements, changing only the number n of multiple readouts and adjusting the weighting function duration accordingly.

We can evaluate in a more analytic and quantitative way the noise figure of a multiple signal processing accomplished by a filter implementing a triangular weighting function for every signal readout. We define an n-fold saw-tooth shaped weighting function, composed of n triangular weighing functions, as follows:

$$Wf(t) = \frac{1}{\tau} \sum_{k=1}^{n} \left\{ \left[t - 2\frac{\tau(k-1)}{n} \right] \cdot \mathcal{H} \left[t - 2\frac{\tau(k-1)}{n} \right] - 2 \left[t - \frac{\tau(2k-1)}{n} \right] \right. \\ \left. \cdot \mathcal{H} \left[t - \frac{\tau(2k-1)}{n} \right] + \left[t - 2\frac{\tau k}{n} \right] \cdot \mathcal{H} \left[t - \frac{\tau k}{n} \right] \right\}$$

(5)

$\mathcal{H}[t]$ represents the Heaviside function. The sum of the amplitude of the n individual signal readouts is normalized to one, *i.e.*, the n maxima Max [Wf(t)] are equal to 1/n. The shaping time τ is defined as the total available measurement time and does not depend on n.

The coefficients A_1 and A_2 can be calculated as follow:

$$A_1 = \int_{-\infty}^{+\infty} \left[Wf'(y) \right]^2 dy$$

(6)

$$A_2 = \int_{-\infty}^{+\infty} \left[Wf^{\frac{1}{2}}(t) \right]^2 dt$$

(7)

where $Wf^{\frac{1}{2}}(y)$ is the derivative of order ½ of the weighting function and $y = t/\tau$ is the time normalized to the shaping time τ.

From Equations (5)–(7) it follows that $A_1 = 4$ for every n, while A_2 decreases as n increases. The A_2 coefficients for different n values are reported in Table 1.

Table 1. 1/f filter coefficients for a saw-tooth weighting functions composed of n triangles. The corresponding ENC due to the 1/f noise is calculated assuming $a_f = 4.5 \times 10^{-12}$ V^2/Hz and an DEPFET input capacitance of 40 fF.

Number of Readouts n	A_2	ENC$_{1/f}$
1	0.88254	1.25
2	0.38287	0.82
4	0.17038	0.55
8	0.07823	0.37
16	0.03695	0.25
32	0.01782	0.18

As an example we can consider the measured noise power spectral density of an existing prototype DEPFET [17]. Typical physical noise sources values are: $a = 1.5 \times 10^{-16}$ V^2/Hz and $af = 4.5 \times 10^{-12}$ V^2. Table 1 reports the ENC component due to the 1/f noise for different number of readouts n, considering an equivalent input capacitance of the DEPFET of 40 fF. These values are independent from the total measurement time (shaping time) τ. Figure 5 shows the different ENC components as a function of τ

for the cases n = 1 and n = 16. As stated before, ENC_{white} is the same for every n. From Figure 7a it is evident that for $\tau < 5$–$10\ \mu s$ the overall ENC is dominated by the white noise and reading out the signal multiple times brings only a negligible improvement.

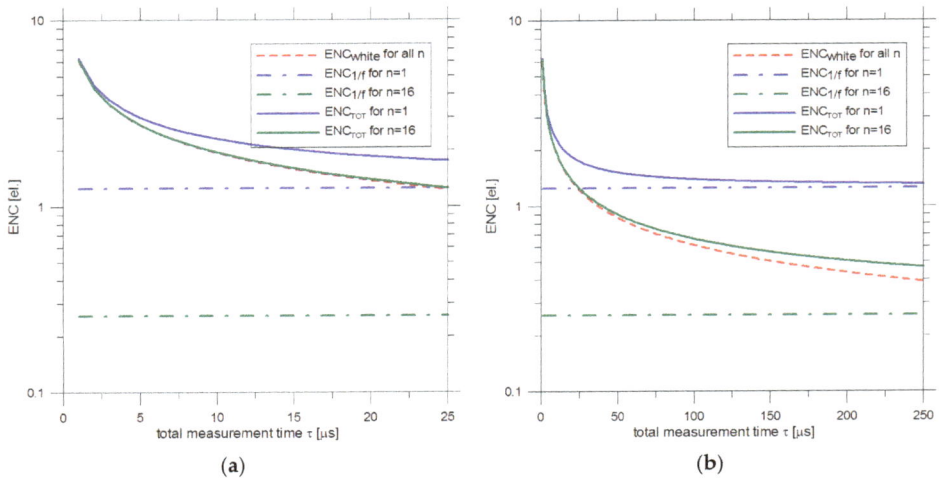

(a)   (b)

Figure 7. Calculated ENC components of a typical DEPFET processed by a triangular filter as a function of the total measurement time for two different number of readouts n = 1 and n = 16 for (**a**) fast readout and (**b**) slow readout.

At shaping times longer than 10 µs, the multiple readout has a consistent impact on the overall noise. For $\tau = 200\ \mu s$ (Figure 7b), for example, it is: $ENC_{TOT} = 1.3$ el. rms for n = 1 and $ENC_{TOT} = 0.5$ el. rms for n = 16.

In system where all the DEPFET pixels in one row of the sensor matrix are read out in parallel, e.g., in the focal plane of the MIXS instrument [18] with the ASTEROID ASIC [10], one can achieve sub-electron noise sensitivity reading out small matrices with a frame rate of some hundreds of Hz. With the mentioned measurement time per row of 200 µs, a 32 × 32 DEPFET array read out from two sides would provide a noise as low as 0.5 el. rms with a frame rate of about 300 Hz.

In a more sophisticated readout approach, e.g., the one adopted for the DEPFET sensor of the DSSC detector [19,20], one can use a dedicated readout channel for every pixel of the array. In this case the readout times needed for a single pixel and for the whole matrix are the same. It follows that a noise of 0.5 el rms is achievable with a frame rate of approximately 5 kHz. In this approach, which requires full parallel readout with an ASIC bump-bonded to the sensor, the frame rate is in principle not limited by the sensor size.

6. Experimental Evidence

Several RNDR-DEPFET structures have been fabricated on a multi-project run and used to experimentally demonstrate the feasibility of the repetitive non-destructive readout of DEPFETs to beat the 1/f noise limit [8]. The impact of the leakage current on the achievable noise level, which for simplicity was neglected in the mathematical treatment, is illustrated in Figure 8. Measurements were done on one cell of a 4 × 4 mini-matrix with a cell size of 75 × 75 µm^2. It shows the achievable noise σ_{end} (r.m.s.) for three different temperatures of $-30\ ^\circ C$, $-40\ ^\circ C$ and $-55\ ^\circ C$ as a function of the number of readouts (n). The required time for a single readout τ_{SINGLE} was 25.5 µs, the total readout time increases linearly with n. Between n = 2 and n = 40 the noise follows the expected 1/sqrt(n) behavior. At the moderate temperature of $-30\ ^\circ C$ the total noise starts to increase again for n > 90 due to leakage current, which is collected in the internal gate during the multiple readouts and thus falsifies the number of stored signal electrons. For lower temperatures this turning point is shifted towards higher n and the curves approach more and more the expected 1/sqrt(n) behavior. Using a Monte-Carlo simulation and taking into account the measured leakage current of 0.02 e per ms and pixel at a temperature of −55 °C, the observed behavior can be well reproduced.

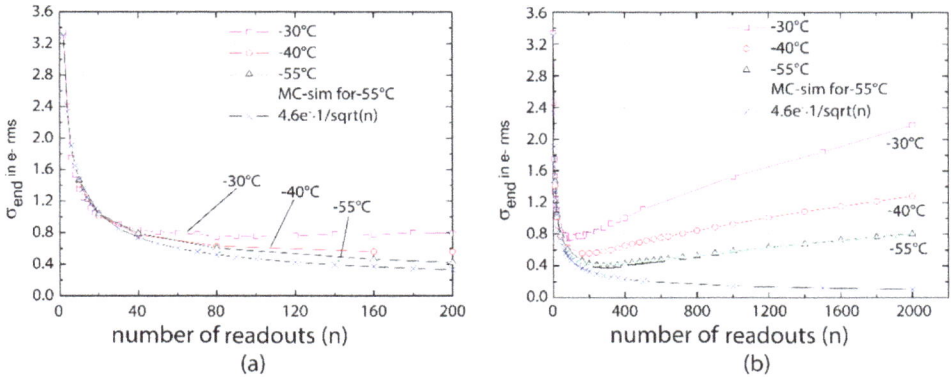

Figure 8. Experimental confirmation of the noise reduction due to multiple readouts using a fixed readout time of τ_{SINGLE} = 25.5 µs. For a single readout (n = 1) the readout noise for was σ_{SINGLE} = 4.6 e⁻ (a) Between 2 and 40 repetitive readouts the read noise follows approximately the expected behavior σ_{end} = σ_{SINGLE}/sqrt(n) = 4.6 e⁻/sqrt(n); (b) For larger n values, the read noise is dominated by the leakage current and σ_{end} increases for higher n.

The lowest read noise so far was measured with a circular RNDR-DEPFET featuring a gate length of 5 µm and thus providing a significantly higher amplification compared to design used for the measurements shown in Figure 9. Due to the higher amplification a read noise of only 3.1 e⁻ r.m.s. was obtained for a single readout and consequently a noise of 0.18 e⁻ r.m.s. at a temperature of −55 °C was achieved after only 300 readout cycles with a readout time of 25.5 µs. In order to demonstrate the deep sub-electron resolution, the DEPFET pixel was illuminated by an optical laser (λ = 672 nm) with a very low intensity, generating only very few signal electrons in the pixel. The number of incident photons was subjected to Poissonian statistics, resulting in the spectrum shown in Figure 9.

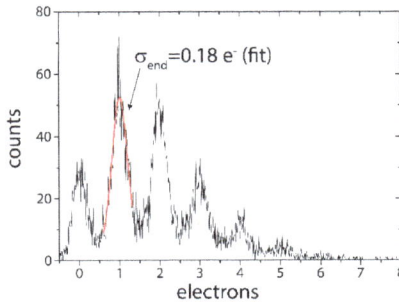

Figure 9. Single photon spectrum measured at low light intensity with a circular RNDR-DEPFET at a temperature of −55 °C. Due to the higher amplification the read noise of a single readout is only 3.1 e⁻ rms and a minimum noise of 0.18 e⁻ was obtained with only 300 readouts.

In contrast to SiPMs, EMCCDs or other avalanche based detectors, the RNDR-DEPFET is a fully linear detector amplifier structure. In order to demonstrate this unique feature, the DEPFET pixel was illuminated with a faint light source so that during the integration time, on average a single photo electron was generated in the internal gate. Subsequently, the exposure time of the DEPFET detector to the light source was increased step by step and for each exposure time a spectrum was taken. By superimposing the various spectra for different exposure times a spectrum, as depicted in Figure 10a, was created. Up to 120 photons, a number which corresponds to the average for photos generated at the maximum exposure time, all photon numbers are equally represented. In the close-up (b) of the red hatched area, well separated peaks can be observed, and by counting the individual peaks, the number of incident photon can be precisely determined.

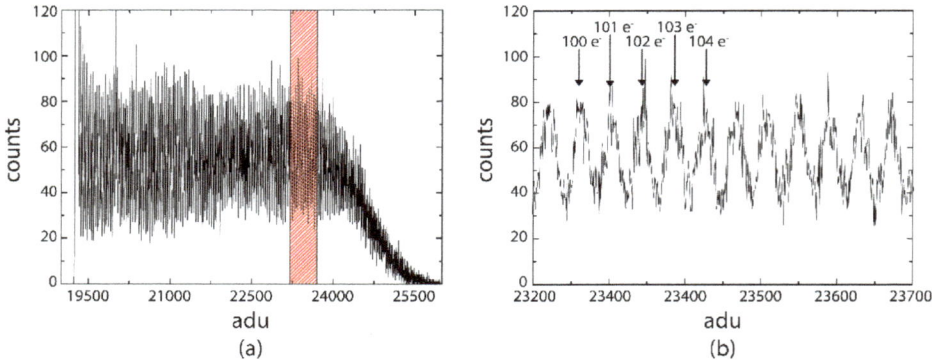

Figure 10. By continuously extending the exposure time of the DEPFET to a faint light source, a spectrum, which includes all photon numbers, can be induced. At an average of 120 electrons the exposure time was not increased any further, resulting in a Gaussian-shaped intensity drop for higher electron numbers (**a**). In the close-up view (**b**) of the red hatched area, well separated peaks can be seen and individual electrons can be counted even for high electron numbers.

7. Summary and Outlook

We have described a sensor developed with the aim of being able to measure charge with sub-electron precision. This is useful for measuring the spectrum of low energy X-rays and especially for photon counting in the visible range. This sensor is based on the low-noise property of the DEPFET structure and its capability to read out the same signal charge repetitively. This RNDR readout method suppresses not only the white serial noise but also the low frequency $1/f$ noise that is normally independent of shaping conditions. A thorough theoretical treatment of noise for RNDR readout has been given. This treatment focusses on serial noise that is due to noise generated in the DEPFET transistors. Shot noise that is due to dark current in the sensor, however, was neglected. Its effect is an occasional increase in the charge that is shifted in and out of the internal gate of the DEPFET in all following read cycles. Thus, the measurement will result in non-integer multiples of the elementary charge. This problem is treated in [8]. Also neglected in the treatment are effects in the DEPFETs due to operation of the charge transfer mechanism.

Measurements on prototype RNDR DEPFET devices have shown excellent results. A sub-electron measurement precision of 0.18 electrons has been achieved with a moderate precision of 3.1 electrons for single readout. Furthermore distinction between n and n + 1 electrons is possible up to several hundred electrons. In the future, increased precision will be possible with improved noise for a single readout and more sophisticated readout electronics allowing for example readout of both DEPFETs of a pixel simultaneously. Furthermore, pixels can be read out in parallel, as is done for example in the DSSC project at the European XFEL, leading to an enormous frame rate, although at increased power consumption. The combination of RNDR-DEPFETs with an intermediate storage device [7] offers additional advantages such as simultaneous gating of all pixels and avoidance of sensitivity to signals during readout.

Acknowledgments: We thank the process engineers and technicians who produced the RNDR-DEPFET prototype sensors.

Author Contributions: All authors have participated in writing the article. G.L. together with the late J. Kemmer has invented the DEPFET and the RNDR device principle. Together with L.S. he participated in all stages of the project. S.W. demonstrated the sub-electron measurement precision in the frame of his PhD thesis work. M.P. derived the RNDR noise analysis formalism and also was involved in the readout electronic development. S.A. has contributed layout variants for recent RNDR DEPFET fabrications.

Conflicts of Interest: The authors declare no conflict of interest.

References

1.   Hartmann, R.; Stephan, K.-H.; Strüder, L. The quantum efficiency of pn-detectors from the near infrared to the soft X-ray region. *Nucl. Instrum. Methods Phys. Res. Sect. A* **2000**, *439*, 216–220. [CrossRef]

2.   Kraft, R.P.; Burrows, D.N.; Garmire, G.P.; Nousek, J.A.; Janesick, J.R.; Vu, P.N. Soft X-ray spectroscopy with sub-electron readnoise charge-coupled devices. *Nucl. Instrum. Methods Phys. Res. Sect. A* **1995**, *361*, 372–383. [CrossRef]

3.   Kemmer, J.; Lutz, G. New semiconductor detector concepts. *Nucl. Instr. Meth. A* **1987**, *253*, 365–377. [CrossRef]

4.   Lutz, G.; Herrmann, S.; Lechner, P.; Porro, M.; Richter, R.H.; Strüder, L.; Treis, J. New DEPFET structures, concepts, simulations and experimental results. *Proc. SPIE* **2008**, *7021*. [CrossRef]

5.   Gatti, E.; Rehak, P. Semiconductor drift chamber—An application of a novel charge transport scheme. *Nucl. Instrum. Methods Phys. Res. Sect. A* **1984**, *225*, 608–621. [CrossRef]

6.   Kemmer, J.; Lutz, G. Low capacitive drift diode. *Nucl. Instr. Meth. A* **1987**, *253*, 378–381. [CrossRef]

7.   Lutz, G. Halbleiterdetektor Mit Einem Zwischenspeicher für Signalladungsträger und Entsprechendes Betriebsverfahren. DE Patent 102011115656 A1, 28 March 2013. (In Germay)

8.   Wölfel, S.; Herrmann, S.; Lechner, P.; Lutz, G.; Porro, M.; Richter, R.H.; Strüder, L.; Treis, J. A Novel Way of Single Optical Photon Detection: Beating the 1/f Noise Limit with Ultra High Resolution DEPFET-RNDR Devices. *IEEE Trans. Nucl. Sci.* **2007**, *54*, 1311–1318. [CrossRef]

9.   Porro, M.; Herrmann, S.; Hörnel, N. Multi correlated double sampling with exponential reset. In *Nuclear Science Symposium Conference Record*, Proceedings of the IEEE NSS'07, Honolulu, HI, USA, 26 October–3 November 2007; pp. 291–298.

10. Porro, M.; de Vita, G.; Herrmann, S.; Lauf, T.; Treis, J.; Wassatsch, A.; Bombelli, L.; Fiorini, C. ASTEROID: A 64 channel ASIC for source follower readout of DEPFET arrays for X-ray astronomy. *Nucl. Instr. Meth. Phys. Res. Sect. A* **2010**, *617*, 351–357. [CrossRef]

11. Porro, M.; Bianchi, D.; De Vita, G.; Hartmann, R.; Hauser, G.; Herrmann, S.; Strüder, L.; Wassatsch, A. VERITAS: A 128-channel ASIC for the readout of pnCCDs and DEPFET arrays for X-ray imaging, spectroscopy and XFEL applications. *IEEE Trans. Nucl. Sci.* **2013**, *60*, 446–455. [CrossRef]

12. Wölfel, S.; Herrmann, S.; Lechner, P.; Lutz, G.; Porro, M.; Richter, R.; Strüder, L.; Treis, J. Sub-electron noise measurements on repetitive non-destructive readout devices. *Nucl. Instr. Meth. Phys. Res. Sect. A* **2006**, *566*, 536–539. [CrossRef]

13. Gatti, E.; Manfredi, P.F.; Sampietro, M.; Speziali, V. Suboptimal filtering of 1/f-noise in detector charge measurements. *Nucl. Instr. Meth. Phys. Res. Sect. A* **1990**, *297*, 467–478. [CrossRef]

14. Gatti, E.; Manfredi, P.F. Processing the signals from solid-state-detector in elementary particle physics. *Riv. Nuovo Cimento* **1986**, *9*, 1–146. [CrossRef]

15. Gatti, E.; Sampietro, M.; Manfredi, P.F. Optimum filters for detector charge measurements in presence of 1f noise. *Nucl. Instr. Meth. Phys. Res. Sect. A* **1990**, *287*, 513–520. [CrossRef]

16. Gatti, E.; Geraci, A.; Guazzoni, C. Multiple read-out of signals in presence of arbitrary noises Optimum filters. *Nucl. Instr. Meth. Phys. Res. Sect. A* **1998**, *417*, 342–353. [CrossRef]

17. Porro, M.; Ferrari, G.; Fischer, P.; Halker, O.; Harter, M.; Herrmann, S.; Hornel, N.; Kohrs, R.; Krueger, H.; Lechner, P.; *et al.* Spectroscopic performance of the DePMOS detector/amplifier device with respect to different filtering techniques and operating conditions. *IEEE Trans Nucl. Sci.* **2006**, *53*, 401–408. [CrossRef]

18. Treis, L.; Andricek, F.; Aschauer, K.; Heinzinger, S.; Herrmann, M.; Hilchenbach, T.; Lauf, P.; Lechner, G.; Lutz, P.; Majewski, M.; *et al.* MIXS on BepiColombo and its DEPFET based focal plane instrumentation. *Nucl. Instr. Meth. Phys. Res. Sect. A* **2010**, *624*, 540–547. [CrossRef]

19. Porro, M.; Andricek, L.; Aschauer, S.; Bayer, M.; Becker, J.; Bombelli, L.; Castoldi, A.; de Vita, G.; Diehl, I.; Erdinger, F.; *et al.* Development of the DEPFET Sensor with Signal Compression: A Large Format X-ray Imager with Mega-Frame Readout Capability for the European XFEL. *IEEE Trans. Nucl. Sci.* **2012**, *59*, 3339–3351. [CrossRef]

20. Lutz, G.; Lechner, P.; Porro, M.; Strüder, L.; de Vita, G. DEPFET sensor with intrinsic signal compression developed for use at the XFEL free electron laser radiation source. *Nucl. Instr. Meth. Phys. Res. Sect. A* **2010**, *624*, 528–532. [CrossRef]

sensors

MDPI

Review

Photon Counting Imaging with an Electron-Bombarded Pixel Image Sensor

Liisa M. Hirvonen and Klaus Suhling *

Department of Physics, King's College London, Strand, London WC2R 2LS, UK; liisa.2.hirvonen@kcl.ac.uk
* Correspondence: klaus.suhling@kcl.ac.uk; Tel.: +44-20-7848-2119

Academic Editor: Edoardo Charbon
Received: 27 January 2016; Accepted: 25 April 2016; Published: 28 April 2016

Abstract: Electron-bombarded pixel image sensors, where a single photoelectron is accelerated directly into a CCD or CMOS sensor, allow wide-field imaging at extremely low light levels as they are sensitive enough to detect single photons. This technology allows the detection of up to hundreds or thousands of photon events per frame, depending on the sensor size, and photon event centroiding can be employed to recover resolution lost in the detection process. Unlike photon events from electron-multiplying sensors, the photon events from electron-bombarded sensors have a narrow, acceleration-voltage-dependent pulse height distribution. Thus a gain voltage sweep during exposure in an electron-bombarded sensor could allow photon arrival time determination from the pulse height with sub-frame exposure time resolution. We give a brief overview of our work with electron-bombarded pixel image sensor technology and recent developments in this field for single photon counting imaging, and examples of some applications.

Keywords: photon counting; electron-bombarded sensor; single photon detection; low light level imaging; EBCCD; EBCMOS

1. Introduction

Photon counting imaging is a well-established low light level imaging technique where an image is assembled from individual photons whose position is recorded during the detection process, usually with a position-sensitive sensor (*i.e.*, a camera). In astronomy, photon counting imaging technology was originally introduced due to its sensitivity, and continues to be used on both ground- and space-based observatories, particularly in the UV [1–4]. Other advantages of photon counting imaging include linearity, high dynamic range, high sensitivity, zero read-out noise and well-defined Poisson statistics, and photon counting imaging is now finding applications in diverse fields of science, such as fluorescence spectroscopy and microscopy, LIDAR, optical tomography and quantum cryptography, as reviewed recently [5–8].

The light detection capability of solid state sensors is based on their ability to convert photons, via electron-hole pair generation, into an electronic signal that can be read out. Despite recent developments in these sensors, especially in CMOS technology which now allow megapixel resolution and up to MHz frame rates [9], these detectors are still not sensitive enough to detect single photons without amplification [10]. Several methods have been developed to produce a detectable output signal from the incoming single photons. Traditionally, microchannel plate (MCP) image intensifiers have been used to amplify the signal, and intensified CCDs (ICCDs), which combine a photon counting MCP-intensifier with a CCD camera in one package, have been commercialised. In these devices, a single photon creates a photoelectron which creates secondary electrons as it travels through the MCPs, before the electrons are converted back into photons with a phosphor screen, coupled to the CCD with by a fibre optic taper or a lens. Electron-multiplying CCDs (EMCCDs) where the signal is amplified in a gain register placed between the shift register and the output amplifier, are also single photon sensitive and commercially available, and an EMCMOS concept has been demonstrated [11].

Single photon detection is also possible with electron-bombarded (EB) sensors, where a photocathode is placed in front of the sensor, and a single photoelectron is accelerated through a high voltage directly into a solid state sensor. EB sensors are conceptually similar to old silicon intensified target television cameras, and have found applications in microscopy [12,13] and biological low-light imaging [14–16], optical spectroscopy [17] and radiography [18,19]. The advantages of EB sensors over intensified camera systems include reduced sensor size and weight, increased sensitivity and dynamic range, faster response time, and better contrast and resolution. They require a high voltage of several kV between the photocathode and the CCD sensor in vacuum, and backscattered photoelectrons can be detected on the low energy side of the pulseheight distribution, making it asymmetric [20]. Note that manufacture of such a device requires skilful incorporation of a wire-bonded silicon chip into a vacuum tube enclosure with cleanliness, low outgassing and vacuum bake requirements. In contrast to intensifiers, it neither requires microchannel plates, nor a phosphor screen.

The EB concept has also been utilised in point detectors. In a hybrid photodetector (HPD), a photocathode is placed in front of an avalanche photodiode (APD) and single photons are accelerated into the APD [21,22]. To provide both spatial and timing resolution, HPD arrays have been built and demonstrated [23] and linear 16 HPDs are now commercially available, e.g., for spectrally-resolved fluorescence lifetime measurements.

We note here that recent developments in single photon avalanche diode (SPAD) detectors, which can be manufactured in large arrays using CMOS technology, show great promise as an alternative to vacuum-based detector technology. They simultaneously deliver single photon sensitivity and picosecond timing resolution in tens of thousands of pixels, and have the potential to significantly advance time-resolved fluorescence microscopy and other fields, see Section 5 for more details.

2. EB-Technology

2.1. How EB-Sensors Work

An EB sensor combines a photocathode with a silicon solid state CCD or CMOS sensor under vacuum, as illustrated in Figure 1a, such that a single photoelectron, ejected from the photocathode by a photon, is directly accelerated into the sensor by a high voltage of several kVs [24]. In the sensor, the electron creates a well-defined number of electron-hole pairs (around one electron-hole pair per 3.7 eV in silicon [25]) and consequently the pulse height distribution of these devices is narrow and strongly dependent on the acceleration voltage.

EB-sensors are usually back-thinned (as the front side electronic layers prevent any low energy particle detection), leading to charge sharing between pixels; during the diffusion of the electrons from the back of the sensor to the front the charge spills over into adjacent pixels, and the photon events typically have a sharp central peak and small wings (Figure 1b). Centroiding methods can be employed to find the photon event location with sub-pixel accuracy; in this case the resolution is governed by the proximity focussing principle [26] (*i.e.*, the photoelectron trajectories from the photocathode to the sensor over a small, typically ~1 mm gap). The backscatter in our EB-sensor is small [20], and, in general, EB-sensors have better contrast and resolution compared to MCP intensified systems, where backscattering of electrons degrades the image quality [27].

Besides photon events, another type of event detected with EB-sensors are ion events, which are caused by an electron hitting a residual gas molecule inside the imperfect vacuum. The gas molecule is ionised and accelerated towards the photocathode, where it causes secondary emission of electrons, and bright, large events when the electrons hit the sensor. In MCP-based intensifiers, the MCPs are placed in a chevron arrangement, *i.e.*, with a small bias angle, to minimise this type of ion feedback, but in EB-sensors, there is no such barrier and a free line of sight between the photocathode and the sensor.

These ion events can usually be discarded during data processing, as their large size and brightness allows them to be easily differentiated from the photon events.

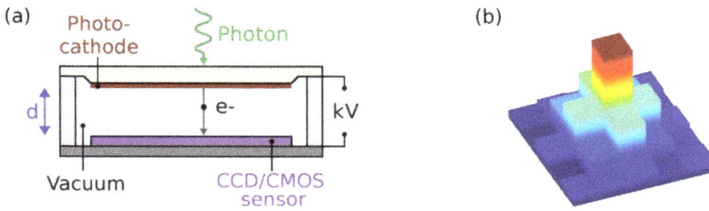

Figure 1. (**a**) Schematic diagram of an electron bombarded sensor. A photon impinging on the photocathode liberates a photoelectron, e⁻. Using a large potential difference of several kV, the photoelectron is accelerated over a distance d into the sensor, where it creates electron-hole pairs; in silicon, one electron-hole pair for each 3.7 eV depending on conditions such as temperature and impurities; (**b**) Schematic of typical photon event charge distribution in the sensor. The event covers several pixels, with a high central peak and small wings.

Using a simple model, the gain of an EB-sensor is determined by [27,28]

$$EB_{gain} = \frac{V - V_{th}}{W} \tag{1}$$

where V is the potential difference between the cathode and the sensor, V_{th} is the threshold voltage, and W is the energy needed to create one electron-hole pair in the sensor. More detailed gain models have been devised, taking backscattered photoelectrons into account [27]. For silicon, W is ~3.7 eV, depending on the local conditions [25,29]. The sensor is covered by a layer of aluminium, and thus only electrons with energy above a threshold energy eV_{th} will be detected. The gain is thus strongly dependent on the acceleration voltage, leading to a narrow pulse height distribution. The variance in gain, σ^2, is expressed as

$$\sigma^2 = F \times EB_{gain} \tag{2}$$

where F is the Fano factor (0.12 for silicon) [30].

The time-of-flight τ of a photoelectron with mass m and charge q from the photocathode to the sensor is given by

$$\tau = \frac{d}{\sqrt{2V}} \sqrt{\frac{m}{q}} \tag{3}$$

where d is the distance between the cathode and CCD and V is the potential. For a typical potential V of a few kV and a typical distance d around 1 mm, the time of flight for electrons is few tens of ps.

2.2. EBCCD Cameras

The first reports in the literature characterising EBCCD cameras appear in the 1980s [31–34]. They were originally developed for their high signal-to-noise ratio under low light levels [28] which was deemed a considerable advantage for astronomical applications [35]. EBCCDs were also developed for military night vision applications [27]. The first commercially available EBCCD was made by Hamamatsu in 2000, and offered 1024 × 2014 pixels, operating voltage of 6–8 kV and a 3 Hz maximum frame rate [36]. Other EBCCD developments have reported frame rates up to 200 Hz [37], and maximum acceleration voltages of 14 kV and 15 kV [18,38,39]. Intensified EBCCDs, comprising a MCP between the photocathode and the CCD, have also been developed [40].

2.3. EBCMOS Cameras

With CMOS cameras each pixel, or row of pixels, has its own amplification and read-out electronics, and can thus achieve faster frame rates than CCD cameras [10]. CMOS sensor technology has developed at a rapid pace over the past two decades, now replacing CCD cameras in consumer electronics and also in scientific research. EBCMOS sensors were first developed for night vision devices in the military [41,42]. For scientific research, the development of EBCMOS cameras originated from applications in particle physics, and Mimosa 5 was demonstrated in 2007, with 1024 × 1024 pixels, 40 Hz frame rate and operating voltage of 6–10 kV [43]. A number of other sensors have been developed [44,45], with the latest development offering a 500 Hz frame rate [46].

A CMOS pixel read-out chip, developed in CERN for particle physics applications, the MediPix2/TimePix ASIC with 256 × 256 pixels and 55 µm pixel size, has been combined with EB concept for single photon detection [47]. Using a clock of 100 MHz and a parallel readout the entire chip can be read out in 266 µs, which makes frame rates of over 3000 fps possible [48].

2.4. Photon Arrival Timing

Unlike MCP-intensifiers, EB sensors cannot directly provide photon arrival timing. Point detectors (HPDs) are often combined with TCSPC timing electronics based on a time-to-amplitude converter (TAC) and used for photon arrival timing, for example, in fluorescence lifetime imaging scanning fluorescence microscopy [21,22], but with pixel image sensors photon arrival timing is less straightforward. Although the Medipix/Timepix chips are in principle capable of high timing resolution, their main drawback is 266 µs frame read-out time which limits the global count rate, and they find more applications in photon counting imaging where the arrival timing of the photons is not required and the photons can be accumulated in each pixel before the frame readout.

With EB-sensors, it could be possible to exploit the dependency between the photon event brightness and the acceleration voltage for photon arrival timing [20,49,50]. By varying the voltage in time, in a similar fashion to varying the voltage on the deflector plates in a cathode ray oscilloscope or a streak camera, the photon event height in the sensor corresponds to the photon arrival time at the photocathode. Thus, by converting the arrival time into an amplitude, each pixel is used as a photoelectronic TAC, see Section 3.3. This approach could parallel-process the arrival time of photons in each pixel of the image simultaneously. This kind of time-tagging is not possible with MCP-based intensified CCD cameras due to the broad pulse height distribution of MCPs [24].

3. Experimental Characterisation

3.1. Single Photon Events & Centroiding

Typical single photon events detected with a Hamamatsu C7190-13 EBCCD at the maximum 8 kV acceleration voltage and maximum read-out gain are shown in Figure 2. The central peak is high with small wings; during the diffusion of the electrons from the back of the sensor to the front, the charge spills over into adjacent pixels, although the pixel's full well capacity is not reached [20]. Brighter, larger ion events are also detected. Single photon events detected with EBCMOS cameras are reported to be very similar to EBCCD photon events (see, for example, Figure 3 in [46]).

In EB pixel image sensors, the pixel's potential wells can be filled by the electrons created by only a few photons. For this reason, EB sensors usually have large pixels [51], to facilitate collection of many photons per pixel in analogue fashion. However, for single photon counting applications, a maximum of one photon is collected per pixel per frame, and with photon counting approaches where the photon events cover an area bigger than one pixel, the resolution of the image does not need to be limited by the sensor pixel size. A characteristic feature of this method is the possibility of employing a centroiding technique, where the position of a photon event can be determined with sub-pixel accuracy [52–56]. With EB-sensors, centroiding can be used to recover the resolution lost in the electron diffusion process, and the resolution of the image is then limited by the photoelectron trajectories from the photocathode to the sensor, governed by the proximity focussing principle [26].

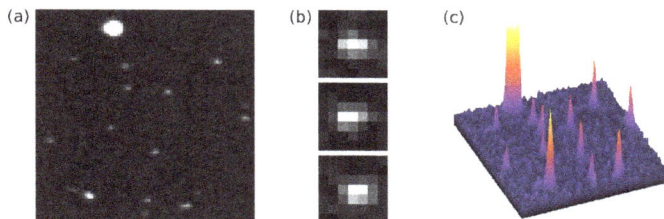

Figure 2. (**a**) 80 × 80 pixel area of a frame with single photon events as detected with a Hamamatsu electron-bombarded CCD at 8 kV acceleration voltage and maximum read-out gain. A bright, large ion event can be seen near the top edge; (**b**) Enlarged areas of three photon events; (**c**) 3D representation of the area in (**a**).

The algorithms employed for event centroiding in photon counting imaging were originally developed for implementation in hardware and are based on a simple center-of-mass calculation [52]. The centroiding is nowadays done in software but the algorithms employed in photon counting imaging are still usually simple, one-iteration algorithms [3]. Recently, we have applied iterative algorithms developed for super-resolution fluorescence microscopy for centroiding of photon events, and found that these algorithms yield excellent results for both MCP-intensified camera systems [57] and EBCCDs (Figure 3a–c) [58], providing efficient photon event recognition, low fixed pattern noise and excellent localisation results. Moreover, multi-emitter fitting algorithms–developed for super-resolution microscopy to separate fluorescent emitters whose point-spread functions overlap partially–allow separation of overlapping photon events with EBCCDs, see Figure 3d,e, an important aspect to facilitate an increased count rate and shorter acquisition times.

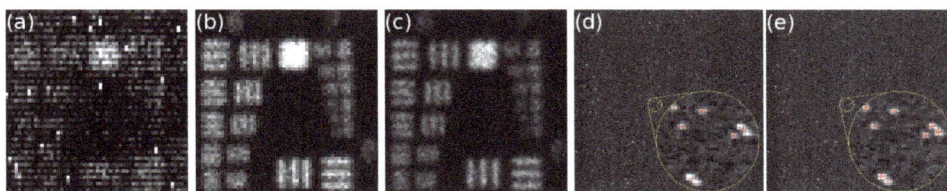

Figure 3. (**a–c**) Images of a USAF test pattern obtained by photon counting imaging with a Hamamatsu EBCCD: (**a**) sum of 30,000 frames; (**b**) 1-pixel centroiding; (**c**) 1/5-pixel centroiding. With 1 pixel centroiding, the photon event is assigned to the to the center pixel of the event and the edges are ignored. With 1/5 pixel centroiding, each pixel is divided into 5 × 5 subpixels, and each photon event is assigned a sub-pixel according to the centroid position calculated by the sub-pixel localisation algorithm; (**d–e**) A raw frame of USAF data with photon positions localised with super-resolution software marked with red crosses. Overlapping events that are normally counted as one event (**d**) are resolved with multi-emitter fitting analysis (**e**) [58].

3.2. Pulse Height Distribution

The pulse height distribution of an EB sensor can provide information about the electron-hole generation process in the sensor. The pulse height distributions of a Hamamatsu C7190-13 EBCCD were measured for different acceleration voltages (Figure 4a). The slight asymmetry is probably due to backscattered photoelectrons [20]. The mean pulse height was plotted against the acceleration voltage (Figure 4b). A straight line fit according to Equation (1) yields a gradient of 266 e/kV, and the energy needed for the creation of one electron-hole pair in silicon can be determined from the inverse gradient: 3.76 ± 0.05 eV. A mean threshold voltage of 2.5 ± 0.1 kV for this device can be found from the x-intercept (Figure 4b); only electrons with an energy above this threshold are detected due to the aluminium layer protecting the CCD. According to Equation (1), this yields a maximum gain of ~1500 at 8 kV for this step [20,50].

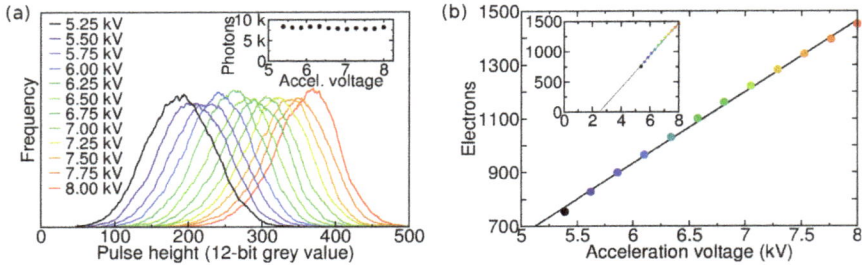

Figure 4. (a) Photon event pulse height distributions for different acceleration voltage settings, measured with a Hamamatsu EBCCD. The count rate and number of acquired frames was the same for each setting, resulting in the integral of the distributions being similar (inset); (b) Mean pulse height in electrons versus acceleration voltage. A straight line fit yields a gradient of 266 e/kV, from which an electron-hole pair generation energy in silicon, 3.76 eV, can be obtained. The inset shows the same data with extended data range; the threshold voltage 2.5 kV is at $y = 0$.

3.3. Photon Arrival Timing

Since the photon event brightness is strongly dependent on the acceleration voltage, it could be possible to use a gain sweep during exposure for photon arrival timing. By sweeping the voltage, each EBCCD frame would consist of photon events of different heights, which represent the arrival time after an excitation pulse, see Figure 5a–d. The frame is read into a computer, where it is analysed and the pulse height and pixel coordinates are stored. By repeating this process many times, *i.e.*, acquiring and analysing many frames, a histogram of photon arrival times is built up in each pixel of the image. This method could be used to measure fluorescence decays. The fluorescence decay can be a function of viscosity, temperature, pH, ion or oxygen concentrations, glucose, refractive index or polarity, and of interaction with other molecules, e.g., via Förster resonance energy transfer [59]. The fluorescence decay is characterised by the fluorescence lifetime, which is the average time a fluorophore remains in the excited state after excitation. By determining the fluorescence lifetime in each pixel of an image, via fluorescence lifetime imaging (FLIM), image contrast according to the fluorescence lifetime is obtained.

As there are currently no devices that allow the acceleration voltage to be changed during exposure, linear sweeps from high to low voltage with 50 ns sweep time and 5, 8 and 20 ns decay times were simulated to test the determination of photon arrival time from the pulse height [20]. The decays were simulated by acquiring sets of frames with Hamamatsu C7190-13 EBCCD at different acceleration voltages. A number of frames from each data set were combined in such a proportion as to yield exponential decays. The photons were thus distributed as if they had arrived at different times during a gain voltage sweep. The frames were processed as a single data set, where the pulse height of each photon was converted to arrival time and added to an arrival time histogram. This was done with the aid of Figure 4b which is effectively a calibration curve to convert photon event brightness into an acceleration voltage (which varies linearly in time). The key point here is not the linearity of the sweep, but the stability and reproducibility of the calibration of brightness versus time. The arrival time histograms, shown in Figure 5e, are in fact the fluorescence decay curves. The histograms were fitted with single-exponential decay law (Figure 5e, lines) using iterative reconvolution and the 8 kV pulse height distribution as an instrumental response function (Figure 5e, black diamonds). This yields decay times of 20.27, 8.78 and 4.72 ns for the 20, 8 and 5 ns simulated decays with chi-squared values of 1.03, 1.08 and 1.31. The residuals are flat, without any systematic deviations. The simulation shows that photon arrival times can be obtained from the photon event pulse heights.

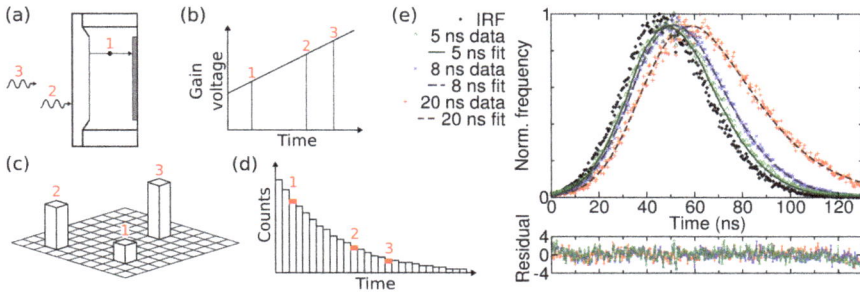

Figure 5. (**a**–**d**) Schematic of proposed gain sweep scheme. Photons arriving at different times during the CCD exposure (**a**) experience different gain voltage which is linearly swept in time (**b**); The pulse heights (**c**) can be converted to arrival times and added to the arrival time histogram (**d**); The gain sweep is repeated many times to build up a histogram for each pixel of the image; (**e**) Simulated fluorescence decays obtained from frames acquired at different acceleration voltages, and the instrument response function (IRF) obtained from measurement with the highest voltage (simulated time 0). The arrival time of each photon was found from its pulse height. Single-exponential fits to the decays yield decay times of 4.72, 8.78 and 20.27 ns for the 5, 8 and 20 simulated decays, respectively [20].

4. Some Applications of EB-Sensors

EB pixel image sensors were developed for high-resolution imaging with high signal-to-noise ratio at extremely low light level, and the low light level imaging capability has been utilised in night vision applications [27,41,42]. The single photon detection capabilities were first used in particle physics applications, for example, in observing neutrino interactions at CERN [39], and in astronomical applications [35]. Although recent developments in EMCCDs and sCMOS sensors have meant that the sensitivity advantage has disappeared [10], EB pixel image sensors continue to find applications in particle physics and in life science imaging.

In life sciences, the sensitivity and low noise of EB-sensors allows imaging of weak luminescence signals that are difficult to detect with other sensors. For microscopy applications, EB-sensors have been demonstrated to be suitable for imaging cells at low light level [20,43,45,50], and EBCMOS cameras have been used for tracking of multiple single-emitters [60]. EBCCDs have been used to visualise protein interactions in plant and animal cells and in tissues with subcellular resolution using bioluminescence resonance energy transfer (BRET) imaging [14], and EBCMOS cameras have been applied to marine bioluminescence imaging [61]. EBCCDs have also been evaluated for use in combination with a spectrometer, where high sensitivity in combination with high spatial resolution is required [17].

EB-sensors could also find applications in clinical use. In X-ray digital radiography and computed tomography, the low light level imaging capability of an EBCCD allows the reduction of the irradiation dose to the patient [18,19]. EBCCDs have also been evaluated for visualising stimulated functional brain areas during surgery [16].

5. Discussion

EB-sensors are more compact, smaller and lighter than intensified camera systems. The single photoelectrons are accelerated directly into the solid state sensor without MCP intensification and without being converted back into light on a phosphor screen, which is then imaged. With GaAsP photocathodes, the quantum efficiency can reach ~50%, and unlike MCP detectors, EB sensors have a fill factor or open area ratio of ~100%. The device lifetime is limited by the damage done to the chip by the high energy electrons striking it and producing x-rays, and ion events reduce the photocathode lifetime. However the relatively small volume and surface area compared to traditional image intensified tubes increase the photocathode lifetime. A lifetime of 10^{12} cnts/mm^2 has been quoted for EBCCDs, an order of magnitude longer than MCP devices [62].

The photon event brightness in EB pixel image sensors is strongly dependent on the acceleration voltage, and as the photon events typically cover an area of a few pixels, resolution lost in the detection process can be recovered by photon event centroiding—both one-iteration centre-of-mass [20] and iterative fitting [58] algorithms have been shown to produce excellent results. The local count rate is given by the frame rate of the camera; EBCMOS cameras with 500 frames per second have been described [46], and 1000 Hz planned. The global count rate depends on the number of pixels in the sensor, and as both CMOS and CCD sensors can be manufactured in large, megapixel arrays, the detection of hundreds or thousands of photons per frame is possible. The imaging speed is usually limited by the CCD or CMOS read-out time, and with photon event centroiding, the localisation time depends on the complexity of the algorithm [57].

However, despite the many applications and advantages of the ideal EBCCD, and being commercially available for over 15 years, there have been drawbacks, such as low frame rates of a few Hz and artefacts in the images, and it seems the development of these sensors has stopped before their full potential has been realised. EBCMOS cameras, on the other hand, are a recent development and not yet widely available, but show great potential, especially regarding the increased frame rate (500 Hz has been demonstrated [46]). A distinctive advantage of EB-sensors is the low dark count due to thermionic emission from the photocathode, in common with other photocathode and MCP-based devices, for which 0.02 events/s/cm^2 have been quoted [63], This would be useful for situations where a good signal to noise ratio is required, e.g., for very weak bioluminescence, or decay measurements of probes with microsecond decay times, for example oxygen sensing, or time-resolved fluorescence anisotropy measurements of large molecular weight proteins for which nanosecond decay times are too short.

A voltage sweep could be used to time photon arrival in EB pixel image sensors. Gated intensifiers can operate with gates as short as 200 ps over ∼1.5 kV [64–66], and some gated optical intensifiers and high rate imagers, which have been used for time-gated FLIM for over a decade [67,68], can operate at 500 ps gate width at 100 MHz. If the EBCCD gain can be swept in 50 ns or so over 4 kV, this approach seems feasible. With a photoelectron time of flight of 25 ps at 8 kV and 35 ps at 4 kV according to Equation (3), the tens of nanosecond sweep times which would typically be needed for nanosecond fluorescence lifetime measurements are a thousand times longer than the time-of-flight of the photoelectron. The length of the time window, *i.e.*, sweep time, would also be easily adjustable, as in a TAC, by adjusting the duration of the voltage gain sweep: it could extend over microseconds to measure decays in that range [9].

Single pixel hybrid detectors, which comprise a photocathode in front of an avalanche photodiode (biased below the breakdown voltage), are excellent for photon arrival timing with picosecond resolution [21,22] and are often used in scanning fluorescence microscopy-based FLIM [69,70]. The single photoelectrons liberated by photons at the photocathode are accelerated across a high voltage (8 kV or so) into the avalanche photodiode. They can have a GaAsP photocathode with a high quantum efficiency of 50% around 500 nm, a large active area, are free of afterpulsing and cost less than a MCP.

SPAD arrays are extremely promising alternative devices, based on all-solid state sensors, to perform photon arrival timing in each pixel with picosecond resolution. At the time of writing, the fill factor (*i.e.*, the light sensitive area compared to the whole pixel area) is low, although this is being addressed by current developments in 3D stacking of integrated circuits [71]. Moreover, microlens arrays can be placed in front of the detector to focus more of the fluorescence signal onto the light-sensitive area [72]. The low fill factor problem can also be circumvented by multibeam scanning fluorescence microscopy, by projecting the fluorescence onto the light sensitive area only [73,74]. The noise levels of SPAD arrays are currently higher than for photocathode based detectors; the dark noise performance of SPAD arrays, typically 100s of counts per pixel (SPAD), depending on the operating voltage and temperature [75], can be improved to 10s of counts per pixel (25 Hz has been quoted [76]), and appropriate cooling could reduce this further. Nevertheless, the outstanding capability of enormous global count rates well into the gigahertz region [77] is a decisive advantage of these devices.

6. Conclusions

In electron-bombarded sensors a single photoelectron is accelerated directly into a CCD or CMOS sensor without multiplication. With a low gain, these devices can be used in analogue mode, and at high gain, they are sensitive enough to detect single photons: they enable wide-field imaging at extremely low light levels, allowing the detection of up to hundreds or thousands of photon events per frame. Photon event centroiding can be employed to recover resolution lost in the detection process, as described in more detail in [58]. Unlike photon counting cameras employing electron-multiplying MCPs, the photon events have a narrow, acceleration-voltage-dependent pulse height distribution. A gain voltage sweep during exposure in an EB-sensor could allow photon arrival time determination from the pulse height with sub-frame exposure time resolution. The low noise performance of EB-sensors may make them suitable for ultra-low intensity measurements, or time-resolved imaging of microsecond decay probes, e.g. for oxygen sensing, or for time-resolved fluorescence anisotropy measurements, or imaging of large molecular weight proteins.

Conflicts of Interest: The authors declare no conflict of interest.

References

1. Roming, P.W.; Kennedy, T.E.; Mason, K.O.M.; Nousek, J.A.; Ahr, L.; Bingham, R.E.; Broos, P.S.; Carter, M.J.; Hancock, B.K.; Huckle, H.E.; *et al.* The Swift Ultra-Violet/Optical Telescope. *Space Sci. Rev.* **2005**, *120*, 95–142.
2. Hutchings, J.B.; Postma, J.; Asquin, D.; Leahy, D. Photon event centroiding with UV photon-counting detectors. *Publ. Astron. Soc. Pac.* **2007**, *119*, 1152–1162.
3. Postma, J.; Hutchings, J.B.; Leahy, D. Calibration and Performance of the Photon-counting Detectors for the Ultraviolet Imaging Telescope (UVIT) of the Astrosat Observatory. *Publ. Astron. Soc. Pac.* **2011**, *123*, 833–843.
4. Fordham, J.L.A.; Bone, D.A.; Read, P.D.; Norton, T.J.; Charles, P.A. Astronomical performance of a micro-channel plate intensified photon counting detector. *Mon. Not. R. Astron. Soc.* **1989**, *237*, 513–521.
5. Buller, G.S.; Collins, R.J. Single-photon generation and detection. *Meas. Sci. Technol.* **2010**, *21*, doi:10.1088/0957-0233/21/1/012002.
6. Hadfield, R.H. Single-photon detectors for optical quantum information applications. *Nat. Photon.* **2009**, *3*, 696–705.
7. Eisaman, M.D.; Fan, J.; Migdall, A.; Polyakov, S.V. Invited Review Article: Single-photon sources and detectors. *Rev. Sci. Instrum.* **2011**, *82*, doi:10.1063/1.3610677.
8. Seitz, P.; Theuwissen, A.J.P. *Single Photon Imaging*; Springer: Heidelberg, Germany 2011.
9. Hirvonen, L.M.; Festy, F.; Suhling, K. Wide-field time-correlated single-photon counting (TCSPC) lifetime microscopy with microsecond time resolution. *Opt. Lett.* **2014**, *39*, 5602–5605.
10. Long, F.; Zeng, S.; Huang, Z.L. Localization-based super-resolution microscopy with an sCMOS camera Part II: Experimental methodology for comparing sCMOS with EMCCD cameras. *Opt. Express* **2012**, *20*, 17741–17759.
11. Brugière, T.; Mayer, F.; Fereyre, P.; Guérin, C.; Dominjon, A.; Barbier, R. First measurement of the in-pixel electron multiplying with a standard imaging CMOS technology: Study of the EMCMOS concept. *Nucl. Instum. Meth. A* **2015**, *787*, 336–339.
12. Berland, K.; Jacobson, K.; French, T.; Rajfur, Z.; Electronic Cameras for Low-Light Level Microscopy. In *Methods in Cell Biology*; Sluder, G., Wolf, D.E., Eds.; Elsevier: Amsterdam, the Netherlands, 2003; Volume 72, pp. 103–132.
13. Levitt, J.A.; Chung, P.H.; Kuimova, M.K.; Yahioglu, G.; Wang, Y.; Qu, J.; Suhling, K. Fluorescence Anisotropy of Molecular Rotors. *ChemPhysChem* **2011**, *12*, 662–672.

14. Xu, X.D.; Soutto, M.; Xie, Q.; Servick, S.; Subramanian, C.; von Arnim, A.G.; Johnson, C.H. Imaging protein interactions with bioluminescence resonance energy transfer (BRET) in plant and mammalian cells and tissues. *Proc. Natl. Acad. Sci. USA* **2007**, *104*, 10264–10269.

15. Mac Raighne, A.; Brownlee, C.; Gebert, U.; Maneuski, D.; Milnes, J.; O'Shea, V.; Rugheimer, T.K. Imaging visible light with Medipix2. *Rev. Sci. Instrum.* **2010**, *81*, doi:10.1063/1.3501385.

16. Sobottka, S.B.; Meyer, T.; Kirsch, M.; Koch, E.; Steinmeier, R.; Morgenstern, U.; Schackert, G. Evaluation of the clinical practicability of intraoperative optical imaging comparing three different camera setups. *Biomed. Tech.* **2013**, *58*, 237–248.

17. Haisch, C.; Becker-Ross, H. An electron bombardment CCD-camera as detection system for an echelle spectrometer. *Spectrochim. Acta B* **2003**, *58*, 1351–1357.

18. Rossi, M.; Casali, F.; Golovkin, S.V.; Covorun, V.N. Digital radiography using an EBCCD-based imaging device. *Appl. Radiat. Isot.* **2000**, *53*, 699–709.

19. Baruffaldi, F.; Bettuzzi, M.; Bianconi, D.; Brancaccio, R.; Cornacchia, S.; Lanconelli, N.; Mancini, L.; Morigi, M.P.; Pasini, A.; Perilli, E.; *et al.* An Innovative CCD-Based High-Resolution CT System for Analysis of Trabecular Bone Tissue. *IEEE Trans. Nucl. Sci.* **2006**, *53*, 2584–2590.

20. Hirvonen, L.M.; Jiggins, S.; Sergent, N.; Zanda, G.; Suhling, K. Photon counting imaging with an electron-bombarded CCD: Towards a parallel-processing photoelectronic time-to-amplitude converter. *Rev. Sci. Instrum.* **2014**, *85*, doi:10.1063/1.4901935.

21. Becker, W.; Su, B.; Holub, O.; Weisshart, K. FLIM and FCS Detection in Laser-Scanning Microscopes: Increased Efficiency by GaAsP Hybrid Detectors. *Microsc. Res. Tech.* **2011**, *74*, 804–811.

22. Michalet, X.; Cheng, A.; Antelman, J.; Suyama, M.; Arisaka, K.; Weiss, S. Hybrid photodetector for single-molecule spectroscopy and microscopy. *Proc. SPIE* **2008**, *6862*, doi:10.1117/12.763449.

23. Suyama, M.; Fukasawa, A.; Haba, J.; Iijima, T.; Iwata, S.; Sakuda, M.; Sumiyoshi, T.; Takasaki, F.; Tanaka, M.; Tsuboyamaothers, T.; *et al.* Development of a multi-pixel photon sensor with single-photon sensitivity. *Nucl. Instrum. Meth. A* **2004**, *523*, 147–157.

24. Howard, N.E. Theoretical comparison between image intensifier tubes using EBCCD and phosphor readout. *Proc. SPIE* **1995**, *2549*, 188–198.

25. Fiebiger, J.R.; Muller, R.S. Pair-production energies in silicon and germanium bombarded with low-energy electrons. *J. Appl. Phys.* **1972**, *43*, 3202–3207.

26. Lyons, A. Design of proximity-focused electron lenses. *J. Phys. E Sci. Instrum.* **1985**, *18*, doi:10.1088/0022-3735/18/2/007.

27. Williams, G.M.; Rheinheimer, A.L.; Aebi, V.W.; Costello, K.A. Electron-bombarded back-illuminated CCD sensors for low-light-level imaging applications. *Proc. SPIE* **1995**, *2415*, doi:10.1117/12.206518.

28. Johnson, C.B. Review of electron-bombarded CCD cameras. *Proc. SPIE* **1998**, *3434*, 45–53.

29. Fraser, G.W.; Abbey, A.F.; Holland, A.; McCarthy, K.; Owens, A.; Wells, A. The X-ray energy response of silicon Part A. Theory. *Nucl. Instrum. Meth. A* **1994**, *350*, 368–378.

30. van Roosbroeck, W. Theory of the Yield and Fano Factor of Electron-Hole Pairs Generated in Semiconductors by High-Energy Particles. *Phys. Rev.* **1965**, *139*, A1702–A1716.

31. Lowrance, J.L.; Carruthers, G.R. Electron bombarded charge-coupled device (CCD) detectors for the vacuum ultraviolet. *Proc. SPIE* **1981**, *279*, 123–128.

32. Lemonier, M.; Piaget, C.; Petit, M. Thinned backside-bombarded RGS-CCD for electron imaging. *Adv. Imaging Electron Phys.* **1985**, *64*, 257–265.

33. Carruthers, G.R.; Heckathorn, H.M.; Opal, C.B.; Jenkins, E.B.; Lowrance, J.L. Development of EBCCD cameras for the far ultraviolet. *Adv. Electron. Electron Phys.* **1988**, *74*, 181–200.

34. Cuby, J.G.; Richard, J.C.; Lemonier, M. Electron bombarded CCD-1st results with a prototype tube. *Proc. SPIE* **1990**, *1235*, 294–304.

35. Auriemma, G.; Errico, L.; Satriano, C.; Vittone, A.A. EBCCD applications in astronomy. *Mem. Della SAIT* **2002**, *73*, 433–438.

36. Hamamatsu Photonics. *Electron Bombardment CCD Cameras C7190*; Hamamatsu Photonics: Hamamatsu, Japan, 2003.

37. Rousset, G.; Beuzit, J.L., The COME-ON/ADONIS Systems. In *Adaptive Optics in Astronomy*; Roddier, F., Ed.; Cambridge University Press: Cambridge, UK, 1999; pp. 171–203.

38. Benussi, L.; Fanti, V.; Frekers, D.; Frenkelc, A.; Gianninid, G.; Golovkine, S.V.; Kozarenkof, E.N.; Kresloc, I.E.; Libertic, B.; Martellottic, G.; *et al.* A multichannel single-photon sensitive detector for high-energy physics: The megapixel EBCCD. *Nucl. Instum. Meth. A* **2000**, *442*, 154–158.

39. Buontempo, S.; Chiodi, G.; Dalinenko, I.N.; Ereditato, A.; Ekimov, A.V.; Fabre, J.P.; Fedorov, V.Y.; Frenkel, A.; Galeazzi, F.; Garufi, F.; *et al.* The Megapixel EBCCD: A high-resolution imaging tube sensitive to single photons. *Nucl. Instum. Meth. A* **1998**, *413*, 255–262.

40. Suyama, M.; Sato, T.; Ema, S.; Ema, S.; Ohba, T.; Inoue, K.; Ito, K.; Ihara, T.; Mizuno, I.; Maruno, T.; Suzuki, H.; Muramatsu, M. Single-photon-sensitive EBCCD with additional multiplication. *Proc. SPIE* **2006**, *6294*, doi:10.1117/12.680381.

41. Aebi, V.; Boyle, J. Electron Bombarded Active Pixel Sensor. US Patent 6285018, 4 September 2001.

42. Aebi, V.W.; Costello, K.A.; Arcuni, P.W.; Genis, P.; Gustafson, S.J. EBAPS®: Next Generation, Low Power, Digital Night Vision. In Proceedings of the OPTRO 2005 International Symposium, Paris, France, 10 May 2005.

43. Baudot, J.; Dulinski, W.; Winter, M.; Barbier, R.; Chabanat, E.; Depasse, P.; Estre, N. Photon detection with CMOS sensors for fast imaging. *Nucl. Instrum. Meth. A* **2009**, *604*, 111–114.

44. Barbier, R.; Cajgfinger, T.; Calabria, P.; Chabanata, E.; Chaizea, D.; Depassea, P.; Doana, Q.T.; Dominjona, A.; Guérina, C.; Houlesa, J.; *et al.* A single-photon sensitive ebCMOS camera: The LUSIPHER prototype. *Nucl. Instrum. Meth. A* **2011**, *648*, 266–274.

45. Barbier, R.; Baudot, J.; Chabanat, E.; Depasse, P.; Dulinski, W.; Estre, N.; Kaiser, C.T.; Laurent, N.; Winter, M. Performance study of a MegaPixel single photon position sensitive photodetector EBCMOS. *Nucl. Instrum. Meth. A* **2009**, *610*, 54–56.

46. Cajgfinger, T.; Dominjon, A.; Barbier, R. Single photon detection and localization accuracy with an ebCMOS camera. *Nucl. Instrum. Meth. A* **2015**, *787*, 176–181.

47. Mac Raighne, A.; Teixeira, A.; Mathot, S.; McPhate, J.; Vallerga, J.; Jarron, P.; Brownlee, C.; O'Shea, V. Development of a high-speed single-photon pixellated detector for visible wavelengths. *Nucl. Instrum. Meth. A* **2009**, *607*, 166 – 168.

48. Fisher-Levine, M.; Nomerotski, A. TimepixCam: A fast optical imager with time-stamping. *J. Instrum.* **2016**, *11*, doi:10.1088/1748-0221/11/03/C03016.

49. Suhling, K. Photon arrival time detection. UK Patent EP1590687, 27 January 2004.

50. Hirvonen, L.M.; Jiggins, S.; Sergent, N.; Zanda, G.; Suhling, K. Photon counting imaging with an electron-bombarded CCD: Towards wide-field time-correlated single photon counting (TCSPC). *Nucl. Instrum. Meth. A* **2015**, *787*, 323–327.

51. Spring, K.R. Cameras for Digital Microscopy. In *Methods in Cell Biology*; Sluder, G., Wolf, D.E., Eds.; Elsevier: Amsterdam, the Netherland 1998; Volume 72, pp. 87–102.

52. Boksenberg, A.; Coleman, C.I.; Fordham, J.; Shortridge, K. Interpolative centroiding in CCD-based image photon counting detectors. *Adv. Electron. Electron Phys.* **1985**, *64A*, 33–47.

53. Bulau, S.E. Simulation of various centroiding algorithms. *Proc. SPIE* **1986**, *627*, 680–687.

54. Jenkins, C.R. The Image Photon Counting System: Performance in detail, and the quest for high accuracy. *Mon. Not. R. Astron. Soc.* **1987**, *226*, 341–360.

55. Suhling, K.; Airey, R.W.; Morgan, B.L. Optimisation of centroiding algorithms for photon event counting imaging. *Nucl. Instrum. Meth. A* **1999**, *437*, 393–418.

56. Suhling, K.; Airey, R.W.; Morgan, B.L. Minimization of fixed pattern noise in photon event counting imaging. *Rev. Sci. Instrum.* **2002**, *73*, 2917–2922.

57. Hirvonen, L.M.; Kilfeather, T.; Suhling, K. Single-molecule localization software applied to photon counting imaging. *Appl. Opt.* **2015**, *54*, 5074–5082.

58. Hirvonen, L.M.; Barber, M.; Suhling, K. Photon counting imaging and centroiding with an EBCCD using single molecule localisation software. *Nucl. Instrum. Meth. A* **2016**, *820*, 121–125.

59. Suhling, K.; Hirvonen, L.M.; Levitt, J.A.; Chung, P.-H.; Tregidgo, C.; Marois, L.A.; Rusakov, D.A.; Zheng, K.; Ameer-Beg, S.; *et al.* Fluorescence lifetime imaging (FLIM): Basic concepts and some recent developments. *Med. Photon.* **2015**, *27*, 3–40.

60. Cajgfinger, T.; Chabanat, E.; Dominjon, A.; Doan, Q.T.; Guerin, C.; Houles, J.; Barbier, R. Single-photon sensitive fast ebCMOS camera system for multiple-target tracking of single fluorophores: Application to nano-biophotonics. *Proc. SPIE* **2011**, *7875*, doi:10.1117/12.872396.

61. Dominjon, A.; Ageron, M.; Barbier, R.; Billault, M.; Brunner, J.; Cajgfinger, T.; Calabria, P.; Chabanat, E.; Chaize, D.; Doan, Q.T.; *et al.* An ebCMOS camera system for marine bioluminescence observation: The LuSEApher prototype. *Nucl. Instrum. Meth. A* **2012**, *695*, 172–178.

62. Blades, J.C. (Ed.) *Ultraviolet and Visible Detectors for Future Space Astrophysics Missions: A Report from the Ad-hoc, UV-Visible Detectors Working Group of NASA's Offics of Space Science*; Office of Space Science, National Aeronautics and Space Administration: Washington, DC, USA; Space Telescope Science Institute: Baltimore, MD, USA, 2002.

63. Siegmund, O.H.W. High-performance microchannel plate detectors for UV/visible astronomy. *Nucl. Instrum. Meth. A* **2004**, *525*, 12–16.

64. Scully, A.D.; Macrobert, A.J.; Botchway, S.; O'Neill, P.; Parker, A.W.; Ostler, R.B.; Phillips, D. Development of a laser-based fluorescence microscope with subnanosecond time resolution. *J. Fluoresc.* **1996**, *6*, 119–125.

65. Dowling, K.; Hyde, S.C.W.; Dainty, J.C.; French, P.M.W.; Hares, J.D. 2-D fluorescence lifetime imaging using a time-gated image intensifier. *Opt. Commun.* **1997**, *135*, 27–31.

66. Blandin, P.; Lévêque-Fort, S.; Lécart, S.; Cossec, J.C.; Potier, M.C.; Lenkei, Z.; Druon, F.; Georges, P. Time-gated total internal reflection fluorescence microscopy with a supercontinuum excitation source. *Appl. Opt.* **2009**, *48*, 553–559.

67. Siegel, J.; Suhling, K.; Lévêque-Fort, S.; Webb, S.E.D.; Davis, D.M.; Phillips, D.; Sabharwal, Y.; French, P.M.W. Wide-field time-resolved fluorescence anisotropy imaging (TR-FAIM)-Imaging the rotational mobility of a fluorophore. *Rev. Sci. Instrum.* **2003**, *74*, 182–192.

68. Suhling, K.; Siegel, J.; Phillips, D.; French, P.M.W.; Lévêque-Fort, S.; Webb, S.E.D.; Davis, D.M. Imaging the environment of green fluorescent protein. *Biophys. J.* **2002**, *83*, 3589–3595.

69. Levitt, J.A.; Chung, P.H.; Suhling, K. Spectrally resolved fluorescence lifetime imaging of Nile red for measurements of intracellular polarity. *J. Biomed. Opt.* **2015**, *20*, doi:10.1117/1.JBO.20.9.096002.

70. Levitt, J.A.; Morton, P.E.; Fruhwirth, G.O.; Santis, G.; Chung, P.H.; Parsons, M.; Suhling, K. Simultaneous FRAP, FLIM and FAIM for measurements of protein mobility and interaction in living cells. *Biomed. Opt. Express* **2015**, *6*, 3842–3854.

71. Garrou, P.; Bower, C.; Ramm, P., Eds. *Handbook of 3D Integration: Volume 1—Technology and Applications of 3D Integrated Circuits*; John Wiley & Sons: Weinheim, Germany, 2011.

72. Pavia, J.M.; Wolf, M.; Charbon, E. Measurement and modeling of microlenses fabricated on single-photon avalanche diode arrays for fill factor recovery. *Opt. Express* **2014**, *22*, doi:10.1364/oe.22.004202.

73. Poland, S.P.; Krstajić, N.; Coelho, S.; Tyndall, D.; Walker, R.J.; Devauges, V.; Morton, P.E.; Nicholas, N.S.; Richardson, J.; Li, D.D.U.; *et al.* Time-resolved multifocal multiphoton microscope for high speed FRET imaging *in vivo*. *Opt. Lett.* **2014**, *39*, 6013–6016.

74. Poland, S.P.; Krstajić, N.; Monypenny, J.; Coelho, S.; Tyndall, D.; Walker, R.J.; Devauges, V.; Richardson, J.; Dutton, N.; Barber, P.; *et al.* A high speed multifocal multiphoton fluorescence lifetime imaging microscope for live-cell FRET imaging. *Biomed. Opt. Express* **2015**, *6*, 277–296.

75. Charbon, E. Single-photon imaging in complementary metal oxide semiconductor processes. *Philos. Trans. A Math. Phys. Eng. Sci.* **2014**, *372*, doi:10.1098/rsta.2013.0100.

76. Richardson, J.A.; Grant, L.A.; Henderson, R.K. Low Dark Count Single-Photon Avalanche Diode Structure Compatible With Standard Nanometer Scale CMOS Technology. *IEEE Photon. Technol. Lett.* **2009**, *21*, 1020–1022.

77. Krstajić, N.; Poland, S.; Levitt, J.; Walker, R.; Erdogan, A.; Ameer-Beg, S.; Henderson, R.K. 0.5 billion events per second time correlated single photon counting using CMOS SPAD arrays. *Opt. Lett.* **2015**, *40*, 4305–4308.

sensors

MDPI

Article

X-ray Photon Counting and Two-Color X-ray Imaging Using Indirect Detection

Bart Dierickx *, Qiang Yao, Nick Witvrouwen, Dirk Uwaerts, Stijn Vandewiele and Peng Gao

Caeleste CVBA, Hendrik Consciencestraat 1 b, 2800 Mechelen, Belgium; Qiang.yao@caeleste.be (Q.Y.); Nick.witvrouwen@caeleste.be (N.W.); Dirk.uwaerts@caeleste.be (D.U.); Stijn.vandewiele@caeleste.be (S.V.); Peng.gao@caeleste.be (P.G.)
* Correspondence: Bart.Dierickx@caeleste.be; Tel.: +32-478-299-757

Academic Editors: Nobukazu Teranishi and Eric R. Fossum
Received: 27 January 2016; Accepted: 23 May 2016; Published: 26 May 2016

Abstract: In this paper, we report on the design and performance of a 1 cm^2, 90 × 92-pixel image sensor. It is made X-ray sensitive by the use of a scintillator. Its pixels have a charge packet counting circuit topology with two channels, each realizing a different charge packet size threshold and analog domain event counting. Here, the sensor's performance was measured in setups representative of a medical X-ray environment. Further, two-energy-level photon counting performance is demonstrated, and its capabilities and limitations are documented. We then provide an outlook on future improvements.

Keywords: X-ray photon counting; scintillator; color X-ray; spectral X-ray

1. Introduction

Photon-counting-based X-ray imaging is assumed to be superior in performance as compared to the more state-of-the-art charge integration X-ray imaging [1]. This is obvious at very low fluxes where photon counting yields quantum limited noise, but also, at high fluxes, photon counting has a DQE (detective quantum efficiency) advantage over integration. A second advantage of photon counting, especially for medical imaging, is that it offers the possibility of extracting spectral information from each photon separately, thus without multiple exposures or an increased X-ray dose. This spectral information reflects the chemical composition of the tissue examined—in this case, the ratio of carbon to oxygen [2,3].

Most, if not all of today's successful photon counting X-ray imagers are based on "direct detection" [4–13], *i.e.*, the X-ray detection happens by absorption of the photon in a high-Z semiconductor photo diode or photo resistor. From a pure detection performance standpoint, this approach is ideal: the photo-electric conversion happens in a very limited volume, and the energy quantum is deposited in a narrow trace or cloud of secondary electron-hole pairs, which are quickly and with little sideward dispersion collected by the electric drift field. As the collection time is in the order of nanoseconds, the direct detector can be operated at very high count rates. As the secondary charges remain confined, the modulation transfer function (MTF), and hence the DQE, is excellent, and the reproducibility of the collected charge packet size is nearly perfect, allowing, if required, an accurate photon energy measurement.

The limiting factor for the widespread use of direct detection in photon counting imaging is the cost and manipulation of the material.

The alternative route, indirect detection, *i.e.*, detection of the X-ray photon indirectly by first absorbing it in a high-Z scintillator and then detecting the secondary, visible light radiation by a visible light image sensor, is economically viable [14]. Many scintillators are inexpensive and easy to co-integrate, and CMOS visible-light event counting is an easily scalable and mature technology.

However, indirect detection has disadvantages as compared to detect detection: the overall indirect process has a significantly lower photon to electron conversion, suffers from slow decay times, may suffer light dispersion and poor MTF, and has poor reproducibility of charge packet sizes, making photon energy measurements unreliable.

In this paper, we report on our experiments to design, manufacture, and evaluate a two-energy-channel indirect detection X-ray image sensor [15–18]. We prove its feasibility, demonstrate that decent performance can be reached in pure counting and in energy resolution, and also show which limitations are present. At the end of the paper, we provide an outlook on a route for future improvements and solutions for the shortcomings.

2. Design of a Photon Counting Image Sensor

Our experimental device "QX2010" (Figure 1 and Table 1) is an image sensor with an array of X-ray photon counting pixels based on indirect (*i.e.*, using a scintillator) X-ray detection. Pixels are capable of detecting charge packets of at least 100 electrons per X-photon. This requirement stems from practice: this is the amount of charge carriers collected by the visible light photodiode for a single event, for photon energies in the range of medical X-ray, *i.e.*, 20 to 100 keV, using typical scintillator materials and scintillator thicknesses for medical X-ray imaging.

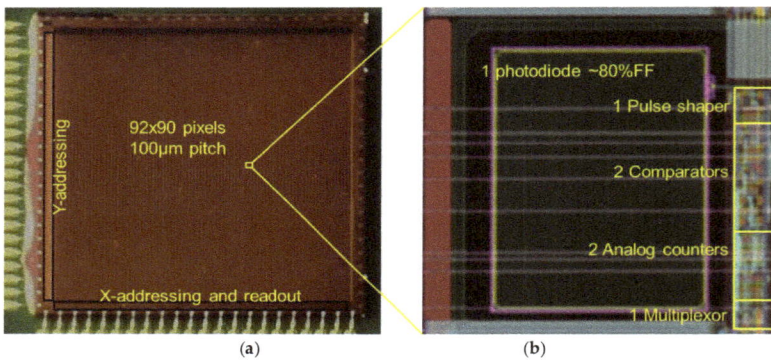

(a)   (b)

Figure 1. (**a**): microphotograph of the QX2010 device wire bonded on the CoB (chip on board); (**b**): Layout detail of one pixel, indicating the relative sizes and positions of the main pixels parts.

Table 1. QX2010 image sensor specifications.

Item	Specification	Item	Specification
Technology	TowerJazz TSL018IS	Detection concept	Photon counting with a scintillator
Die size	1 × 1 cm	Sense node capacitance	2 ... 3 fF
Array size	X: 91 + 1 test column Y:90	Wavelength spectrum	400 ... 900 nm (typical Si)
Pixel pitch	100 μm	QE × FF (quantum efficiency × fill factor)	~50% assuming optical glue between scintillator and image sensor
Analog counter step height	20 mV, exponentially decaying (see further)	FF (Fill factor)	75%, metal limited
Number of energy channels	2	Smallest charge packet that can be counted	~50 electrons estimated
Test pixels	In column 92	Q_N noise on threshold	~15 e-$_{RMS}$ estimated
Acquisition scheme	Global shutter (*i.e.*, global reset of counters)	Q_N variability	Not measured

Table 1. *Cont.*

Item	Specification	Item	Specification
Array readout scheme	X/Y addressing	Dark current and dark current variability	Not considered (DC current does not affect a pulse shaper)
#transistors per pixel	45 (53 in some test pixels)	MTF	Not measured or not relevant on imager part only.
Full frame readout time	8 ms Most measurements done with frame time, including count time 80 ms.	FPN (threshold voltage accuracy & reproducibility)	15 e-$_{RMS}$ estimated
Maximum count rate (separating two events)	~1000 kHz max. Most measurements with setting allowing up to 100 kHz.	PRNU (photo response non-uniformity)	No data
Pulse shaper band	Adjustable by current mirrors	Threshold of comparators	Adjusted by voltage
Number of IO pins	40 at two edges	Power consumption at 5 fps	30 mW

2.1. Pixel Topology

The pixel topology is shown in Figure 2. It is essentially a two-channel counting pixel. As transistor number and area is expensive, we had to reduce the complexity of each component to the minimum. The concept of the charge packet sense amplifier or "pulse shaper" is shown in Figure 3.

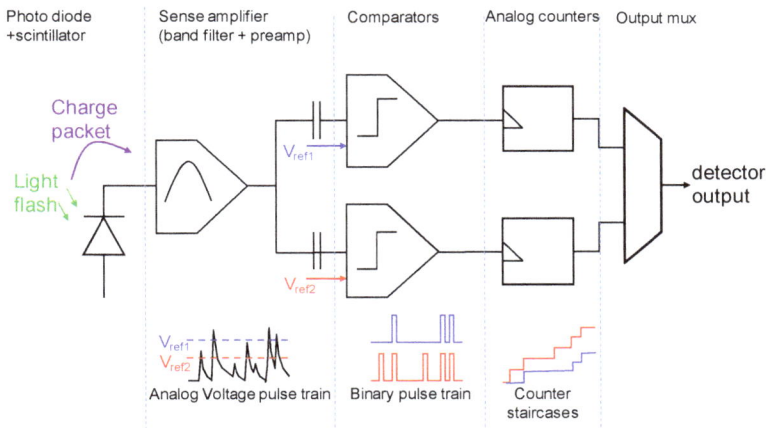

Figure 2. Topology of the QX2010 pixel.

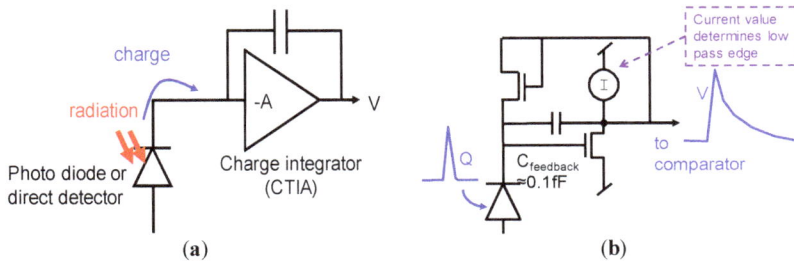

Figure 3. (**a**): General concept of a charge sensitive "charge trans-impedance amplifier" (CTIA); (**b**): CTIA with resistive feedback implemented as a MOSFET, becoming a "pulse shaper."

The comparator concept of this pixel is shown in Figure 4. The single pulse shaper feeds two comparators, each having different threshold "Vref." Both the bias currents in the pulse shaper and comparators are set through current mirrors. By choosing a proper value for the off-chip bias resistors of these current mirrors, one can independently set the overall high pass and low pass edge of the band-pass characteristic of the combination pulse-shaper + comparator, as exemplified in Figure 5.

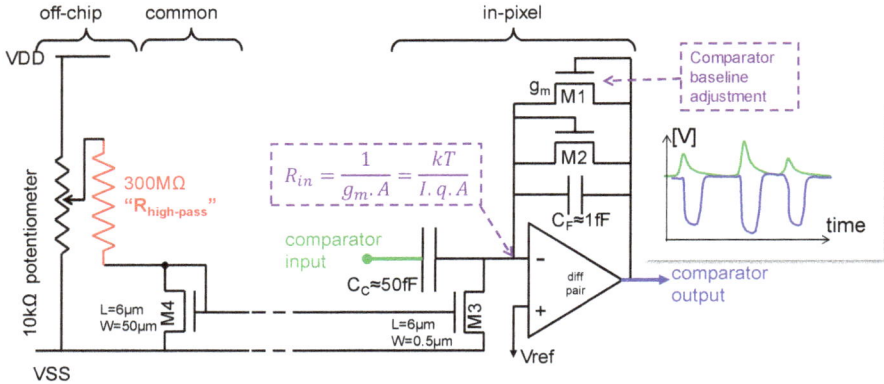

Figure 4. Comparator concept.

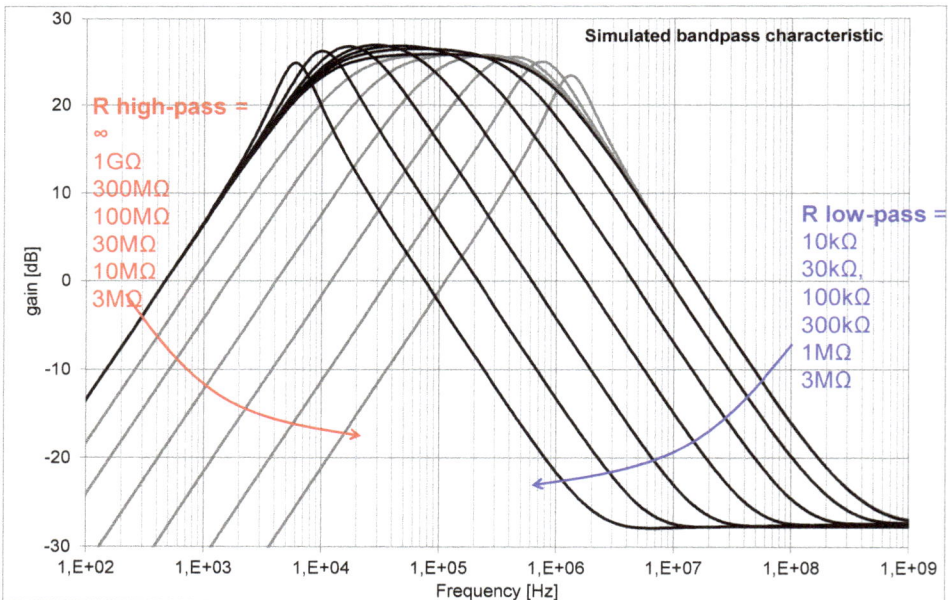

Figure 5. Simulated bandpass characteristics of the combination of pulse-shaper and comparator.

An important feature of the pixel is the introduction of what we call "non-linear analog domain counting." A classic, digital domain counter consisting of logic gates such as DFFs and NANDs would result in a sub circuit containing hundreds of transistors. This would not only eat up the available area in the pixel, it would also result in poor yield if the pixel array size would be scaled to the size desired in X-ray imaging, being several cm² up to wafer-scale. The solution that we proposed in [15,16] is to replace the digital counter by a circuit that realizes an analog signal staircase, which one may interpret as an extreme example of multi-level logic. Such a circuit can be realized in a very compact fashion, as in Figure 6. Here, an initial DC voltage on the large capacitor C2 is gradually and step-wise decreased by the switched capacitor network around the capacitor C1. Such an approach is acceptable as long as one can unambiguously retrieve the digital count by converting the staircase signal with an analog to digital convertor (ADC), *i.e.*, as long as the steps of the staircase are sufficiently large. As the step height is proportional to the "analog count" value itself, the step height decreases as the number of events increases, in a decaying exponential fashion. The non-linear transfer function of the analog counter helps to extend the range of the counter (Figure 7).

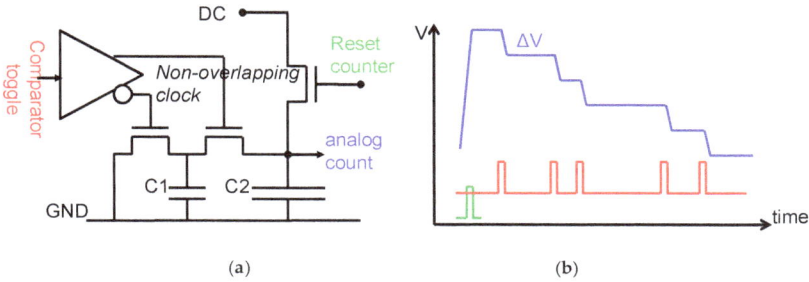

(a) (b)

Figure 6. Analog domain counter. (**a**): circuit concept; (**b**): signals as expected on key nodes of the circuit concept.

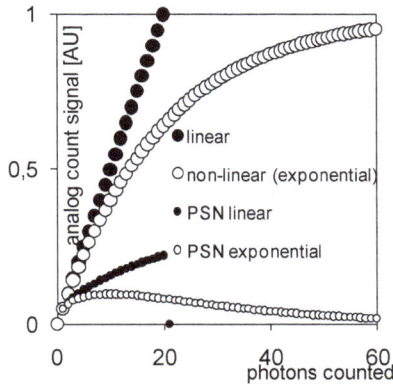

Figure 7. "Non-linear analog counting" is valid as long as the voltage step can be discriminated by the external ADC; yet this may be relaxed as long as the PSN (photon shot noise) expressed in voltage steps can be discriminated by the ADC. PSN is calculated as the square root of the number of photons counted.

Using such a counter and other, equally compact sub-circuits, we managed to keep the MOSFET circuit area below 20% of the area of the 100 × 100 μm² pixel, leaving almost 75% fill factor for the optical photodiode.

2.2. Pixel Measurements with Visible Light

The X/Y addressing structure of the QX2010 imager allows us to observe each analog counter in real-time. Figure 8 shows the analog counter response for a periodic visible light LED pulse train.

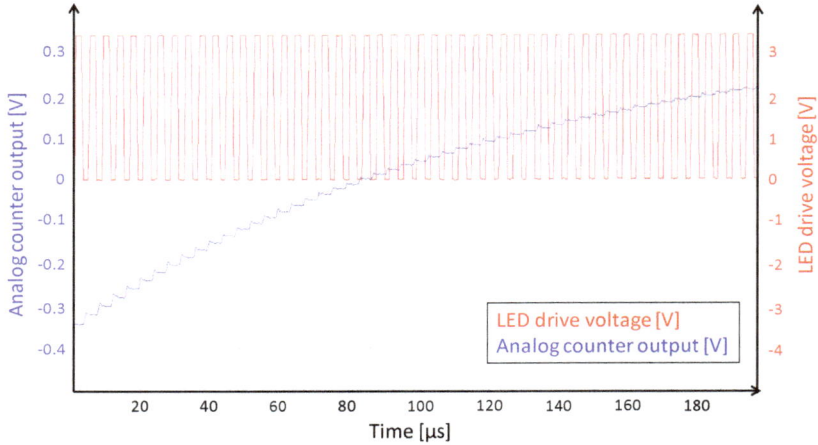

Figure 8. Signal of an analog counter during illumination of the QX2010 by a pulse train of LED pulses. X-axis: Time 0–1 ms; Y-axis: *Red trace* LED forward bias voltage; *Blue trace* analog counter voltage, measured real-time inside the circuit.

In Figure 9, we show the response of the analog counter on a pseudo-random LED pulse train. Short pulses create small charge packets, and wide LED pulses create large charge packets, thus mimicking the situation of an X-ray illuminated scintillator that would output a random population of stronger and weaker secondary light flashes. Of interest here is that we clearly see the effect of a different reference threshold voltage of each comparator; we also see a pretty good reproducibility of the counter pattern over time.

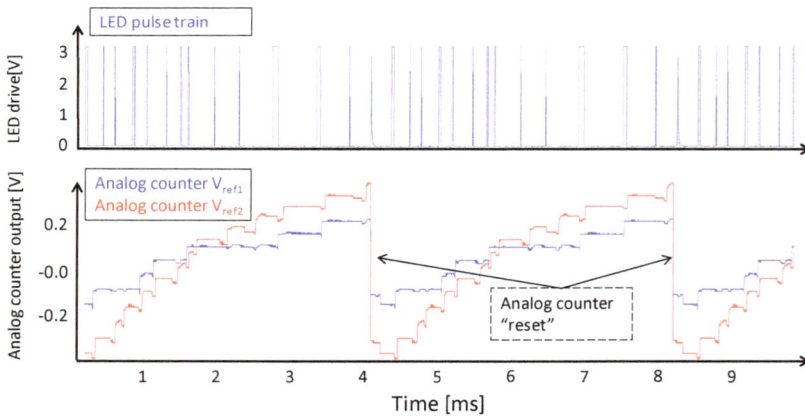

Figure 9. Demonstration of analog counting, emulating the variation of charge packet size by means of LED pulse width modulation. For the evaluation of noise and reproducibility, we defined a repeated random pattern of long and short LED pulses (blue top trace). The bottom traces are the two real-time observed analog counter outputs (with different Vref thresholds) of one pixel.

3. Experimental Results under X-ray Illumination and Discussion

In this paragraph we report on the experimental results of the QX2010 device with a CsI scintillator in an X-ray beam. A picture of one of the setups and a detail of the QX2010 device is shown in Figure 10.

Figure 10. Photograph of an in-beam measurement setup.

3.1. Beam Experiments with the QX2010

In such experiments, we recorded images under widely different conditions of beam voltages and current—various objects, various bias conditions of the QX2010, and various reference threshold voltages. It is beyond the scope of this paper to report on these. As an example of what such a photon counting imager can do, we show Figure 11. These are very low flux images with on average 10 and 30 counts (high/low threshold) per frame in the brightest parts of the image. In both images, metal parts have very low transmission, but the plastic part shows a distinct difference in transmission. These two "black and white" images can be combined in a color image for rendering.

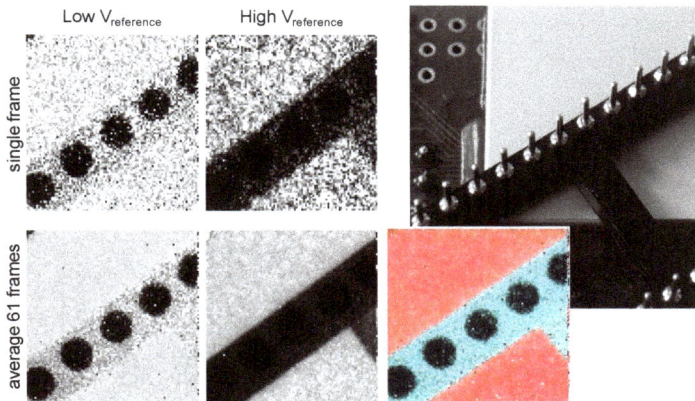

Figure 11. Recorded 92 × 90 pixel images of the object shown at the top right (a DIL socket on top of the scintillator glued in the QX2010). Top left/middle: simultaneous low reference and high reference threshold images from the QX2010. Bottom left/middle: the same, averaged 61 frames to smooth the photon shot noise. Color image: A "color-matrixed" combination of the two bottom left/middle images.

As two count values per pixel per frame are recorded, one can plot these as a series for a certain pixel over time. Such a series is shown in Figure 12. The non-linear analog counter values are properly linearized and scaled on the Y-axis, so that they correspond as closely as possible with the discrete integer counts. One clearly sees the photon shot noise in the X-ray detection. The histogram on the right is compared to a theoretical Poisson distribution with the same average count.

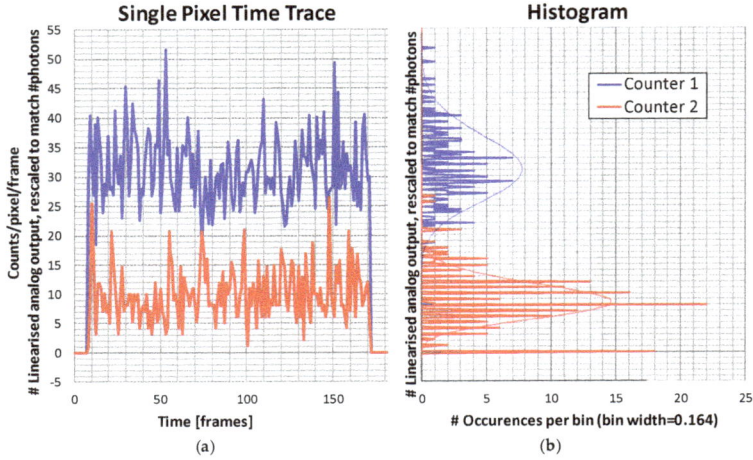

Figure 12. One pixel, two thresholds, observed over a time comprising 180 frames. Beam powers on at frame 9 and powers off after frame 170. (**a**) Counts *versus* frame number; (**b**) Horizontal histogram of the occurrences of analog count values. Thin red and blue lines are theoretical Poisson distributions with the same average counts.

Figure 13 displays images taken with a lead resolution target. When recording images of the target with two thresholds, one observes that the image with the higher reference voltage ($V_{reference}$) threshold, *i.e.*, the one recording the largest charge packets only, is sharper and thus has a better MTF. It also has a distinctly lower average count than the lower threshold image. This cannot be explained by any spectral sensitivity effect, as the lead in the target has a practically 100% absorption; thus, both images should be identical. The explanation of this observation is given in the next paragraph.

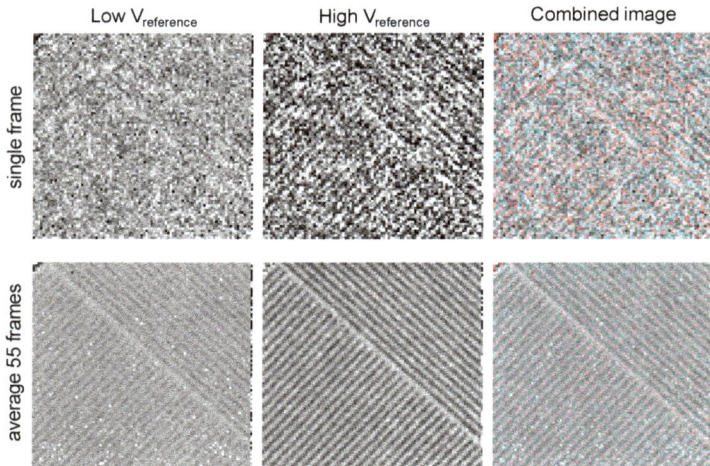

Figure 13. Images of a lead resolution target. Simultaneously recorded frames with two threshold levels (left, middle). Right: "color matrix-like" linear combination of the two frames for color rendering. Upper row: single frames; lower row: average of 55 frames.

3.2. Discussion on Issues Found

In these experiments, we demonstrated that scintillator-based photon counting is feasible. At the same time, however, we were able to pinpoint the major shortcomings of scintillator-based photon counting, all related to the well-known *scintillator depth effect* or to *Lubbert's effect* [19]: the amount of secondary radiation that falls on a single Silicon pixel depends significantly on the depth in the scintillator where the photo-electric conversion from X-ray photo to secondary light flash takes place, as illustrated in Figure 14.

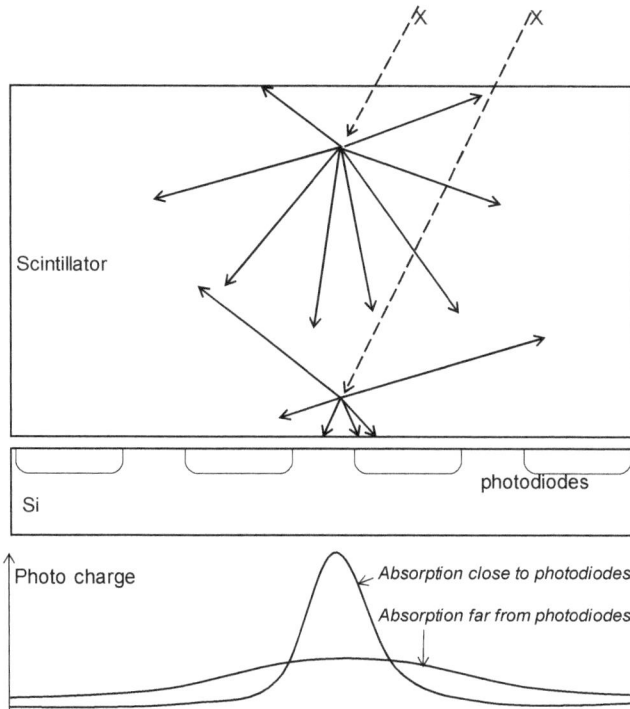

Figure 14. Schematic drawing of Lubbert's effect in a scintillator-covered Si pixel array.

Shallow absorption (far from the silicon photodiodes) *versus deep* absorption (close to the silicon photodiodes) results in larger or lesser optical diffusion of photo charges to neighboring pixels. As a consequence, one will see

- that the MTF depends on the depth of absorption and hence on the energy of the photon;
- that the number of visible photons per event per pixel depends on depth of absorption, therefore resulting in missed counts if the photo charge is diluted too much;
- that double or multiple (false) counts may occur if the event acts on multiple pixels;
- that these combined effects affect the DQE adversely; and
- that the effect deteriorates the assumed spectral energy sensitivity: the *observed* larger charge packets are not only charge packets originating from higher energy photons, but also X-ray photons absorbed close to the Si. The observed smaller charge packets may as well come from the higher energy photons that are absorbed far from the Si.

4. Next-Generation Devices

In this paragraph, we will discuss solutions to the issues encountered, as may be implemented in future devices.

4.1. Scintillators with Optical Confinement

Although we see a clear advantage of CsI *versus* most other scintillators, the anisotropic nature of CsI crystals is far from sufficient to cancel Lubbert's effect. The root cause solution of the light diffusion problem may be in the optical confinement of the light in a volume aligned with the pixel [20,21]. By confining the light in a volume that is equal in size to, or smaller than, the pixel (Figure 15), one will prevent the MFT degradation and thus rival the MTF performance of direct detection.

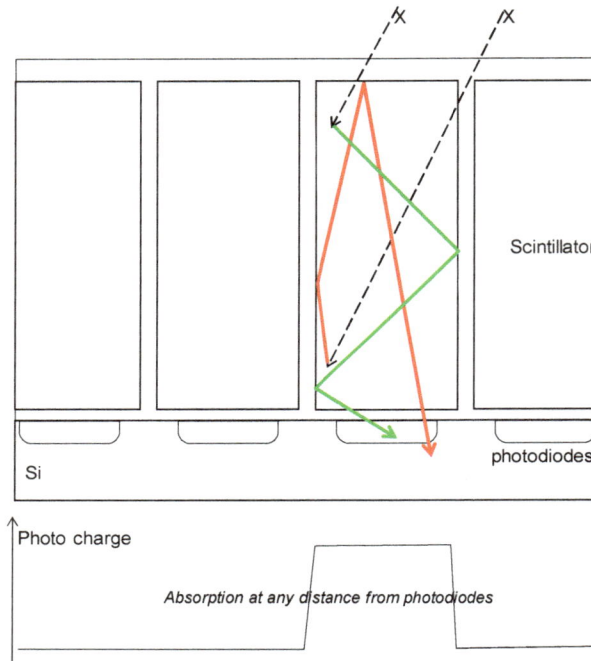

Figure 15. Similar drawing as Figure 14, now with optically confined scintillators on top of the silicon pixels.

4.2. Combining Photon Counting and Charge Integration in One Pixel

Two other weak points of photon counting in general, and analog domain photon counting specifically, are the limitations of count range and count speed. This also limits the more widespread use of photon counting X-ray imagers.

The count range may be virtually unlimited in digital domain, as long as the number of bits in the counter is high enough. In the analog domain, our concept allows us to reliably count up to about 100, and optimizations might result in counts up to 1000.

The count rates are limited by circuit performance, circuit power, and scintillator decay times. Failure here may lead to missed counts or even to counter paralysis.

Both issues are not present in the classic integrating X-ray imager. It has been proposed before by several groups and users to combine both concepts. In this paragraph, we propose a pixel concept that realizes photon counting and charge integration in the same pixel, on the same photo charge.

Consider the pulse shaper circuit of Figure 3. In this circuit, both the AC and DC photo currents from the photodiode end up in the output node of the feedback amplifier. By separating the drain and gate of the feedback MOSFET as in Figure 16, one can maintain the pulse shaping properties and add the capability of integrating the DC photo current with a capacitor.

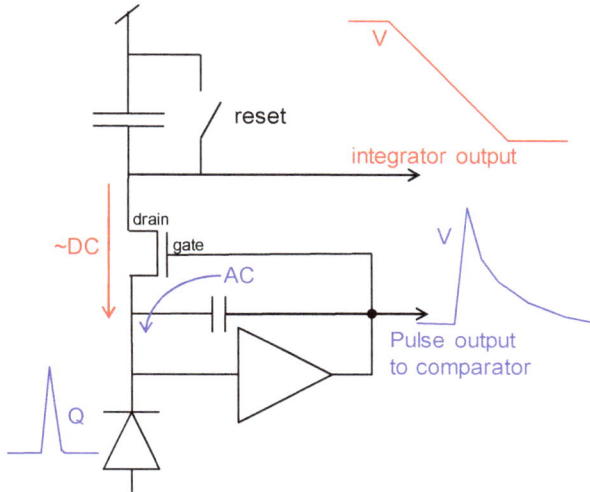

Figure 16. Concept circuit of a pulse shaper that feeds its photocurrent to a charge integrator.

This integrator in itself reminds us of the well-known integrating 3T pixel. The extra circuitry compared to the photon counter is small. The overall pixel may look like the illustration in Figure 17.

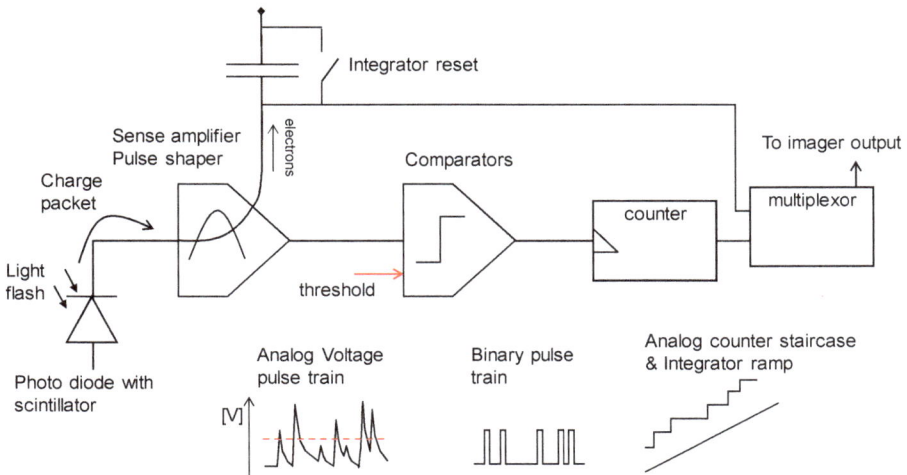

Figure 17. Overall pixel topology allowing simultaneous photon counting and charge integration.

This pixel is significantly more compact than the two-channel pixel of Figure 2. Notwithstanding, it has clear advantages:

- It does genuine photon counting. Its low count range and count rate make it suitable for low flux applications such as fluoroscopy. At medium fluxes, this signal may suffer from counter paralysis.
- It does genuine charge integration and can handle any charge quantity as required in medical imaging, limited only by the integration capacitor. These are applications such as mammography and high SNR imaging. In low flux conditions, the charge integration signal will be read noise limited and thus perform worse than the photon counting signal.
- There is a flux range where both signals are available and of good quality. In that range, the ratio of the two signals is a form of spectral information, as treated hereafter.

One can express both pixel signals as voltages in this "pseudo code":

$$V_{counter} = \#photons \bullet V_{step} \text{ and} \tag{1}$$

$$V_{integrator} = \sum\nolimits_{all\ photons} LO \tag{2}$$

where $V_{counter}$ is the linearized and normalized analog counter signal, $V_{integrator}$ is the offset corrected integrator signal, #photons is the number of photons absorbed during the integration time, V_{step} is the analog counter normalized voltage step, QE is the photodiode's quantum efficiency at the emission wavelength of the scintillator; the summation is over all primary photons; LO is the light output expressed as secondary photons for each primary photon, which, in good approximation, is proportional to the photon energy $h\nu$:

$$V_{counter} \sim \sum\nolimits_{all\ photons} (1) \text{ and} \tag{3}$$

$$V_{integrator} \sim \sum\nolimits_{all\ photons} (h\nu) \tag{4}$$

where α is the proportionality factor between photon energy and the effective number of photoelectrons in the photodiode.

$$\frac{V_{integrator}}{V_{counter}} \sim \frac{\sum_{all\ photons} (h\nu)}{\sum_{all\ photons} (1)} \text{ and} \tag{5}$$

$$\frac{V_{integrator}}{V_{counter}} \sim \overline{h\nu} \tag{6}$$

Hence, ratio (Equation (6)) is roughly proportional to the average photon energy during this integration time. The ratio between the counting channel and the integration channel represents the average photo charge per X-photon, which is a direct measure of the average photon energy on that pixel during the integration time and hence contains spectral information relating to the chemical composition of the tissue being imaged.

4.3. Photon Shot Noise-Free Spectral Information

A peculiar property of the combination of photon counting and charge integration on the same photo charge is that the above ratio (Equation (6)) is free of photon shot noise. The ratio, even in the hypothetical case of a single (X-ray) photon integrated, yields the energy of that photon.

5. Conclusions and Outlook

In this paper, we demonstrated that it is feasible to realize an X-ray photon counting image sensor using indirect detection. The presumed hurdles can all be overcome: the low light output of scintillators is not a showstopper. For the scintillators studied, GOS and CsI, charge packets in the order of 100 to 500 electrons for medical X-ray energies, are perfectly detectable and can be discriminated with low error rates, at least if there is no dilution by optical crosstalk. Although we experienced that in practice, the decay time of the scintillators is not as short as claimed by the suppliers or in textbooks. The actual delay times of CsI are suitable for count rates up to about 100,000 kHz. We managed to design 100-µm pitch pixels with a fill factor close to 75% by using very compact electronics, including "analog domain" counters.

Outlook

We expect the concept herein to be scalable to pixel sizes smaller than 50 µm and to array sizes that may go up to the full wafer size. Alternative and variant concepts may reach even smaller pixels pitches by using smaller linewidth CMOS technologies, or by using backside illumination, so that the fill factor remains high notwithstanding a small photodiode junction surface, or hybrid configurations of the indirect and even direct detection type.

We pointed out that a combination of photon counting and charge integration on the same photo charge expands the usability of the concept; an interesting side effect is the capability of obtaining photon shot noise—free spectral information.

Concerning the basic performance limitations due to Lubbert's effect or the scintillator absorption depth effect, we expect that the root cause solution must come from CMOS-process compatible optical confinement of the scintillator's emitted light inside the boundaries of the pixel.

Acknowledgments: The authors acknowledge the support and fruitful discussions of and with the Flemish IWT, the UZ Brussel team of Nico Buls; Paul De Keyser and colleagues, and several other fine scientists and engineers at other companies whose names we are not allowed to disclose here.

Author Contributions: B.D.: circuit and device concepts; Q.Y.: IC design; N.W.: simulations and test software; D.U.: test manager; S.V.: testing and visualization, data handling algorithms; P.G.: IC design.

Conflicts of Interest: The authors declare no conflict of interest.

References

1. Cahn, R.N.; Cederström, B.; Danielsson, M.; Hall, A.; Lundqvist, M.; Nygren, D. Detective quantum efficiency dependence on x-ray energy weighting in mammography. *Med. Phys.* **1999**, *26*, 2680–2683. [CrossRef] [PubMed]
2. Bourgain, C.; Dierickx, B.; Willekens, I.; Buls, N.; Breucq, C.; Schiettecatte, A.; de Mey, J. A new technique for enhanced radiological-pathological correlation in breast cancer: multi-energy color X-ray. In Proceedings of the RSNA 2011, Chicago, IL, USA, 27 November–2 December 2011.
3. Barber, W.C.; Nygard, E.; Wessel, J.C.; Malakhov, N. Large Area Photon Counting X-Ray Imaging Arrays for Clinical Dual-Energy Applications. In Proceedings of the 2009 IEEE Nuclear Science Symposium Conference Record (NSS/MIC), Orlando, FL, USA, 24 Octobrt–1 November 2009; pp. 3029–3031.
4. Llopart, X.; Campbell, M.; Dinapoli, R.; San Segundo, D.; Pernigotti, E. Medipix2: A 64-k Pixel Readout Chip With 55-µm Square Elements Working in Single Photon Counting Mode. *IEEE Trans. Nucl. Sci.* **2002**, *49*, 2279–2283. [CrossRef]
5. Ballabriga, R.; Campbell, M.; Heijne, E.H.M.; Llopart, X.; Tlustos, L. The Medipix3 Prototype, a Pixel Readout Chip Working in Single Photon Counting Mode with Improved Spectrometric Performance. In Proceedings of the 2006 IEEE Nuclear Science Symposium Conference Record, San Diego, CA, USA, 29 October 2006–1 November 2006; pp. 3557–3561.
6. Koenig, T. Charge Summing in Spectroscopic X-Ray Detectors With High-Z. *IEEE Trans. Nucl. Sci.* **2013**, *60*, 4713–4718. [CrossRef]
7. Zuber, M.; Hamann, E.; Ballabriga, R.; Campbell, M.; Fiederle, M.; Baumbach, T.; Koenig, T. An investigation into the temporal stability of CdTe-based photon counting detectors during spectral micro-CT acquisitions. *Biomed. Phys. Eng. Express* **2005**, *1*, 025205. [CrossRef]

8. Iwanczyk, J.S.; Nygard, E.; Meirav, O.; Arenson, J.; Barber, W.C.; Hartsough, N.E.; Malakhov, N.; Wessel, J.C. Photon Counting Energy Dispersive Detector Arrays for X-ray Imaging. *IEEE Trans. Nucl. Sci.* **2009**, *56*, 535–542. [CrossRef] [PubMed]

9. Spartiotis, K.; Leppänen, A.; Pantsar, T.; Pyyhtiä, J.; Laukka, P.; Muukkonen, K.; Männistö, O.; Kinnari, J.; Schulman, T. A photon counting CdTe gamma- and X-ray camera. *Nucl. Instrum. Methods Phys. Res.* **2005**, *550*, 267–277. [CrossRef]

10. Lotto, C.; Seitz, P. Charge Pulse Detection with Minimum Noise for Energy-Sensitive Single-Photon X-Ray Sensing. In Proceedings of the EOS Conference on the Frontiers in Electronic Imaging, Munich, Germany, 15–16 June 2009.

11. Perenzoni, M.; Stoppa, D.; Malfatti, M.; Simoni, A. A Multi-Spectral Analog Photon Counting Readout Circuit for X-Ray Hybrid Pixel Detectors. *IEEE Trans. Instrum. Meas.* **2008**, *57*, 1438–1444. [CrossRef]

12. User_Manual_PILATUS3_RSX_V3 (2).pdf. Avaliable online: http://www.dectris.com (accessed on 12 January 2016).

13. MicroDose_White_Paper,_Proven_clinical_effectiveness_at_low_radiation_dose.pdf. Avaliable online: http://incenter.medical.philips.com/doclib (accessed on 12 January 2016).

14. Miyata, E.; Tawa, N.; Mukai, K.; Tsunemi, H. High resolution X-ray photon-counting detector with scintillator-deposited charge-coupled device. In Proceedings of the 2004 IEEE Nuclear Science Symposium Conference Record, Rome, Italy, 16–22 October 2004.

15. Dierickx, B.; Dupont, B.; Defernez, A.; Henckes, P. *Towards Photon Counting X-ray Image Sensors*; OSA Symposium: Tucson, AZ, USA, 2010.

16. Dierickx, B.; Dupont, B.; Defernez, A.; Ahmed, N. Indirect X-ray Photon-Counting Image Sensor with 27T Pixel and $15e^-_{rms}$ Accurate Threshold. In Proceedings of the 2011 IEEE International Solid-State Circuits Conference Digest of Technical Papers (ISSCC), San Francisco, CA, USA, 20–24 February 2011.

17. Dierickx, B.; Vandewiele, S.; Dupont, B.; Defernez, A.; Witvrouwen, N.; Uwaerts, D. Scintillator based color X-ray photon counting imager. In Proceedings of the CERN workshop, Geneva, Switzerland, 23 April 2013.

18. Dierickx, B.; Vandewiele, S.; Dupont, B.; Defernez, A.; Witvrouwen, N.; Uwaerts, D. Two-color indirect X-ray photon counting imager. In Proceedings of the IISW, Snowbird, UT, USA, 12–16 June 2013.

19. Lubberts, G. Random Noise Produced by X-Ray Fluorescent Screens. *J. Opt. Soc. Am.* **1968**, *58*, 1475–1482. [CrossRef]

20. Scint-x.com. Avaliable online: http://www.scint-x.com/technology/ (accessed on 31 December 2015).

21. Rolf Kaufmann, R.; Seitz, P. High-Sensitivity X-ray Detector. U.S. Patent US9086493 B2, 21 July 2015.

sensors

MDPI

Review

Single Photon Counting UV Solar-Blind Detectors Using Silicon and III-Nitride Materials

Shouleh Nikzad [1,*], Michael Hoenk [1], April D. Jewell [1], John J. Hennessy [1], Alexander G. Carver [1], Todd J. Jones [1], Timothy M. Goodsall [1], Erika T. Hamden [2], Puneet Suvarna [3], J. Bulmer [3], F. Shahedipour-Sandvik [3], Edoardo Charbon [4], Preethi Padmanabhan [4], Bruce Hancock [1] and L. Douglas Bell [1]

[1] Jet Propulsion Laboratory, California Institute of Technology, Pasadena, CA 91109, USA;
    michael.hoenk@jpl.nasa.gov (M.H.); April.D.Jewell@jpl.nasa.gov (A.D.J.);
    John.J.Hennessy@jpl.nasa.gov (J.J.H.); Alexander.G.Carver@jpl.nasa.gov (A.G.C.);
    todd.jones@jpl.nasa.gov (T.J.J.); timothy.goodsall@jpl.nasa.gov (T.M.G.); bruce.hancock@jpl.nasa.gov (B.H.);
    lloyddoug.bell@jpl.nasa.gov (L.D.B.)
[2] Department of Physics, Mathematics and Astronomy, California Institute of Technology, Pasadena,
    CA 91125, USA; hamden@caltech.edu
[3] College of Nanoscale Science and Engineering, SUNY Polytechnic Institute, Albany, NY 12203, USA;
    psuvarna@albany.edu (P.S.); jbulmer@albany.edu (J.B.); sshahedipour-sandvik@albany.edu (F.S.-S.)
[4] Department of Microelectronics, Delft University of Technology, Delft, The Netherlands;
    e.charbon@tudelft.nl (E.C.); PreethiPadmanabhan@student.tudelft.nl (P.P.)
* Correspondence: shouleh.nikzad@jpl.nasa.gov; Tel.: +1-818-354-7496

Academic Editor: Eric R. Fossum
Received: 31 March 2016; Accepted: 7 June 2016; Published: 21 June 2016

Abstract: Ultraviolet (UV) studies in astronomy, cosmology, planetary studies, biological and medical applications often require precision detection of faint objects and in many cases require photon-counting detection. We present an overview of two approaches for achieving photon counting in the UV. The first approach involves UV enhancement of photon-counting silicon detectors, including electron multiplying charge-coupled devices and avalanche photodiodes. The approach used here employs molecular beam epitaxy for delta doping and superlattice doping for surface passivation and high UV quantum efficiency. Additional UV enhancements include antireflection (AR) and solar-blind UV bandpass coatings prepared by atomic layer deposition. Quantum efficiency (QE) measurements show QE > 50% in the 100–300 nm range for detectors with simple AR coatings, and QE $\cong$ 80% at ~206 nm has been shown when more complex AR coatings are used. The second approach is based on avalanche photodiodes in III-nitride materials with high QE and intrinsic solar blindness.

Keywords: ultraviolet; quantum efficiency; MBE; ALD; EMCCD; APD; ROIC; Avalanche; visible rejection; MOCVD; GaN

1. Introduction

The ultraviolet (UV) spectral range is populated with atomic and molecular lines that are highly relevant for studying planetary bodies, including solar system planets, exoplanets, comets and asteroids as well as stars, supernovae, black holes, galaxies, and the cosmos. In recent years, Hubble Space Telescope (HST) [1–4], Galaxy Evolution Explorer (GALEX) [5,6], Rosetta [7–11], and the Cassini mission [12] have shown exciting and intriguing results, hinting at and sometimes leading to new discoveries. In depth studies of these phenomena will require further observations with more powerful UV instruments. Discoveries of plumes on Europa, oceans on Enceladus, and theories of intergalactic medium (IGM) and circumgalactic medium (CGM) have opened new scientific questions and windows of study that require improved UV detection capabilities. Using UV spectroscopy and imaging spectrometry, thin atmospheres and surface composition of primitive bodies can be examined. UV emission lines and bands from H, C, O, N, S, OH and CO; UV absorption lines by CO_2, H_2O, NH_3, N_2; and UV surface reflectance spectra are all essential for the detection of ice, iron oxides, organics, and other compounds on planetary bodies. All of these are used as diagnostic tools for understanding the nature and habitability of these bodies. In addition to space applications, UV is used in defense applications, cancer detection, bacterial detection, machine vision, wafer inspection, lithography, and electrical safety inspection.

Photon counting is a key capability enabling faint object detection with NASA's UV instruments. In addition, visible-blindness, or more generally speaking, out-of-band rejection, is often required to detect UV signals in the presence of a significant visible background. UV instruments have traditionally addressed these requirements using image-tube technologies, such as photomultiplier tubes (PMTs) and microchannel plates (MCPs).

Silicon-based imaging detectors with single photon counting capability in the UV represents a significant leap forward in detector manufacturability, accessibility, and reliability. The enormous investment by industry and defense organizations has led to the development of large-format, high-resolution silicon imaging arrays. Relatively recent advances in solid-state imaging technology have produced detector architectures with high efficiency and built-in gain [13–15]. Challenges are posed, however, in producing silicon detectors with high UV sensitivity, stability with respect to illumination history, photon counting ability, and out-of-band rejection. Nanoscale surface and interface engineering technologies for surface passivation, together with progress in large-scale production, address these challenges and are leading to solid-state imagers being highly competitive and even superior in performance and cost to replace image-tube technologies in UV instrumentation [16–18].

In this paper we discuss two different approaches to solid-state, photon-counting, UV imaging and detection. In the first approach, we employ back-illumination processes developed in our laboratory, including Molecular Beam Epitaxy (MBE)-based superlattice and delta doping as well as Atomic Layer Deposition (ALD)-based custom antireflection (AR) coatings on electron multiplying charge-coupled devices (EMCCDs) and integrated filters on avalanche photodiodes (APDs). Our recent results show record high UV quantum efficiency (QE) of 60%–80%. Integrated metal dielectric filters (MDF) are used for visible rejection. We have measured dark current and clock-induced charge at low enough levels to make these detectors attractive and competitive for photon counting applications. In the second approach, we use gallium nitride and its alloys in a hybrid APD design. Gallium nitride and its alloy gallium aluminum nitride are wide bandgap materials that, depending on the fraction of aluminum in the alloy, span a wide range of direct bandgaps with tailorable out of band cutoff from 3.4 eV to 6.2 eV. We have achieved 50% QE in GaN and AlGaN APDs. Due to lack of native substrates, III-N's suffer from defects and leakage. ALD's nanoscale precision and conformal coating capability were used for sidewall passivation against leakage, resulting in consistent improvement over detectors coated using Plasma Enhanced Chemical Vapor Deposition (PECVD). Both approaches have resulted in better than 10^4 out-of-band rejection, which is at the same level as the traditional image tube-based UV detectors.

We present results for both of these promising approaches as part of our overall focused effort in developing technologies and instrumentation for challenging UV science where single photon counting is required. We discuss concepts, designs, fabrication and processing, and characterization techniques.

2. Materials and Methods: Silicon and Gallium Nitride/Gallium Aluminum Nitride Detector Designs with Avalanche Gain

2.1. Single Photon Counting in the UV with Silicon

2.1.1. Silicon Detectors with Gain

Avalanche multiplication has been used in various silicon detector architectures to achieve signal gain, including APDs, APD arrays, single photon avalanche photodiode (SPAD) arrays, and EMCCDs [15]. It is now possible to achieve single photon counting in silicon arrays rivaling that achieved with high-voltage photocathode-based imaging technologies, provided that the noise levels are still kept low. The focus of this work is the combination of delta doping and custom AR coatings to achieve high and stable QE for single photon counting imaging detectors, particularly in the UV. Two types of silicon detectors with avalanche gain were used. A brief description of each detector follows.

Electron Multiplying CCDs

EMCCDs leverage the advantages of the mature CCD technology while enabling single photon detection. Charge amplification is achieved in EMCCDs by appending a gain serial shift register to the serial register of a conventional CCD. At each stage of this gain serial shift register, a higher gate voltage (~40 V) in the second serial clock phase causes an impact ionization effect (Figure 1). The impact ionization effect produces a small gain in each charge transfer, enabling a cumulative (mean) charge gain of more than one thousand [13,14].

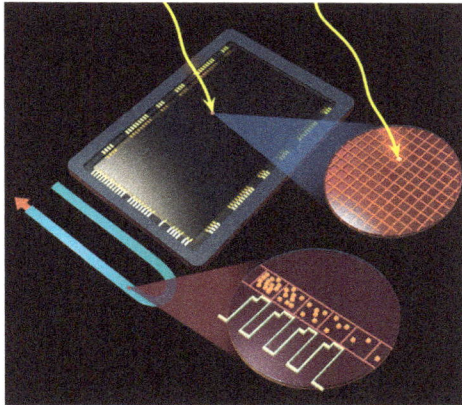

Figure 1. Artist's concept of the avalanche gain process in an Electron Multiplying Charge Coupled Device (EMCCD). The EMCCD architecture is the same as a regular two-phase CCD but with an added serial register stage where additional voltage is applied at each transfer that induces probability of avalanche. Photons impinge on CCD pixels resulting in the generation of photoelectrons as depicted in the CCD and the blow up circle on top right. The photoelectron is transferred through normal CCD charge transfer process until it reaches the special gain register of the EMCCD. In this gain register, avalanche multiplication is induced by applying higher than normal clock voltages (~40 V) as shown in the figure blow up. The CCD shown in the figure is an artist's recreation of e2v's (Chelmsford, UK) CCD201.

Charge gain in EMCCDs mitigates output read noise. Low light level sensitivity in EMCCDs is limited by low-level sources of spurious signal, such as dark current and "clock induced charge" (CIC). Clock induced charge, which is generated in the detector pixels during the readout process, is commonly attributed to electric field-induced hole transport between the detector surface and column stops. The generation rates are tied to bias voltages and surface inversion during clocking. Although relatively insignificant in most CCDs, CIC can be a source of spurious signal in EMCCDs [13,14,19]. As demonstrated by Hamden *et al.*, these remaining spurious signal sources are manageable by conventional techniques and have recently been measured for the Faint Intergalactic Red-shifted Emission Balloon (FIREBall-2) experiment [20]. Cooling (<-120 °C) and reducing exposure from array saturating sources can reduce dark current to levels well below the typically reported 1 e$^-$-pixel^{-1}-hr^{-1} [21], leaving CIC as the limiting noise source. There are simple yet powerful techniques to reduce the effects of the CIC such as optimal wave shaping filters applied to the CCD clocks [22,23]. Furthermore, because CIC results from charge transfer, its impact increases proportionally with frame rate. Many scientific applications do not require a high frame rate, and therefore can be read out only as often as necessary, for example, to minimize the effects of cosmic rays (~1000 s).

Avalanche Photodiodes

Unlike conventional photodiodes, APDs are designed to sustain high bias voltages without breaking down. In large area linear-mode APDs, high electric fields in the region of the p-n junction can generate multiplication gains as high as 1000×. The UV sensitivity and response time of standard linear-mode APDs are limited by diffusion and recombination in the undepleted silicon near the surface. In a collaboration between Radiation Monitoring Devices, Inc. (RMD, Watertown, MA, USA), the California Institute of Technology (Caltech) and the Jet Propulsion Laboratory (JPL), we have demonstrated superlattice-doped APDs in which the depletion region can approach the surface, thereby enabling high UV sensitivity and fast response. Superlattice doping in these devices has also enabled the development of visible-blind, UV band pass filters, with high QE in the deep ultraviolet, and excellent out-of-band rejection.

2.1.2. Silicon Passivation

The concept of backside illumination (BSI) was created in the 1970s to reduce the losses in the front-side device circuitry, effectively increasing the QE and expanding the spectral response into the UV range [24–26]. Today, BSI detectors have become prevalent even in commercial applications; however, many of the processes needed for BSI were initially developed for scientific detectors, with HST's Wide Field/Planetary Camera-1 (WF/PC-1) leading the way as the first implementation of the idea in space. As was quickly discovered, these early BSI silicon detectors had poor due UV efficiency due to the shallow absorption depth of UV photons and limitations of then-available technologies for thinning and surface passivation. To circumvent these limitations and enable the detection of UV photons in HST's WF/PC-1 instrument, backside-thinned CCDs were coated with phosphors such as coronene or lumogen that convert UV photons into visible photons that could be detected more efficiently. Unfortunately, the performance of WF/PC-1 CCDs was severely compromised by instabilities known as QE hysteresis (QEH). QEH in BSI detectors is characterized by low and unstable sensitivity, especially at the blue end of spectrum. QEH is a consequence of poor surface passivation, as unpassivated surface defects trap photogenerated electrons and holes, resulting in time-variable surface charge and surface depletion depth. Despite another two decades of development, state-of-the-art ion-implanted CCDs currently flying in Hubble's Wide Field Camera 3 instrument still exhibit QEH at a level of several percent, which is a testament to the importance and difficulty of surface passivation [27].

The surface of thinned, BSI detectors comprises high purity silicon, which is extremely sensitive to charge in surface oxides and at interfaces. Charge transfer from silicon to interface traps causes the Si-SiO$_2$ interface to acquire a net charge. At thermal equilibrium, p-type silicon surfaces acquire a net positive charge due to trapping of majority carriers. The physical location of traps in the oxide plays an important role in the dynamics of the illuminated surface. Defects and contaminants create a fixed charge in the oxide, while traps at the Si-SiO$_2$ interface acquire a variable charge depending on the near-surface Fermi level and the time-dependent illumination of the surface. Traps located within ~1 nm of the silicon surface interact with mobile charge in the underlying silicon, and are therefore capable of changing charge state in response to transient changes in the surface potential. Traps located deeper in the oxide change state much more slowly.

The net charge trapped in oxide and interface states generates an electric field that depletes the nearby silicon. From a device physics standpoint, interface traps "pin" the surface Fermi level in the middle of the bandgap. In general, the silicon surface is transformed into a "surface dead layer" that renders the detector insensitive to light that is absorbed near the surface. In order to achieve high efficiency and stable response, it is therefore necessary to combine thinning of BSI detectors with a surface passivation technology that reduces or eliminates the surface dead layer while at the same time suppressing the surface-generated dark current associated with a high density of interface traps.

Since the first BSI CCDs were demonstrated in 1974 [24], many methods have been developed for back surface passivation. These can be roughly divided into backside charging and back surface doping technologies, which are distinguished by the location and distribution of charge in the detector and its oxide. Backside charging methods stabilize the detector by introducing negative charge into the oxide. Provided that the density of interface traps is not too high, the negatively charged passivation layer will bias the surface into accumulation. Backside doping methods take the opposite approach by introducing an impurity profile that stabilizes the surface and creates an internal bias in the detector.

The advantage of surface doping lies in the potentially greater stability that can be achieved. In particular, surface doping promises greater resistance to radiation-induced traps. In p-type surfaces, holes are trapped in interface states, depleting the surface of holes and forming a depletion layer with a negative net charge density. The surface is therefore characterized by a charge dipole, bounded on one side by positive charge in the surface/oxide and on the other by the negatively charged depletion layer. To a first approximation, the depth of the depletion layer is given by the ratio of surface charge density to silicon dopant density. Thus, variations in the surface charge density are reflected by variations in the depletion layer depth. Within the depletion layer, photogenerated electrons experience a force directed toward the surface where they can be lost to trapping and recombination. Back surface doping methods seek to create a near-surface doping profile that simultaneously shrinks the surface depletion layer and creates a strong, built-in electric field. Controlling the doping profile can stabilize the detector against variations in surface charge density. Ideally, surface doping achieves a high near-surface dopant concentration with a strong gradient, thereby creating a surface that is insensitive to charge trapped at the surface. Surface doping is potentially more stable than back surface charging in challenging environments, including high radiation environments in space.

Unlike conventional "3D" doping methods, in which dopants are randomly distributed in the silicon lattice, 2D doping (also known as delta-doping) incorporates dopant atoms in highly ordered, self-organized two-dimensional layers. For p-type doping with boron, at concentrations >3 × 10^{20} B/cm^3 the conventional (3D) approach suffers from poor quality crystals, in which dopant atoms are only partially activated due to clustering and incorporation into interstitial sites. In contrast, 2D doping is marked by high crystal quality with nearly 100% electrical activation at concentrations exceeding the 3D doping limit by an order of magnitude. The growth of 2D-doped silicon is based on self-organized surface phases that form during sub-monolayer deposition of dopant atoms on silicon. Once formed, these self-organized surface phases are stabilized by covalent bonds in the silicon lattice. In our work, we have routinely used 2D sheet densities of 2 × 10^{14} B/cm^2; however, 2D-doped surfaces with single-layer sheet densities as high as 3 × 10^{14} B/cm^2 have been demonstrated.

An additional major challenge in fabricating BSI detectors is that standard surface passivation processes require high temperatures that would damage the detector. Low-temperature surface passivation technologies have been developed, but many—such as ion implantation and laser anneal—still suffer from lower QE or QE instability as a function of environment or illumination history. JPL-invented low-temperature MBE has been developed as a method of passivating BSI detectors [28,29]. This approach enables control over the surface doping profile with nearly atomic-scale precision (Figure 2). The precision control of surface band structure engineering and interface engineering afforded by 2D doping by MBE results in high QE without any QEH [16,17,30]. The first time JPL passivated a CCD using MBE, the passivation layer comprised 5 nm of silicon doped at 3×10^{20} B/cm^{-3}, *i.e.*, 3-D doping, plus a sacrificial 1-nm un-doped silicon cap layer to form the surface oxide [29]. In all subsequent devices, JPL has used 2D doping for surface passivation [17,18,28,30–38]. During MBE growth, a thin layer of un-doped silicon is grown on the substrate to form an atomically clean, uniform silicon surface. The silicon flux is interrupted by closing a shutter, and boron is deposited on the atomically clean silicon surface, where it spontaneously forms a self-organized surface phase, as described above. Once the surface coverage reaches a pre-defined level, the boron shutter is closed and the silicon shutter is opened in order to grow a thin layer of crystalline silicon on the boron-doped surface. This process is termed delta doping, because the resulting vertical dopant profile resembles the mathematical delta function. Occasionally we will refer to "superlattice-doping" in which more than one delta layer is included in the MBE structure.

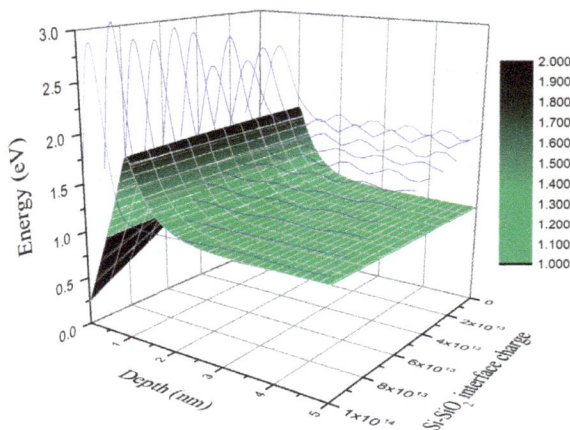

Figure 2. Three-dimensional plot of conduction band edge *vs.* depth and silicon-silicon oxide interface charge density for band structure engineered, 2D-doped BSI silicon array (shown for a delta-doped device). The figure shows that the placement of a high density of boron in a single atomic sheet creates a delta function change at the edge of the conduction band (green plot). The electron wave functions (shown in blue) are unconfined, illustrating that the placement of a delta layer within 1 nm of the surface reduces the probability of photoelectron trapping. The figure illustrates the stability of the 2D-doped surface against variable surface charge, which is associated with interface traps and radiation damaged surfaces.

2.1.3. Atomic Layer Deposition for Antireflection Coatings & Detector-Integrated Visible-Rejection Filters

Many of the concepts of AR coatings and visible rejection filters for the far UV spectral range have been developed over the years, but could not be implemented on live devices with accuracy or fidelity [35]. ALD is a technique that is a close spinoff of chemical vapor deposition (CVD). ALD films are created a single atomic layer at a time through a series of self-limiting chemical reactions with the substrate surface; different reactants designed for layer-by-layer growth are introduced sequentially, followed by purging (e.g., N$_2$, Ar) to remove reaction byproducts after each layer is formed. This chemical-reaction-driven process allows for the precise control of stoichiometry, thickness, and uniformity, while producing highly dense and pinhole free films with sharp and well-defined interfaces. These characteristics make ALD ideal for preparing optical coatings based on single and multilayer films, and we have employed this technique to develop highly effective AR coatings in the challenging far UV spectral range [34,37,39,40].

There are cases in which it is highly desirable to extend a UV detector's sensitivity to visible and infrared wavelengths. Silicon detectors are ideal for such cases, for example spectroscopic applications where broadband response is needed. However, there are also cases in which a detector's sensitivity to visible photons introduces an undesirable signal that interferes with UV measurement. This type of out-of-band background, also known as "red leak", is particularly problematic in environments with weak UV signals masked by presence of a strong visible signal. For these applications, some degree of visible-blindness or out-of-band rejection is required.

To achieve visible rejection, we turn to metal dielectric filters (MDFs). As stand-alone filters, these Fabry-Perot structures, which are also referred to as photonic bandgaps, have been used in the past as bandpass filters for spectral ranges from the UV to the infrared [41–44]. Briefly, the metal layers are separated by transparent dielectric spacer layers. The layered structure is designed to resonantly transmit light within a specific wavelength band, while out-of-band light is strongly reflected by the metal layers. The in-band light suffers absorption losses in the metal layer; therefore, it is necessary to choose metals with a large absorption coefficient to index of refraction (k/n) ratio in the band of interest. In the UV the primary choice is aluminum due to its high plasma frequency and relative lack of significant interband transitions in this spectral range [45,46].

Designs of this type have been extended to Si substrates for potential use as a filter directly integrated on a photodetector, making it possible to at once optimize the in-band sensitivity together with the out-of-band rejection [47]. The complexity of the filter design influences several performance metrics, including peak transmission percentage, out-of-band rejection ratio, passband width, and target wavelength. Depending on the target wavelengths these filter designs may be based variety of dielectric materials, including MgF_2 as illustrated in Figure 3. ALD remains a critical aspect of this work, especially for complex multilayer stacks, which requires accurate thickness control, layer uniformity, and precise control of the interface chemistry between the metal and dielectric layers.

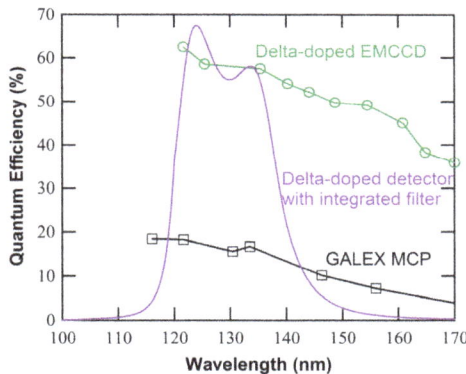

Figure 3. Model performance of a 2D-doped detector with an integrated visible-blind bandpass filter for far UV wavelengths. The multilayer MDF was designed to provide high in-band QE, and high out-of-band rejection (>10^4). As designed the MDF includes layers of MgF_2 (20 nm thick), Al (26 nm), MgF_2 (18 nm), Al (20 nm), and MgF_2 (11 nm), starting from the silicon interface up to the air interface.

2.1.4. Large-Scale, High Throughput Affordable Production of High Efficiency Single Photon Counting Silicon Imagers for Missions and Commercial Applications

With the need for high throughput processing of scientific detector arrays as well as high throughput processing for commercial use apparent, a scaled up version of our post fabrication end-to end processing was developed over the last few years. The scaling up process began with upgrading the MBE capability by design, procurement, and commissioning of a multiple wafer batch mode, wafer-scale MBE machine. All the steps of the process can be performed at wafer-scale or die level and can be customized based on project needs.

Figure 4 shows JPL's end-to-end post-fabrication processing as applied to a "generic" device type. This example is illustrative of a process flow that can be customized and adapted according to the requirements of a wide variety of silicon detectors. The process begins with fully-fabricated detector arrays, complete with final metallization. The VLSI (Very-Large-Scale Integration)-fabricated circuitry and pixel structure are protected by direct oxide-oxide bonding of the device wafer's front surface to a blank silicon wafer that has an identical diameter and is a few hundred microns thick. The blank wafer also serves as a mechanical support once the device wafer is thinned down to the epitaxial silicon layer thickness, *i.e.*, a few micrometers to tens of micrometers thick. The oxide-oxide bond must be devoid of any epoxy or organics, in order for the bond to survive subsequent high temperature processing steps, including MBE and ALD. Following direct wafer bonding, the bulk of the detector substrate is removed by grinding and chemical mechanical polishing, which reduces the device wafer thickness from hundreds to ~50 μm. The remainder of the substrate is removed by a selective thinning process, using an isotropic chemical etchant that removes the bulk P+ Si at a rate of >100:1 compared to the P-epilayer. This dopant-sensitive selectivity enables the use of the epilayer as an etch stop. Final polishing produces a smooth mirror finish surface that will be subjected to a simple series of cleaning steps to prepare the surface for epitaxial growth.

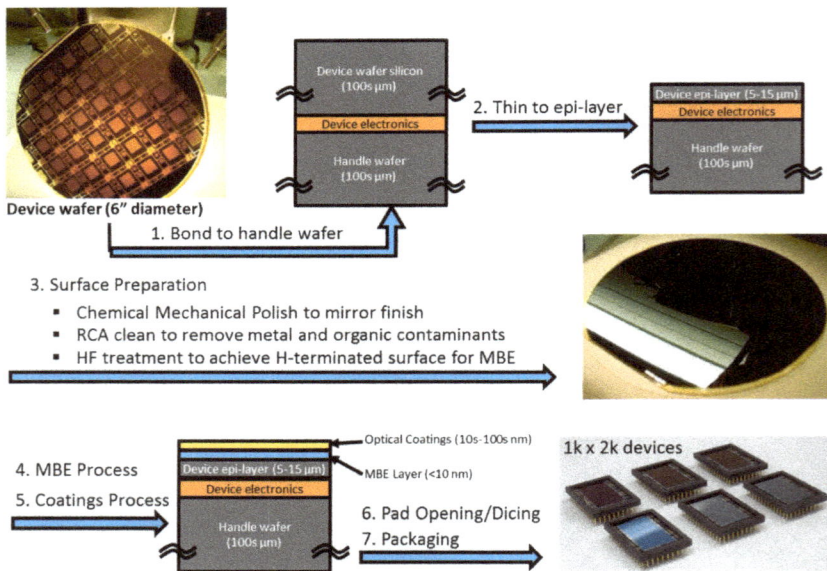

Figure 4. Process flow for end-to-end post fabrication processing along with schematic diagram of a superlattice-doped or delta-doped arrays (summarized as MBE layers), and the photographs of bare and AR-coated BSI silicon detector arrays. The coated arrays can be discerned from the reflective hue seen in photograph.

The backside-thinned detector wafer is then passivated by MBE growth of 2D-doped Si layers, comprising single or multilayer stacks of delta-doped Si as described in Section 2.1.2. MBE growth provides the precise band structure engineering of the Si surface to allow detection of higher energy photons (100 nm < λ < 400 nm) that are absorbed in shallow depths (4–10 nm) with minimal to no loss of photoelectrons [16,18].

After passivation, wafers are moved into the ALD machine. A thin oxide layer is grown to protect the 2D-doped surface. AR coatings or filters can be applied to the entire wafer at this time; alternatively, the wafer can be diced and coatings applied to individual or multiple die. As described in Section 2.1.3, our optical coatings are typically prepared by ALD, which enables atomic-scale control of structure and composition, including multilayer coatings with embedded metal films for bandpass coatings with high in-band QE and excellent out-of-band rejection [36].

2.2. Single Photon Counting in the UV with III-Nitride APDs

The III-Nitride material family can be alloyed to span the group of direct bandgap semiconductors spanning in bandgap range from 3.4 to 6.2 eV. Due to their wide bandgaps, they are naturally insensitive to visible photons, the cutoff wavelength can be tailored and the resulting devices can be operated at higher temperature. Most III-Nitride materials are grown on sapphire or silicon substrates. This is because there are no native substrates except for bulk substrates produced by hydride vapor phase epitaxy (HVPE) which are small and costly. The growth on mismatched substrates leads to defects and consequently leakage when devices are processed.

Our approach in developing III-Nitride-based single photon counting detectors in the UV has been to use novel growth techniques in order to start from the optimized material quality; to use design and processing techniques for reduction of leakage and improvement of efficiency; and finally to design readouts that can address the need for the high avalanche voltage and the quenching of breakdown possibility. Here we discuss some special aspects of growth, processing and readout and in the results section we show quantum efficiency, out-of-band rejection, tailorability, and gain of APDs formed of GaN and AlGaN material.

2.2.1. Brief Description of the Special Features of III-Nitride Materials Growth

For the work described here, the team from the College of Nanoscale Science and Engineering (CNSE) at State University of New York (SUNY)-Polytechnic performed all device growth development and contributed to device design, fabrication and processing. Growths were performed using Metal Organic Chemical Vapor Deposition (MOCVD). Novel features of the growth are briefly described here in experiments using two different substrates.

Here we have used HVPE GaN substrates for the growth that creates smooth and high quality material. GaN p-i-n APDs were grown on HVPE bulk substrate using MOCVD. The high quality of the material grown on bulk substrates is evident from the smooth atomic force microscopy (AFM) image of the GaN surface showing uniform parallel step edges, in comparison to the surface on sapphire that shows step terminations and defects (Figure 5).

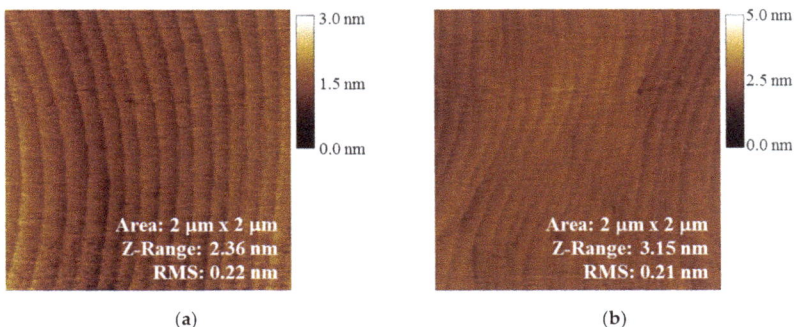

(a)   (b)

Figure 5. AFM images of MOCVD GaN grown on (a) HVPE GaN and (b) GaN template on sapphire showing high quality materials growth on the native substrate.

For some applications, including those that require solar blind operation, AlGaN APDs with [Al] > 40% need to be developed; this brings with it many more challenges in addition to the problems already discussed for GaN APDs. Conventional growth of $Al_xGa_{1-x}N$ with x > 0.4 (required for cut-off wavelengths below 280 nm) is problematic, producing highly defective films due to gas phase reactions and low surface adatom mobility. We have recently developed a new technique for growth of high quality, high Al composition AlGaN films, where the Al and Ga precursors and ammonia are sequentially pulsed during the MOCVD film growth. The process has some similarities to the growth protocol used for ALD, and improves the material for similar reasons: growth species have time to saturate the surface and move to stable sites before the next species is introduced. Pulsed growth also addresses another MOCVD limitation: in some cases, MOCVD precursors partially react in the gas phase before reaching the growth surface. By sequentially introducing the precursors and purging the system between pulses, the precursors cannot interact in gas phase. Using these new pulsed growth methods, we have maintained material quality while increasing doping levels by about an order of magnitude. This gives us very high quality films, with X-ray diffraction from the 0002 crystal orientation yielding peak widths (FWHM) of less than 100 arcsec.

This is especially important for higher doping levels in AlGaN, since doping becomes more difficult as the Al fraction increases, eventually resulting in a decrease in material quality. Due to the low activation of dopants in AlGaN (~1%), much more dopant must be incorporated in AlGaN than in GaN to achieve the same carrier concentration, leading to difficulty in maintaining the crystalline quality of the AlGaN. Pulsed growth allows higher carrier concentrations while maintaining sufficient material quality.

Growth parameters were varied to get-high quality Si-doped $Al_{0.6}Ga_{0.4}N$ films, which show a carrier concentration of 2×10^{18} cm^{-3} and mobility of 90 cm$^2 \cdot$V$^{-1} \cdot$s^{-1}. The crystalline quality and electrical properties achieved by our team are the state of the art for high aluminum percentage AlGaN. Using pulsed growth technique, we have been able to grow crack-free AlGaN films with thickness greater than 500 nm on sapphire substrates. The values of carrier concentration and mobility of the pulsed n-AlGaN films are among the best reported results for high aluminum percentage n-AlGaN on sapphire. Values of electrical characteristics are given in Figure 6.

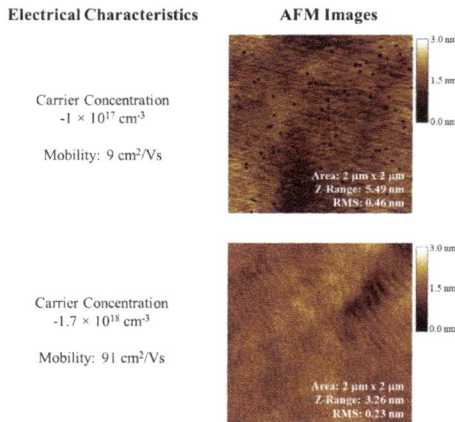

Electrical Characteristics   **AFM Images**

Carrier Concentration
-1 × 10^{17} cm^{-3}

Mobility: 9 cm^2/Vs

Carrier Concentration
-1.7 × 10^{18} cm^{-3}

Mobility: 91 cm^2/Vs

Figure 6. Pulsed growth in MOCVD shows marked improvement in the quality of AlGaN films as shown in the AFM images. The higher quality of both films is the result of the pulsed growth. Introduction of Si during the pulsed sequence is shown to greatly impact defect formation and density as well as carrier concentration. Si dopants were introduced during the Al pulse in the AFM image (**top**), whereas introduction of Si during the Ga pulse (**bottom**) increases both the free electron concentration as well as marked improvement in the quality of the AlGaN film.

2.2.2. Processing Features

Some of the typical challenges facing APD production are device yield, robustness, and dark current. In particular, sidewall-related defects in GaN APDs have often been observed to contribute to undesirable current components, such as those produced by defect-related microplasmas. SiO_2 is most commonly used for passivation due to its availability and simplicity of growth; Figure 7 shows an example device structure and one implementation of the sidewall passivation using SiO_2. SiO_2 is not the ideal passivation material for AlGaN materials; thus, we are exploring other passivation materials based on ALD processes. ALD is particularly well suited for passivation because the technique results in conformal coatings with monolayer uniformity, even for complex geometries.

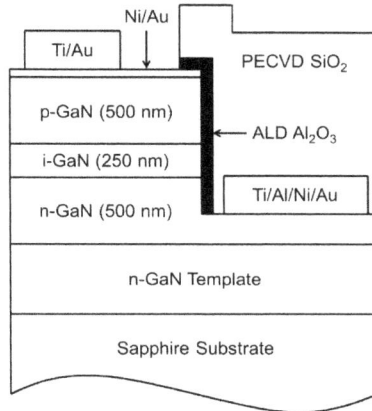

Figure 7. Schematic of device geometry showing the application of the ALD Al_2O_3 layers as sidewall passivation and incorporating PECVD SiO_2 as a contact isolation layer. In devices not receiving the ALD treatment, the mesa sidewalls are passivated by the PECVD film only.

2.2.3. Readout Design and Fabrication for III-Nitride APDs

The III-Nitride sensor discussed here is a hybrid structure where the detector is processed in III-Nitride material and the readout integrated circuit (ROIC) is designed and fabricated in silicon; the two pieces are hybridized using indium bump bonding. The major part of the effort up to this point has been focused on detector design and fabrication. Recently, in collaboration with the Delft University of Technology (TU Delft), we have begun the work on ROICs. Here we describe some of the major parameters considered for the readout and the particular challenges for this detector. In the results section, simulation plots and preliminary fabrication results are discussed.

Based on the detector characteristics, an equivalent diode circuit model was chosen, consisting of the photodiode current along with its equivalent capacitance. One of the goals was to keep the design as simple as possible while obtaining a suitable match with the detector characteristics and adequate readability at the readout output. These considerations led to a Capacitive TransImpedance Amplifier (CTIA) as a feasible solution, as shown in Figure 8. Initially, the feedback capacitor (C_{fb}) is discharged by the reset transistor and the output equals V_{ref}. Once the reset is released, incoming photodiode current (I_{pd}) passing through the high voltage N-type Metal Oxide Semiconductor (NMOS) transistor is integrated on C_{fb}. The amplifier generates whatever output is necessary to keep the input voltage zero as the capacitor charges up. Thus, the output swings negative by a voltage equal to the integrated charge divided by the feedback capacitance.

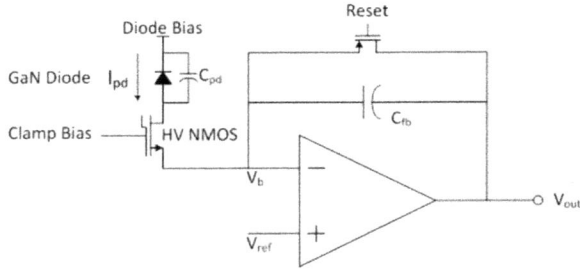

Figure 8. Block diagram of a capacitive transimpedence amplifier basic concept used for the readout of the GaN detector.

One challenge with these detectors is that the voltage required for electron multiplication is high, on the order of 60–100 V. Readout circuits must be protected from these high voltages. A convenient solution, available in the chosen Complementary Metal Oxide Semiconductor (CMOS) process, is the use of the high voltage NMOS clamp transistor. The clamp bias is set so that, initially, the clamp transistor is in its ohmic regime. There is very little voltage drop across it and the photodiode sees its full bias. Eventually, however, the amplifier output will saturate. Thereafter, the current charging C_{fb} will begin to raise the voltage of the positive input. Left unchecked, it could reach damaging levels. Instead, with this circuit, as the amplifier input voltage rises the gate-to-source voltage of the clamp transistor is reduced and the transistor begins to shut off. The bias is chosen so that shutoff occurs before the input exceeds a safe voltage. Once the transistor shuts off, the photodiode current (I_{pd}) no longer flows through it to charge C_{fb}, instead charging the photodiode capacitor (C_{pd}) and reducing the diode bias. In the case of avalanche breakdown the diode current will be high and the processes will happen quickly: saturation of the amplifier, shutoff of the clamp transistor, and debiasing of the photodiode. The debiasing of the diode, in turn, will quench the avalanche. A high voltage will remain on the drain of the clamp transistor, but it is designed for this. In order to resume operation, the reset transistor must be operated. This will pull down the source of the clamp transistor, turning it on. I_{pd}, no longer high, will sink through the amplifier output, along with the current needed to charge C_{pd} back to its full bias voltage. When the reset is released, the process begins anew.

Our implementation of this concept uses a simple single-ended common-source PMOS transistor for the amplifier. An NMOS source follower was added to buffer the output of the feedback amplifier for monitoring, and another to allow monitoring the amplifier input voltage. This should be useful for observing the increase in voltage when saturation occurs, and should be especially valuable in detecting avalanche breakdown.

3. Results

This section describes the results of superlattice-doped EMCCDs and APDs—with AR coatings and visible-blind filters followed by the results of the III-Nitride APDs.

3.1. Silicon Detectors

3.1.1. Quantum Efficiency in the Ultraviolet Spectral Range

Quantum efficiency results of 2-D growth as single layer (delta doping) and multilayer (superlattice doping) are well documented showing 100% internal QE from soft X-ray to near infrared in various silicon detector architecture and designs [16,28,31,34,48,49]. Addition of AR coatings has shown high external QE tailored for various parts of the spectrum. Here we show two examples of results obtained from back illumination and passivation with 2D growth combined with custom coatings in the FUV and NUV for EMCCDs. Results of this process as manifested in QE are shown in Figure 9 [34,37,40]. Figure 9a shows the QE results of coated delta doped arrays for an imaging spectrometer that covers the challenging part of the spectrum from 100–300 nm. Coatings are simple, single layer designs that result in better than 50% external QE. Figure 9b shows the QE measurements of two superlattice-doped, two-megapixel EMCCDs with two different ALD multilayer AR coatings designed for the atmospheric window of a balloon, aiming to maximize the QE at 205 nm. The higher number of multilayers result in improvement in the QE and narrowing of the peak.

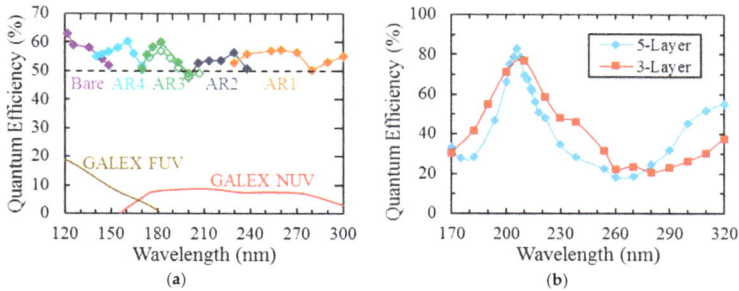

Figure 9. (**a**) QE data of delta-doped conventional (closed diamonds) and EMCCDs (open diamonds) enhanced with single layer AR coatings [34]. (**b**) QE data from two superlattice-doped CCD201s (e2v's 1k × 2k EMCCD) optimized for the 200–220 nm wavelength range; the device designs included a three-layer AR coating (red squares) and a five-layer AR coating (blue diamonds).

3.1.2. Visible Rejection Using Metal Dielectric Films

The high QE of 2D-doped silicon imagers combined with integrated red-rejection filters is a potentially disruptive sensor technology for applications currently dominated by low QE MCPs. Preliminary results were recently reported based on die-level coating of superlattice-doped APD detectors [16]. Figure 10 shows QE measurements from these superlattice-doped APDs with both MDFs and AR coatings. The MDFs were formed by ALD-grown aluminum oxide and *ex-situ* e-beam evaporated metallic aluminum, while the dielectric-only coatings comprise ALD-grown thermal Al_2O_3. A comparison of the dielectric-only coating with the metal-dielectric multilayer stack clearly demonstrates the concept. With a single embedded metal layer, a rejection ratio of near an order of magnitude is achieved. With multiple embedded metal layers, rejection ratios from 10^4 to 10^7 are projected. This significant achievement now makes possible UV-sensitive Si detectors with tailorable out-of-band rejection.

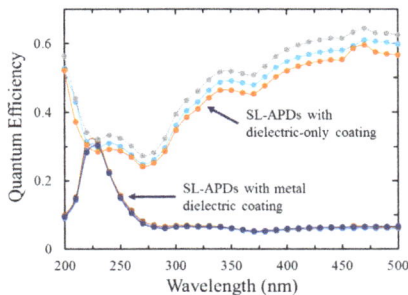

Figure 10. An example of ALD multilayer stacks of metal-dielectrics films, shown in figure is a single metal layer sandwiched between two dielectric layers. The results shown here represent the application of superlattice doping and AR coating technologies to APDs fabricated and characterized by RMD.

This initial demonstration involves an intermediate air exposure of the metallic Al layer prior to ALD Al$_2$O$_3$ encapsulation which is estimated to result in the formation of 1–2 nm of interfacial Al oxide. For this application at λ > 200 nm, the impact of the oxidation layer is minimal on the final performance, but for applications below 200 nm the oxide absorption losses will begin to limit the predicted peak transmission. For these shorter wavelength applications (which utilize ALD metal fluoride dielectric layers) we are exploring vacuum transfer approaches to limit environmental exposure, as well as methods incorporating atomic layer etching to chemically remove the Al surface oxide immediately prior to ALD encapsulation.

3.2. III-Nitride APDs

Photodiodes and photodiode arrays were processed from the III-Nitride material and QE was measured in comparison with a calibrated diode. Photodiode arrays with 50% QE, four orders of magnitude out-of-band rejection ratio, and low leakage current (<<1 nA@15 V reverse bias for 250-um pixel size) were demonstrated. Figure 11 shows measured external QE for these devices with no applied voltage; the QE increases with voltage due to voltage-enhanced collection of carriers. At higher voltages (not shown), carrier multiplication (gain) increases the measured photocurrent.

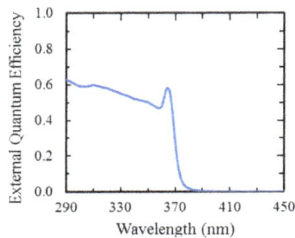

Figure 11. External QE as a function of photon wavelength for an Al$_2$O$_3$ –passivated GaN p-i-n APD with zero applied voltage. With no applied voltage, there is unity internal gain.

The use of ALD Al$_2$O$_3$ as a sidewall passivation layer has resulted in the reduced occurrence of premature breakdown in mesa p-i-n GaN APDs when compared to devices fabricated with a more common plasma-enhanced chemical vapor deposition (PECVD) SiO$_2$ passivation. Mesa APDs with diameters ranging from 25 to 100 µm show a significant reduction in median dark current for the ALD-passivated devices (Figure 12). The reduction in median dark current was most significant for the smallest devices, showing an order of magnitude improvement at reverse biases near avalanche. The interfacial effect of ALD Al$_2$O$_3$ was investigated by fabricating MOS capacitors, which show a large reduction in both slow trapping and faster interface states compared to SiO$_2$-passivated devices (Figure 13).

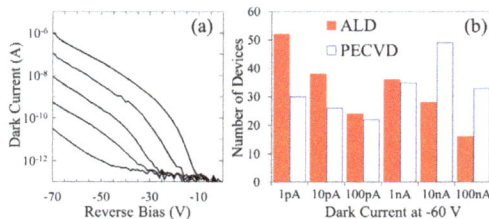

Figure 12. (a) Example of the variation in reverse bias behavior observed for ALD-passivated GaN APDs; **(b)** Histogram of ~200 devices with 25-µm diameter samples receiving a sidewall passivation layer of ALD Al$_2$O$_3$ or PECVD SiO$_2$. First appeared in *MRS Proceedings* [50].

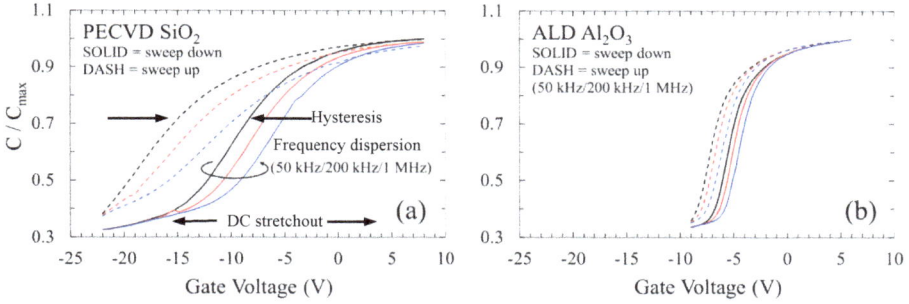

Figure 13. High-frequency CV characteristics for MOS capacitors fabricated on n-type GaN. A large reduction in the qualitative indicators of charge trapping—including voltage hysteresis, frequency dispersion, and stretchout—is observed for devices fabricated with (**b**) ALD Al_2O_3 compared to those with (**a**) PECVD SiO_2.

We have also shown that the high-field region around the top device contact is a significant source of avalanche dark current. One method for reducing this dark current component is ion implantation. Implantation around the contact perimeter creates high-resistance region that spreads the potential drop at the contact edge over a larger area, greatly reducing the electric field at the edges. It also creates a disordered region of heavy scattering that prevents electrons from acquiring the energy necessary for multiplication. This reduced field prevents premature breakdown at the contact edges. We have shown a large reduction in dark current and voltage-induced damage using this method (Figure 14) [51]. Innovative passivation methods using other materials can be explored to further reduce dark current in our APD devices.

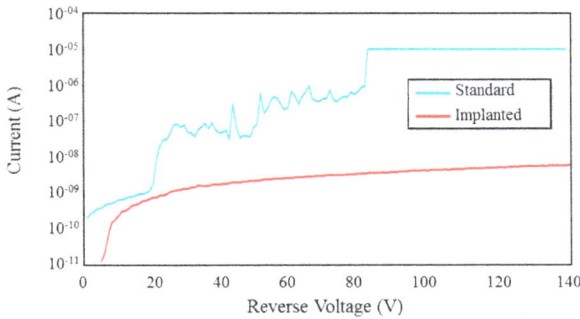

Figure 14. Current-voltage characteristics for an APD without contact-edge ion implantation (blue) and with implantation (red). Dark current and irreversible damage present in the unimplanted sample are reduced or eliminated by the implantation. First published in *IEEE Photonic Technology Letters* [51].

Avalanche operation has been demonstrated in GaN p-i-n APDs. Figure 15 shows characteristics from these devices, with large avalanche gain and low dark current. The gain values are competitive with the current state of the art for GaN APDs. These data demonstrate high-efficiency APDs resulting from our ability to grow low-defect-density III-N material.

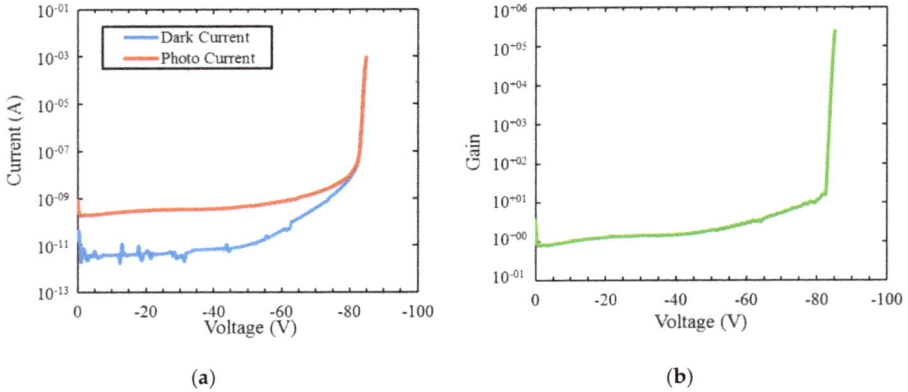

Figure 15. (a) Dark current (blue) and photocurrent (red) for a GaN p-i-n APD. (b) Gain derived from Figure 16a is shown in green. Avalanche gains of $>10^5$ have been measured on these devices.

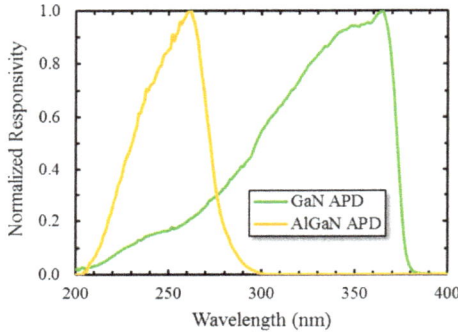

Figure 16. Comparison of the spectral response of GaN and $Al_{.4}Ga_{.6}N$ APDs. First published in *Journal of Electronic Materials* [52].

Properties of the III-Nitride family of materials can be tuned by varying the concentration of Al in $Al_xGa_{1-x}N$; capitalizing on this phenomenon we have recently demonstrated avalanche breakdown with high gain in the solar blind region. Figure 16 shows a comparison of the spectral response of GaN and $Al_{.4}Ga_{.6}N$ APDs; the latter shows a detection cutoff at 280 nm in the solar blind region, a dramatic blue shift relative to the former.

As previously mentioned, work has also been done to develop a readout device for AlGaN-based APDs. Among the specialized requirements for these readouts is the ability to operate at high device bias voltages (in some cases V > 100V). Measurements from recently-fabricated readout devices have confirmed the functionality of the chip. Here we present transient and AC simulation results performed using the Cadence 6.1.5 modeling software. The transient modeling of the CTIA starts with the "reset" switch initially closed and released at t = 0. When the switch is opened, the photodiode current starts to be integrated and the voltage at the amplifier inverting input, V_b, begins to rise, causing the amplifier output voltage, V_{out1}, to fall. The small rise in V_b is due to the finite gain of the amplifier, with $\Delta V_b = -\Delta V_{out1}/A$ and $\Delta V_{out1} \approx -Q/C_{fb}$, where A is the open loop gain and Q is the integrated input charge. V_{out1} drops from its initial value down to 0 V, as reflected in the output of the source follower, shown in Figure 17a). The source follower saturates somewhat sooner than the amplifier, and at this point the capacitor must be reset for another integration cycle.

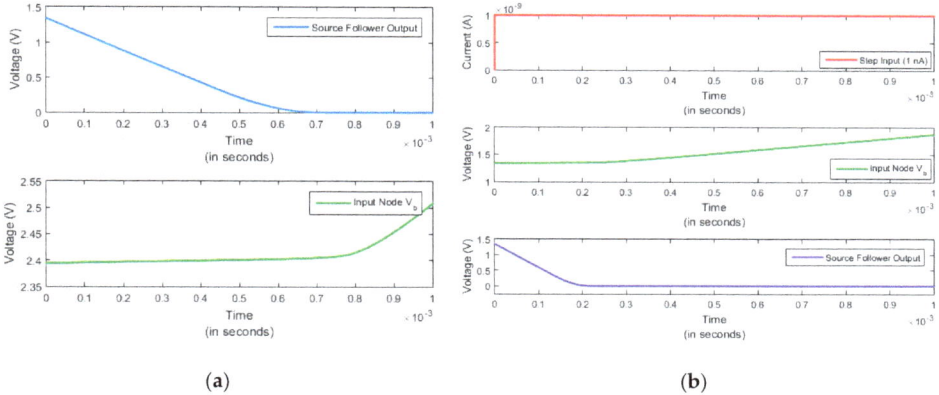

(a)   (b)

Figure 17. (**a**) As photodiode current starts to flow, the voltage at the node V_b rises from its initial value. As this occurs, the output voltage (source follower output) drops to 0. (**b**) A step response is simulated for a sudden increase in the input current from 0 to 1 nA. As soon as the input current rises to 1 nA, the output nodes settle to their steady state value without any ringing or overshoot.

We have also shown that the system is stable. Figure 17b shows a simulated step response for a sudden rise in the input current from 0 to 1 nA during which the output nodes were observed and plotted. The moment the input current rises to 1 nA, the output nodes settle to their steady state value without any ringing or overshoot.

Finally, Figure 18 shows a simulation of the open loop gain. Here the gain obtained is ~45 dB, while the 3 dB frequency is 10 MHz, close our target approximation. Furthermore, the system is stable with a phase margin of 66°.

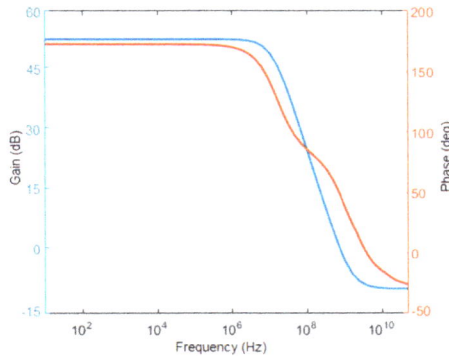

Figure 18. Frequency response of the readout circuit. The gain is around 45 dB with a 3 dB frequency of 10 MHz, close our target value. The modeling results indicate that the system operation is stable within the designed bandwidth.

4. Summary and Conclusions

Silicon-based and III-Nitride-based single photon counting ultraviolet detectors are described. In silicon, photon counting is achieved using EMCCD architecture. Although APD structure are capable of photon counting operation, the APDs used here are operated in the linear mode and are discussed because of their use to demonstrate detector-integrated visible rejection filters. EMCCDs are back illuminated and passivated using superlattice doping and delta doping. Antireflection coatings are developed using standard techniques for single and multilayerd coatings and are implemented using atomic layer deposition. Out-of-band rejection in silicon is achieved through detector-integrated metal-dielectric stacks as visible rejection filters. Avalanche photodiodes structures are grown and processed using GaN and AlGaN materials to take advantage of the wide bandgap of these materials for visible blindness. Readout is achieved by hybridization through a CMOS readout integrated circuit. A simple CTIA ROIC designed for this project has been fabricated. Both approaches show excellent promise. Silicon-based devices are ready for deployment and III-N arrays will have applications in longer-term missions. The first approach takes advantage of the maturity and mass production of silicon imaging while the second approach takes advantage of intrinsic materials characteristics. In both cases, surface and interface engineering and nano-scale materials growth and processing play key roles in achieving high quantum efficiency and low dark current.

Supplementary Materials: More information can be obtained regarding the techniques, equipment, and related projects described here at http://microdevices.jpl.nasa.gov/infrastructure/.

Acknowledgments: We gratefully acknowledge the collaborative effort with e2v, Inc. (Chelmsford, UK) and helpful discussions with P. Pool, P. Fochi, A. Reinheimer, P. Jorden, and P. Jerram of e2v. We acknowledge excellent collaborative effort and support from M. McClish at Radiation Monitoring Devices (RMD, Watertown, MA, USA) and D. Hitlin of Caltech. The authors thank S. Riccardi and R. Myers of RMD for QE measurements on superlattice-doped avalanche photodiodes. This work was carried out at Jet Propulsion Laboratory, California Institute of Technology under a contract with NASA. We also acknowledge the support of W.M. Keck Institute (Pasadena, CA, USA) for Space Studies for partial support of this work. One of us, E.T.H acknowledges the support of NASA Earth and Space Science Fellowship, NSF Fellowship, and R.M & G.B. Millikan Prize Fellowship at different stages of this project. P. P. was supported by an internship supported through JPL Visiting Research Student Program (JVRSP) and TU Delft Faculty of Electrical Engineering.

Conflicts of Interest: The authors declare no conflict of interest. The funding sponsors had no role in the design of the study; in the collection, analyses, or interpretation of data; in the writing of the manuscript, and in the decision to publish the results.

Abbreviations

The following abbreviations are used in this manuscript:

AFM	Atomic Force Microscopy
ALD	Atomic Layer Deposition
APD	Avalanche PhotoDiode
ARC	AntiReflection Coatings
BSI	BackSide Illumination
CCD	Charge Coupled Detector
CGM	CircumGalactic Medium
CIC	Clock Induced Charge
CMOS	Complementary Metal Oxide Semiconductor
CTIA	Capacitive TransImpedance Amplifier
CV	Capacitance-Voltage
EMCCD	Electron Multiplying CCD
GALEX	GALaxy Evolution eXplorer
HST	Hubble Space Telescope
HVPE	Hydride Vapor Phase Epitaxy
IGM	InterGalactic Medium
MBE	Molecular Beam Epitaxy
MCP	MicroChannel Plate
MDF	Metal Dielectric Filter
MOCVD	Metal Organic Chemical Vapor Deposition
MOS	Metal Oxide Semiconductor
NMOS	N-type Metal Oxide Semiconductor

PECVD	Plasma Enhanced Chemical Vapor Deposition
PMOS	P-type Metal Oxide Semiconductor
PMT	PhotoMultiplier Tube
QE	Quantum Efficiency
QEH	QE Hysteresis
RMD	Radiation Monitoring Devices, Inc.
ROIC	Read Out Integrated Circuit
VLSI	Very-large-scale Integration
WF/PC	Wide Field/Planetary Camera

References

1. Clarke, J.T.; Ajello, J.; Ballester, G.E.; Ben Jaffel, L.; Connerney, J.E.P.; Gerard, J.-C.; Gladstone, G.R.; Pryor, W.R.; Tobiska, K.; Trauger, J.; *et al.* HST/STIS images of uv auroral footprints from Io, Europa, and Ganymede. *Bull. Am. Astron. Soc.* **1999**, *31*, 1185.

2. McGrath, M.A.; Feldman, P.D.; Strobel, D.F.; Retherford, K.; Wolven, B.; Moos, H.W. HST/STIS ultraviolet imaging of Europa. *Bull. Am. Astron. Soc.* **2000**, *32*, 1056.

3. Woodgate, B.E.; Kimble, R.A.; Bowers, C.W.; Kraemer, S.; Kaiser, M.E.; Grady, J.F.; Loiacono, J.J.; Brumfield, M.; Feinberg, L.D.; Gull, T.R.; *et al.* The space telescope imaging spectrograph design. *Publ. Astron. Soc. Pacific* **1998**, *110*, 1183–1204. [CrossRef]

4. Green, J.C.; Froning, C.S.; Osterman, S.; Ebbets, D.; Heap, S.H.; Leitherer, C.; Linsky, J.L.; Savage, B.D.; Sembach, K.; Michael Shull, J.; *et al.* The cosmic origins spectrograph. *Astrophys. J.* **2012**, *744*, 60. [CrossRef]

5. Morrissey, P. A GALEX instrument overview and lessons learned. *Proc. SPIE* **2006**, *6266*. [CrossRef]

6. Schiminovich, D.; Ilbert, O.; Arnouts, S.; Milliard, B.; Tresse, L.; Le Fèvre, O.; Treyer, M.; Wyder, T.K.; Budavári, T.; Zucca, E.; *et al.* The GALEX -VVDS measurement of the evolution of the far-ultraviolet luminosity density and the cosmic star formation rate. *Astrophys. J.* **2005**, *619*, L47–L50. [CrossRef]

7. Feldman, P.D.; Steffl, A.J.; Parker, J.W.; A'Hearn, M.F.; Bertaux, J.L.; Stern, S.A.; Weaver, H.A.; Slater, D.C.; Versteeg, M.; Throop, H.B.; *et al.* Rosetta-Alice observations of exospheric hydrogen and oxygen on Mars. *Icarus* **2011**, *214*, 394–399. [CrossRef]

8. Feldman, P.D.; A'Hearn, M.F.; Bertaux, J.-L.; Feaga, L.M.; Parker, J.W.; Schindhelm, E.; Steffl, A.J.; Stern, S.A.; Weaver, H.A.; Sierks, H.; *et al.* Measurements of the near-nucleus coma of comet 67P/Churyumov-Gerasimenko with the Alice far-ultraviolet spectrograph on Rosetta. *Astron. Astrophys.* **2015**, *583*, A8. [CrossRef]

9. Stern, S.A.; Feaga, L.M.; Schindhelm, E.; Steffl, A.; Parker, J.W.; Feldman, P.D.; Weaver, H.A.; A'Hearn, M.F.; Cook, J.; Bertaux, J.-L. First extreme and far ultraviolet spectrum of a Comet Nucleus: Results from 67P/Churyumov-Gerasimenko. *Icarus* **2015**, *256*, 117–119. [CrossRef]

10. Stern, S.A.; Scherrer, J.; Slater, D.C.; Gladstone, G.R.; Dirks, G.; Stone, J.; Davis, M.; Versteeg, M.; Siegmund, O.H.W. ALICE: The ultraviolet imaging spectrograph aboard the New Horizons Pluto mission spacecraft. *Proc. SPIE* **2005**, *5906*. [CrossRef]

11. Slater, D.C.; Davis, M.W.; Olkin, C.B.; Scherrer, J.; Stern, S.A. Radiometric performance results of the New Horizons' ALICE UV imaging spectrograph. *Proc. SPIE* **2005**, *5906*. [CrossRef]

12. Esposito, L.W.; Barth, C.A.; Colwell, J.E.; Lawrence, G.M.; Mcclintock, W.E.; Stewart, A.I.F.; Keller, H.U.; Korth, A.; Lauche, H.; Festou, M.C.; *et al.* The Cassini ultraviolet imaging spectrograph investigation. *Space Sci. Rev.* **2004**, *115*, 299–361. [CrossRef]

13. Jerram, P.; Pool, P.J.; Bell, R.; Burt, D.J.; Bowring, S.; Spencer, S.; Hazelwood, M.; Moody, I.; Catlett, N.; Heyes, P.S. The LLCCD: Low-light imaging without the need for an intensifier. *Proc. SPIE* **2001**, *4306*, 178–186.

14. Hynecek, J. Impactron—A new solid state image intensifier. *IEEE Trans. Electron. Devices* **2001**, *48*, 2238–2241. [CrossRef]

15. Niclass, C.; Favi, C.; Kluter, T.; Gersbach, M.; Charbon, E. A 128 × 128 single-photon image sensor with column-level 10-bit time-to-digital converter array. *IEEE J. Solid State Circuits* **2008**, *43*, 2977–2989. [CrossRef]

16. Hoenk, M.E.; Nikzad, S.; Carver, A.G.; Jones, T.J.; Hennessy, J.; Jewell, A.D.; Sgro, J.; Tsur, S.; McClish, M.; Farrell, R. Superlattice-doped silicon detectors: Progress and prospects. *Proc. SPIE* **2014**, *9154*. [CrossRef]

17. Hoenk, M.E.; Carver, A.G.; Jones, T.J.; Dickie, M.; Cheng, P.; Greer, F.; Nikzad, S.; Sgro, J.; Tsur, S. The DUV stability of superlattice-doped CMOS detector arrays. In Proceedings of the International Image Sensor Workshop, Snowbird, UT, USA, 12–16 June 2013.

18. Nikzad, S.; Hoenk, M.E.; Carver, A.G.; Jones, T.J.; Greer, F.; Hamden, E.; Goodsall, T. High Throughput, High Yield Fabrication of High Quantum Efficiency Backilluminated Photon Counting, Far UV, UV, and Visible Detector Arrays. In Proceedings of the International Image Sensor Workshop, Snowbird, UT, USA, 12–16 June 2013.

19. Harding, L.K.; Demers, R.T.; Hoenk, M.; Nemati, B.; Cherng, M.; Michaels, D.; Peddada, P.; Loc, A.; Bush, N.; Hall, D.; *et al.* Technology Advancement of the CCD201-20 EMCCD for the WFIRST-AFTA Coronograph Instrument: Sensor characteriation and radiation damage. *J. Astron. Telesc. Instrum. Syst.* **2015**, *2*, 011007. [CrossRef]

20. Hamden, E.T.; Lingner, N.; Kyne, G.; Morrissey, P.; Martin, D.C. Noise and dark performance for FIREBall-2 EMCCD delta-doped CCD detector. *Proc. SPIE* **2015**, *9601*. [CrossRef]

21. Reinheimer, A. *Personal Communication*; e2v: Chelmsford, UK, 2012.

22. Daigle, O.; Gach, J.-L.; Guillaume, C.; Carignan, C.; Balard, P.; Boisin, O. L3CCD results in pure photon-counting mode. *Proc. SPIE* **2004**, *5499*. [CrossRef]

23. Daigle, O.; Djazovski, O.; Laurin, D.; Doyon, R.; Artigau, É. Characterization results of EMCCDs for extreme low-light imaging. *Proc. SPIE* **2012**, *8453*. [CrossRef]

24. Shortes, S.R.; Chan, W.W.; Rhines, W.C.; Barton, J.B.; Collines, D.R. Characteristics of thinned backside-illuminated charge-coupled device imagers. *Appl. Phys. Lett.* **1974**, *24*, 565. [CrossRef]

25. Stoller, A.; Speers, R.; Opresko, S. A new technique for etch thinning silicon wafers. *RCA Rev.* **1970**, *31*, 265–270.

26. Kern, W. Chemical etching of silicon, germanium, gallium arsenide, and gallium phosphide. *RCA Rev.* **1978**, *39*, 278–308.

27. Collins, N.R.; Boehm, N.; Delo, G.; Foltz, R.D.; Hill, R.J.; Kan, E.; Kimble, R.A.; Malumuth, E.; Rosenberry, R.; Waczynski, A.; *et al.* Wide field camera 3 CCD quantum efficiency hysteresis: Characterization and mitigation. *Proc. SPIE* **2009**, *7439*. [CrossRef]

28. Hoenk, M.E.; Grunthaner, P.J.; Grunthaner, F.J.; Terhune, R.W.; Fattahi, M.; Tseng, H.-F. Growth of a delta-doped silicon layer by molecular beam epitaxy on a charge-coupled device for reflection-limited ultraviolet quantum efficiency. *Appl. Phys. Lett.* **1992**, *61*, 1084–1086. [CrossRef]

29. Hoenk, M.E.; Grunthaner, P.J.; Grunthaner, F.J.; Terhune, R.W.; Fattahi, M.M. Epitaxial Growth of p+ Silicon on a Backside-thinned CCD for Enhanced UV Response. *Proc. SPIE* **1992**, *1656*, 488–496.

30. Hoenk, M.E.; Carver, A.G.; Jones, T.J.; Dickie, M.R.; Sgro, J.; Tsur, S. Superlattice-doped imaging detectors: Structure, physics and performance. In Proceedings of the Scientific Detectors Workshop, Florence, Italy, 14 October 2013.

31. Nikzad, S.; Hoenk, M.E.; Grunthaner, P.J.; Terhune, R.W.; Grunthaner, F.J.; Winzenread, R.; Fattahi, M.; Tseng, H.-F.; Lesser, M. Delta-doped CCDs: High QE with long-term stability at UV and visible wavelengths. *Proc. SPIE* **1994**, *2198*, 907–915.

32. Nikzad, S.; Jones, T.J.; Elliott, S.T.; Cunningham, T.J.; Deelman, P.W.; Walker, A.B.C.; Oluseyi, H.M. Ultrastable and uniform EUV and UV detectors. *Proc. SPIE* **2000**, *4139*, 250–258.

33. Hoenk, M.E.; Jones, T.J.; Dickie, M.R.; Greer, F.; Cunningham, T.J.; Blazejewski, E.R.; Nikzad, S. Delta-doped back-illuminated CMOS imaging arrays: Progress and prospects. *Proc. SPIE* **2009**, *7419*. [CrossRef]

34. Nikzad, S.; Hoenk, M.E.; Greer, F.; Jacquot, B.; Monacos, S.; Jones, T.J.; Blacksberg, J.; Hamden, E.; Schiminovich, D.; Martin, D.C.; *et al.* Delta doped electron multiplies CCD with absolute quantum efficiency over 50% in the near to far ultraviolet range for single photon counting applications. *Appl. Opt.* **2012**, *51*, 365–369. [CrossRef] [PubMed]

35. Greer, F.; Hamden, E.; Jacquot, B.C.; Hoenk, M.E.; Jones, T.J.; Dickie, M.R.; Monacos, S.P.; Nikzad, S. Atomically precise surface engineering of silicon CCDs for enhanced UV quantum efficiency. *J. Vac. Sci. Technol. A* **2013**, *31*, 01A103. [CrossRef]

36. Jewell, A.D.; Hennessy, J.; Hoenk, M.E.; Nikzad, S. Wide band antireflection coatings deposited by atomic layer deposition. *Proc. SPIE* **2013**, *8820*. [CrossRef]

37. Jewell, A.D.; Hamden, E.T.; Ong, H.R.; Hennessy, J.; Goodsall, T.; Shapiro, C.; Cheng, S.; Carver, A.; Hoenk, M.; Schiminovich, D.; *et al.* Detector performance for the FIREBall-2 UV experiment. *Proc. SPIE* **2015**, *9601*. [CrossRef]

38. Hoenk, M.E. Surface Passivation by Quantum Exclusion Using Multiple Layers. U.S. Patent No. 8,395,243 B2, 12 March 2013.

39. Hamden, E.T.; Greer, F.; Hoenk, M.E.; Blacksberg, J.; Dickie, M.R.; Nikzad, S.; Martin, D.C.; Schiminovich, D. Ultraviolet antireflection coatings for use in silicon detector design. *Appl. Opt.* **2011**, *50*, 4180–4188. [CrossRef] [PubMed]

40. Hamden, E.T.; Jewell, A.D.; Shapiro, C.A.; Cheng, S.R.; Goodsall, T.M.; Hennessy, J.; Nikzad, S.; Hoenk, M.E.; Jones, T.J.; Gordon, S.; *et al.* CCD detectors with greater than 80% QE at UV wavelengths. Under review.

41. Bates, B.; Bradley, D.J. Interference filters for the far ultraviolet (1700 A to 2400 A). *Appl. Opt.* **1966**, *5*, 971–975. [CrossRef] [PubMed]

42. Scalora, M.; Bloemer, M.J.; Pethel, A.S.; Dowling, J.P.; Bowden, C.M.; Manka, A.S. Transparent, metallo-dielectric, one-dimensional, photonic band-gap structures. *J. Appl. Phys.* **1998**, *83*, 2377–2383. [CrossRef]

43. Renk, K.F.; Genzel, L. Interference filters and Fabry-Perot interferometers for the far infrared. *Appl. Opt.* **1962**, *1*, 643–648. [CrossRef]

44. Sigalas, M.M.; Chan, C.T.; Ho, K.M.; Soukoulis, C.M. Metallic photonic band-gap materials. *Phys. Rev. B* **1995**, *52*, 11744–11751. [CrossRef]

45. Piegari, A.; Bulir, J. Variable narrowband transmission filters with a wide rejection band for spectrometry. *Appl. Opt.* **2006**, *45*, 3768–3773. [CrossRef] [PubMed]

46. Bloemer, M.J.; Scalora, M. Transmissive properties of Ag/MgF_2 photonic band gaps. *Appl. Phys. Lett.* **1998**, *72*, 1676–1678. [CrossRef]

47. Hennessy, J.; Jewell, A.D.; Hoenk, M.E.; Nikzad, S. Metal-dielectric filters for solar-blind silicon ultraviolet detectors. *Appl. Opt.* **2015**, *54*, 3507–3512. [CrossRef] [PubMed]

48. Blacksberg, J.; Hoenk, M.E.; Elliott, S.T.; Holland, S.E.; Nikzad, S. Enhanced quantum efficiency of high-purity silicon imaging detectors by ultralow temperature surface modification using Sb doping. *Appl. Phys. Lett.* **2005**, *87*, 254101. [CrossRef]

49. Blacksberg, J.; Nikzad, S.; Hoenk, M.E.; Holland, S.E.; Kolbe, W.F. Near-100% quantum efficiency of delta doped large-format UV-NIR silicon imagers. *IEEE Trans. Electron Devices* **2008**, *55*, 3402–3406. [CrossRef]

50. Hennessy, J.; Bell, L.D.; Nikzad, S.; Suvarna, P.; Leathersich, J.M.; Marini, J.; Shahedipour-Sandvik, F.S. Atomic-layer Deposition for Improved Performance of III-N Avalanche Photodiodes. In *MRS Online Proceeding Library*; Materials Research Society: Warrendale, PA, USA; Cambridge University Press: Cambridge, UK, 2014; Volume 1635, pp. 23–28.

51. Suvarna, P.; Bulmer, J.; Leathersich, J.M.; Marini, J.; Mahaboob, I.; Hennessy, J.; Bell, L.D.; Nikzad, S.; Shahedipour-Sandvik, F. Ion implantation-based edge termination to improve III-N APD reliability and performance. *IEEE Photonics Technol. Lett.* **2015**, *27*, 498–501. [CrossRef]

52. Suvarna, P.; Tungare, M.; Leathersich, J.M.; Agnihotri, P.; Shahedipour-Sandvik, F.; Bell, L.D.; Nikzad, S. Design and growth of visible-blind and solar-blind III-N APDs on sapphire substrates. *J. Electron. Mater.* **2013**, *42*, 854–858. [CrossRef]

sensors

MDPI

Article

Quantum Random Number Generation Using a Quanta Image Sensor

Emna Amri [1,*], Yacine Felk [1], Damien Stucki [1], Jiaju Ma [2] and Eric R. Fossum [2]

[1]   ID Quantique SA, Ch. de la Marbrerie 3, 1227 Carouge, Switzerland; yacine.felk@idquantique.com (Y.F.); damien.stucki@idquantique.com (D.S.)

[2]   Thayer Engineering School at Dartmouth College, Hanover, NH, USA; Jiaju.Ma.TH@dartmouth.edu (J.M.); eric.r.fossum@dartmouth.edu (E.R.F.)

*   Correspondence: emna.amri@idquantique.com; Tel. +41-22-301-83-71; Fax: +41-22-301-83-79

Academic Editor: Albert Theuwissen
Received: 6 April 2016; Accepted: 23 June 2016; Published: 29 June 2016

Abstract: A new quantum random number generation method is proposed. The method is based on the randomness of the photon emission process and the single photon counting capability of the Quanta Image Sensor (QIS). It has the potential to generate high-quality random numbers with remarkable data output rate. In this paper, the principle of photon statistics and theory of entropy are discussed. Sample data were collected with QIS jot device, and its randomness quality was analyzed. The randomness assessment method and results are discussed.

Keywords: QRNG; random number generator; QIS; quanta image sensor; photon counting; jot; entropy; randomness

1. Introduction

The generation of high-quality random numbers is becoming more and more important for several applications such as cryptography, scientific calculations (Monte-Carlo numerical simulations) and gambling. With the expansion of computers' fields of use and the rapid development of electronic communication networks, the number of such applications has been growing quickly. Cryptography, for example, is one of the most demanding applications. It consists of algorithms and protocols that can be used to ensure the confidentiality, the authenticity and the integrity of communications and it requires true random numbers to generate the keys to be used for encoding. However, high-quality random numbers cannot be obtained with deterministic algorithms (pseudo random number generator); instead, we can rely on an actual physical process to generate numbers. The most reliable processes are quantum physical processes which are fundamentally random. In fact, the intrinsic randomness of subatomic particles' behavior at the quantum level is one of the few completely random processes in nature. By tying the outcome of a random number generator (RNG) to the random behavior of a quantum particle, it is possible to guarantee a truly unbiased and unpredictable system that we call a Quantum Random Number Generator (QRNG).

Several hardware solutions have been used for true random number generation, and some of them are exploiting randomness in photon emission process. This class of QRNG includes beam splitters and single-photon avalanche diodes (SPADs) [1–3], homodyne detection mechanisms [4,5] and conventional CMOS image sensors (CIS) [6]. Although it has been demonstrated that these devices produce data of a satisfactory randomness quality, more work needs to be done to enhance the generation process, especially on the improvement of output data rate and device scalability. Practically, in an RNG utilizing image sensors, the photon emission is not the only source of randomness, and some noise sources in the detector, such as dark current and $1/f$ noise, will act as extra randomness sources and reduce the randomness quality since they have a strong thermal dependency. Therefore, an ideal detector should have high photon-counting accuracy with low read noise and low dark current to completely realize quantum-based randomness.

The Quanta Image Sensor (QIS) can be regarded as a possible solution to meet these goals because of its high-accuracy photon-counting capability, high output-data rate, small pixel-device size, and strong compatibility with the CIS fabrication process.

Proposed in 2005 as a "digital film sensor" [7], QIS can consist of over one billion pixels. Each pixel in QIS is called a "jot". A jot may have sub-micron pitch, and is specialized for photon-counting capability. A QIS with hundreds of millions of jots will work at high speed, e.g., 1000 fps, with extremely low power consumption, e.g., 2.5 pJ/bit [8]. In each frame, each jot counts incident photons and outputs single-bit or multi-bit digital signal reflecting the number of photoelectrons [9]. The realization of QIS concept relies on the photon-counting capability of a jot device. As photons are quantized particles in nature, the signal generated by photons is also naturally quantized. However, with the presence of noise in the read out electronics, the quantization effect is weakened or eliminated. To realize photon-counting capability, deep sub-electron read noise (DSERN) is a prerequisite, which refers to read noise less than 0.5 e– r.m.s. But, high-accuracy photon-counting requires read noise of 0.15 e– r.m.s. or lower [10,11].

The pump-gate (PG) jot device designed by the Dartmouth group achieved 0.22 e– r.m.s. read noise with single correlated double sampling (CDS) read out at room temperature [12,13]. The low read noise of PG jot devices was fulfilled with improvements in conversion gain (CG) [14], and the photoelectron counting capability was demonstrated with quantization effects in the photon counting histogram (PCH) [15].

2. Randomness Generation Concept

To quantify the randomness in a sequence of bits, we refer to the concept of entropy, first introduced by Shannon [16]. Entropy measures the uncertainty associated with a random variable and is expressed in bits. For instance, a fair coin toss has an entropy of 1 bit, as the exact outcome—head or tail—cannot be predicted. If the coin is unfair, the uncertainty is lower and so is the entropy. And when tossing a two-headed coin, there is no uncertainty which leads to 0 bit of entropy.

To compute the value of the entropy, we need to have full information about the random number generation process. In a photon source, the photon emission process obeys the principle of Poisson statistics [10], and the probability $P[k]$ of k photoelectron arrivals in a QIS jot is given by:

$$P[k] = \frac{e^{-H}H^k}{k!} \tag{1}$$

where the quanta exposure H is defined as the average number of photoelectrons collected in each jot per frame. So under the illumination of a stable light source, randomness exists in the number of photoelectrons arriving in each frame.

During readout, the photoelectron signal from the jot is both converted to a voltage signal through the conversion gain (V/e–) and corrupted by noise. Let the readout signal U be normalized by the conversion gain and thus measured in electrons. The readout signal probability distribution function (PDF) becomes a convolution of the Poisson distribution for quanta exposure H and a normal distribution with read noise u_n (e– r.m.s.). The result is a sum of constituent PDF components, one for each possible value of k and weighted by the Poisson probability for that k [11]:

$$P[U] = \sum_{k=0}^{\infty} \frac{1}{\sqrt{2\pi u_n^2}} \exp\left[-\frac{(U-k)^2}{2u_n^2}\right] \cdot \frac{e^{-H}H^k}{k!} \tag{2}$$

An example of a Poisson distribution corrupted with read noise is shown in Figure 1. While in practice the photodetector may be sensitive to multiple photoelectrons, subsequent circuitry can be used to discriminate the output to two binary states (either a "0" meaning no photoelectron, or a

"1" meaning at least one photoelectron) by setting a threshold U_t between 0 and 1, typically 0.5 and comparing U to this threshold. From a stability perspective, it is better to choose the threshold U_t at a valley of the readout signal PDF, such as at a 0.50 e− when H = 0.7, so that small fluctuations in light intensity have minimal impact on the value of entropy. The probability of the "0" state is given by:

$$P[U < U_t] = \sum_{k=0}^{\infty} \frac{1}{2}\left[1 + \mathrm{erf}\left(\frac{U_t - k}{u_n\sqrt{2}}\right)\right] \cdot \frac{e^{-H}H^k}{k!} \tag{3}$$

and the probability of the "1" state is just:

$$P[U \geqslant U_t] = 1 - P[U < U_t] \tag{4}$$

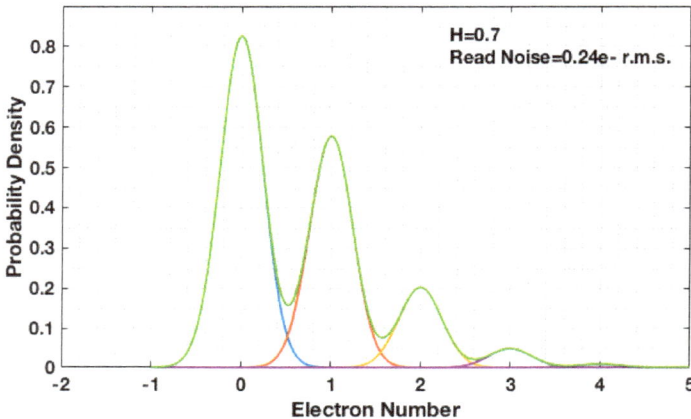

Figure 1. Readout signal probability distribution function (PDF) from Poisson distribution corrupted with read noise. Quanta exposure H = 0.7 and read noise u_n = 0.24 e− r.m.s.

The minimum quantum entropy of this distribution is given by [6]:

$$S_{min} = -\log_2[\max(P[U \geqslant U_t], P[U < U_t])] \tag{5}$$

If the measured value U will be encoded over b bits, the quantum entropy per bit of output will be, on average, equal to:

$$\overline{S} = \frac{S_{min}}{b} < 1 \tag{6}$$

where b = 1 for the single-bit QIS. It is, therefore, optimal to choose a quanta exposure H such that $P[U < U_t] = P[U \geqslant U_t] = 0.5$. These two conditions of stability and entropy lead to a preferred quanta exposure $H \cong 0.7$. An example of the cumulative probability function for the readout signal for H = 0.7 is shown in Figure 2. It should be noted that other combinations of H and U_t such as H = 2.67 and U_t = 2.5 e− are also viable options. For read noise u_n above 1 e− r.m.s., where the photon-counting peaks of Figure 1 are fully "blurred" by noise (e.g., conventional CMOS image sensors), the optimum settings of U_t and H converge so that the resultant Gaussian readout signal PDF is split in half at the peak, as one might deduce intuitively.

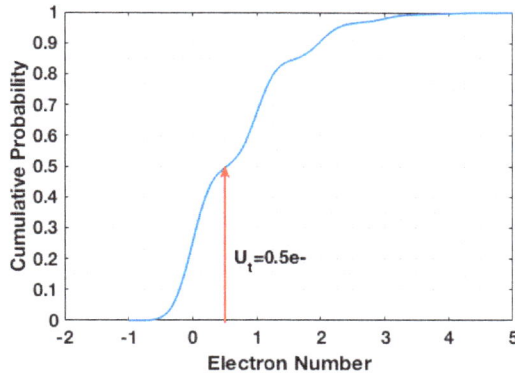

Figure 2. Cumulative probability of readout signal with read noise u_n = 0.24 e− r.m.s. and quanta exposure H = 0.7.

Stability is illustrated by comparing two cases with different quanta exposures and respective thresholds: H = 0.7 and H = 1.2. As shown in Figure 3, the thresholds for each case were selected to maximize the binary data entropy: U_t = 0.5 is located at a valley of PCH for H = 0.7, and U_t = 1 is located at a peak of PCH for H = 1.2. With 2% variation of quanta exposure in both cases, the output data of H = 0.7 showed better stability in entropy.

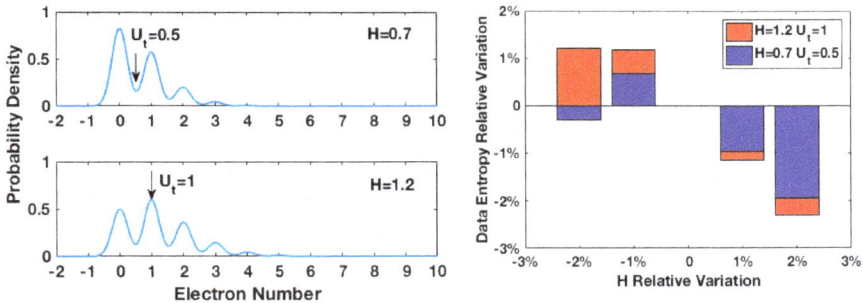

Figure 3. Binary data entropy variation caused by quanta exposure fluctuation during data collection.

It should be noted that only perfectly random bits will have unity quantum entropy, otherwise an extractor is required. A randomness extractor is a mathematical tool used to post-process an imperfect sequence of random bits (with an entropy less than 1) into a compressed but more random sequence. The quality of a randomness extractor is defined by the probability that the output deviates from a perfectly uniform bit string. This probability can be made arbitrarily very small by increasing the compression factor. The value of this factor depends on the entropy of the raw sequence and the targeted deviation probability and must be adjusted accordingly.

In this paper, we used a non-deterministic randomness extractor based on Universal-2 hash functions [17]. This extractor computes a number q of high-entropy output bits from a number $n > q$ of lower-entropy (raw) input bits. This is done by performing a vector-matrix multiplication between the vector formed by the raw bit values and a random $n \times q$ matrix M generated using multiple entropy sources. The compression ratio is thus equal to the number of lines divided by the number of columns of M. After extraction, statistical tests are run in order to make sure that randomness specifications are fulfilled.

3. Data Collection

The feasibility of applying the QIS to the QRNG application was tested with PG jot devices. In the PG jot test chip, an analog readout approach is adopted. The output signal from 32 columns is selected by a multiplexer and then amplified by a switch-capacitor programmable gain amplifier (PGA) with a gain of 24. The output signal from the PGA is sent off-chip and digitized through a digital CDS implemented with an off-chip 14-bit ADC. A complete description of readout electronics can be found in [13]. A 3 × 3 array of green LEDs was used as light source, located in front of the test chip. The distance from the light source to the sensor was 2 cm, and the intensity of the light source was controlled by a precision voltage source. During the data collection, a single jot with 0.24 e− r.m.s. read noise was selected and read out repeatedly, and a 14-bit raw digital output was collected. Under the limitation of the readout electronics on this test chip, the single jot was readout at a speed of 10 ksample/s. The testing environment was calibrated with 20,000 testing samples, and the quanta exposure H was obtained using the PCH method. In order to improve the randomness entropy of the data, the threshold U_t was determined as the median of the testing samples and then used with later samples to generate binary random numbers. The experimental PCH created by 200,000,000 samples is shown in Figure 4, which shows quanta exposure H of 0.7, and a read noise of 0.24 e− r.m.s. The threshold was set to 27.5DN, or 0.5 e−. The binary random numbers generated by first 10,000 samples are shown in Figure 5.

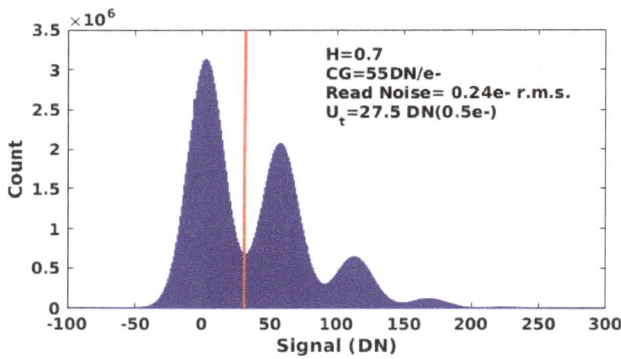

Figure 4. Photon counting histogram (PCH) of the first 200,000,000 samples.

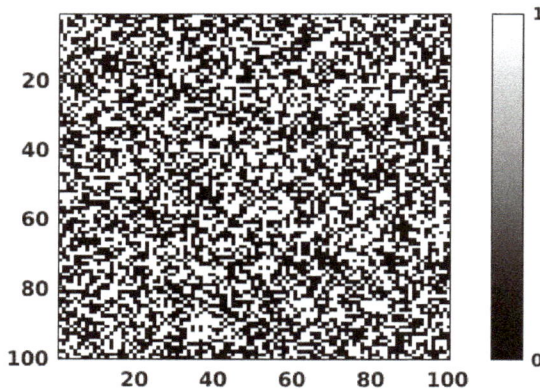

Figure 5. The binary output of the first 10,000 samples.

Although the light source was controlled by a stable voltage source, there was still a small fluctuation inferred in the light intensity. As shown in Figure 6, the quanta exposure H of 200 datasets is depicted, in which each dataset contains 1,000,000 samples and H is determined for each data set using its PCH. During the data collection, about 2.1% variation in quanta exposure was observed. To minimize the impact of light source fluctuation, the testing environment was calibrated to have an average quanta exposure H close to 0.7, for which the threshold U_t is located at a valley between two quantized peaks in the PCH.

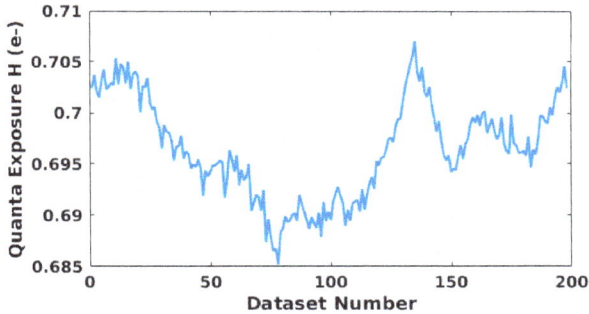

Figure 6. Quanta exposure fluctuation during data collection. Each dataset contains 1,000,000 samples.

4. Results

For a first test, we collected 500 Mbyte of raw random numbers by reading the jot at 5 ksamples/s (200 h of data collection). Using Equation (5), we were able to compute a minimum quantum entropy per output bit equal to 0.9845 for $H = 0.7$ and $u_n = 0.24$ e− r.m.s. Then we used the obtained value in the formula of the probability that the extractor output will deviate from a perfectly uniform q-bit string:

$$\varepsilon_{hash} = 2^{-(\overline{S}n-m)/2} \tag{7}$$

where n is the number of raw bits and m the number of extracted random bits.

Since a value of $\varepsilon_{hash} = 0$ is generally unachievable, we try to keep ε_{hash} below 2^{-100} implying that even using millions of jots one will not see any deviation from perfect uniform randomness in a time longer than the age of the universe. This gave a compression factor for $n = 1024$ equal to 1.23 which corresponds to losing only 18% of the input raw bits.

After extraction, we perform NIST tests [18] on the obtained random bits. This set of statistical tests evaluate inter alia, the proportion of 0 s and 1 s in the entire sequence, the presence of periodic or non-periodic patterns and the possibility of compression without loss of information. The QIS-based QRNG passed all these tests.

5. Comparison with Other Technologies

The idea of using an optical detector for random number generation is not new and has been driven by the intrinsic quantum nature of light. Single Photon Avalanche Diode (SPAD) arrays illuminated by a photon source and operating in Geiger mode have been widely used for this purpose [19,20]. Besides the single photon detection capability and technology maturity, SPAD matrices offer high-quality random data and can be fabricated in standard CMOS manufacturing line. However, these SPAD sensors require high supply voltage (22–27 V) for biasing above breakdown, suffer from after-pulsing phenomena, and have lower throughput per unit area than other optical detectors because of larger pixel size (600 Mbits/s for a matrix size of 2.5 mm^2 [19] and 200 Mbits/s for a matrix size of 3.2 mm^2 [20]).

Another technology exploiting optical quantum process has been recently introduced by the University of Geneva [6] and it consists of extracting random numbers of a quantum origin from an illuminated CIS. This low-power technology is more compatible with consumer and portable electronics since cameras are currently integrated in many common devices. Unfortunately, conventional image sensors are not capable of single-photon detection and provide lower randomness quality [6], which requires higher compression factor and hence lower output data rate. The choice of using QIS for random number generation was driven by the results obtained with SPADs and CIS since we noticed that QIS covers the advantages of both technologies (best tradeoff between data rate and scalability, single photon detection and CMOS manufacturing line) while providing solutions for most of their problems (speed, dark count rate, detection efficiency). Table 1 summarizes the comparison of the three techniques performances under the assumption of being used as RNGs. Note that the generation processes are different which limits the comparison points.

Table 1. The three technologies main comparison points.

Criteria	QIS	CIS	SPADs Matrix
Data Rate [1]	5–12 Gb/s	0.3–1 Gb/s	0.1–0.6 Gb/s
Read Noise	<0.25 e− r.m.s.	>1 e− r.m.s.	<0.15 e− r.m.s.
Dark Current/Count Rate [2]	0.1 e−/(jot·s)	10–500 e−/(pix·s)	200 counts/(pix·s)
Power Supply	2.5/3.3 V	2.5/3.3/5 V	22–27 V
Single Photon Counting	YES	NO	YES

[1] For a device with 2.5 mm^2 area size; [2] We define Dark Current for QIS/CIS and Dark Count Rate for SPADs, these values are measured at room temperature.

6. Summary

A new quantum random number generation method based on the QIS is proposed. Taking advantage of the randomness in photon emission and the photon counting capability of the Quanta Image Sensor, it shows promising advantages over previous QRNG technologies. Testing data was collected with QIS pump-gate jot device, and the randomness quality was assessed. Both randomness assessment method and data collection process are discussed, and the results show good randomness quality.

Acknowledgments: ID Quantique work has been sponsored by the Swiss State Secretariat for Education, Research, and Innovation (SERI) grants received for IDQ participation to European Marie Skłodowska-Curie Actions (MSCA), Innovative Training Network (ITN), Postgraduate Research on Dilute Metamorphic Nanostructures and Metamaterials in Semiconductor Photonics (PROMIS) and Eurostars project Quantum Random Number Generator (QRANGER). The QIS project at Dartmouth is sponsored by Rambus Inc. (Sunnyvale, CA, USA).

Author Contributions: Emna Amri and Damien Stucki co-conceived the random number data assessment experiments; Yacine Felk provided data for comparing technologies; Emna Amri performed the randomness experiments on the QIS data and co-wrote the paper; Jiaju Ma and Eric R. Fossum co-conceived, co-designed and performed the data collection experiments and co-wrote the paper.

Conflicts of Interest: The authors declare no conflict of interest.

References

1. Stefanov, A.; Gisin, N.; Guinnard, O.; Guinnard, L.; Zbinden, H. Optical quantum random number generator. *J. Modern Opt.* **2000**, *47*, 595–598. [CrossRef]
2. Dultz, W.; Hidlebrandt, E. Optical Random-Number Generator Based on Single-Photon Statistics at the Optical Beam Splitter. U.S. Patent No. 6,393,448, 21 May 2002.
3. Wei, W.; Guo, H. Bias-Free true random-number generator. *Opt. Lett.* **2009**, *34*, 1876–1878. [CrossRef] [PubMed]
4. Gabriel, C.; Wittmann, C.; Sych, D.; Dong, R.; Mauerer, W.; Andersen, U.L.; Marquardt, C.; Leuchs, G. A generator for unique quantum random numbers based on vacuum states. *Nat. Photonics* **2010**, *4*, 711–715. [CrossRef]

5.   Shen, Y.; Tian, L.A.; Zou, H.X. Practical quantum random number generator based on measuring the shot noise of vacuum states. *Phys. Rev.* **2010**, *61*. [CrossRef]

6.   Sanguinetti, B.; Martin, A.; Zbinden, H.; Gisin, N. Quantum random number generation on a mobile phone. *Phys. Rev.* **2014**, *4*. [CrossRef]

7.   Fossum, E.R. The quanta image sensor (QIS): Concepts and challenges. In Proceedings of the 2011 Optical Society of America Topical Meeting on Computational Optical Sensing and Imaging, Toronto, ON, Canada, 10–14 July 2011.

8.   Masoodian, S.; Rao, A.; Ma, J.; Odame, K.; Fossum, E.R. A 2.5 pJ/b binary image sensor as a pathfinder for quanta image sensors. *IEEE Trans. Electron. Devices* **2015**, *63*, 100–105. [CrossRef]

9.   Fossum, E.R. Modeling the performance of single-bit and multi-bit quanta image sensors. *IEEE J. Electron. Devices Soc.* **2013**, *1*, 166–174. [CrossRef]

10. Fossum, E.R. Application of photon statistics to the quanta image sensor. In Proceedings of the International Image Sensor Workshop (IISW), Snowbird Resort, UT, USA, 12–16 June 2013.

11. Fossum, E.R. Photon counting error rates in single-bit and multi-bit quanta image sensors. *IEEE J. Electron. Devices Soc.* **2016**. [CrossRef]

12. Ma, J.; Fossum, E.R. Quanta image sensor jot with sub 0.3 e− r.m.s. read noise. *IEEE Electron. Device Lett.* **2015**, *36*, 926–928. [CrossRef]

13. Ma, J.; Starkey, D.; Rao, A.; Odame, K.; Fossum, E.R. Characterization of quanta image sensor pump-gate jots with deep sub-electron read noise. *IEEE J. Electron. Devices Soc.* **2015**, *3*, 472–480. [CrossRef]

14. Ma, J.; Fossum, E.R. A pump-gate jot device with high conversion gain for a Quanta Image Sensor. *IEEE J. Electron. Devices Soc.* **2015**, *3*, 73–77. [CrossRef]

15. Starkey, D.; Fossum, E.R. Determining conversion gain and read noise using a photon-counting histogram method for deep sub-electron read noise image sensors. *IEEE J. Electron. Devices Soc.* **2016**. [CrossRef]

16. Shannon, C.E. A mathematical theory of communication. *Bell Syst. Tech. J.* **1948**, *3*, 379–423. [CrossRef]

17. Troyer, M.; Renner, R. A Randomness Extractor for the Quantis Device, ID Quantique. Available online: http://www.idquantique.com/wordpress/wp-content/uploads/quantis-rndextract-techpaper.pdf (accessed on 27 June 2016).

18. Rukhin, A.; Soto, J.; Nechvatal, J.; Smid, M.; Barker, E. A Statistical Rest Suite for Random and Pseudorandom Number Generators for Cryptographic Applications. National Institute of Standards and Technology (NIST), Special Pub. 800-22, 15 May 2001. Available online: http://oai.dtic.mil/oai/oai?verb=getRecord& metadataPrefix=html&identifier=ADA393366 (accessed on 27 June 2016).

19. Stucki, D.; Burri, S.; Charbon, E.; Chunnilall, C.; Meneghetti, A.; Regazzoni, F. Towards a high-speed quantum random number generator. *Proc. SPIE* **2013**, *8899*. [CrossRef]

20. Tisa, S.; Villa, F.; Giudice, A.; Simmerle, G.; Zappa, F. High-Speed quantum random number generation using CMOS photon counting detectors. *IEEE J. Sel. Top. Quant. Electron.* **2015**, *21*. [CrossRef]

Review

Towards a Graphene-Based Low Intensity Photon Counting Photodetector

Jamie O. D. Williams [1,*], **Jack A. Alexander-Webber** [2], **Jon S. Lapington** [1], **Mervyn Roy** [1], **Ian B. Hutchinson** [1], **Abhay A. Sagade** [2], **Marie-Blandine Martin** [2], **Philipp Braeuninger-Weimer** [2], **Andrea Cabrero-Vilatela** [2], **Ruizhi Wang** [2], **Andrea De Luca** [2], **Florin Udrea** [2] and **Stephan Hofmann** [2]

[1] Department of Physics and Astronomy, University of Leicester, University Road, Leicester LE1 7RH, UK; jsl12@le.ac.uk (J.S.L.); mr6@le.ac.uk (M.R.); ibh1@le.ac.uk (I.B.H.)
[2] Department of Engineering, University of Cambridge, 9 JJ Thomson Avenue, Cambridge CB3 0FA, UK; jaa59@cam.ac.uk (J.A.A.-W.); aas73@cam.ac.uk (A.A.S.); mbcbm2@cam.ac.uk (M.-B.M.); pab96@cam.ac.uk (P.B.-W.); ac769@cam.ac.uk (A.C.-V.); rw520@cam.ac.uk (R.W.); ad597@cam.ac.uk (A.D.L.); fu10000@hermes.cam.ac.uk (F.U.); sh315@cam.ac.uk (S.H.)
* Correspondence: jodw1@le.ac.uk; Tel.: +44-161-229-7729

Academic Editor: David Stoppa
Received: 19 February 2016; Accepted: 15 August 2016; Published: 23 August 2016

Abstract: Graphene is a highly promising material in the development of new photodetector technologies, in particular due its tunable optoelectronic properties, high mobilities and fast relaxation times coupled to its atomic thinness and other unique electrical, thermal and mechanical properties. Optoelectronic applications and graphene-based photodetector technology are still in their infancy, but with a range of device integration and manufacturing approaches emerging this field is progressing quickly. In this review we explore the potential of graphene in the context of existing single photon counting technologies by comparing their performance to simulations of graphene-based single photon counting and low photon intensity photodetection technologies operating in the visible, terahertz and X-ray energy regimes. We highlight the theoretical predictions and current graphene manufacturing processes for these detectors. We show initial experimental implementations and discuss the key challenges and next steps in the development of these technologies.

Keywords: graphene; single photon; photodetector; visible; terahertz; cryogenic; X-ray

1. Introduction

Single photon counting photodetectors require an incident single photon to be absorbed and to give a measurable signal. A number of different photodetector technologies have been developed for optical single photon counting with a wide range of specifications such as energy and time resolution, and operating temperature. For instance photomultipliers, avalanche diodes [1] and transition edge sensors [2] are able to operate with single photon resolution but without wavelength specificity in the optical range. Other detector technologies do exist that allow for single photon counting with optical wavelength specificity [3], but mostly operate at extreme cryogenic temperatures [4].

These detectors have many different applications, in areas as diverse as medical and space sciences or security applications. For instance a photon counting photodetector has applications on a satellite for the detection of faint, distant stars, or in fluorescence spectroscopy for use in characterizing biological samples. Single photon counting photodetectors also have quantum information applications, ranging from quantum key distribution (QKD) [5,6] to time-correlated fluorescence spectroscopy of quantum wells [7]. These new quantum applications are making significant demands on existing technologies due to the required signal to noise ratio, detection efficiency, spectral range and photon number resolution [8,9] .

Graphene is an allotrope of carbon, specifically arranged in a 2D hexagonal lattice structure with sp^2 bonded carbon atoms. It has captured the world's attention since it was first isolated in 2004 [10,11] due to a unique combination of mechanical and optoelectronic properties [11–16]. Graphene provides an interesting solution for single photon counting photodetection [17] with many potential applications; graphene has already been used for ultrafast photodetection on a femtosecond timescale [18] for pulsed lasers, its high carrier mobility enabling greater operational bandwidth. In addition, the tuneable band gap in bilayer graphene may enable sensitive photon counting photodetectors to operate with a trade off between resolution and operational temperatures, with resulting operational benefits.

2. Existing Technologies

A number of different techniques are currently utilised for single photon counting photodetection over a wide range of photon energies. For instance, a microwave kinetic inductance detector (MKID) passes a microwave through a circuit with a given frequency resulting in an inductance impedance through the circuit related to the frequency. A photon incident on a superconducting film (typically TiN) breaks Cooper pairs, creating additional charge carriers and changing the resonant frequency within the range 1–10 GHz [19]. To observe the change in phase and amplitude, very sensitive measurements are made before charge carriers recombine in time periods of order, 10^{-3}–10^{-6} s. This technique has been used in detectors built into a 1000 pixel array [20]. MKIDs operate at temperatures ~ 100 mK [21] and have demonstrated position sensitivity with a noise equivalent power (NEP) of ~10^{-17} W·Hz$^{-1/2}$ [22–25]. Ongoing research activities are being performed to investigate the use of graphene as an MKID [26,27].

Like the MKID, a superconducting tunnelling junction (STJ) can also be used for single photon counting at cryogenic temperatures. An STJ works by the absorbed photon energy breaking Cooper Pairs in a superconducting film, typically tantalum [22]. STJs have an effective band gap of order 1 meV, and operate at a low temperature, typically 300 mK, to ensure low dark noise. They have a time resolution of order microseconds and a typical resolution of order 1 eV for soft X-ray photons, and 0.1–0.2 eV for near-infrared and visible photons, with the Fano limit as the inherent energy resolution [22,23].

A number of different techniques have been proposed to allow low intensity photodetection at terahertz photon frequencies. Terahertz photodetection has been demonstrated using techniques such as bolometry [28], but many of these are at sub-THz frequencies. A technique using Photon Counting Terahertz Interferometry (PCTI) utilises the pulsed nature of photons at sub-far infrared frequencies, whereby detection on two or more telescopes can be used to measure the intensity correlation, enabling a wide bandwidth [29–32]. This technique requires detectors with a high count rate of 1–100 MHz and a time resolution better than 1 ps [31].

Table 1 provides a summary of the existing state of the art photodetectors for low intensity photon source illumination and for photon counting. Existing techniques, such as STJs and MKIDs, are able to count single photons, but have a timing resolution that is limited to approximately 1 µs. At similar photon wavelengths covered by the STJ, other detectors such as Avalanche Photodiodes and Transition Edge Sensors provide solutions. The Avalanche Photodiodes provide improved timing resolution but with compromised energy resolution. Transistion Edge Sensors provide less time resolution but improved energy resolution and very good responsivity. Across a wide range of wavelengths, microchannel plate photomultipler tubes provide an alternative to an STJ, with improved timing resolution up to ~25 ps, but with no energy resolution at optical wavelengths, and only very poor energy resolution at soft X-ray wavelengths. No detector exists that has the required combination of features for the current application demands of single photon counting photodetectors, such as high detection efficiency with wavelength specificity, high temporal resolution and low dark count [33].

Table 1. Brief summary of a selection of photodetector technologies, with up and coming graphene-based technologies highlighted in grey followed by other potential solutions.

Detector Type [34]	Operating Temperature	Operational Wavelength	Timing Resolution	Energy Resolution $\frac{E}{\delta E}$	Responsivity	Size of Active Area	Photon Intensity
Superconducting Tunnelling Junction [23,35]	<1 K	1 nm–100 μm	1 μs	<20 (for E = 1.8 eV); <6 (for E = 3.1 eV); ~200 (for E = 0.4 keV); ~500 (for E = 5.9 keV)	>~100 AW^{-1}	~1 mm²	Single photon
Microwave Kinetic Inductance Detector [22–25,36]	0.1 K–1 K	Sub-mm and mm	~1 μs	>20	10^{-7} rad per quasi-particle	>1000 pixel array.	Single photon
Avalanche Photodiodes [37–39]	−20 °C / −90 °C	<1 μm	40 ps+	~16 (for E = 5.9 keV); ~45 (for E = 5.9 keV)	~50 AW^{-1}	<~25 mm²	Single photon
Transition Edge Sensors [40–42]	0.1 K	~1 nm	0.5 ms	~70 (for E = 0.1 keV); ~7000 (for E = 10 keV)	~100,000 AW^{-1} on transition region	~5 cm²	Single photon
Microchannel plate photomultiplier tube [33]	300 K	X-ray to IR	25 + ps	None across most of the spectrum, very poor at soft X-ray.	5–1000 mAW^{-1}	>1000 mm²	Single photon
Ultrafast Graphene-based Photodetector. Photothermoelectric effect [18]	40–300 K	500–1500 nm	~50 fs	Photovoltage greater for lower temperatures.	~100 μAW^{-1}	~10 μm	50 μW
X-ray GFET on SiC substrate. Field effect. [43–49]	300 K	~0.01–0.03 nm	-	10,000 (for E = 15 keV)	0.1 AW^{-1}	20 μm × 4 μm	15 kV, 15 μA →
X-ray GFET on Si substrate. Field effect. [43–48]	4.3 K	~0.01–0.03 nm	-		-	~10 μm	40 kV, 80 μA
Ultrafast GFET [50]. Photovoltaic effect.	300 K	1.55 μm	~25 ps (2 ps theory)	-	0.5 mAW^{-1}	1 μm × 2.5 μm	3 mW
THz GFET. Dyakanov–Shur effect [51,52]	300 K	100 μm	~1 s	-	100 mVW^{-1}	10 μm	-
Quantum Dot (Field Effect Transistor) [8,53]	4 K	805 nm	1 μs–1 ms	-	650 AW^{-1}	15 μm	~3.5 mW
Black Phosphorus FET [54,55]	323–383 K	<940 nm	~1 ms	-	4.8 mAW^{-1}	~10 μm	~500 μW

Graphene-based photodetector techniques have been an exciting topic of research in recent years, with many potential applications in a number of different areas. The main detector techniques investigated are the photovoltaic effect, photo-thermoelectric effect, bolometric effect and the Dyakanov-Shur effect [24]. The photovoltaic effect exploits the separation of electron-hole pairs, with a resulting generation of a photocurrent between p and n doped areas. For the photo-thermoelectric effect, a photon absorption excites an e-h pair that leads to the ultrafast heating of the lattice, as this relaxes it induces a measurable photovoltage [56]. The increased temperature of the lattice can also be used for detection through bolometry due to a change in carrier conductance. The change in temperature is measured, with the thermal resistance also related to the power of the incident radiation [57]. Terahertz detection also exploits the Dyakanov-Shur effect, whereby radiation couples to the antennae, and excites a plasmon resonance between the contacts that generates a measurable DC photocurrent.

Field effect transistor detectors have been developed to exploit these detection mechanisms; for instance, graphene-based terahertz detectors have been developed by a number of groups [51,52,58], utilising many different photodetection techniques which usually require the coupling of the terahertz photon to the detector resulting in heating of the lattice or a plasmon resonance leading to a measurable photocurrent. These detectors have demonstrated excellent noise equivalent power (NEP) in the 10^{-10}–10^{-11} W·Hz$^{-1/2}$ range [52]. In addition, the Jovanovic group showed the development of a graphene field effect transistor (GFET) sensitive to X-ray photons, with silicon and silicon carbide absorbers and an applied back gate voltage [43–48]. These often require the photon to be absorbed in an absorber exciting multiple charge carriers that modulate the field applied to the graphene and resulting in a measurable change in the resistance. The Jovanovic group found that it was not possible to obtain an X-ray signal at room temperature for highly resistive silicon, only at 4.3 K [44]. Additionally it can be shown that a significant energy is required for a measurable change in resistance, with a signal rise time of order of seconds, which makes this technique currently not suitable for a single photon counting photodetector. However the change of measured resistance of a graphene field effect transistor-like structure has already been shown to enable sensitive detection of single molecules [59] suggesting that single photon sensitivity is feasible. In addition, work by Xia et al. [50] has shown sensitivity to 1.55 μm laser illumination with a 3 mW energy deposition, leading to an experimentally determined bandwidth of 40 GHz, compared to the theoretically predicted maximum of 500 GHz. Other novel field effect detectors have potential, such as a black phosphorus-zinc oxide nanomaterial heterojunction with a reported on/off ratio of 10^4 and no time delay [54].

Detectors with wavelength specificity such as the MKID and STJ detectors require cryogenic cooling to prevent dark noise that is critically dependent on the energy gap in the Cooper pairs for both techniques. Varying this energy gap by means of graphene's tuneable band-gap would enable potential operation at higher temperatures, overcoming cost and operational issues of cryogenic cooling. Scope also exists to exploit graphene to develop further high speed photodetectors for different photon energies with possibility for femtosecond photodetection [18], and to enable PCTI with smaller pixel sizes to allow for greater resolution resulting from a greater pixel density [29–32].

Table 1 highlights the already impressive characteristics of graphene-based photodetectors using a number of different techniques, suggesting that it may provide a potentially interesting and viable solution to future technologies. Throughout the rest of this paper we will outline how graphene can be applied to such future single photon counting technologies, with a particular focus on the devices that we are developing. In Section 3 we outline the critical properties of single and bi-layer graphene for photodetection. In Section 4 we consider our theoretical study of bilayer graphene as a single photon counting photodetector at visible wavelengths, and in Section 5 we discuss our studies working towards a detector optimised for operation at a frequency of 1.2 THz. In Section 6 we discuss our progress to develop an X-ray detector at room temperature and suggest potential iterations to the design. This motivates our discussion in Section 7, where we consider the latest state of the art for graphene device fabrication, its limitations, and possible future solutions.

3. Properties of Single and Bilayer Graphene

Graphene has many properties that make it promising to the development of new photodetector technologies and potentially outperform other existing materials. The low energy band structure of graphene is dictated by π states which form symmetrical cones touching at the so called Dirac point (Figure 1a). Graphene is therefore usually described as zero-bandgap semiconductor. The electron dispersion in this region is linear (Figure 1b), reminiscent to that of light and unlike conventional parabolic dispersions in semiconductors. The band structure is symmetric about the Dirac point, i.e., electrons and holes should have the same properties. The Fermi velocity is calculated to be approximately 10^6 ms^{-1} [12,16,60]. Graphene can support very high carrier mobilities (10^6 cm$^2\cdot$V$^{-1}\cdot$s^{-1} for suspended graphene at temperatures ~5 K [13] to higher temperatures [14]) but, as with most of its properties, this strongly depends on the environment and support. Fully encapsulated graphene devices on silicon/silicon dioxide support show mobilities in the order of 10^3 cm$^2\cdot$V$^{-1}\cdot$s^{-1} at room temperature [61]. High carrier mobilities offer the potential for an ultrafast detector; photodetection has been demonstrated at femtosecond resolution [62], with GFETs developed with a theoretical bandwidth up to 500 GHz [50].

The carrier density (or doping level) of graphene is continously tunable from p-type to n-type through charge transfer, often unintentionally due to external factors such as air exposure and substrate effects. Due to this high sensitivity, reproducibility of electrical characteristics is a key challenge which may be addressed by considering techniques such as encapsulation [61–63] to reduce atmospheric effects or controlled doping [64,65]. We can also exploit the change of doping through the field effect, whereby a field applied to the graphene shifts its Fermi level [43] and hence changes the number of charge carriers and therefore the conductivity of the graphene [47,48,66–69]. In Figure 1c we see the change in conductivity resulting from the application of a gate voltage for four different samples, with hole transport and electron transport at negative and positive gate voltages respectively. At gate voltages far from the Dirac point we obtain a linear conductivity-gate voltage relationship, with the gradient related to the carrier mobility of the sample [70]. Employing the field effect has enabled detection of X-rays with a relatively simple device fabrication and detector measurements [43–48].

Graphene has a wideband absorption of 2.3% [15] per layer at visible frequencies, although this leads to low photoresponsivity and low external quantum efficiency (EQE) [71–73]. However it is possible to exploit plasmonic nanostructures to improve this EQE, a technique that has been shown to enhance the photocurrent by up to 1500% [73]. Interestingly, we can exploit the production of plasmons to enable terahertz photodetection by utilising the Dyakanov-Shur effect [51,52]. In this technique terahertz radiation is coupled into an antennae resulting in the excitation of plasmon waves in a graphene channel and the generation of a measurable DC photocurrent.

Flexible graphene-based photodetectors using centimetre-scale grown samples have also been developed. In [76] the authors report an internal responsivity of 45.5 AW^{-1} and internal responsivity of 570 AW^{-1} for a laser source intensity of 0.1 nW$\cdot\mu$m^{-2} and maintain this photodetection down to a bending radius of 6 cm.

Bilayer graphene is also of interest in the development of photodetector technologies. For bilayer graphene the crucial additional parameter is the stacking of the two layers [75]. For instance AA stacked graphene has the two layers directly above each other, whereas AB (Bernal) stacking has an offset in the arrangement as shown in Figure 1d. The layer interactions change the band structure, as highlighted in Figure 1e for AB-stacking, which shows a hyperbolic (non-linear) bandstructure. An approach for opening a tunable band gap for such bilayer graphene is to apply an electric field perpendicular to the layers (Figure 1e) [75], a technique that shows no hysteresis and also allows tuning of the Fermi level. The band gap magnitude is given by $U_g = \frac{|U|\gamma_1}{\sqrt{\gamma_1^2 + U^2}}$, where U_g is the band gap, U is the interlayer asymmetry and γ_1 is the interlayer hopping parameter; the magnitude of the band gap saturates $U_g \rightarrow \gamma_1$ for large U [77]. Other techniques that have been reported to open a band gap include the controlled adsorption of water [78] or hydrogen [79], applying strain [80], and molecular doping [81].

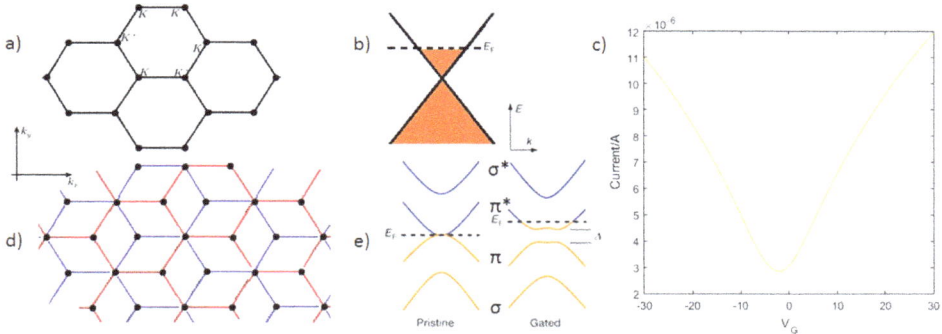

Figure 1. Showing (**a**) the honeycomb structure of single layer graphene with the K and K' points in the first Brillouin Zone (reproduced by permission of Cambridge University Press, subject to cambridge.org/uk/information/rights/permission.htm); (**b**) the linear energy-wavenumber relationship close to the Dirac point with a Fermi level that we can change through the application of a electric field; (**c**) the drain-source current of graphene against gate voltage [74] with a sample dependent Dirac point and electron and hole mobilities; (**d**) the structure of AB stacked bilayer graphene, with the two layers marked in red and blue respectively and a hopping parameter of ~0.4 eV between the layers; and (**e**) the band structure for bilayer graphene showing pristine bilayer graphene and the opening of a band gap for gated bilayer graphene with AB stacking (reproduced with permission of Nature Publishing Group) [75]. In the gated we see "trigonality" at very low energies [16].

4. Bilayer Graphene Single Photon Counting Photodetector—Simulations and Design

Our work considers the application of a potential, V, applied perpendicularly to the lattice [16,75]. This breaks the interlayer symmetry and leads to the electron energy spectrum [16] given by:

$$E^2 = \gamma_0^2 |S(k)|^2 + \frac{\gamma_1^2}{2} + \left(\frac{V}{2}\right)^2 \pm \sqrt{\left(\frac{\gamma_1^2}{2}\right)^2 + \left(\gamma_1^2 + V^2\right) \gamma_0^2 |S(k)|^2} \tag{1}$$

as described in Figure 1e [75], where $\gamma_0 = 2.97$ eV and $\gamma_1 = 0.4$ eV [16] are the intralayer and interlayer hopping parameters respectively and:

$$S(k) = \sum_\delta e^{ik\delta} = 2\exp\left(\frac{ik_x a}{2}\right) \cos\left(\frac{k_y a \sqrt{3}}{2}\right) + \exp\left(-ik_x a\right) \tag{2}$$

where k is the wavevector and $a = 1.42$ A is the near neighbour distance [16].

As bilayer graphene possesses a variable band gap [75], unlike many other materials including single layer graphene, it allows the potential for a detector that can exploit this tuneability to vary the resolution for optimal performance.

Initially, we developed a number of simulations for our bilayer graphene single photon counting photodetector, which indicate the fundamental operational properties and parameters of the detector. We firstly calculate the density of states and investigate the optimum operational window [82]. We then use a Monte Carlo simulation using a Gillespie Algorithm [83] to simulate the absorption of an incident photon on the graphene lattice, the excitation of a photoelectron and its subsequent relaxation in the conduction band.

4.1. Density of States and Optimum Operational Window

Firstly we calculate the density of states, $n(E)$, numerically (Figure 2a) and integrate the Fermi-Dirac distribution over the first Brillouin zone to determine the number of charge carriers in the conduction band per unit area given by:

$$N = \int_0^{\frac{E_{photon}}{2}} dE \frac{1}{\exp\left(\frac{E}{k_b T}\right) + 1} n(E) \tag{3}$$

where E is the electron energy and T is the temperature. The integration limit given by $\frac{E_{photon}}{2}$ arises from the possible photon excitations from the valence band to the conduction band at energies we are interested in.

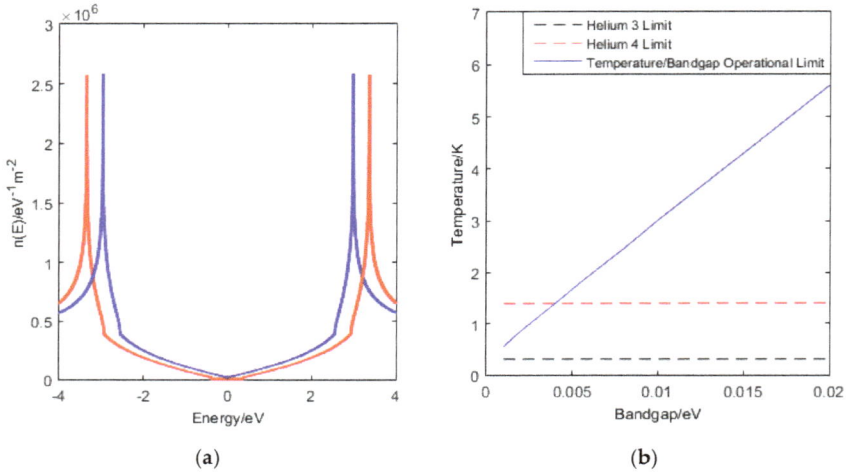

Figure 2. Showing (**a**) the density of states for bilayer graphene with a band gap of 5 meV; red is the σ band, blue is the π band. (**b**) shows the o perational limit of a bilayer graphene photodetector. In this simulation, A = 1 mm². Helium-4 cooling limit is 1.4 K and Helium-3 limit is 0.3 K.

For a single photon counting photodetector we require it to be statistically unlikely that electrons are thermally excited into the conduction band. We therefore calculate numerically NA, where A is the sample area, and look for cases where NA = 1, as plotted in Figure 2b. Below this line, NA < 1, is the regime where there is theoretically no dark current. This is critically dependent on the bilayer graphene density of states. The tuneable band gap in bilayer graphene allows us to exploit this operational limit, as this approach allows us to run our device at higher temperatures, with a larger band gap, but with a trade off against energy resolution.

4.2. Monte Carlo Simulations

We have developed a Monte Carlo simulation to determine the likely properties of our photodetector [82]. Our model assumes that we operate within the limit shown in Figure 2b, i.e., electrons in the conduction band result solely from the initial photoexcitation (or subsequent relaxations). Furthermore, excitation occurs when the photon energy is equal to the energy difference between two bands in the valence and conduction bands respectively shown in Figure 1e.

After the initial excitation, the electron can relax through a number of different relaxation paths. For instance electron-electron scattering (EES) is the inelastic scattering between two electrons in the conduction band (CB) and does not affect the total energy or the number of electrons in the CB. Another possibility is electron-phonon scattering (EPS) which is the scattering of an electron due to the emission (absorption) of a phonon to (from) the lattice [84], resulting in energy lost (gained) from the electrons. Alternatively the electron may relax through impact ionisation (II) or Auger recombination (AR); II is the excitation of an electron from the valence band (VB) to the CB due to the loss of energy from a CB electron. In this model II is the only process which results in an increase in the number of electrons in the conduction band [85–87]. AR is the reverse process, where an electron relaxes from CB to VB, when another CB electron becomes more excited. At low temperatures, the rates of electron-phonon scattering, σ_{Phonon}, and electron-electron scattering, σ_{E-E}, are given respectively by [88]:

$$\sigma_{Phonon} = \sigma_{Acoustic} + \sigma_{Optical}$$

$$\approx \frac{D_0^2}{\rho_m \omega_0 (\hbar v_F)^2} \left[(E_k - \hbar\omega_0) \left[\frac{1}{e^{\frac{\hbar\omega_0}{k_B T}} - 1} + 1 \right] \theta (E_k - \hbar\omega_0) + (E_k + \hbar\omega_0) \left(\frac{1}{e^{\frac{\hbar\omega_0}{k_B T}} - 1} \right) \right] \tag{4}$$

$$\sigma_{E-E} = \frac{1}{\tau_{MFT}} = \frac{v_F}{\lambda} = 2k_f \frac{\hbar k_f}{m_e} = \pi n \frac{2\hbar}{m_e} \tag{5}$$

where ω_0 is the phonon frequency, E_k is the electron energy, T is the temperature, ρ_m is the mass density, D_0 is the deformation potential constant, λ is the wavelength, v_F is the Fermi velocity, k_f is the Fermi wavenumber and n is the density of charge carriers.

In the literature, little work has been done on the analytical II and AR rates for low CB electron density at low temperature. However, as we start with only one conduction band electron following the photoexcitation, we assume that EES, EPS and AR relaxation rates will be significantly lower than II as the former are CB density dependent, whereas II is VB density dependent [86]. Furthermore, relaxation rates at lower energies such that electrons relax out of CB altogether are low, due to the necessity to conserve energy and momentum whilst filling vacant holes in the VB from previous electron excitations. Therefore in the low electron density, low temperature limit, II highly dominates. To run simulations we choose a ratio, μ, of phonon scattering rate to impact ionisation rate, where II dominates. We run simulations with each of the relaxation events chosen randomly, weighted based on the relevant rates, and solve numerically to find solutions where energy and momentum are conserved. We test the dependence of the number of charge carriers produced as a function of time, initial photon energy, band gap and $\mu = \frac{\sigma_{II}}{\sigma_{Phonon}}$. In our simulations we use the interlayer hopping parameter $\gamma_1 = 0.4$ eV. A schematic of this is shown in Figure 3a.

We ran our first simulations over a given time, at different initial energies and different band gaps, as shown in Figure 3b. The results show, as anticipated, that the number of electrons produced increase with initial energy. Additionally, as the band gap is increased the number of electrons produced is significantly reduced.

By simulating with different size band gaps and photon energies we calculate the average electron-hole pair creation energy, $W = \frac{E_{photon}}{N}$, as shown in Figure 3b. This gives a W to band gap ratio of 3–4, similar to that of semiconductors such as silicon and germanium (Figure 3d) [89].

A plot of the dependence on the initial photon energy of the distribution of charge carriers produced is shown in Figure 4a. Clearly, for a more energetic photon, more electrons will be produced. We observe wavelength specificity as the difference in the distributions at each wavelength. Additionally, in Figure 4a we see four peaks in the simulations, caused by the four alternative excitations from the π and σ bands to the π^* and σ^* bands respectively. The gap between the centre of the peaks is equal to $\Delta N = \frac{\gamma_1}{W}$, where γ_1 is the hopping parameter between layers, and W is the average ionisation energy. The characteristic peak of an event is highly dependent on the initial transition between the bands, and the initial relaxation step. The presence of the four peaks makes energy resolution of the incident photon problematic.

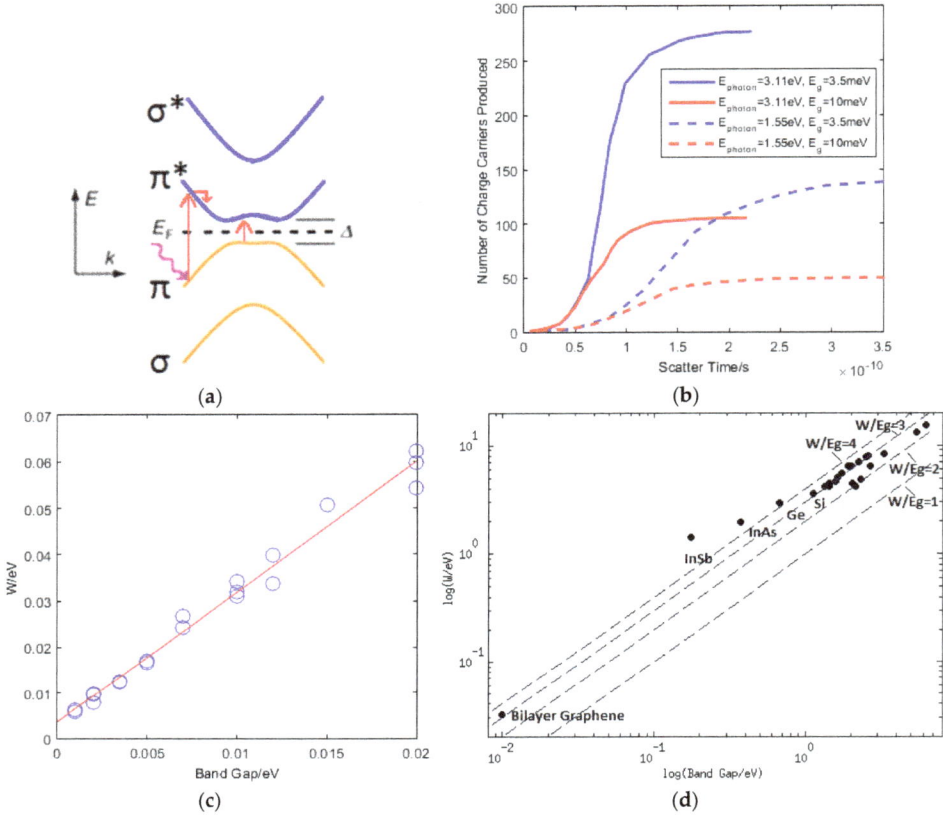

Figure 3. (**a**) Schematic to show the absorption of a photon and the excitation and subsequent relaxation the hot photoelectron; (**b**) The distribution N (t) at E_{gap} = 1 meV and 3.5 meV for photons with energy 3.11 eV and 1.55 eV, and with μ = 100; (**c**) Electron-hole pair creation energy as a function of band gap with μ = 100. The circles show simulation results, and the red line is best fit straight line; (**d**) Comparison of W vs. band gap for bilayer graphene with other semiconductors.

However for a photon energy less than γ_1 = 0.4 eV (i.e., in the IR spectrum), we obtain only one peak since the lower available energy allows only one possible transition. Figure 4b shows, with λ = 3500 nm and a band gap of 3.5 meV, that we get one large peak in the distribution, with a W value still in the range, 3–4, as also seen at visible wavelengths.

Initially we arbitrarily picked the II rate by using a ratio to the phonon rate, μ. We then tested the effect of changing the ratio to ensure that the total number of charge carriers produced tends towards the same value, but at an increased time, with decreasing values of the ratio. The results are shown in Figure 5 for a photon with λ = 400 nm and a 3.5 meV band gap.

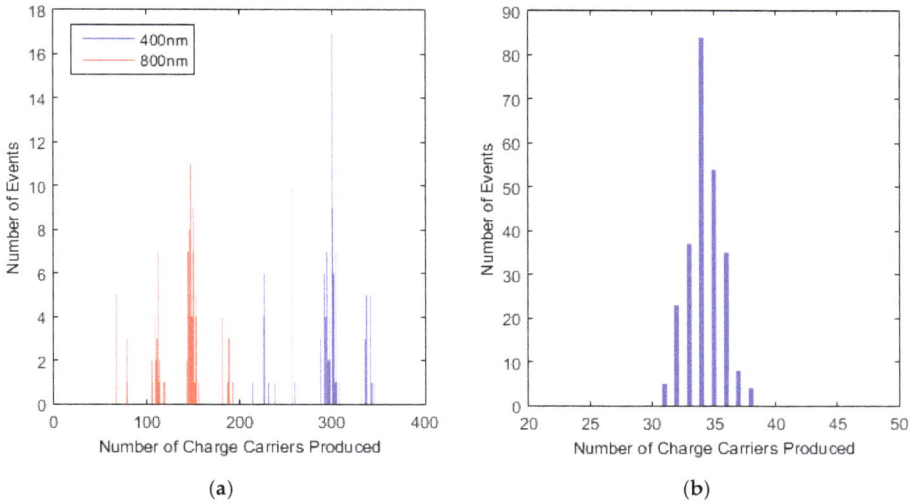

Figure 4. Showing (**a**) the number distribution as a function of photon energy for μ = 100; and (**b**) the distribution of events for λ = 3500 nm photon. μ = 100.

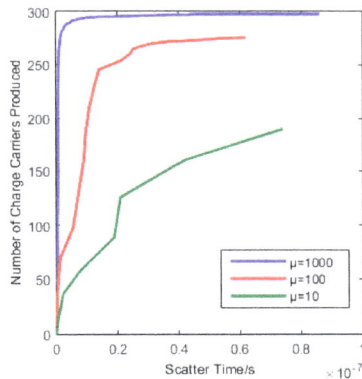

Figure 5. A plot indicating the number of charge carriers versus time showing the effect of changing the II rate.

If we integrate over the entire active scattering time (i.e., the time during which electrons continue to relax and collect at the bottom of the conduction band) then this gives us an estimate of the total number of charge carriers produced. The II rate is then indicative of the active scattering time, with an active scattering time of order 10^{-8} s illustrated in Figure 5.

Our results enabled us to design our prototype detector, based on the schematic from [75], Figure 6, where they first demonstrated a tuneable band gap in AB-stacked bilayer graphene. Our prototype single pixel detector design has a silicon substrate with a 300 nm thick silicon dioxide insulating layer. Ni-Al contacts are deposited on top of the graphene in order to provide electrical connections to the graphene, with a top gate dielectric of alumina deposited through atomic layer deposition (ALD). For the top gate, indium tin oxide (ITO) contacts are deposited; indium tin oxide is typically used in transparent electronics and is opaque at UV photon energies but is transparent at visible photon energies [90].

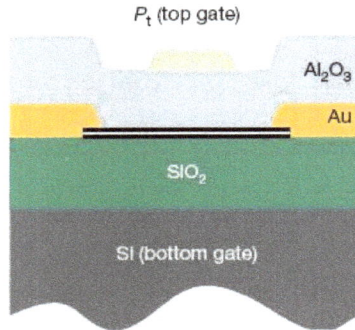

Figure 6. Schematic used in [75] that has been used to show opening of a tuneable bandgap in bilayer graphene. Our bilayer graphene detector design is based on this schematic (reproduced with permission of Nature Publishing Group).

In summary, our results demonstrate the feasibility of a new type of ultrafast photon counter operating at optical and IR wavelengths. Such a device can be operated at approximately 100 MHz, although higher frequencies may be possible with improved calculations of the impact ionisation rate to give our detector comparable or superior results to other detectors. We obtain a value of the electron-hole pair creation energy, W, as a function of the band gap. The ratio between W and the band gap is found to be comparable to that of other detectors such as Si and Ge [89]. The detector has scope to enable a trade-off between operating temperature and energy resolution, allowing for a cryogenic single photon counting photodetector to operate at temperatures that do not require helium-3 cooling albeit with reduced energy resolution. This approach could enable a lower cost detector to be developed for space science where extreme levels of cooling are complex and expensive.

5. Dyakonov-Shur GFET Optimised for 1.2 THz—Simulations and Design

A number of different techniques can be used for photodetection at terahertz frequencies, such as the photothermoelectric effect and bolometry [62]. Another technique is the Dyakonov-Shur effect, whereby a terahertz photon impinges on GFET contacts, designed as antennae, and excites a plasma wave that resonates between the source and the drain of the channel that gives a non-linear photoresponse as a DC voltage [91]. We base our detector on this technique, and utilise simulations discussed in [51] to design our detector and optimise the parameters for the regime that we are interested in.

Tomadin [51] discusses a THz detector in a FET structure, with an AC potential U_a generated between the source and drain and the back gate, kept at a voltage U_0 relative to this. The graphene is of length L between the gates and width W, with a substrate thickness, d. This design, illustrated in Figure 7a, measures the generated photocurrent I which is related to the energy of the incident photon. The photocurrent generated between the contacts is given by:

$$\frac{I}{I_d} = 1 + 2\beta\left(\omega\tau\right)F\left(\omega, \tau\right) \tag{6}$$

where I_d is the diffusive current, $\beta\left(x\right) = \frac{2x}{\sqrt{1+x^2}}$, ω is the frequency of the incident photon, τ is the momentum relaxation time:

$$F\left(\omega, \tau\right) = \frac{\cosh\left(\left(2K_2L\right)\right) + \cos\left(2K_1L\right) - 2}{\cosh\left(2K_2L\right) - \cos\left(2K_1L\right)} \tag{7}$$

and K_1 and K_2 are the real and imaginary parts of the wave number K respectively.

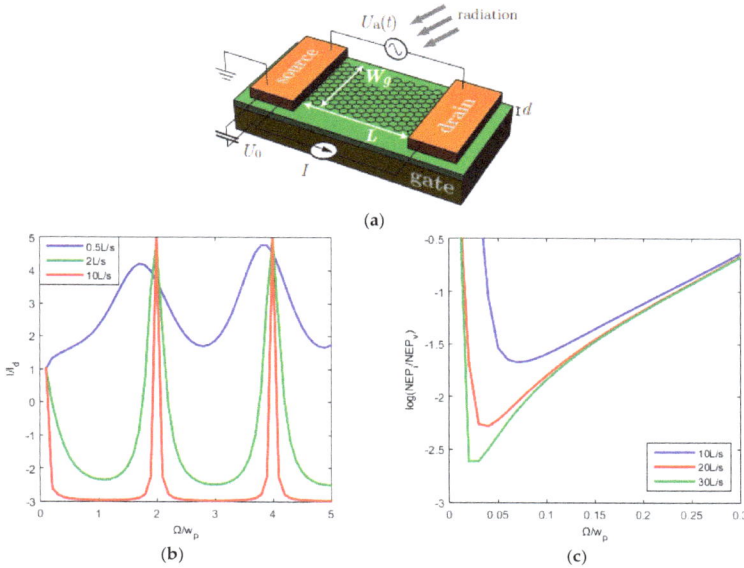

Figure 7. (a) The layout of a GFET utilising the Dyakonov Shur Instability (reproduced from [51] with permission of AIP Publishing under a Creative Commons license subject to https://publishing.aip.org/authors/rights-and-permissions); (b) the photocurrent against the photon frequency for different momentum relaxation time, where s is the plasma wave speed; and (c) $\log\left(\frac{NEP_I}{NEP_V}\right)$ against photon frequency with lower noise at higher momentum relaxation time.

In Figure 7b,c, $\frac{I}{I_d}$ and $\log\left(\frac{NEP_I}{NEP_V}\right)$ are plotted against $\frac{\Omega}{\omega_P} = \frac{2L\Omega}{\pi s}$ respectively, where Ω is the frequency of the incident THz radiation, $\omega_P = \frac{\pi s}{2L}$ is the resonant plasma angular frequency, s is the plasma wave velocity, L is the length of the graphene channel, NEP is the noise equivalent power and $\frac{NEP_I}{NEP_V}$ is the ratio between the current noise and voltage noise [51]. Figure 7b shows that we see a peak in the I/I_d which becomes increasingly sharper with increasing momentum relaxation time, and at regular values of Ω/ω_p. For larger momentum relaxation time we also see a lower noise, Figure 7c. These plots show that we can pick a number of solutions for the parameters of our detector designed for detection of photons with a frequency of 1.2 THz and potentially provide results which are measurable and realistic. In addition, as outlined in [51], by varying U_0 it is possible to control the Fermi level, plasma wave speed, fundamental plasma angular frequency and diffusive photocurrent. Therefore, by changing U_0, we can maximize the photocurrent for a given photon frequency, trade off the noise for optimised device response, and enable a degree of tuneability to maximize the response of the detector over the wide frequency range of interest.

Across our devices the graphene channels were coupled to a number of different antennae, either a bowtie (or a variant "beetle" antenna) as shown in Figure 8a, or a log periodic circular toothed antenna shown in Figure 8b. These were optimised using Sonnet Lite simulation software to resonate at the required frequency range. The schematics show a silicon back gate with a 300 nm thick silicon dioxide insulator between the silicon and a graphene channel of 10 μm × 5 μm, with the graphene channel etched to the required dimensions and nickel-aluminium deposited for the contacts and antennae. The antennae are of order 100 μm from the graphene to the edge of the antennae, with the ratio between the arms of the electrode set to 1.5. This means we operate in the long gate regime, discussed further in [92], where plasma waves have been shown to be excited.

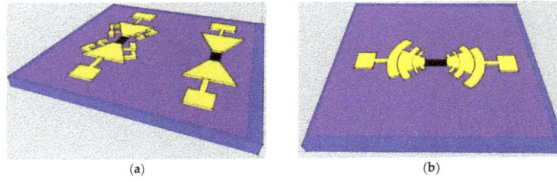

Figure 8. (**a**) The "beetle" (left) and bowtie (right) antennae and (**b**) the log-periodic, circular toothed antenna (not to scale). Both these were designed to have a resonating frequency at the required 1.2 THz. Here blue is Si, light purple is SiO$_2$, yellow is the Ni-Al contact and black is the graphene.

6. X-ray Graphene Field Effect Transistor

A number of groups are working on the development of graphene-based X-ray detectors using a number of different techniques. The most promising developments are from the Jovanovic group [43–49], where they have showed a graphene field effect transistor on a silicon carbide structure at room temperature. They have also demonstrated sensitivity to an X-ray photon beam (15 kV, 15 µA and 40 kV, 80 µA) for an undoped silicon substrate, but only at 4.3 K [48]. This has shown good energy resolution, of order $\frac{E}{\delta E} \sim 10,000$ with contributions from the number of charge carriers produced and limitations due to device design [43]. They have also shown a responsivity of 0.1 AW^{-1} but has presented difficulties with regards to the speed of detection. As shown in Figure 8a the illumination time is ~40 s, with a signal decay time of seconds for both the silicon carbide and silicon respectively. This technique works by modulating the charge carrier density in the substrate, with a resulting change in the resistance of the graphene.

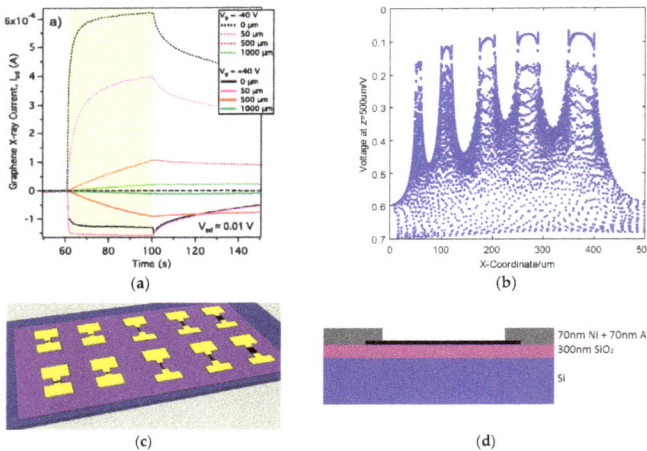

Figure 8. (**a**) The signal with illumination by a 10.1 µW X-ray source over time for a graphene FET based on a silicon carbide substrate (reproduced from [49] with permission of AIP Publishing under a Creative Commons license subject to https://publishing.aip.org/authors/rights-and-permissions). This shows slow illumination and slow decay for an X-ray photon beam, The same group have also shown the sensitivity of a graphene FET based on a silicon substrate at 4.3 K [48]; (**b**) the funnelling of charge carriers towards the substrate dielectric with the application of a gate voltage. In this simulation the charge carriers are funnelled towards 5 contacts on the substrate surface of increasing size; (**c**) shows the the design of an X-ray GFET test device; here blue is Si, light purple is SiO$_2$, yellow is the Ni-Al contact and black is the graphene of different channel sizes for each device, with another, side-on, schematic shown in Figure 8d.

The detectors developed by the Jovanovic group would be unsuitable for high speed, low intensity single photon counting photodetection, but the graphene channel resistance technique, which has also been shown capable of single molecule sensing [59], potentially provides a good basis for our X-ray single photon counting photodetector. For our prototype detectors we have used a silicon substrate with a conductivity of ~100 Ω·cm and operating at room temperature. An incident X-ray photon is absorbed by the silicon, with the resulting electron-hole pair scattering through the silicon directed by the application of a field to funnel the charge carriers to the substrate dielectric as shown in Figure 8b. This build up of charge develops a field across the substrate and applies a field to the graphene resulting in a change in the channel resistance. Our test chip consists of CVD grown graphene that was transferred onto a silicon substrate with resistivity ~100 Ω·cm and a 300 nm thick silicon dioxide insulating layer.

The graphene channels were etched to different sizes from 5 µm × 10 µm up to ~50 µm × 100 µm. These were connected to nickel-aluminium source and drain pads, as shown in Figure 8c,d; nickel obtaining low contact resistance with the graphene and aluminium for better wirebonding.

Whilst our eventual aim is to detect low intensity or single photon sources, we chose to undertake initial experiments using illumination from a pulsed optical laser to characterise the behaviour of the detector and, in particular, its likely sensitivity. These pulsed lasers were calibrated using an Excelitas C30742-33 Series silicon photomultiplier (SiPM). The pulsed laser offers many advantages for initial characterisation including simple control of the deposited energy via variation of pulse width or by attenuation with filters, as well as providing a periodic strobe signal with which the detector output pulse, if present, will be synchronised cf. the unknown random arrival time of X-ray events from an X-ray source. The latter capability is critical when trying to measure the sensitivity while looking for the smallest detectable pulse above the noise. In addition the laser pulse can be used to generate a deposited energy at equivalent depths in the substrate to UV and soft X-ray single photons. Figure 9a shows the wavelength dependence of the photon absorption depth in silicon, the red and blue horizontal lines indicating the absorption depths at 650 nm and 405 nm respectively, showing that the red laser absorption depth is analogous to soft X-rays ~1–4 keV. The device was characterised by applying a 10 mV source-drain voltage, and varying the back gate voltage whilst measuring the source-drain current. The device has a Dirac point at approximately 10 V gate voltage, as shown in Figure 9b.

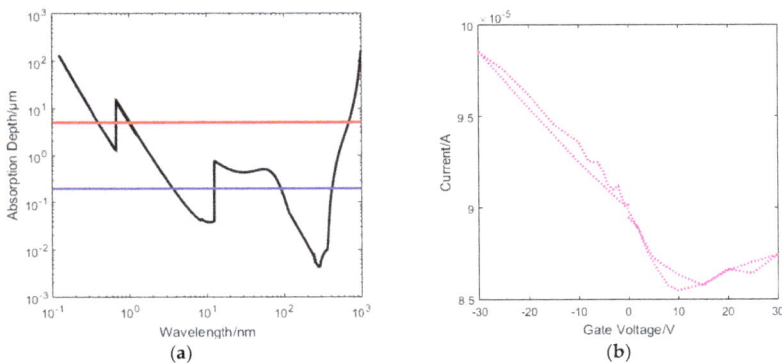

Figure 9. (**a**) the wavelength dependence of photon absorption length for silicon. The red and blue lines show the absorption depth at 650 nm and 405 nm laser wavelengths respectively; and (**b**) showing the current as a function of the gate voltage with the Dirac point at approximately 10 V.

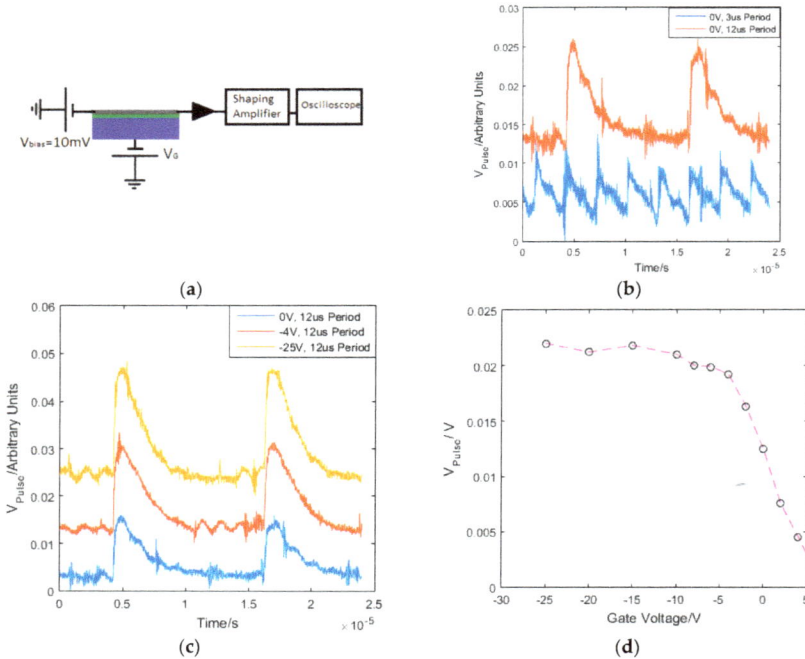

Figure 10. (**a**) the current sensitive preamplifier arrangement with the graphene (grey), SiO$_2$ (green) and Si (blue), with a 10 mV bias between the source and drain of the graphene and the output from the current sensitive preamplifier output on the oscilloscope (**b**) the dependence of the detector signal on the laser pulse frequency and (**c**) the gate voltage applied. N.B. periodic noise at ~200 kHz is also apparent. The detector signal has a very fast rise time and a fall time linked to the recombination time of the charge carriers in the silicon. (**d**) shows the dependence of the signal pulse height on the gate voltage, with a saturation point at approximately −10 V attributed to limits on carrier transport in the Si given by SRH recombination.

The device was initially connected to an Analog Devices ASA4817-1 amplifier in transimpedance mode, with V_{bias} = 10 mV voltage applied between the source and drain contacts, as shown in Figure 10a. The device was then illuminated by a pulsed optical laser with a wavelength of 650 nm and pulse width down to 40 ps. Following the illumination of the detector, the current sensitive preamplifier detects the change in source-drain current and provides a voltage output, V_{Pulse}, captured on an oscilloscope. Figure 10b,c show the dependence of the detector signal on pulse frequency and back gate voltage respectively. Figure 10d shows that the peak amplitude increases for increasing negative gate voltages, until saturating at approximately −15 V.

In order to reduce the noise for higher sensitivity we rearranged the GFET measurement circuit. The device was connected to two low noise, high gain, Canberra 2001 charge sensitive preamplifiers on each contact coupled through two 1 pF capacitors, with two 1 kΩ resistors in series with the graphene, as shown in Figure 11a. Conceptually, when the detector is illuminated, the resistance of the graphene and therefore the voltage across the graphene varies creating a voltage pulse between graphene and the resistor, which in the presence of the capacitor creates a charge pulse which is measured by the the charge sensitive preamplifier, and outputs a voltage pulse, V_{Pulse}. In this arrangement we identified the same saturation as in Figure 10d, which is shown in Figure 11b.

(a)

(b)

(c)

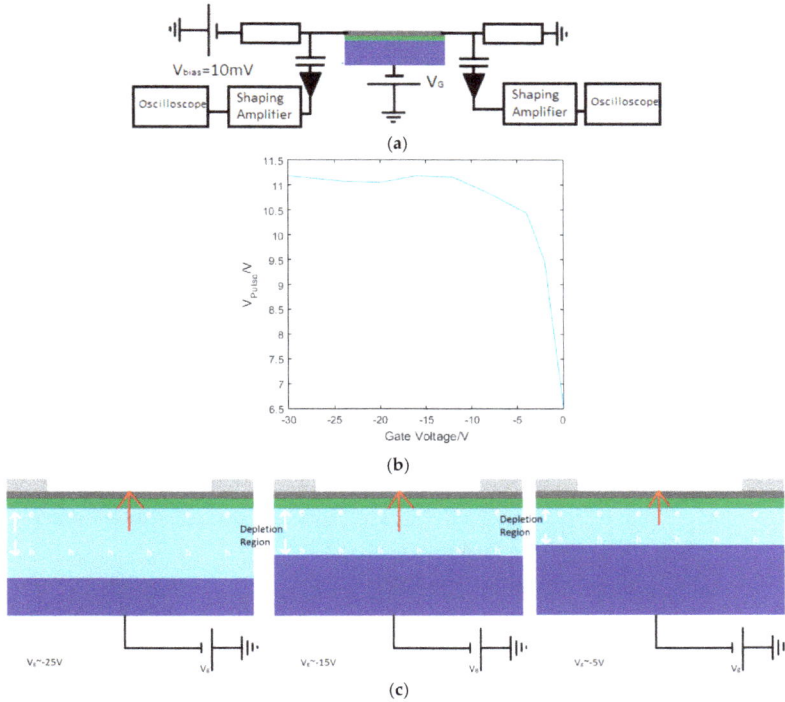

Figure 11. (**a**) shows the charge sensitive preamplifier arrangement with a resistor each side of the graphene to give a voltage change that gives a measurable change in charge due to the presence of the capacitor; and (**b**) show the dependence on the gate voltage in the charge sensitive preamplifier arrangement, with saturation in the pulse amplitude that we again attributed to limits on carrier transport in the silicon given by SRH recombination. (**c**) shows a schematic for the detection mechanism, with the dipole between the electron and hole pair larger for larger depletion regions until they become limited by the SRH recombination time. The dipole created causes a field that changes the graphene conductivity.

We attribute this saturation to the generation of a depletion region in the silicon by the application of a negative gate voltage, extracting the majority carriers, holes, from the silicon gate with electrons travelling towards the insulating dielectric and the graphene. Photons are absorbed in this depletion region and generate an electron hole pair which creates a dipole aligned with the field across the silicon [93,94], whose generation controls the rise time in the signal. The charge carriers generated by the absorption of the photon scatter through the silicon in a region limited by the size of the depletion region. When this becomes large for increasingly negative gate voltages the limiting factor becomes the Shockley-Hall-Read (SRH) recombination time [95]. The V_{Pulse} peak occurs before the charge carriers recombine, with the recombination time driving the fall time of the signal. The calibrated pulsed laser was attenuated using a set of ND filters to simulate a range of X-ray energies. As expected, with increased attenuation we measured a smaller V_{Pulse}, as shown in Figure 11a. The equivalent energy absorbed by the silicon absorber was then calculated, indicating the sensitivity of the detector in terms of equivalent energy of X-ray photons; this is shown in Figure 11b indicating a sensitivity equivalent to ~100 keV. When we applied the graphene drain-source current we observed a change in V_{pulse} that suggests there are two contributions to the signal. Figure 11c shows the two detector output pulses, one with, and one without the drain-source voltage (for V_{bias} = 10 mV and 0 mV respectively). The graphene peak due to the change in resistance is given by the difference between the two peaks. The invariant, larger component of the signal results from charge carriers that accumulate at the dielectric interface and are capacitively coupled to the contact. Figure 11d,e show the varying contribution that the graphene resistance change makes to the total signal with, and without a source-drain current. A schematic for the two contributions to the signal is presented in Figure 12.

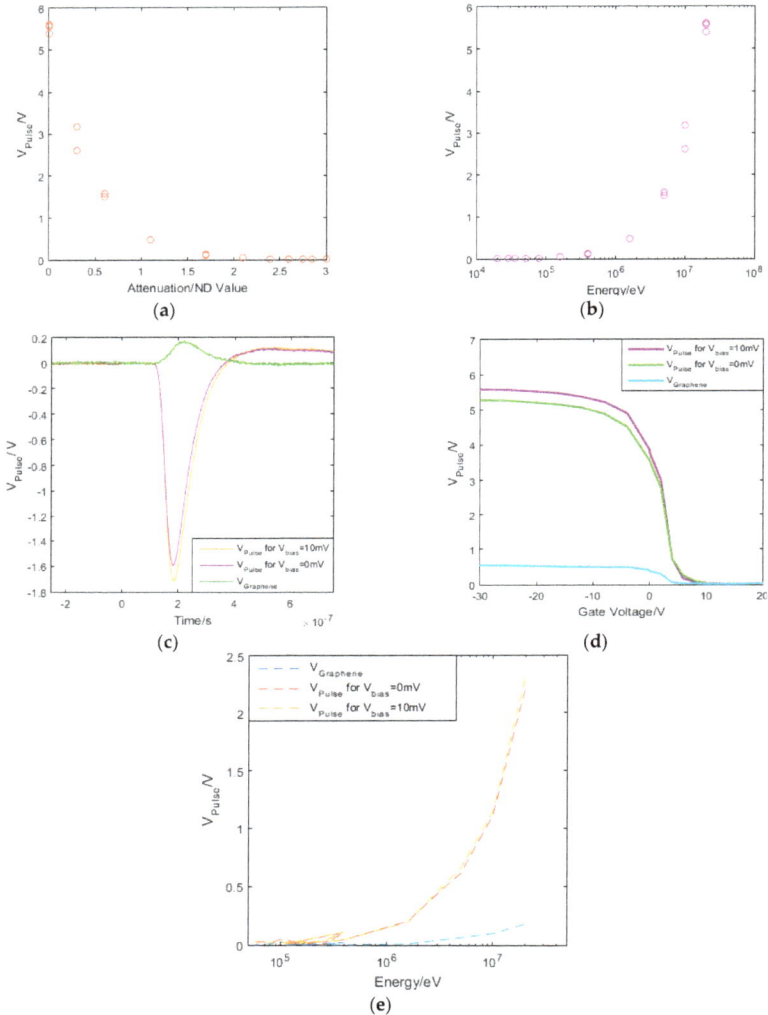

Figure 11. Showing (**a**) the exponential decay on the signal with increasing attenuation of the laser input; (**b**) the measured pulse energy collected in the absorber, indicating an energy sensitivity to ~ 100 keV; (**c**) the difference in the pulse with Vbias = 10 mV and 0 mV for a gate voltage of −15 V showing a small peak attributed to the change in graphene resistance; (**d**) the magnitude of V_{Pulse} for V_{bias} = 10 mV and 0 mV and the signal from the graphene for increasing gate voltage, again showing the saturation described previously; and (**e**) the magnitude of V_{Pulse} for different energies deposited in the absorber by attenuating the incident laser signal to indicate our detector's energy sensitivity, for V_{bias} = 10 mV and 0 mV and identifying the signal attributed to the graphene.

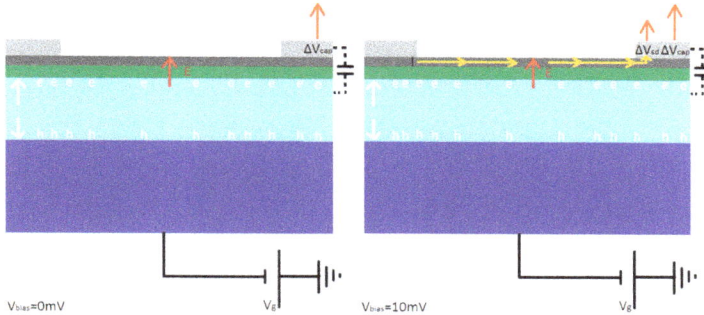

Figure 12. showing the different contributions to the signal that we obtain, with only a contribution from the capacitive coupling between the absorber and contact for V_{bias} = 0 mV, and the capacitative coupling contribution and from the change in voltage across the graphene providing an additional voltage pulse when V_{bias} = 10 mV.

Our results thus far suggest that our current devices have the potential for single photon counting at X-ray energies above 100 keV. Further work is required to improve the signal to noise ratio to improve the sensitivity to lower energies. Our detector shows a significant contribution from capacitively coupled charge carriers in the silicon, and the next step is a redesign to enhance the contribution from the change in the resistance of the graphene which, at the moment, only contributes ~10%. We are currently looking to improve this contribution by increasing the charge carrier dipole field at the graphene using a thinner insulator dielectric, or an intrinsic absorber not requiring a separate insulator, and/or by encapsulating the device to give more reliable and enhanced electrical characteristics. For instance, with an increase in the carrier mobility of the graphene, and the Dirac point located such that we obtained the maximum mobility and operated at gate voltages where we have previously observed V_{Pulse} saturation, we would operate the device where the current-gate voltage curve has a larger gradient and therefore we would expect a proportional increase in the contribution of the signal attributed to the graphene to the overall signal. In addition, including a top gate would enable variation in the drain-source current and enable the depletion region in the silicon to be created to maximise V_{Pulse}.

7. Device Fabrication, Challenges and Progress

The devices discussed in the previous sections have specific requirements that create challenges in the fabrication process, such as graphene coverage, stacking and homogeneity. Micromechanical exfoliation provides easy access to graphene flakes and has been widely used experimentally to explore the properties for single-crystalline graphene and related device structures on the nano- to micrometre scale [96]. The main technical barrier to commercialisation is the development of manufacturing and processing techniques that fulfill the industrial demands for quality, quantity, reliability, and low cost [96,97]. A plethora of diverse fabrication methods have emerged to produce different types of graphene material. For the discussed photodetector applications the requirement is for "electronic-grade" material, in particular continuous films with detailed structural control that support high mobilities. The two main routes for manufacturing "electronic-grade" graphene films are epitaxial growth on silicon carbide (SiC) and chemical vapour deposition (CVD) [97,98]. The former is based on thermal decomposition of high-quality SiC wafers at high temperature (>1300 C) in a controlled atmosphere to control the Si sublimation [98]. As grown graphene-SiC interfaces can be modified by passivation and intercalation [99,100], which allows detailed interfacial tuning. However, the SiC route is severly limited by cost and allows no flexibility in substrate (limited to max 4" high quality SiC). Hence CVD has emerged as main industrial technique to scalably and economically synthesise high-quality graphene films [96,97]. The CVD process typically utilizes a planar catalyst film/foil, on which upon exposure to a gaseous carbon precursor at elevated temperatures a graphitic layer forms. Most widely used are transistion metal catalysts, such as Cu, Ni, Co and Pt, but also semiconductor surfaces such as Ge [101–103]. An increasingly detailed understanding of the CVD growth mechanisms [96,101,104,105] allows increasingly better structural control of the film microstructure, with single-crystal domains of cm dimensions already being achieved (Figure 13, [106]) and also CVD-grown AB-stacked graphene bi-layer films in the order of half-millimetre size recently reported on Cu via oxygen activation [107]. Figure 13b shows a >100 μm grain of bilayer graphene on device fabrication substrate that is readily achievable by CVD.

CVD is rapidly emerging also as industrially preferred technique for other 2D materials, such as h-BN and TMDs [108–113]. Particularly, CVD not only allows the growth of individual 2D layers but potentially also the direct growth of 2D heterostructures [114].

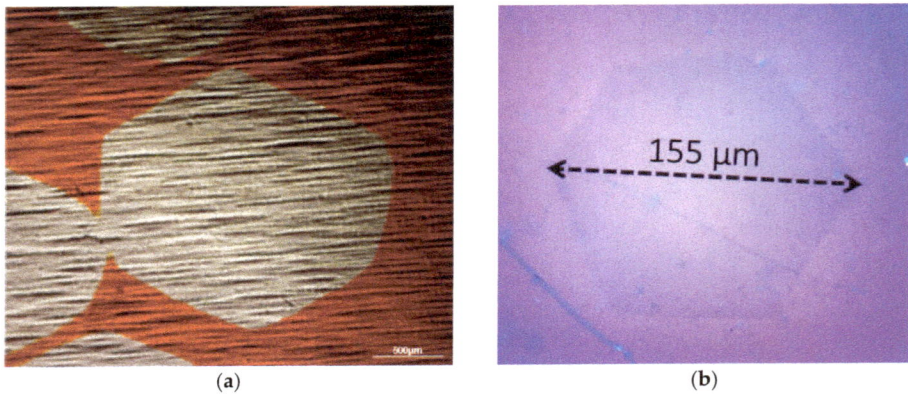

(a)   (b)

Figure 13. Optical micrograph of (**a**) an individual domain of mm sized CVD grown single layer graphene on Cu and (**b**) bilayer graphene on SiO_2/Si substrates.

Using a high quality bilayer graphene sample it is practical now to obtain mobilities of 60,000 $cm^2 \cdot V^{-1} \cdot s^{-1}$ at 1.7 K when encapsulated between h-BN [107]. When processing individual as-grown graphene layers, however, the adsorption of contaminants remains a critical issue. The operation of a GFET in air, Figure 14, results in trap states forming at the surface of graphene and at the graphene/SiO_2 interface from moisture and OH^- states respectively. These trap states cause charge carriers to be trapped in these states resulting in two different Dirac points. This detriments the reliablility and instability of devices during measurements.

To overcome such issues en route to scalable future technology it is desirable to encapsulate the device; hBN is the most promising 2D insulator for this purpose and has showed promising results in proof-of-concept devices [115–117]. Realising its suitability for large area CVD growth is a challenging path and hence difficult to implement in current technology. Another technique that can be used is atomic layer deposition (ALD), an industrially viable large area process that has shown complete passivation and encapsulation of large area graphene devices [61]. Recently Sagade et al. [36] has demonstrated viability of 90 nm alumina layer grown by ALD on GFET that demonstrated highly consistant device operation, Figure 15.

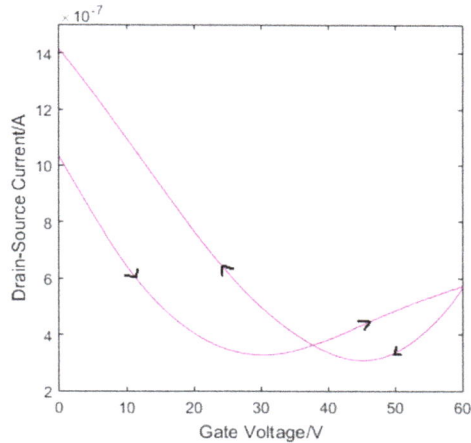

Figure 14. The transfer characteristics of a GFET operated in air. The large hysteresis at two Dirac points is due to trapping of charge carriers. The arrows denote the sweep direction.

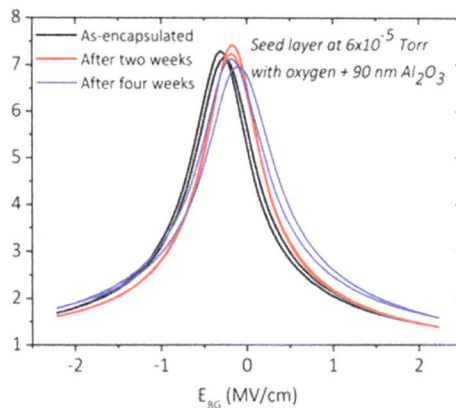

Figure 15. Endurance of electrical properties of encapsulated GFET measured over several weeks in ambient conditions (reproduced from [61] with permission of the Royal Society of Chemistry). The reproducibility in the characteristics is very important in photon counting applications.

8. Conclusions

In this review we have considered the various challenges facing graphene-based single photon counting photodetectors and their prospects at a technological level. The future applications of single photon counting photodetectors requires high detection efficiency with wavelength specificity, good temporal resolution and low dark counts. Graphene's high mobility, tunable band gap (in bilayer graphene), strong dependence of conductivity on electric field, and other properties make it particularly suitable for this application. Here graphene acts as an (indirect) photoconductor with a high gain of transconductance due to the sharp field effect in graphene. Compared with more conventional detector architectures based on charge sensing, the effective decoupling of the detector (graphene) and the absorber (substrate/electrode for X-ray/THz regime respectively) could offer potential benefits. This will also allow more flexibility in the choice of the absorber material. Graphene, therefore, provides an interesting solution for single photon counting applications due to its unique properties, which will make it more favorable than other 2D materials such as metal chalcogenides. The other advantages which graphene can provide are the ultralow noise and high speed of operation.

At visible wavelengths, current detector technologies are able to count single photons such as MKIDs and STJs, but are limited by a temporal resolution of ~1 µs. By contrast graphene photodetectors have shown detection on a femtosecond timescale. In addition, MKIDs are required to be operated at very low temperatures requiring expensive cryogenic techniques. Our simulations of bilayer graphene devices demonstrate wavelength specificity for a photon counter that can be operated over a wide range of temperatures; which can reduce the cost as well as size of an operating system, two factors crucial for implementation in space science. This may also enable more sensitive detectors, owing to the avoidance of wavelength dispersive elements, with potential applications in single photon fluorescence spectroscopy and the ability to sense multiple fluorophores simultaneously.

In this review we have also discussed future graphene-based THz detectors that have applications in areas such as security, astronomy and medical sciences. The lack of sensitive commercial devices currently limit opportunities for detection at 1.2 THz, a regime where significant scientific research could be enabled by graphene THz detectors. We have shown simulations and designs of our proposed detector that exploits the Dyakanov-Shur principle and have identified various antennae designs optimized to these frequencies. We have also discussed critical properties of graphene which may provide a future solution required for PCTI.

A number of options are available for detection of single X-ray photons, such as STJs and microchannel plate photomultipler tubes. STJs have good energy resolution, but must be operated at cryogenic temperatures, whilst MCP-PMTs have a timing resolution on the order of picoseconds, but provide very poor to no energy resolution. For graphene X-ray detectors, our experimental research with pulsed optical lasers, which simulate X-ray absorption, suggest a potential energy sensitivity of the detector equivalent to ~100 keV X-ray photons. We have also discussed the ample scope for the improvement in the design and operation of the detector to improve future sensitivity.

Effective integration of graphene at industrial scale in these different types of photodetectors critically depends on the development of integrated manufacturing pathways, in particular progress in CVD graphene (single- and bi-layer) growth technologies in terms of control over homogeneity of layers, defect density, doping and transfer to device relevant substrates. We have also highlighted the importance of interfacial control and graphene encapsulation to ensure reproducible and reliable device characteristics. Graphene photosensors have the unique capability to cover an energy range from THz to X-rays. Our simulations and experimental results, combined with continuing advances in growth and fabrication techniques suggest that graphene-based new photodetector technologies have a highly promising future.

Acknowledgments: J.O.D. Williams is funded by an STFC PhD studentship. Simulations were done using SPECTRE High Power Computing at the University of Leicester. Experimental work funded by the European Space Agency (AO/1-8070/14/F/MOS), with acknowledgements to Alan Owens and Elena Saenz for their input. CVD growth and device work in the Hofmann group was supported by EPSRC (Grant No. EP/K016636/1, GRAPHTED). J.A. Alexander-Webber acknowledges a Research Fellowship from Churchill College Cambridge.

Author Contributions: J.O.D. Williams, J.S. Lapington, M. Roy and I.B. Hutchinson initiated the project; J.O.D. Williams conducted the simulations and testing at the University of Leicester, with design work done with discussions with those at the University of Cambridge. All CVD work was carried out by the Hofmann group at the University of Cambridge. J.O.D. Williams and J.A. Alexander-Webber fabricated the devices. All authors contributed to analyzing the results and writing the paper.

Conflicts of Interest: The authors declare no conflict of interest.

References

1. Santavicca, D.F.; Carter, F.W.; Prober, D.E. Proposal for a GHz Count Rate Near-IR Single-Photon Detector Based on a Nanoscale Superconducting Transition Edge Sensor. Available online: http://arxiv.org/ftp/arxiv/papers/1202/1202.4722.pdf (accessed on 17 August 2016).

2. Rajteri, M.; Taralli, E.; Portesi, C.; Monticone, E. Single Photon Light Detection with Transition Edge Sensors. *Nuovo Cimento C* **2009**, *31*, 549–555.

3. Dierickx, B.; Yao, Q.; Witrouwen, N.; Uwaerts, D.; Vandewiele, S.; Gao, P. X-ray Photon Counting and Two-Colour X-ray Imaging Using Indirect Detection. *Sensors* **2016**, *16*, 764. [CrossRef] [PubMed]

4. Fraser, G.W.; Heslop-Harrison, J.S.; Schwarzacher, T.; Holland, A.D.; Verhoeve, P.; Peacock, A. Detection of multiple fluorescent labels using superconducting tunnel junctions. *Rev. Sci. Instrum.* **2003**, *74*. [CrossRef]

5. Comandar, L.C.; Frohlich, B.; Lucamarini, M.; Patel, K.A.; Sharpe, A.W.; Dynes, J.F.; Yuan, Z.L.; Penty, R.V.; Shields, A.J. Room temperature single-photon detectors for high bit rate quantum key distribution. *Appl. Phys. Lett.* **2014**, *104*. [CrossRef]

6. Eisaman, M.D.; Fan, J.; Migdall, A.; Polyakov, S.V. Single-photon sources and detectors. *Rev. Sci. Instrum.* **2011**, *82*. [CrossRef] [PubMed]

7. Korneev, A.; Vachtomin, Y.; Minaeva, O.D.; Smirnov, K.; Okunev, O.; Gol'tsman, G.; Zinoni, C.; Chauvin, N.; Balet, L.; Marsili, F.; et al. Single-photon detection system for quantum optics applications. *IEEE J. Sel. Top. Quantum Electron.* **2007**, *13*, 944–951. [CrossRef]

8. Hadfield, R.H. Single-photon detectors for optical quantum information applications. *Nat. Photonics* **2009**, *3*, 696–705. [CrossRef]

9. Varnava, M.; Browne, D.E.; Rudolph, T. How good must single photon sources and detectors be for efficient linear optical quantum computation? *Phys. Rev. Lett.* **2008**, *100*, 060502. [CrossRef] [PubMed]

10. Novoselov, K.S.; Geim, A.K.; Morozov, S.V.; Jiang, D.; Zhang, Y.; Dubonos, S.V.; Grigorieva, I.V.; Firsov, A.A. Electric field in atomically thin carbon films. *Science* **2004**, *306*, 666–669. [CrossRef] [PubMed]

11. Geim, A.K. Graphene: Status and prospects. *Science* **2009**, *324*, 1530–1534. [CrossRef] [PubMed]

12. Castro Neto, A.H.; Guinea, F.; Peres, N.M.; Novoselov, K.S.; Geim, A.K. The electronic properties of graphene. *Rev. Mod. Phys.* **2009**, *81*, 109. [CrossRef]

13. Bolotin, K.I.; Sikes, K.J.; Jiang, Z.; Klima, M.; Fudenberg, G.; Hone, J.; Kim, P.; Stormer, H.L. Ultrahigh electron mobility in suspended graphene. *Solid State Commun.* **2008**, *146*, 351–355. [CrossRef]

14. Hwang, E.H.; Adam, S.; Das Sarma, S. Carrier Transport in Two-Dimensional Graphene Layers. *Phys. Rev. Lett.* **2007**, *98*, 186806. [CrossRef] [PubMed]

15. Nair, R.R.; Blake, P.; Grigorenko, A.N.; Novoselov, K.S.; Booth, T.J.; Stauber, T.; Peres, N.M.R.; Geim, A.K. Fine Structure Constant Defines Visual Transparency of Graphene. *Science* **2008**, *320*, 1308. [CrossRef] [PubMed]

16. Katsnelson, M.I. *Graphene: Carbon in Two Dimensions*; Cambridge University Press: Cambridge, UK, 2012.

17. McKitterick, C.B.; Prober, D.E.; Karasik, B.S. Performance of Graphene Thermal Photon Detectors. *J. Appl. Phys.* **2013**, *113*. [CrossRef]

18. Tielrooij, K.J.; Piatkowski, L.; Massicotte, M.; Woessner, A.; Ma, Q.; Lee, Y.; Myrho, K.S.; Lau, C.N.; Jarillo-Herrero, P.; van Hulst, N.F.; et al. Generation of photovoltage in graphene on a femtosecond timescale through efficient carrier heating. *Nat. Nanotechnol.* **2015**. [CrossRef] [PubMed]

19. Monfardini, A.; Benoit, A.; Bideaud, A.; Swenson, C.; Cruciani, A.; Camus, P.; Hoffmann, C.; Deser, F.X.; Doyle, S.; Ade, P.; et al. A Dual-Band Millimeter-Wave Kinetic Inductance Camerca for the IRAM 30 m Telescope. *Astrophys. J. Suppl. Ser.* **2011**, *194*. [CrossRef]

20. Van Rantwijk, J.; Grim, M.; van Loon, D.; Yates, S.; Baryshev, A.; Baselmans, J. Multiplexed readout for 1000-pixel arrays of microwave kinetic inductance detectors. *IEEE Trans. Microw. Theory Tech.* **2016**, *64*, 1876–1883. [CrossRef]

21. Day, P.K.; LeDuc, H.G.; Mazin, B.A.; Vayonakis, A.; Zmuidzinas, J. A broadband superconducting detector suitable for use in large arrays. *Nature* **2003**, *425*, 817–821. [CrossRef] [PubMed]

22. Princeton. Josephson Junctions. Available online: http://www.princeton.edu/~romalis/PHYS210/stj.htm (accessed on 17 August 2016).

23. Martin, D.D.; Verhoeve, P. Superconducting Tunnel Junctions. *ISSI SR* **2010**, *9*, 441–457.

24. Czakon, N.G.; Vaykonakis, A.; Schlaerth, J.; Hollister, M.I.; Golwala, S.; Day, P.K.; Gao, J.S.; Glenn, J.; LeDuc, H.; Maloney, P.R.; et al. Microwave Kinetic Inductance Detector (MKID) Camera Testing for Submillimeter Astronomy. Available online: http://web.physics.ucsb.edu/~bmazin/Papers/preprint/czakon_LTD13.pdf (accessed on 17 August 2016).

25. Mazin, B.A. Microwave Kinetic Inductance Detectors: The First Decade. In Proceedings of the AIP Thirteenth International Workshop on Low Temperature Detectors, Stanford, CA, USA, 20–24 July 2009.

26. Xia, F. Electrons en masse. *Nat. Nanotechnol.* **2014**, *9*, 575–576. [CrossRef] [PubMed]

27. Yoon, H.; Forsythe, C.; Wang, L.; Tombros, N.; Watanabe, K.; Taniguchi, T.; Hone, J.; Kim, P.; Ham, D. Measurement of Collective Dynamical Mass of Dirac Fermions in Graphene. *Nat. Nanotechnol.* **2014**, *9*, 594–599. [CrossRef] [PubMed]

28. Karasik, B.S.; Sergeev, A.V.; Prober, D.E. Nanobolometers for THz Photon Detection. Available online: https://arxiv.org/ftp/arxiv/papers/1208/1208.5803.pdf (accessed on 17 August 2016).

29. Matsuo, H. Photon Statistics for Space Terahertz Astronomy. 2014. Available online: http://www.nrao.edu/meetings/isstt/papers/2014/2014067000.pdf (accessed on 17 August 2016).

30. Matsuo, H. Photon Counting Terahertz Interferometry (PCTI). Available online: http://fisica.iaps.inaf.it/2014-02-15-Fisica-Presentations/pres_FISICA-Matsuo.pdf (accessed on 17 August 2016).

31. Matsuo, H. Requirements on Photon Counting Detectors for Terahertz Interferometry. *J. Low Temp. Phys.* **2012**, *167*, 840–845. [CrossRef]

32. Hajime, E.; Go, F.; Hiroshi, M.; Shigetomo, S.; Masahiro, U. SIS Detectors for Terahertz Photon Counting System. In Proceedings of the 16th International Workshop on Low Temperature Detectors, Grenoble, France, 20–24 July 2015.

33. Bulter, A. Single-Photon Counting Detectors for the Visible Range Between 300 and 1000 nm. In *Advanced Photon Counting*; Springer International Publishing: Cham, Switzerland, 2014; pp. 23–42.

34. Wahl, M. Time Correlated Single Photon Counting. Available online: https://www.picoquant.com/images/uploads/page/files/7253/technote_tcspc.pdf (accessed on 17 August 2016).

35. STJ X-ray Spectrometers. Available online: http://starcryo.com/stj-x-ray-spectrometers/ (accessed on 17 August 2016).

36. Doyle, S. Lumped Element Kinetic Inductance Detectors. Available online: http://www.astro.cardiff.ac.uk/~spxsmd/Lumped_Element_Kinetic_Inductance_Detectors.pdf (accessed on 17 August 2016).

37. White, S.; Chiu, M.; Diwan, M.; Atoyan, G.; Issakov, V. Design of a 10 picosecond Time of Flight Detector using Avalanche Photodiodes. 2009, arXiv:0901.2530.

38. Tan, C.H.; Gomes, R.B.; David, J.P.; Barnett, A.M.; Bassford, D.J.; Lees, J.E.; Shien, J. Avalanche Gain and Energy Resolution of Semiconductor X-Ray Detectors. *IEEE Trans. Electron Dev.* **2011**, *58*, 1696–1701.

39. Excelitas. Avalanche Photodiode—A User Guide. Available online: http://www.excelitas.com/downloads/app_apd_a_user_guide.pdf (accessed on 17 August 2016).

40. Oshima, T.; Yamakawa, Y.; Kurabayashi, H.; Hoshnino, A.; Ishisaki, Y.; Ohashi, T.; Mitsuda, K.; Tanaka, K. A High Energy Resolution Gamma-Ray TES Microcalorimeter with Fast Response Time. *J. Low Temp. Phys.* **2008**, *151*, 430–435. [CrossRef]

41. Figueroa Group, MIT. Transition Edge Sensors (TES). Available online: http://web.mit.edu/figueroagroup/ucal/ucal_tes/ (accessed on 17 August 2016).

42. Lee, S.-F.; Gildemeister, J.M.; Holmes, W.; Lee, A.T.; Richards, P.L. Voltage-viased superconducting transition-edge bolometer with strong electrothermal feedback operated at 370 mK. *Appl. Opt.* **1998**, *37*, 3391–3397. [CrossRef] [PubMed]

43. Foxe, M.; Lopez, G.; Childres, I.; Patil, A.; Roecker, C.; Boguski, J.; Jovanovic, I.; Chen, Y.P. Graphene Field-Effect Transistors on Undoped Semiconductor Substrates for Radiation Detectors. *IEEE Trans. Nanotechnol.* **2012**, *11*, 581–587. [CrossRef]

44. Jovanovic, I.; Cazalas, E.; Childres, I.; Patil, A.; Koybasi, O.; Chen, Y.P. Graphene field effect transistor-based detectors for detection of ionising radiation. In Proceedings of the 2013 3rd International Conference on Advancements in Nuclear Instrumentation, Measurement Methods and Their Applications, Marseille, France, 23–27 June 2013.

45. Koybasi, O.; Cazalas, E.; Childres, I.; Jovanovic, I.; Chen, Y.P. Detection of light, X-rays and gamma rays using graphene field effect transistors fabricated on SiC, CdTe and AlGaAs/GaAs substrates. In Proceedings of the IEEE Nuclear Science Symposium Conference Record, Seoul, Korea, 27 October–2 November 2013.

46. Koybasi, O.; Childres, I.; Jovanovic, I.; Chen, Y.P. Graphene field effect transistor as a radiation and photo detector. *Proc. SPIE* **2012**, *8373*. [CrossRef]

47. Koybasi, O.; Childres, I.; Jovanovic, I.; Chen, Y. Design and Simulation of a Graphene DEPFET Detector. In Proceedings of the IEEE Nuclear Science Symposium and Medical Imaging Conference Record, Anaheim, CA, USA, 27 October–3 November 2012; pp. 4249–4254.

48. Patil, A.; Koybasi, O.; Lopez, G.; Foxe, M.; Childres, I.; Roecker, C.; Boguski, J.; Gu, J.; Bolen, M.L.; Capano, M.A.; et al. Graphene Field Effect Transistor as Radiation Sensor. In Proceedings of the 2011 IEEE Nuclear Science Symposium and Medical Imaging Conference (NSS/MIC), Valencia, Spain, 23–29 October 2011.

49. Cazalas, E.; Sarker, B.K.; Moore, M.E.; Childres, I.; Chen, Y.P.; Jovanovic, I. Position sensitivity of graphene field effect transistors to X-rays. *Appl. Phys. Lett.* **2015**, *106*, 223503. [CrossRef]

50. Xia, F.; Mueller, T.; Lin, Y.-M.; Valdes-Garcia, A.; Avouris, P. Ultrafast Graphene Photodetector. *Nature* **2009**, *4*, 839–843. [CrossRef] [PubMed]

51. Tomadin, A.; Tredicucci, A.; Pellegrini, V.; Vitiello, M.S.; Polini, M. Photocurrent-Based detection of Terahertz radiation in graphene. *Appl. Phys. Lett.* **2013**, *103*, 211120. [CrossRef]

52. Vicarelli, L.; Vitiello, M.S.; Coquillat, D.; Lombardo, A.; Ferrari, A.C.; Knap, W.; Polini, M.; Pellegrini, V.; Tredicucci, A. Graphene field effect transistors as room-temperature Terahertz detectors. *Nat. Mater.* **2012**, *11*, 865–871. [CrossRef] [PubMed]

53. Hetsch, F.; Zhao, N.; Kershaw, S.V.; Rogach, A.L. Quantum Dot Field Effect Transistors. *Mater. Today* **2013**, *16*, 312–325. [CrossRef]

54. Jeon, P.J.; Lee, Y.T.; Lim, J.Y.; Kim, J.S.; Hwang, D.K.; Im, S. Black Phosphorus-Zinc Oxide Nanomaterial Heterojunction for p-n Diode and Junction Field Effect Transistor. *Nano Lett.* **2016**, *16*, 1293–1298. [CrossRef] [PubMed]

55. Bucsema, M.; Groenendijk, D.J.; Blanter, S.L.; Steele, G.A.; van den Zant, H.S.; Castellanos-Gomez, A. Fast and Broadband Photoresponse of Few-Layer Black Phosphorus Field Effect Transistors. *Nano Lett.* **2014**, *14*, 3347–3352.

56. Song, J.C.W.; Rudner, M.S.; Marcus, C.M.; Levitov, L.S. Hot Carrier Transport and Photocurrent Response in Graphene. *Nano Lett.* **2011**, *11*, 4688–4692. [CrossRef] [PubMed]

57. Yan, J.; Kim, M.H.; Elle, J.A.; Sushkov, A.B.; Jenkins, G.S.; Milchberg, H.M.; Fuhrer, M.S.; Drew, H.D. Dual-gated bilayer graphene hot-electron bolometer. *Nat. Nanotechnol.* **2012**, *7*, 472–478. [CrossRef] [PubMed]

58. Jessop, D.S.; Kindness, S.J.; Xiao, L.; Braeuninger-Weimer, P.; Lin, H.; Ren, Y.; Ren, C.X.; Hofmann, S.; Zeitler, J.A.; Beere, H.E.; et al. Graphene based plasmonic terahertz amplitude modulator operating above 100 MHz. *Appl. Phys. Lett.* **2016**, *108*, 171101. [CrossRef]

59. Schedin, F.; Geim, A.K.; Morozov, S.V.; Hill, E.W.; Blake, P.; Katsnelson, M.I.; Novoselov, K.S. Detection of individual gas molecules absorbed on graphene. *Nat. Mater.* **2007**, *6*, 652–655. [CrossRef] [PubMed]

60. Deacon, R.S.; Chuang, K.C.; Nicholas, R.J.; Novoselov, K.S.; Geim, A.K. Cyclotron resonance study of the electron and hole velocity in graphene monolayers. *Phys. Rev. B* **2007**, *76*, 081406(R). [CrossRef]

61. Sagade, A.A.; Neumaier, D.; Schall, D.; Otto, M.; Pesquera, A.; Centeno, A.; Zurutuza Elorza, A.; Kurz, H. Highly air stable passivation of graphene based field effect devices. *Nanoscale* **2015**, *7*, 3558–3564. [CrossRef] [PubMed]

62. Koppens, F.H.L.; Mueller, T.; Avouris, P.; Ferrari, A.C.; Vitiello, M.S.; Polini, M. Photodetectors based on graphene, other two-dimensional materials and hybrid systems. *Nat. Nanotechnol.* **2014**, *9*, 780–793. [CrossRef] [PubMed]

63. Kretinin, A.V.; Cao, Y.; Tu, J.S.; Yu, G.L.; Jalil, R.; Novoselov, K.S.; Haigh, S.J.; Gholinia, A.; Mishchenko, A.; Lozada, M.; et al. Electronic Properties of Graphene Encapsulated with Different Two-Dimensional Atomic Crystals. *Nano Lett.* **2014**, *14*, 3270–3276. [CrossRef] [PubMed]

64. Sanders, S.; Cabrero-Vilatela, A.; Kidambi, P.R.; Alexander-Webber, J.A.; Weijtens, C.; Braeuninger-Weimer, P.; Aria, A.I.; Qasim, M.M.; Wilkinson, T.D.; Robertson, J.; et al. Engineering high charge transfer n-doping of graphene electrodes and its application to organic electronics. *Nanoscale* **2015**, *7*, 13135–13142. [CrossRef] [PubMed]

65. Meyer, J.; Kidambi, P.R.; Bayer, B.C.; Weijtens, C.; Kuhn, A.; Centeno, A.; Pesquera, A.; Zurutuza, A.; Robertson, J.; Hofmann, S. Metal Oxide Induced Charge Transfer Doping and Band Alignment of Graphene Electrodes for Efficient Organic Light Emitting Diodes. *Sci. Rep.* **2014**, *4*. [CrossRef] [PubMed]

66. Novoselov, K.S.; Geim, A.K.; Morozov, S.V.; Jiang, D.; Katsnelson, M.I.; Grigorieva, I.V.; Dubonos, S.V.; Firsov, A.A. Two Dimensional Gas of Massless Dirac Fermions in Graphene. *Nature* **2005**, *438*, 197–200. [CrossRef] [PubMed]

67. Zhan, B.; Li, C.; Yang, J.; Jenkins, G.; Huang, W.; Dong, X. Graphene field-effect transistor and its application for electronic sensing. *Small* **2014**, *10*, 4042–4065. [CrossRef] [PubMed]
68. Foxe, M.; Cazalas, E.; Lamm, H.; Majcher, A.; Piotrowski, C.; Childres, I.; Patil, A.; Chen, Y.P.; Jovanovic, I. Graphene-Based Neutron Detectors. In Proceedings of the 2011 IEEE Nuclear Science Symposium and Medical Imaging Conference (NSS/MIC), Valencia, Spain, 23–29 October 2011; pp. 352–355.
69. Foxe, M.; Lopez, G.; Childres, I.; Jalilian, R.; Roecker, C.; Boguski, J.; Jovanovic, I.; Chen, Y.P. Detection of Ionizing Radiation Using Graphene Field Effect Transistors. In Proceedings of 2009 IEEE Nuclear Science Symposium Conference Record (NSS/MIC), Orlando, FL, USA, 25–31 October 2009; pp. 90–95.
70. Zou, K.; Hong, X.; Zhu, J. Effective mass of electrons and holes in bilayer graphene: Electron-Hole asymmetry and electron-electron interaction. *Phys. Rev. B* **2011**, *84*, 085408. [CrossRef]
71. Xia, F.; Mueller, T.; Golizadeh-Mojarad, R.; Freitag, M.; Lin, Y.-M.; Tsang, J.; Perebeinos, V.; Avouris, P. Photocurrent Imaging and Efficient Photon Detection in a Graphene Transistor. *Nano Lett.* **2009**, *9*, 1039–1044. [CrossRef] [PubMed]
72. Mueller, T.; Xia, F.; Avouris, P. Graphene photodetectors for high-speed optical communications. *Nat. Photonics* **2010**, *4*, 297–301. [CrossRef]
73. Liu, Y.; Cheng, R.; Liao, L.; Zhou, H.; Bai, J.; Liu, G.; Liu, L.; Huang, Y.; Duan, X. Plasmon resonance enhance multicolour photodetection by graphene. *Nat. Commun.* **2011**, *2*. [CrossRef] [PubMed]
74. Van Veldhoven, Z.A.; Alexander-Webber, J.A.; Sagade, A.A.; Braeuninger-Weimer, P.; Hofmann, S. Electronic properties of CVD graphene: The role of grain boundaries, atmospheric doping and encapsulation by ALD. *Phys. Stat. Sol. B* **2016**. [CrossRef]
75. Zhang, Y.; Tang, T.-T.; Girit, C.; Hao, Z.; Martin, M.C.; Zettl, A.; Crommie, M.F.; Shen, Y.R.; Wang, F. Direct Observation of a Widely Tunable Bandgap in Bilayer Graphene. *Nature* **2009**, *459*, 820–823. [CrossRef] [PubMed]
76. De Fazio, D.; Goykhman, I.; Bruna, M.; Eiden, A.; Milana, S.; Yoon, D.; Sassi, U.; Barbone, M.; Dumcenco, D.; Marinov, K.; et al. High Responsivity, Large-Area Graphene/MoS2 Flexible Photodetectors. Available online: http://arxiv.org/pdf/1512.08312v1.pdf (accessed on 17 August 2016).
77. McCann, E.; Koshino, M. The electronic properties of bilayer graphene. *Rep. Prog. Phys.* **2013**, *76*. [CrossRef] [PubMed]
78. Yavari, F.; Kritzinger, C.; Gaire, C.; Song, L.; Gullapalli, H.; Borca-Tasciuc, T.; Ajayan, P.M.; Koratkar, N. Tunable Bandgap in Graphene by the Controlled Adsorption of Water Molecules. *Small* **2010**, *6*, 2535–2538. [CrossRef] [PubMed]
79. Balog, R.; Jorgensen, B.; Nilsson, L.; Andersen, M.; Rienks, E.; Bianchi, M.; Fanetti, M.; Laegsgaard, E.; Baraldi, A.; Lizzit, S.; et al. Bandgap Opening in Graphene Induced by Patterned Hydrogen Adsorption. *Nat. Mater.* **2010**, *9*, 315–319. [CrossRef] [PubMed]
80. Lee, J.-K.; Yamazaki, S.; Yun, H.; Park, J.; Kennedy, G.P.; Kim, G.-T.; Pietzsch, O.; Wiesendanger, R.; Lee, S.; Hong, S.; et al. Modification of Electrical Properties of Graphene by Substrate-Induced Nanomodulation. *Nano* **2013**, *13*, 3494–3500. [CrossRef] [PubMed]
81. Zhang, W.; Lin, C.-T.; Liu, K.-K.; Tite, T.; Su, C.-Y.; Chang, C.-H.; Lee, Y.-H.; Chu, C.-W.; Wei, K.-H.; Kuo, J.-L.; et al. Opening an Electrical Band Gap of Bilayer Graphene with Molecular Doping. *Nano* **2011**, *5*, 7517–7524.
82. Williams, J.O.D.; Lapington, J.S.; Roy, M.; Hutchinson, I.B. Graphene as a Novel Single Photon Counting Optical and IR Photodetector. *IET* **2015**. [CrossRef]
83. Gillespie, D.T. Exact Stochastic Simulation of Coupled Chemical Reactions. *J. Phys. Chem.* **1977**, *81*, 2340–2361. [CrossRef]
84. Pop, E.; Varshney, V.; Roy, A.K. Thermal Properties of Graphene: Fundamentals and Applications. *MRS Bull.* **2012**, *37*, 1273–1280. [CrossRef]
85. Zhang, Y.; Liu, T.; Meng, B.; Li, X.; Liang, G.; Hu, X.; Wang, Q.J. Broadband High Photoresponse from Pure Monolayer Graphene Photodetector. *Nat. Commun.* **2013**, *4*. [CrossRef] [PubMed]
86. Winzer, T.; Knorr, A.; Malic, E. Carrier Multiplication in Graphene. *Nano Lett.* **2010**, *10*, 4839–4843. [CrossRef] [PubMed]
87. Winzer, T.; Malic, E. Impact of Auger Processes on Carrier Dynamics in Graphene. *Phys. Rev. B* **2012**, *85*, 241404. [CrossRef]

88. Borysenko, K.M.; Mullen, J.T.; Li, X.; Semenov, Y.G.; Zavada, J.M.; Nardelli, M.B.; Kim, K.W. Electron-Phonon interactions in bilayer graphene. *Phys. Rev. B* **2011**, *83*, 161402. [CrossRef]

89. European Space Agency. Future Missions Preparations Office. Available online: http://sci.esa.int/science-e/www/object/index.cfm?fobjectid=41034 (accessed on 17 August 2016).

90. Chen, Z.; Li, W.; Li, R.; Zhang, Y.; Xu, G.; Cheng, H. Fabrication of Highly Transparent and Conductive Indium-Tin Oxide Thin Films with a High Figure of Merit via Solution Processing. *Am. Chem. Soc.* **2013**, *29*, 13836–13842. [CrossRef] [PubMed]

91. Knap, W.; Rumyantsev, S.; Vitiello, M.S.; Coquillat, D.; Blin, S.; Dyakanova, N.; Shur, M.; Teppe, F.; Tredicucci, A.; Nagatsuma, T. Nanometer size field effect transistors for terahertz detectors. *Nanotechnology* **2013**, *24*. [CrossRef] [PubMed]

92. Knap, W.; Dyakonov, M.I.; Coquillat, D.; Teppe, F.; Dyakanova, N.; Lusakowski, J.; Karpierz, K.; Sakowicz, M.; Valusis, G.; Seluita, D.; et al. Field Effect Transistors for Terahertz Detection: Physics and First Imaging Applications. *J. Infrared Millim. Terahertz Waves* **2009**, *30*. [CrossRef]

93. Tuan Trinh, M.; Sfeir, M.Y.; Choi, J.J.; Owen, J.S.; Zhu, X. A Hot Electron-Hole Pair Breaks the Symmetry of a Semiconductor Quantum Dot. *Nanoletters* **2013**, *13*, 6091–6097. [CrossRef] [PubMed]

94. Trinh, M.T.; Wu, X.; Niesner, D.; Zhu, X.-Y. Many-Body interactiosn in photoexcited lead iodide perovskite. *J. Mater. Chem. A* **2015**, *3*, 9285–9290. [CrossRef]

95. Schroder, D.K. Carrier Lifetimes in Silicon. *IEEE Trans. Electron Dev.* **1997**, *44*, 160–170. [CrossRef]

96. Hofmann, S.; Braeuninger-Weimer, P.; Weatherup, R.S. CVD-Enabled Graphene Manufacture and Technology. *J. Phys. Chem. Lett.* **2015**, *6*, 2714–2721. [CrossRef] [PubMed]

97. Nature Nanotechnology. Ten years in two dimensions. *Nat. Nanotechnol.* **2014**, *9*, 725.

98. Ruan, M.; Hu, Y.; Guo, Z.; Dong, R.; Palmer, J.; Hankinson, J.; Berger, C.; de Heer, W.A. Epitaxial graphene on silicon carbide: Introduction to structured graphene. *MRS Bull.* **2012**, *37*, 1138–1147. [CrossRef]

99. Melios, C.; Panchal, V.; Giusca, C.E.; Strupinski, W.; Silva, S.R.P.; Kazakova, O. Carrier type inversion in quasi-free standing graphene: Studies of local electronic and structural properties. *Sci. Rep.* **2015**, *5*. [CrossRef] [PubMed]

100. Huang, J.; Alexander-Webber, J.A.; Janssen, T.J.B.M.; Tzakenchuk, A.; Yager, T.; Lara-Avils, S.; Kubatkin, S.; Myers-Ward, R.L.; Wheeler, V.D.; Gaskill, D.K. Hot carrier relaxation of Dirac fermions in bilayer epitaxial graphene. *J. Phys. Condensed Matter* **2015**, *27*, 164202. [CrossRef] [PubMed]

101. Cabrero-Vilatela, A.; Weatherup, R.S.; Braeuninger-Weimer, P.; Caneva, S.; Hofmann, S. Towards a general growth model for graphene CVD on transition metal catalysts. *Nanoscale* **2016**, *8*, 2149–2158. [CrossRef] [PubMed]

102. Lee, J.-H.; Lee, E.K.; Joo, W.-J.; Jang, Y.; Kim, B.-S.; Lim, J.Y.; Choi, S.-H.; Ahn, S.J.; Ahn, J.R.; Park, M.-H.; et al. Wafer-Scale Growth of Single-Crystal Monolayer Graphene on Reusable Hydrogen-Terminated Germanium. *Science* **2014**, *344*, 286–289. [CrossRef] [PubMed]

103. Babenko, V.; Murdock, A.T.; Koos, A.A.; Britton, J.; Crossley, A.; Holdway, P.; Moffat, J.; Huang, J.; Alexander-Webber, J.A.; Nicholas, R.J.; et al. Rapid epitaxy-free graphene synthesis on silicidated polycrystalline platinum. *Nat. Commun.* **2015**, *6*, 7536. [CrossRef] [PubMed]

104. Degl'Innocenti, R.; Jessop, D.S.; Sol, C.W.; Xiao, L.; Kindness, S.J.; Lin, H.; Zeitler, J.A.; Braeuninger-Weimer, P.; Hofmann, S.; Ren, Y.; et al. Fast Modulation of Terahertz Quantum Cascade Lasers Using Graphene Loaded Plasmonic Antennas. *ACS Photonics* **2016**, *3*, 464–470. [CrossRef]

105. Batzill, M. The surface science of graphene: Metal interfaces, CVD synthesis, nanoribbons, chemical modifications and defects. *Surf. Sci. Rep.* **2012**, *67*, 83–115. [CrossRef]

106. Wu, T.; Zhang, X.; Yuan, Q.; Xue, J.; Lu, G.; Liu, Z.; Wang, H.; Wang, H.; Ding, F.; Yu, Q.; et al. Fast growth of inch-sized single crystalline graphene from a controlled single nucleus on Cu-Ni alloys. *Nat. Mater.* **2015**, *15*, 43–47. [CrossRef] [PubMed]

107. Hao, Y.; Wang, L.; Liu, Y.; Chen, H.; Wang, X.; Tan, C.; Nie, S.; Suk, J.W.; Jiang, T.; Liang, T.; Xiao, J.; Ye, W.; Dean, C.R.; et al. Oxygen-activated growth and bandgap tunability of large single-crystal bilayer graphene. *Nat. Nanotechnol.* **2016**, *11*, 426–431. [CrossRef] [PubMed]

108. Caneva, S.; Weatherup, R.S.; Bayer, B.C.; Brennan, B.; Spencer, S.J.; Mingard, K.; Cabrero-Vilatela, A.; Baehtz, C.; Pollard, A.J.; Hofmann, S. Nucleation Control for Large, Single Crystalline Domains of Monolayer Hexagonal Boron Nitride via Si-Doped Fe Catalysts. *Nanoletters* **2015**, *15*, 1867–1875. [CrossRef] [PubMed]

109. Caneva, S.; Weatherup, R.S.; Bayer, B.C.; Blume, R.; Cabrero-Vilatela, A.; Braeuninger-Weimer, P.; Martin, M.-B.; Wang, R.; Baehtz, C.; Schloegl, R.; et al. Controlling Catalyst Bulk Reservoir Effects for Monolayer Hexagonal Boron Nitride CVD. *Nano Lett.* **2016**, *16*, 1250–1261. [CrossRef] [PubMed]

110. Piquemal-Banci, M.; Galceran, R.; Caneva, S.; Martin, M.-B.; Weatherup, R.S.; Kidambi, P.R.; Bouzehouane, K.; Xavier, S.; Anane, A.; Petroff, F.; et al. Magnetic tunnel junctions with monolayer hexagonal boron nitride tunnel barriers. *Appl. Phys. Lett.* **2016**, *108*, 102404. [CrossRef]

111. Kang, K.; Xie, S.; Huang, L.; Han, Y.; Huang, P.Y.; Mak, K.F.; Kim, C.-J.; Muller, D.; Park, J. High-mobility three-atom-thick semiconducting films with wafer-scale homogeneity. *Nature* **2015**, *520*, 656–660. [CrossRef] [PubMed]

112. Bhimanapati, G.R.; Lin, Z.; Meunier, V.; Jung, Y.; Cha, J.; Das, S.; Xiao, D.; Son, Y.; Strano, M.S.; Cooper, V.R.; et al. Recent Advances in Two-Dimensional Materials Beyond Graphene. *ACS Nano* **2015**, *9*, 11509–11539. [CrossRef] [PubMed]

113. Kidambi, P.R.; Blume, R.; Kling, J.; Wagner, J.B.; Baehtz, C.; Weatherup, R.S.; Schloegl, R.; Bayer, B.C.; Hofmann, S. In Situ Observations during Chemical Vapour Deposition of Hexagonal Boron Nitride on Polycrystalline Copper. *Chem. Mater.* **2014**, *26*, 6380–6392. [CrossRef] [PubMed]

114. Fu, L.; Sun, Y.; Wu, N.; Mendes, R.G.; Chen, L.; Xu, Z.; Zhang, T.; Rummeli, M.H.; Rellinghaus, B.; Pohl, D.; et al. Direct Growth of MoS2/h-BN Heterostructures via a Sulfide-Resistance Alloy. *ACS Nano* **2016**, *10*, 2063–2070. [CrossRef] [PubMed]

115. Jain, N.; Durcan, C.A.; Jacobs-Gedrim, R.; Xu, Y.; Yu, B. Graphene interconnects fully encapsulated in layered insulator hexagonal boron nitride. *Nanotechnology* **2013**, *24*, 355202. [CrossRef] [PubMed]

116. Barnard, H.R.; Zossimova, E.; Mahlmeister, N.H.; Lawton, L.M.; Luxmoore, I.J.; Nash, G.R. Boron nitride encapsulated graphene infrared emitters. *Appl. Phys. Lett.* **2016**, *108*, 131110. [CrossRef]

117. Petrone, N.; Chari, T.; Meric, I.; Wang, L.; Shepard, K.L.; Hone, J. Flexible Graphene Field-Effect Transistors Encapsulated in Hexagonal Boron Nitride. *ACS Nano* **2015**, *9*, 8953–8959. [CrossRef] [PubMed]

sensors

MDPI

Review

A 72 × 60 Angle-Sensitive SPAD Imaging Array for Lens-less FLIM

Changhyuk Lee [1,2], Ben Johnson [1,3], TaeSung Jung [1,2] and Alyosha Molnar [1,*]

[1]   School of Electrical and Computer Engineering, Cornell University, Ithaca, NY 14853, USA;
      cl678@cornell.edu (C.L.); bcj25@cornell.edu (B.J.); tj85@cornell.edu (T.J.)
[2]   Department of Electrical Engineering, Columbia University, New York, NY 10027, USA
[3]   Cortera Neurotechnologies, 2150 Shattuck Ave., PH, Berkeley, CA 94704, USA
[*]   Correspondence: molnar@ece.cornell.edu; Tel.: +1-607-254-8257

Academic Editor: Edoardo Charbon
Received: 31 May 2016; Accepted: 26 August 2016; Published: 2 September 2016

Abstract: We present a 72 × 60, angle-sensitive single photon avalanche diode (A-SPAD) array for lens-less 3D fluorescence lifetime imaging. An A-SPAD pixel consists of (1) a SPAD to provide precise photon arrival time where a time-resolved operation is utilized to avoid stimulus-induced saturation, and (2) integrated diffraction gratings on top of the SPAD to extract incident angles of the incoming light. The combination enables mapping of fluorescent sources with different lifetimes in 3D space down to micrometer scale. Futhermore, the chip presented herein integrates pixel-level counters to reduce output data-rate and to enable a precise timing control. The array is implemented in standard 180 nm complementary metal-oxide-semiconductor (CMOS) technology and characterized without any post-processing.

Keywords: CMOS avalanche photodiodes; highly sensitivity photodetectors; fluorescence imaging; photon timing; rangefinder; single photon detectors; SPAD arrays; time-correlated measurements; 3-D image sensor; lifetime microscopy; low power imaging; point-of-care; lab-on-chip; in-vitro; exponential decay; fill-factor; SPAD; photo-detector

1. Introduction

1.1. Overview

Fluorescence imaging has become one of major enabling techniques of modern biology, allowing sub-micron resolution of multiple markers (fluorophores) within a single sample in three dimensions, to monitor dynamic cellular behaviors. Its broader implementation, however, has been limited by fluorescent microscopy's requirement of bulky optical components, hindering low-cost, miniaturized, and/or implantable imaging systems. Conventionally, resolving fluorophores in 3D requires both (1) high precision scanning and focusing optics and (2) a set of optical filters, such as excitation, dichroic beam splitter, and emission filters to distinguish light emission from different fluorophores and also from (much brighter) stimulus light which are bulky and expensive. In this work, we propose a complementary metal-oxide-semiconductor (CMOS) compatible fluorescence imaging system as a cost effective alternative. We demonstrate a CMOS-based fluorescence lifetime imaging microscopy (FLIM) that is much cheaper and smaller than conventional FLIM systems, at the cost of reduced functional flexibility, sensitivity and 2D spatial resolution. Yet, for many applications such as an in-vivo neural interface, CMOS-based approach possesses several advantages—it can be thinned down (~10 μm) and made flexible. Furthermore, CMOS can also serve as a versatile platform to incorporate multitudes of modalities to realize a system that can faithfully monitor bio-activities expressed by fluorescence or bio-luminescence [1,2] or by electrical activities [3] with cellular level resolution. To date, lens-less imaging [4–6] has been able to expand the field of view without the use of microscope optics, but such efforts require bright-field imaging with structured illumination, which is generally incompatible with fluorescent imaging.

Figure 1a depicts a conventional fluorescent imaging microscope with bulky optical components such as filters and lenses. In this work, we propose the lensless, filterless alternative as depicted in Figure 1b. Such a system not only lowers manufacturing cost by orders of magnitude but also miniaturizes the system while still maintaining core functionality (albeit with reduced resolution and signal-to-noise ratio (SNR)). This approach is amenable to a wide range of applications poorly suited to conventional FLIM systems, including: implantable applications where small size and weight are critical, surgical and endoscopic instruments, cost-sensitive field deployments, and massively parallel assays.

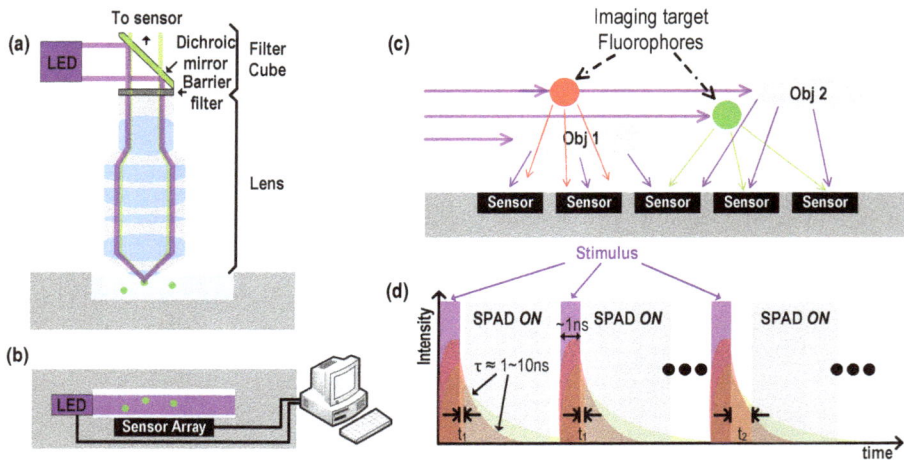

Figure 1. 3D imaging without lens and filter (**a**) conventional fluorescent imaging, (**b**) on-chip fluorescent imaging. (**c**) Stimulus light is illuminated in parallel to the sensor surface, the fluorescence emission and scattered from the imaging space (Obj1 and Obj2) reach the sensor surface. (**d**) Time-resolved fluorescent imaging for stimulus rejection and time-gated lifetime detection.

The largest and most expensive components in the aforementioned conventional fluorescent imaging setup are the stimulus light, optical filters, and lens. Stimulus light cannot be eliminated because it is the source of excitation energy; however, many efficient solutions have been proposed to reduce its cost and size, such as the use of a µLED [7] or an on-chip optical waveguides that couples solid-state excitation through the chip from a shared off-chip source [8].

1.2. Replacing Filters: Time-Gated SPAD Drive

In all cases, the excitation light source has orders of magnitude higher power (>5 mW/mm^2) than the fluorescence emission of interest (power $\sim$10 nW/mm^2), which makes it essential for the sensor to be able to significantly reject the excitation signal. The geometry by which stimulus light is projected onto a sample can also be arranged parallel to the sensor surface to minimize direct illumination of the sensor. But, as shown in Figure 1c, even when the stimulus field is aligned so as to minimally irradiate the sensor array, scattering of excitation by the sample medium itself may cause saturation due to the finite dynamic range of the sensor.

Optical band-pass/notch filters in the range of the excitation wavelength are widely used for this purpose. For an on-chip imaging configuration, however, interference-based filters are too thick, and provided limited rejection to wide-angle illumination from scattering, while relatively thin ($\sim$30 µm) absorption-based filters have an insufficient rejection ratio of $\sim$30 dB [9]. Thus, an alternative rejection technique is required that is more compatible with an integrated CMOS solution. An alternative to wavelength specific filters is to use a synchronized pulsed stimulus and high speed solid state circuitry to time-gate sensing to times when only fluorescent emission is present: fluorophores decay from an excited to ground state with a half-life on the order of nanoseconds, meaning that significant numbers of fluorescent photons may be detected after the end of a brief stimulus pulse as depicted in Figure 1d. Single photon avalanche diodes (SPADs) co-integrated with high speed CMOS electronics are ideally suited to this functionality.

1.3. Fluorescence Lifetime Imaging

In order to distinguish between fluorescent markers, or detect dynamic changes in their fluorescence due to physiological signals such as calcium concentration, one must resolve changes in the intensity, wavelength and/or temporal statistics of fluorescent emission. Intensity based fluorescent imaging of physiological signals uses changes in fluorescent emission intensity to detect changes in biological state or events in the interior of cells. Unfortunately, accurate measurement of fluorescent emission intensity can be quite challenging due to randomness of a probe concentration and changes in light path loss in dynamic, scattering medium. Wavelength-based approaches, and especially ratio-metric wavelength measurements normalize these effects, providing much better results by providing the proportion of fluorophores in a given state (i.e., bound vs. unbound to calcium). Such approaches, however, require multiple, closely spaced wavelength filters. Like a wavelength reporter, a fluorescent lifetime reporter is independent of the probe intensity, but instead of the wavelength of emission, lifetime imaging is sensitive to the nanosecond-scale changes in emission intensity after excitation. As with wavelength measures, lifetime measurement enable quantitative ratio-metric measurements of fluorophore state, but rely on high-speed sensors and circuits rather than selective optical filters. Extensive search for lifetime probes is on-going and expanding to enable sensing of changes in Ca^{2+}, Mg^{2+}, K^+ concentration as well as pH [10–12]. In FLIM, fluorophores are distinguished by their decay constants (τ's) rather than by their wavelengths (color) or absolute intensity. Previous work has demonstrated FLIM with arrays of SPADs in standard CMOS with standard microscope optics [13,14]. Lens-less FLIM with SPAD arrays has also been demonstrated [15], but it did not have the ability to localize fluorophores in 3D space and had limited 3D resolution for objects not in direct contact with chip surface.

1.4. Replacing Lenses: Angle Sensitive Pixels

Due to the absence of collimation of imaged light, spatial resolution when imaging without focusing optics is inherently inferior to that possible with a conventional focusing system. Nevertheless, significant progress has been made on bright field lens-less imaging, where two techniques are most common. The first is *contact mode imaging* which minimizes light spreading between imaged object and sensor array by placing them close together and collimating the illumination. To compensate for finite pixel size, this method repetitively scans the imaging object while shifting either position of the imaging object relative to the sensor, and/or shifts the position/direction of the light source relative to the object [5] which also can provide parallax for 3-D imaging. The second approach utilizes partially coherent illumination and measures interference patterns between non-scattered and scattered illumination through the imaging object [16]. Unfortunately, neither of these techniques are suitable for on-chip volumetric fluorescence lifetime imaging, where all detected light is emitted isotropically.

One way to resolve this drawback is through lens-less light field imaging, where the image sensors resolves incoming light rays in both space and incident angle. Light-field imagers have shown promising results by utilizing computational re-focusing [17], lens-less far-field imaging [18], and on-chip imaging [19]. Among approaches to light-field capture, angle-sensitive pixels (ASP) provide an easily integrated, effective solution by extracting angle information about the rays arriving at each pixel.

ASPs uses a set of two μm-scale gratings to create an incident-angle-dependent diffraction patterns and then filtering the light based on the offset of this pattern, as shown in Figure 2 [17]. The grating structure is compatible with a standard CMOS process and does not require any post-processing steps such as mounting micro-lenses/lens on the image sensor surface which are required for a light-field imager [20].

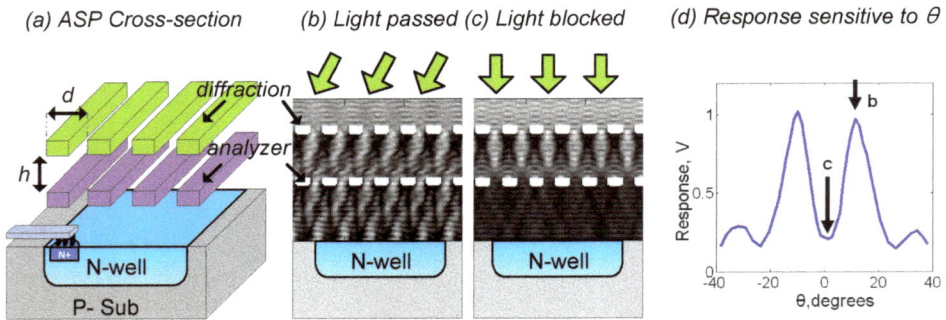

Figure 2. (**a**) Structure of simple angle sensitive pixel (ASP) [17] showing metal gratings and photodiode, grating pitch, *d* is order 1 μm, while vertical separation, *h* is order 3 μm, (**b**,**c**) finite-difference time-domain (FDTD) simulation of field intensity in response to incident plane waves of light: (**b**) diffraction pattern from top grating aligns with gaps in bottom grating, light passes, (**c**) pattern aligns with bars, light is blocked. (**d**) simulated intensity at detector as angle is swept. FDTD simulation is performed using MATLAB. Inter connect metal gratings (Al) is modeled as perfect electrical conductor (PEC) and the inter dielectric layer (SiO$_2$) is assumed lossless. The incident angle of the plane wave are swept.

1.5. Angle Sensitive Time Resolved Fluorescence Lifetime Imaging

This work combines area-efficient, time-gated SPADs with CMOS-compatible integrated optical structures, similar to those used in ASPs [17–19] to replace conventional optical filters and lenses. The 2-D incident angle information not only improves non-contact 2-D spatial resolution of light sources above the plane of the chip but also expands the system's imaging resolution into 3D for volumetric localization. Here we present a 72 × 60, angle-sensitive SPAD (A-SPAD) array fabricated in a conventional 180 nm CMOS process. The pixels temporally reject high-powered UV stimulus pulses [21] while successfully performing 3D localization of different fluorescent sources with different lifetimes through reconstruction utilizing angle information and lifetime measurements, all without the use of lenses or wavelength filters.

In the next section (Section 2) we describe the design of the angle-sensitive SPAD image sensor. Section 3 provides a system-level overview of the architecture of the image sensor and associated challenges. Section 4 shows experimental results demonstrating the function of our image sensor and supporting circuitry. Finally, Section 5 describes the 3D reconstruction algorithm tailored to the unique challenges associated with lens-less 3D FLIM imaging and summarize our work.

2. Angle-Sensitive Single Photon Avalanche Diode

2.1. A-SPAD Pixel Structure and Circuitry

Figure 3 shows a cross-section of A-SPAD structure. The SPAD is formed by the junction between the P+ implant and N-Well with P-epi layer used as a guard ring [13,21]. The two sets of metal gratings for ASP are implemented using conventional CMOS metal stacks, arranged to modulate light based on azimuth and altitude incident angles (θ and ϕ) [17]. As light strikes the top grating, an incident angle dependent diffraction pattern is generated beneath it, similar to what is shown in Figure 2b,c. The lower, analyzer grating, then selectively blocks or passes the generated diffraction pattern depending on relative lateral position. The response of an angle-sensitive pixel to light of intensity I_0 and incident angle θ is proportional to $I_0 \cdot (1 + m\cos(\beta\theta + \alpha))$, where modulation depth m, angular frequency β, and angle offset α are design parameters dictated by the geometry of the gratings (d and h in Figure 2). In order to provide maximum 3D information, our array uses six types of A-SPAD pixels, resulting from combination of two angular frequencies ($\beta = 8, 15$) [22] and three phases ($\alpha = -180°, -60°, 60°$). Note that prior work in ASPs used four phases (differential sampling of I and Q: $\alpha = 0°, 90°, 180°, 270°$), but only resolved three parameters describing incident angle distribution: sinusoidal phase and amplitude, and a background offset, thus conventional four phase sampling has one phase redundancy. To remove the redundancy and increase the spatial sampling efficiency we used three phase sampling where instead of merely removing single phase from the quadrature sampling all three samplings are evenly distributed in phase ($\alpha = -180°, -60°, 60°$) which simplifies the amplitude computation: a numerical average of three output. Both the horizontal and vertical grating orientation are required to make the A-SPAD array sensitive to angle in both x-z or y-z planes, θ and ϕ respectively. Thus the overall array is made up of tiles of 12 A-SPADs each.

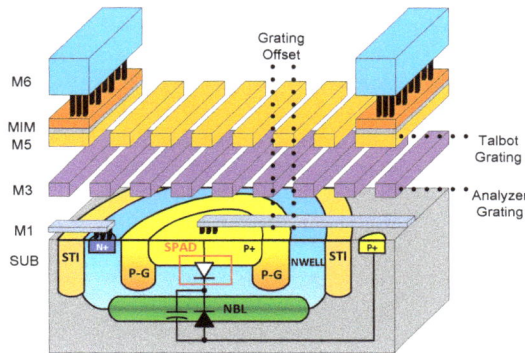

Figure 3. A cross-section of the proposed angle-sensitive single photon avalanche diode (A-SPAD) structure. Reproduced from [22], with the permission of AIP Publishing.

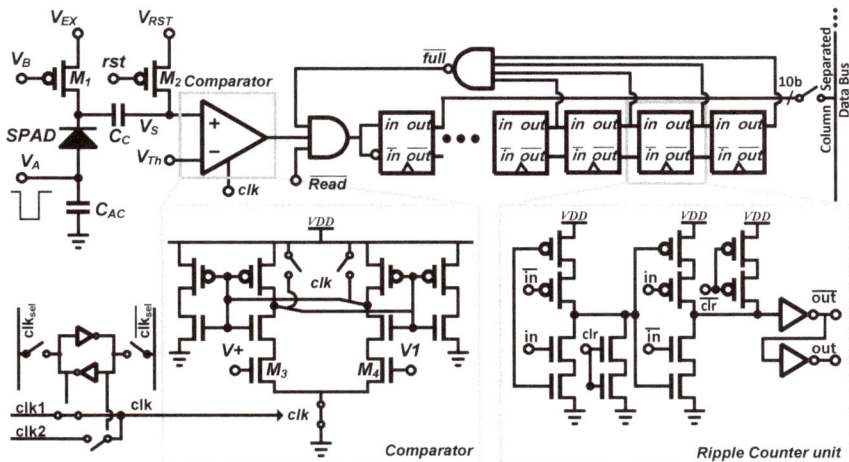

Figure 4. Pixel level schematic of A-SPAD.

Figure 4 shows the schematic of a single A-SPAD pixel with circuitry for time-gated passive quenching (PQ) using transistor M_1, an AC-coupled comparator to detect photon arrival, and a 10-bit ripple counter for Time Correlated Single Photon Counting (TCSPC). For the time-gated operation, every SPAD's anode is connected to a global off-chip driver, V_A, which drives the SPADs beyond their breakdown voltage (V_{BD}) into Geiger mode (ON) by the excess voltage ($V_E = V_A - V_{BD}$). V_A is synchronized to the timing of the stimulus light source such that the SPAD remains insensitive ($V_A < V_{BD}$) until after UV excitation, when it is turned "ON". Detection of a photon results in a step voltage drop across the SPAD by V_{EX} which couples through a MIM capacitor (C_C) to the V_S node, on the input of the comparator.

2.2. Readout Circuitry: Comparator and Counter

The sensitivity of the SPAD is increased as the reverse bias rises above the (V_{BD}), as is dark count rate. Thus, the globally controlled reverse bias needs to be adjusted in order to regulate and optimize array-wide photon counting to avoid desensitization and saturation. A fixed voltage pulse detector such as an inverter [13,21] has a lower limit on pulse amplitude detection when the supply voltage is fixed, and lowering supply voltage degrades temporal precision in the detection. Therefore a differential comparator, rather than an inverter, was implemented for pulse detection in this work. The schematic of the comparator is shown in Figure 4. The comparator's detection threshold (V_{Th}) can be adjusted to track the coupled voltage step on V_S which is linearly dependent on the excess bias voltage (V_E). The freedom to control threshold voltage also allows compensation of the process,voltage, and temperature (PVT) variation in SPAD breakdown voltage (V_{BD}). The device size of the input pair in the comparator is designed to minimize DC offset($\sigma_{V_{th}} < 4.76$ mV) and node capacitance to minimize detection uncertainty and quenching current. Furthermore, a positive feedback (latch) load with reset switch provides high gain to reduce timing uncertainty.

Data readout of SPAD array can be approached in several ways: (1) a single-bit readout can be stored in a single bit memory in pixel [21,23], (2) a single bit multiplexed to a shared off-pixel time interval counter (TDC) [24] or (3) a multi-bit type that uses an in-pixel TDC [14], (4) a multi-bit counter/analog integrator that accumulate photon detection events. The first type is efficient for a small scale arrays and provides good fill factor but requires all SPADs be read out after every excitation pulses limiting frame rate as the number of pixels increases. The second type achieves similar fill factor, and much better single-frame time resolution, but can only capture from a small subset of SPADs at once, requiring many excitation pulses to capture a full frame. The third approach allows simultaneous reading from many SPADs, but at a cost in fill factor, and still requires high data bandwidth to sustain steady frame rate [25]. The last approach, which is implemented in the proposed imager, has a reduced data bandwidth thanks to an in-pixel counter/integrator however at the cost of reduced fill factor [26,27].

The data bandwidth is especially critical in a moderate or low figure imaging environment, since acquiring an accurate time-histogram requires averaging over a large number of measurements. The minimum number of photon detections necessary for fluorescence lifetime measurements (monoexponential decay analysis) can be derived [28], and a typical estimation of fluorescence lifetime histogram requires averaging over 500 photon detection events [24]. At the same time, the average photon detection activity must be controlled to be on the order of 1% $\sim$ 5% to avoid non-linearity due to photon pile-up [29] or local saturation/desensitization. This means that with 1% average photon detection activity, over 50,000 measurements are required to produce a single lifetime histogram [24]. If such TDC measurements were to be exported off-chip for a large scale array, the data bandwidth requirement can easily become impractical. For instance, full histogram reconstruction of a 72×60 system with 10 bit TDC and 10ns laser repetition rate, at worst case, would require an approximate bandwidth of 540 GB/s ($=10 \times 72 \times 60/10$ ns/byte). This is even more true for A-SPADs, since meaningful angular information (which is derived from diffraction, and so the wave-description of light) can only be extracted by measuring the average of discrete photon detection events (derived from the particle description of light) [22]. On-chip data compression or decimation may partly resolve this issue (event-driven readout [30], embedded FIFO [31]), but still suffers from inefficiencies due to moving inherently redundant data from the array core to the compression circuitry [24]. For the A-SPAD array, described here, which operates in a low figure environment while collecting extra dimensions of information, we took the simple yet efficient alternative approach of pixel level averaging combined with tightly controlled detection windows. This allows the read-out rate from each pixel to be reduced by the number of measurements being averaged, while increasing the number of bits to be read out by $\log_2$ ($\sharp$ measurements $\times P_{activity}$) where $P_{activity}$ (a probability of positive output) is limited to ~1% to avoid pileup. For example, with 10 bit in-pixel accumulator and ~1% activity rate we can increase the number of averaged measurement up to 100 ksamples before the in-pixel memory over flows.

To increase the frame rate and provide sufficient figures with pixel level averaging, a 10 bit asynchronous ripple counter was used to integrate output of the comparator over multiple excitation pulses. As shown in Figure 5, the counter consists of tri-state inverters with cascaded asynchronous reset switches and output buffers. Using a fixed window accumulation method significantly reduced the overall data rate compared to the TDC method, albeit requiring more total frames (as much as a factor of TDC bit resolution when time-gate does not overlap) to generate a histogram with equivalent SNR (discussed in Section 3).

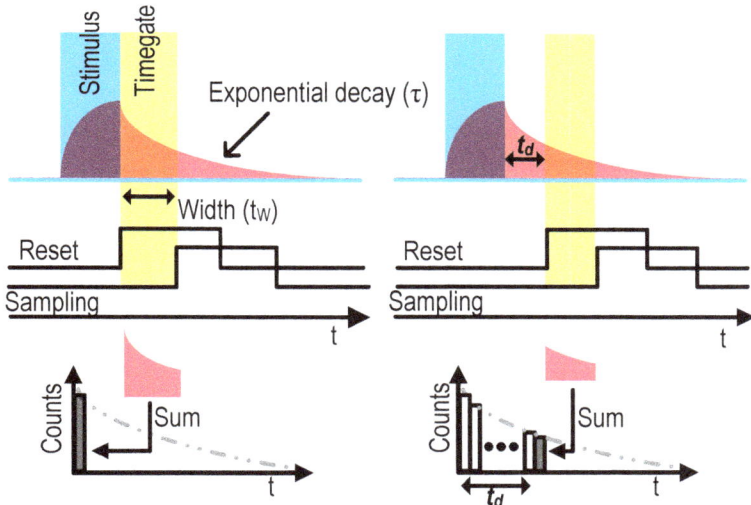

Figure 5. Statistical sampling with detection window width (t_W), detection window shift (t_d), lifetime (τ) for lifetime estimation. Note that the proposed system is able to achieve $t_W = t_d \geq 72$ ps whereas, $t_W \sim 2.4 \times \tau$ is used in this work.

A layout of two A-SPAD pixels (θ and ϕ sensitive) are shown in Figure 6. A fill factor of 14.4% (15 μm N-well diameter) was achieved through a simple pixel-circuit design, which is dominated by a 10 bit ripple counter and overlaid MIM capacitors which double as a light shield.

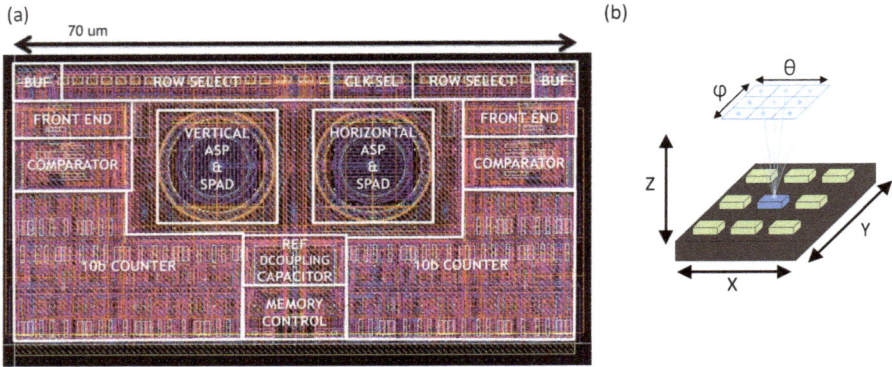

Figure 6. (**a**) Layout of two (θ and φ sensitive) A-SPAD pixels. (**b**) illustration of θ and φ.

3. Large Scale A-SPAD Array Design

3.1. Power and Area Efficient Lifetime Estimation Approach

FLIM depends upon the ability to resolve fine time differences in photon arrival time statistics (≤1 ns). Conventional methods for lifetime estimation can be classified into two categories: (1) time gating (TG) method using a digital counter [32–35], and (2) TCSPC method using time-to-digital converter (TDC) [14,24,25,36].

Time-gating methods are efficient when a large number of samples are needed to provide *"analog like"* intensity information [37]. When using time-gating methods, algorithms such as rapid lifetime determination (RLD) and its variants such as overlap RLD (ORLD) and center-of-mass method (CMM) based RLD are widely used for their simplicity. Unfortunately, this method extracts the average lifetime of multi-exponential decays when they coexist, introducing fixed pattern noise/extension to the estimation [38], and the gate position and interval have to be adjusted based on assumption of lifetime. The TCSPC method combined with an in-pixel TDC (minimum of eight time bins [38]) is close to ideal for producing an unbiased histogram, but comes at the cost of drastic increase in hardware complexity, area, power consumption and high I/O data bandwidth for off chip curve fitting.

A real-time time gating method requires a minimum of two, four and eight bins to estimate single-exponential, bi-exponential and multi-exponential decays, respectively. Although the multi-exponential decay time constant can be extract by small number (eight) of time bin, high temporal resolution raw histogram data is often preferred [35]. For the proposed design, we use a hybrid of the two methods. We use a time gating method for its simplicity and compactness of circuitry, but introduce a dynamically programmable time shift on the gate, with fine temporal accuracy (~100 ps), to enable sweeping of the time-gate, enabling extraction of multi-exponent demixing. Specifically, we repeat the laser excitation and measurement cycles many times for a given gating time offset, counting the photon detection within the time-gate. Lifetime estimation is acquired by shifting the time-gate with fine temporal resolution (>72 ps) relative to the laser pulse. Here the resulting photon count creates an average sum of inhomogeneous Poisson random variables over the gating time. The accumulated count within the detection window can be interpreted as an integration of the combined probability density functions of all light sources illuminating the SPAD (See Figure 5). The time-gate is formed with three global timing references generated by on chip digital-to-time converters (DTC). Short time-gate width (t_w) can provide simple and intuitive histogram for lifetime estimation where the proposed system can reduce the width as short as DTC LSB (>72 ps); however, increases required number of stimulus and sampling to achieve reasonable SNR. Thus, instead of using the minimum width (LSB), we have used a 10 ns time-gate width analogues to the optimal time-gate width of the RLD method for $\tau \geq 4$ ns where $t_w \geq= 2.4 \times \tau$ [35]. The start of the detection window is applied to the gate of the p-FET (M1) reset switch. When it is high, the positive input of the comparator (V_S) sees a high impedance and follows the AC coupled avalanche induced voltage drop from the SPAD. The end of the time-gate is defined by the comparator clock.

3.2. Digital-to-Time Converter (DTC)

The fine resolution time-gate shifting and counting can benefit from a multi-channel synchronous clock reference on chip. For continuous imaging, we used a rolling shutter based technique to read out *"analog like"* accumulated digital counts from each pixel. In a conventional rolling shutter, each sensor's data accumulation occurs synchronously with a global time-gated detection window, and the number of photons counted per pixel can be controlled by the number of cycles between memory reset and read. The time histogram, as shown earlier in Figure 5, can be acquired by using a sliding time-gated window. Efficiently combining a sliding window with rolling shutter requires pixel-by-pixel recording and readout, where each pixel, as it is read out and reset, is switched from one timing window to the next. This requires two precisely controlled windows, which are generated using digital-to-time converters (DTC).

Figure 7 shows the schematic of the DTC used in this work. For minimal area and optimal power efficiency, the open-loop delay-line utilizes a ring oscillator instead of a long single line or a binary weighted programmable delay. For a synchronous open-loop operation, a single reference clock is provided from off-chip. Both positive and negative edges of the reference clock are extracted by positive/negative edge detector which initiates a positive/negative edge ring oscillator whose output is connected to three delay lines to set the relative timing between the excitation pulse and the two time-gates for rolling shutter operation. Each delay line has a coarse control (5-MSBs) that selects the number of cycles around the oscillator. The oscillator consist of 16 differential delay units, and nodes between each delay unit are connected to a thermometer-coded multiplexer which feeds each counter to provide 5 LSBs of finer control. Each delay unit is a current mode differential exclusive or (XOR) logic gate whose tail current can be adjusted to control the LSB delay (<80 ps). To counter the effects of changing the tail current, the p-FET biasing is maintained by an off-chip negative feedback loop. This structure allows flexible and accurate control of both duration and delay of the detection window relative to the reference. Furthermore, to conserve power, it is possible to power down the ring oscillator after all three time-gates are generated by putting it in a low power (En_{LP}) mode.

Time-gates generated by the DTC, in conjunction with an in-pixel clock pointer and a single bit static random-access memory (SRAM) to select the on-going detection window, are used to implement the rolling shutter scheme. A pixel readout is followed by thecounter memory reset.

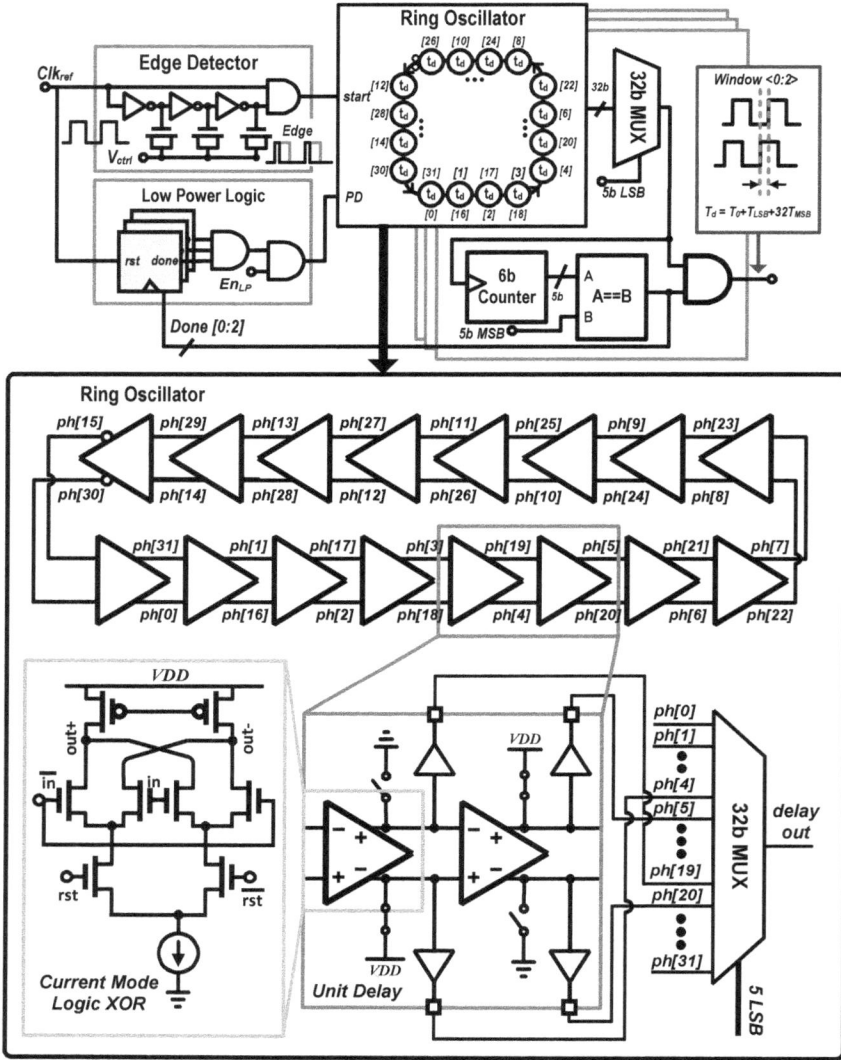

Figure 7. Schematic of high dynamic range, area efficient DTC, where *ph[0:31]* are nodes between each delay unit, the supply voltage *VDD* is 1.8 V, *rst* denotes reset signal to stop the ring oscillator (RO), *Done[0:2]* is a digital bit to power gate the RO to enable the low power DTC operation.

3.3. System Architecture

Figure 8 shows the architecture of imager system. The system synchronizes an off-chip UV stimulus to the internal delay lines which in turn generate detection windows. Other supporting circuits around the A-SPAD imaging core include a column and row decoder for pixel access, data MUX for a readout, SPI for serial communication to off-chip FPGA, the two DTCs with three outputs, and a system state machine.

This simple system architecture is robust; during three-months of continuous testing, the system reliably operated without a system reset.

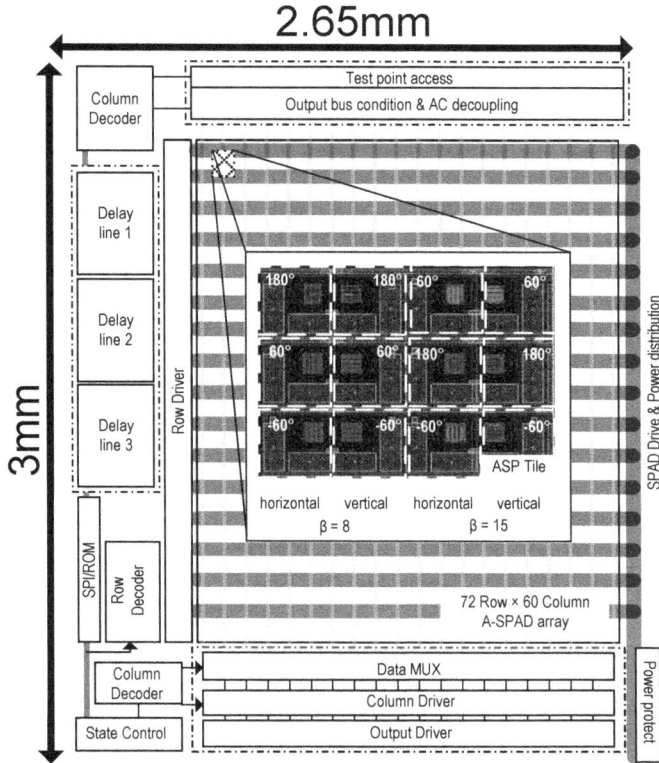

Figure 8. System architecture of a 72 × 60 A-SPAD array.

The Photon detection probability (PDP) for an A-SPAD (with Talbot and analyzer gratings) is 2.72% at 540 nm with V_{EX} = 1.2 V whereas SPAD without gratings resulted in PDP of 18.7 % indicating a factor of 7 loss due to the reflection from gratings [22]. The array wide (72 × 60 pixels) median dark count rate across the array was 404 Hz [22]. Including local the comparator and the 10 bit ripple counter, the pixel pitch is 35 μm × 35 μm. The chip achieves a fill-factor of 14.4% across the active area of the array, and 9.6% including support circuits and pads. The system consumes 73.8 mW or 83.8 mW power in dark or tested illumination conditions respectively.

The sensitivity, or detection efficiency, is one of the photon detector's (SPAD) most important performance factors. Numerous studies have worked to enhance the quantum efficiency or/and fill factor of SPADs [39]. However, in many cases, the sensitivity measurement is performed at the single device level and underestimates the various challenges that may arise in a large scale array. A notable example is fluctuations in local sensitivity caused by the large avalanche current pulse of adjacent pixels, which is a binary response containing very sharp edges. To put a SPAD into Geiger mode (ON), its reverse bias ($\geq$ 10 V) is supplied from off-chip due to the complexity and limitations of generating such high voltage using standard CMOS process. Because the various interconnects, including wire-bond (L_{bond}) and on-chip routing, introduce parasitic resistance and inductance, they present high impedance to the high frequency components of the avalanche current and cause local voltage drops. Despite the use of wide traces for current return path and multiple wire-bonds (6 total) to stabilize V_A, avalanche events inevitably cause local and global fluctuations. Since reducing V_A reduces sensitivity these fluctuations in V_A modulate pixel sensitivity across the array and degrade the overall image quality.

To evaluate the effects of internal bias line impedance on the avalanche count of a single pixel, an array scale simulation illustrated in Figure 9a was performed using a simple lumped pixel model in Figure 9b. The integrated SPAD array may experience global transient drops in bias voltage when one or more SPADs are triggered in a short time window. The combined avalanche current through finite bond-wire impedance lowers V_A, reducing sensitivity and limiting the number of photons that can be detected in a short period of time. Simulation result of the reverse bias V_A variation is depicted in Figure 9c. The local voltage drop from the avalanche current couples to the reverse bias of adjacent SPAD pixels, desensitizing them, and so distorting or even saturating their response. To mitigate this effect, several steps were taken; six parallel wire-bonds that supply V_A were interleaved with current return path bonds to minimize effective inductance, and each pixel contained a decoupling capacitor ($C_{AC} = 375$ fF/SPAD) to absorb a portion of the avalanche transient current. As a result, the amplitude of the local V_A drop was reduced by a factor of 2.5 as shown in Figure 9c. The decoupling capacitor (two series connected MIM) and AC coupling MIM cap C_C also serve as optical shield for active circuitry against any undesirable photo-current generation, especially suppressing leakage from dynamic nodes in the in-pixel counter.

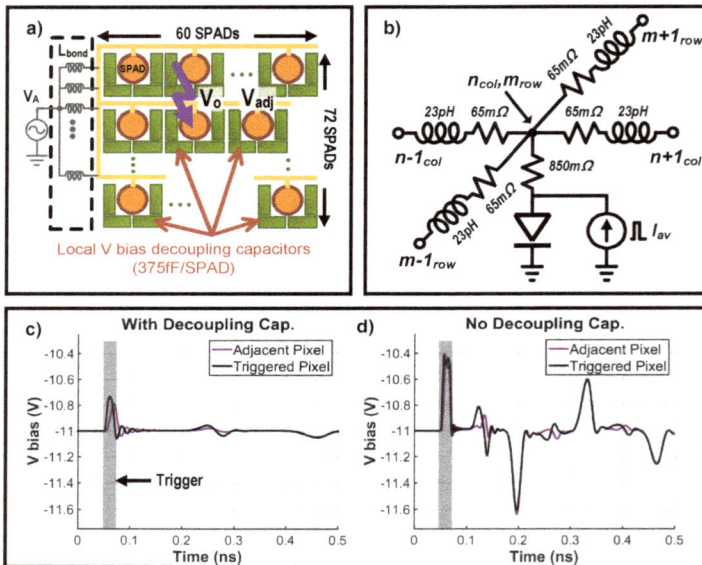

Figure 9. (a) Schematic of SPAD reverse bias network with local decoupling capacitors, (b) SPAD bias network impedance model. (c) Simulated SPAD reverse bias voltage with decoupling capacitor (d) Simulated SPAD reverse bias voltage without decoupling capacitors where red and black lines depict avalanche pixel and neighboring pixel respectively.

4. Results

4.1. Measurement Setup

Figure 10 shows the micro-photograph of the A-SPAD FLIM integrated circuit (IC), fabricated in a standard 180 nm CMOS process. The IC was interfaced with a field-programmable gate array (FPGA) (Cyclone II, Altera, San Jose, CA, USA) for serial peripheral interface (SPI) control, and 20 low-voltage differential signaling (LVDS) (2 × 10 bit) output channels were connected to a high speed digital data acquisition module (NI PXIe-6555, National Instruments, Austin, TX, USA). The lifetime histogram and image reconstruction were post-processed in MATLAB.

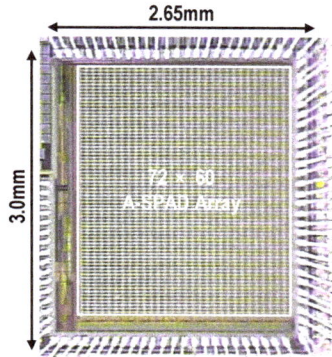

Figure 10. Micro photograph of the A-SPAD FLIM IC.

4.2. Angular Sensitivity Response

A-SPAD's response to change in incident angle (θ, ϕ) was measured by rotating the sensor surface relative to a collimated beam of light generated by green LED (λ = 540 nm). Figure 11 shows measured angle response curves for all six A-SPAD pixel types. Again, this is a combination of three different phases ($\alpha = 60°, 180°, -60°$) and two spatial frequencies ($\beta = 8, 15$) where $\beta = 8$ in Figure 11a refers to low frequency (sparse grating) and $\beta = 15$ in Figure 11b refers to high frequency (dense grating) pixels. The agreement between measurement (accumulation of single/few photon measurements) and simulation in Figure 2 where binary SPAD readout represents particle-like behavior of light and the FDTD simulation represents wave-like behavior of light complies with the particle/wave duality of light and verifies that A-SPAD is capable of detecting angle dimensions, making its sensing dimension a total of 5: X, Y, ϕ, θ, and time [22].

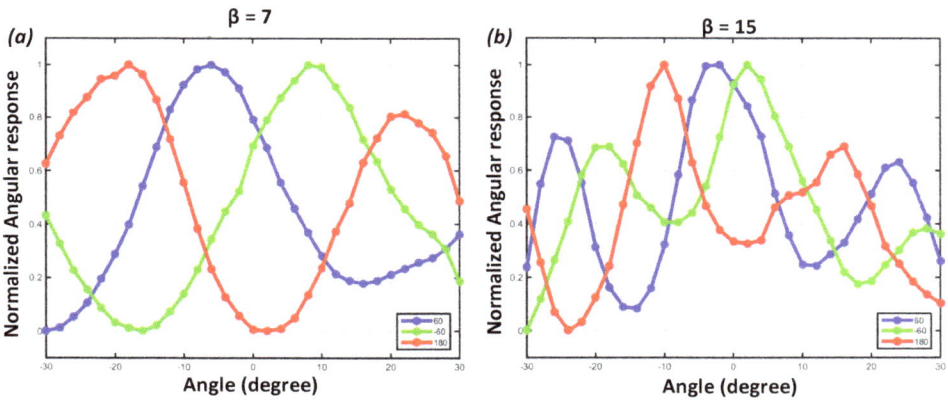

Figure 11. Measured A-SPAD responses to change in incident angle. (a) $\beta = 8$; (b) $\beta = 15$.

4.3. Digital to Time Converter (DTC) Performance

Static performance was measured for each of the three global DTCs. They were independently provided with a 20 MHz external reference clock (equivalent to highest laser repetition rate), and the delay between input reference and DTC output was measured with a 350 MHz Universal Counter/Timer (Agilent 53230A, Agilent Technologies, Santa Clara, CA, USA). The outputs of the DTC were connected to test mode I/O pads to characterize the linearity (integral nonlinearity (INL) and differential nonlinearity (DNL)). Ten measurements were collected and averaged for each delay code. The LSB delay can be adjusted by controlling the tail current of the XOR delay unit shown in Figure 7. The change of voltage swing of the current mode logic (CML) due to change in tail current was compensated by an off-chip voltage regulator [40] and pFET load. The CML voltage swing was set to 1.6V minimize jitter and the LSB delay to 72 ps. Figure 12a,b show the resulting DNL and INL of the DTC. The maximum measured DNL is less than 0.54 LSB, and INL is less than 1.27 LSB. The measured high jump around the MSB switch of INL is similar to our layout extracted simulation results, which implies that this abruptness may have been caused by the parasitic capacitance in layout routing. The root-mean-square (RMS) jitter averaged over the full dynamic range was 13.2 ps, and power consumption was 24.1 mW with 1.8 V supply voltage.

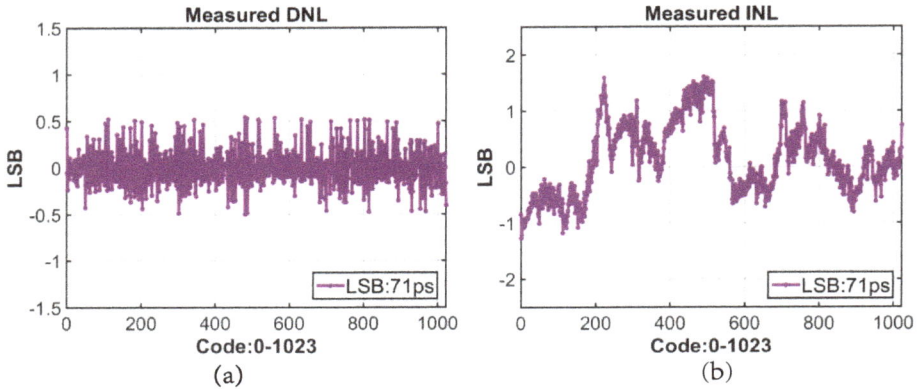

Figure 12. (a) DNL of the DTC; (b) INL of the DTC.

5. 3D Localization Using FLIM

5.1. Measurement Setup

Two sets of measurement were taken: 3D localization based on angle sensitivity and lifetime information, and fluorescence lifetime extraction. Figure 13 shows the setup for both measurements. For each measurement, images were taken with two fluorescent micro-spheres suspended above the A-SPAD array. For 3D localization, two types of solid fluorophore grains—Muscimol Bodipy TMR-X conjugate and Fluoresbrite YG—with diameters around 100 µm were used. For fluorescence lifetime measurement, two types of quantum dot (QPP-450 and QSP-560, OceanNanotech, San Diego, CA, USA) were cured in epoxy to emulate/encapsulate imaging targets.

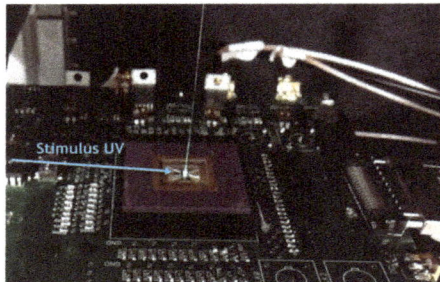

Figure 13. Measurement setup for on-chip 3D localization and fluorescence lifetime extraction.

The imaging targets were made as small as possible to generate response close to point spread functions. For each case, the two fluorescent beads were positioned above the A-SPAD array using a micro-manipulator. The stimulus light was provided by a UV (λ = 385 nm) laser diode (DeltaDiode, HORIBA Scientific, Kyoto, Japan) with an average optical power of 5 mW/mm². For each case, the fluorophores were stimulated by 50 ps wide pulses at a rate of 2 MHz by UV beam parallel to the plate.

5.2. 3D Localization Based on Angle Sensitivity and Lifetime Information

Prior to performing 3D localization, the array-wide angular response to a point light source was verified. This response was first computed considering the total 12 configuration (six pixel types in *X-Y* alignment. See Figure 8) per tile of the A-SPAD array. To measure the actual response, an LED light passing through a pinhole was placed far away from the array to generate a beam from a point light source. This beam was passed through a high aperture lens placed on top of the array so that the resulting light spread would look like it originated from a point source. To extract an angular response, average intensity data of three different phase samplings, is subtracted from each individual pixel response. Figure 14 shows the comparison between computed and measured angular response of the array.

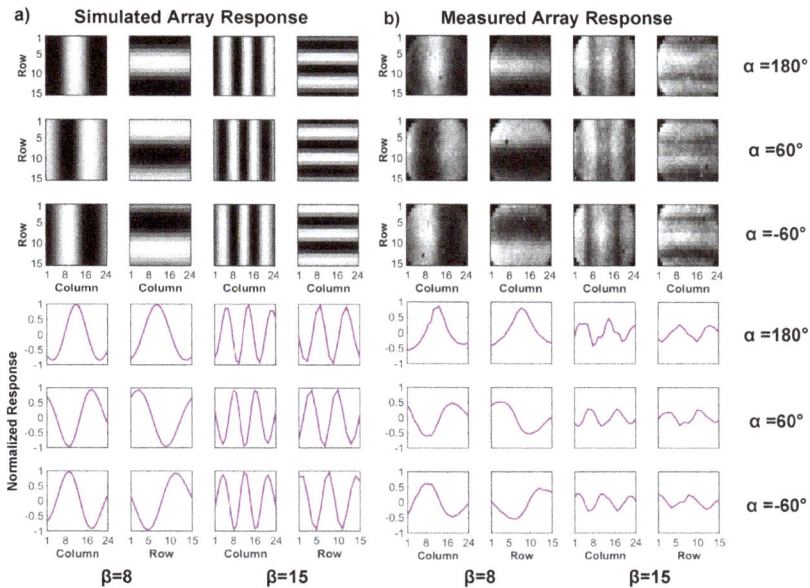

Figure 14. (**a**) Ideal array-wide angular response to a point source with computed *A* matrix; (**b**) Actual angular response to a point-like light source. Note: Ideal angular response is derived from $I_{out}(\theta) = I_0(1 + m\cos(\beta\theta + \alpha)) \times F(\theta)$, where m = 1 [22].

Localization was performed based on the array-wide angular response, with intensity information (from the average response of ASPs with different angular sensitivities) normalized out. This did not degrade the accuracy of 3D localization suggesting that there is more information in incident angle than intensity data. Furthermore, suppressing intensity information provides better tolerance against measurement artifacts such as fixed pattern noise.

3D localization algorithm was implemented by extracting the sparse set of source locations ($\vec{x}$) that best explain the A-SPAD outputs ($\vec{y}$). When an output of the array $\vec{y}$ and its back-projected location $\vec{x}$ have matched dimensions, *A* becomes an invertible matrix, leading to a definitive localized solution

$\vec{x} = A^{-1}\vec{y}$. However, due to the dimension mismatch between two vectors $\vec{x}$ and $\vec{y}$, ($\vec{x}$ describes a volume whereas $\vec{y}$ denotes a plane) the mapping matrix A is not square and therefore not invertible. Fortunately, localizing fluorophore in 3D can be formulated as a quadratic program ($\parallel \vec{y} - A\vec{x} \parallel$) when the number of fluorophore clusters are limited to a low number (i.e., they are sparse). Figure 15 illustrates the overall process as well as specific dimensions of $\vec{x}$, $\vec{y}$, and A.

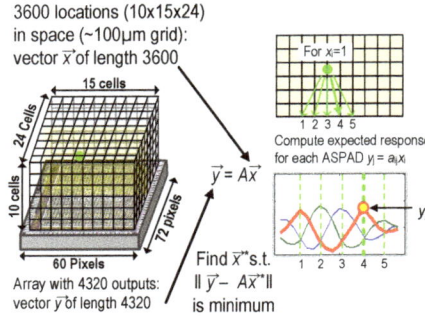

Figure 15. Illustration of 3D localization method based on angular response.

To be more specific, our localization algorithm made use of pseudo-matrix A^+, combined with iterative search similar to the bootstrapping method. Starting with initial response $\vec{y}$, likely positions of fluorophores, $\vec{x}^*$, were found with back-projection $A^+\vec{y}$ and thresholding. Then, $\vec{x}^*$ was multiplied with A to generate a simulated response $\vec{y}^* = A\vec{x}^*$ where entries in $\vec{x}*$ that lead to a large discrepancy between $\vec{y}^*$ and $\vec{y}$ were discarded. Finally, $\vec{y}^*$ was used to re-start the reconstruction cycle by making a second guess on likely positions of fluorophores. Such iterations eventually lead to a converged solution that reconstructs the fluorophore positions. In addition, we have controlled the fluorophore movements with micro-manipulators, which of resulting movement vectors are then compared to the reconstructed images to confirm the validity of estimation.

Compared to lensed systems, the proposed system here suffers from reduced lateral resolution: a lensed system is limited to a lateral resolution set by pixel size and magnification, and or by the diffraction limit of the light itself (100s of nm). The lensless approach described here, does not benefit from magnification, and so for objects close to the chip is limited by the pixel pitch itself (35 μm). For more distant objects, this resolution is limited by the effective angular resolution of the A-SPADS, which is, in the angle domain, $\delta x, y = \frac{\pi}{\beta_{MAX}}$ and translates to $\frac{2\pi \times z}{\beta_{MAX}}$ in the spatial domain, where z is the vertical distance from the chip to the plane being resolved. Vertical resolution is this same term, scaled by the cosine of the angular aperture (θ_{app}) of the A-SPADs: $\delta z = \delta x, y \times cos(\theta_{app}/2)$. Compared to a typical lensless contact imager, this resolution is improved by a factor of $\frac{2\times\pi}{\theta_{app}}$. Thus, resolution wise, the proposed approach falls between full, high resolution lensed systems, and typical lens-less systems. While a quantitative study to analyze key metrics of the reconstruction such as spatial resolution, estimation error, robustness, convergence rate, effect of scattering, number of sources and computational overhead in a greater detail is certainly needed to better understand the system and such performance parameter is inevitably a strong function of algorithms which of details will merit a paper of its own [41].

For this measurement, a 10 ns wide time-gate was shifted in steps of 139 ps, corresponding to two LSB bits of DTC, with 3D localization performed for each time step. The distinctive lifetimes of the two fluorophores—Muscimol Bodipy TMR-X (red) and Fluoresbrite YG (green)—make the solution sparser and easier to reconstruct. The reconstructed peak location for each shifted time-gate are shown in Figure 16. The distinct temporal responses of the two voxels whose response was above threshold reconstruction correspond to each lifetime of the two fluorophores. The global sensitivity which is a function of the SPAD reverse bias was adjusted to meet the SNR requirements to localize the

Muscimol Bodipy TMR-X conjugate (Red voxel) due to the poor absorption efficiency at single photon excitation wavelength (385 nm). As a result, the laser power and the array sensitivity was excessive for the Fluoresbrite YG microspheres (Green voxel), resulting in partially saturated response from many SPADs avalanching at once due to photons from the Fluoresbrite YG. This can be alleviated by simply adding an additional excitation source where each of fluorophore types can be more efficiently stimulated. Nevertheless, even with single excitation source a 3D location of two distinct fluorophore voxels are accurately reconstructed as shown in Figure 17.

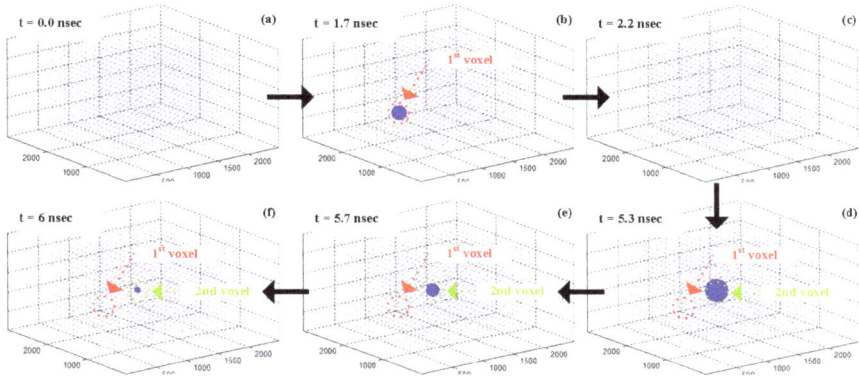

Figure 16. Example snap-shot frames for reconstructing the fluorophores.

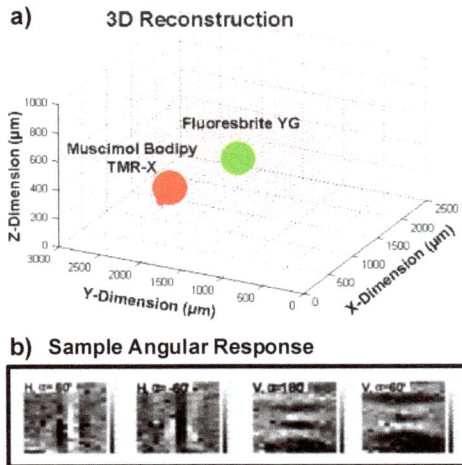

Figure 17. (**a**) 3D localization of two types of fluorophore micro-spheres. (**b**) Sample angular response from these micro-spheres.

5.3. Fluorescence Lifetime Measurement

The two fluorophores used above exist as fine solid grains in μm scale and were thus suitable for 3D localization. Because their excitation wavelength is far away from UV regime, however, it was difficult to obtain sufficient intensity data for their lifetime extraction. To properly test the DTC's capability to resolve shorter lifetimes, two quantum dots whose lifetimes are in scale of nanosecond were measured. Because quantum dots are not functional in solid state, however, they were first dissolved in toluene and then mixed with curable, transparent epoxy (Vitralit 1688, Epoxy Technology, Billerica, MA, USA) to form minuscule beads. These beads, however, were difficult to localize in 3D because they were significantly bigger than solid fluorophore grains, which violated sparsity condition. Furthermore, epoxy's uneven surface caused irregularities in photons' incident angles on the pixel surface. Quantum dot lifetimes alone, however, could be clearly distinguished despite the small difference ~2.5 ns between them. Figure 18 shows the resulting lifetime curves extracted for each quantum dot. Intensity data was collected by shifting the 10 ns wide detection window by 71 ps, by averaging three measurements per time bin. The collected histogram data was curve-fitted to produce lifetimes of 9.23 and 6.74 ns for QPP-450 and QSP-560, respectively.

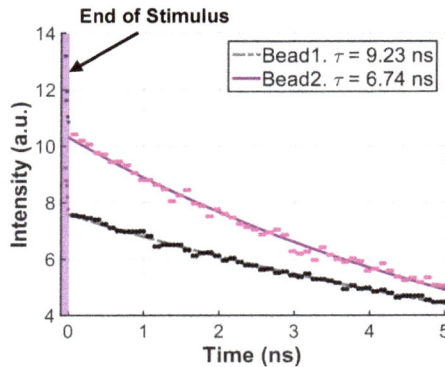

Figure 18. Lifetime extraction for quantum dots (Bead1: QSP-560, Bead2: QPP-450).

The first measurement shows that with combination of lifetime signature and the pre-localized 3D position, the system can localize distinct fluorescent sources in 3D without any lens or filter. The second measurement shows that with the control of detection window formulated by the system's DTC combined with in-pixel averaging, it is possible to resolve nanosecond fluorescence lifetimes with low off-chip data bandwidth.

6. Conclusions

Apart from the powerful conventional fluorescence microscopy, there still is a significant need for low cost and compact imagers that can potentially become implantable devices for in vivo biological applications in the future. Heading towards that goal, in this paper we have built a first reported 3D on-chip fluorescence lifetime imaging system without the use of any lenses or filters both of which are critical to the conventional FLIM system. To eliminate those components, we used a CMOS compatible optical structure, an angle-sensitive pixel, to extract angle information for localization of fluorophores without lens, and circuit techniques to generate accurately controllable detection window and perform on-chip data compression to obtain fluorescence lifetime histogram without any filter. In essence, we have combined SPAD's high sensitivity with metal gratings much like the recent works in angle sensitive pixel (ASP), to make an image sensor closer to the ideal 8D plenoptic function sensor [42]. The simple system architecture is optimized for robustness and high fill-factor with uniform sensitivity across the array of 4230 A-SPAD pixels. To generate global timing window for photon detection, a 10 bit DTC with LSB of 71 ps, RMS jitter of 13.2 ps, and 24 mW power consumption was used. The total power consumption of the IC is 83.8 mW. Other performance highlights are summarized in Table 1.

Table 1. Comparison with other state-of-the-art SPAD FLIM image sensors.

	This Work	[14]	[43]	[24]	[44]	[45]
PDP (Max %) @Ve	2.72 †@1.2 Ve	25@1 Ve	20@3 Ve	30@1.5 Ve	-	40@5 Ve
Pixel Size	35 µm × 35 µm	50 µm × 50 µm	25 µm × 25 µm	48 µm × 48 µm	8 µm × 8 µm	100 µm × 100 µm
Fill Factor	14.4%	2%	4.5%	0.77%	19.63%	3.14%
DCR (kHz)	$0.404@1.2\ V_E$	$0.1@1\ V_E$	$0.078@3\ V_E$	$0.54@2.5\ V_E$	$0.05@2\ V_E$	$4@5\ V_E$
Array Format	72×60	32 × 32	128 × 128	64 × 64	256 × 256	32 × 32
Chip size	2.64 mm × 3 mm	4.8 mm × 3.2 mm	25 mm × 25 mm	9.1 mm × 4.2 mm	3.5 mm × 3.1 mm	3.5 mm × 3.5 mm
Power	83.8 mW	90 mW	363 mW	26.4 W	-	-
Process	180 nm	CIS 130 nm	350 nm	130 nm	CIS 130 nm	350 nm
Lifetime Method	Time-Gated (DTC)	TDC	Time-Gated	TDC	Time-Gated (TAC)	Time-Gated
Resolution (LSB)	<71 ps	119 ps	200 ps	62.5 ps	6.66 ps	-
Max Range	72.7 ns	100 ns	9.6 ns	64 ns	50 ns	-
DNL (LSB)	0.54	0.4	-	4	3.5	-
INL (LSB)	1.27	1.2	-	8	-	-
Application	3D FLIM	FLIM	FLIM	FLIM	FLIM	FLIM
Lens	No	Yes	Yes	-	Yes	Yes
Filter	No	Yes	No	Yes	-	Yes

† includes losses due to ASP gratings. V_E: excess voltage. *TAC*: time-to-amplitude converter. *TDC*: time-to-digital converter. *DTC*: digital-to-time converter.

Acknowledgments: The authors thank Sunwoo Lee, Min Chul Shin, SungYun Park, and Jaeyoon Kim for critical discussions. The authors acknowledge the support by NSF (Grant ECS-0335765).

Author Contributions: Changhyuk Lee conceived and designed the sensor array, performed the experiments, analyzed the data, and wrote the paper; Ben Johnson supported the sensor array design and the experiments; Tae-Sung Jung supported the measurement and analyzed the data; PI. Alyosha Molnar advised the entire project.

Conflicts of Interest: The authors declare no conflict of interest.

Abbreviations

SPAD	Single Photon Avalanche Diode
FLIM	Fluorescence Lifetime Imaging Microscopy
CMOS	Complementary metal-oxide-semiconductor
TCSPC	Time Correlated Single Photon Counting
SRAM	Static random-access memory

References

1. Ghosh, K.K.; Burns, L.D.; Cocker, E.D.; Nimmerjahn, A.; Ziv, Y.; Gamal, A.E.; Schnitzer, M.J. Miniaturized integration of a fluorescence microscope. *Nat. Methods* **2011**, *8*, 871–878.
2. Jin, L.; Han, Z.; Platisa, J.; Wooltorton, J.R.A.; Cohen, L.B.; Pieribone, V.A. Single action potentials and subthreshold electrical events imaged in neurons with a novel fluorescent protein voltage probe. *Neuron* **2012**, *75*, 779–785.
3. Johnson, B.; Peace, S.T.; Cleland, T.A.; Molnar, A. A 50 µm pitch, 1120-channel, 20 kHz frame rate microelectrode array for slice recording. In Proceedings of the 2013 IEEE Biomedical Circuits and Systems Conference (BioCAS), Rotterdam, The Netherlands, 31 October–2 November 2013; pp. 109–112.
4. Greenbaum, A.; Luo, W.; Su, T.W.; Gorocs, Z.; Xue, L.; Isikman, S.O.; Coskun, A.F.; Mudanyali, O.; Ozcan, A. Imaging without lenses: Achievements and remaining challenges of wide-field on-chip microscopy. *Nat. Methods* **2012**, *9*, 889–895.
5. Zheng, G.; Lee, S.A.; Antebi, Y.; Elowitz, M.B.; Yang, C. The ePetri dish, an on-chip cell imaging platform based on subpixel perspective sweeping microscopy (SPSM). *Proc. Natl. Acad. Sci. USA* **2011**, *108*, 16889–16894.
6. Monjur, M.; Spinoulas, L.; Gill, P.R.; Stork, D.G. Ultra-miniature, computationally efficient diffractive visual-bar-position sensor. In Proceedings of the 9th International Conference on Sensor Technologies and Applications (SENSORCOMM 2015), Venice, Italy, 24–29 August 2015; pp. 51–56.

315

7. Kim, T.i.; McCall, J.G.; Jung, Y.H.; Huang, X.; Siuda, E.R.; Li, Y.; Song, J.; Song, Y.M.; Pao, H.A.; Kim, R.H.; et al. Injectable, Cellular-Scale Optoelectronics with Applications for Wireless Optogenetics. *Science* **2013**, *340*, 211–216.
8. Zorzos, A.N.; Boyden, E.S.; Fonstad, C.G. Multiwaveguide implantable probe for light delivery to sets of distributed brain targets. *Opt. Lett.* **2010**, *35*, 4133–4135.
9. Coskun, A.F.; Sencan, I.; Su, T.W.; Ozcan, A. Lensfree Fluorescent On-Chip Imaging of Transgenic *Caenorhabditis elegans* Over an Ultra-Wide Field-of-View. *PLoS ONE* **2011**, *6*, e15955.
10. Szmacinski, H.; Lakowicz, J.R. Fluorescence lifetime-based sensing and imaging. *Sens. Actuators B Chem.* **1995**, *29*, 16–24.
11. Bastiaens, P.I.; Squire, A. Fluorescence lifetime imaging microscopy: Spatial resolution of biochemical processes in the cell. *Trends Cell Biol.* **1999**, *9*, 48–52.
12. Hum, J.M.; Siegel, A.P.; Pavalko, F.M.; Day, R.N. Monitoring Biosensor Activity in Living Cells with Fluorescence Lifetime Imaging Microscopy. *Int. J. Mol. Sci.* **2012**, *13*, 14385.
13. Tyndall, D.; Rae, B.; Li, D.; Richardson, J.; Arlt, J.; Henderson, R. A 100 Mphoton/s time-resolved mini-silicon photomultiplier with on-chip fluorescence lifetime estimation in 0.13 μm CMOS imaging technology. In Proceedings of the 2012 IEEE International Solid-State Circuits Conference (ISSCC), Digest of Technical Papers, San Francisco, CA, USA, 19–23 February 2012; pp. 122–124.
14. Gersbach, M.; Maruyama, Y.; Trimananda, R.; Fishburn, M.W.; Stoppa, D.; Richardson, J.A.; Walker, R.; Henderson, R.; Charbon, E. A Time-Resolved, Low-Noise Single-Photon Image Sensor Fabricated in Deep-Submicron CMOS Technology. *IEEE J. Solid-State Circuits* **2012**, *47*, 1394–1407.
15. Field, R.M.; Shepard, K. A 100-fps fluorescence lifetime imager in standard 0.13-μm CMOS. In Proceedings of the 2013 Symposium on VLSI Circuits (VLSIC), Kyoto, Japan, 11–14 June 2013; pp. C10–C11.
16. Bishara, W.; Sikora, U.; Mudanyali, O.; Su, T.W.; Yaglidere, O.; Luckhart, S.; Ozcan, A. Holographic pixel super-resolution in portable lensless on-chip microscopy using a fiber-optic array. *Lab Chip* **2011**, *11*, 1276–1279.
17. Wang, A.; Molnar, A. A Light-Field Image Sensor in 180 nm CMOS. *IEEE J. Solid-State Circuits* **2012**, *47*, 257–271.
18. Gill, P.R.; Lee, C.; Lee, D.G.; Wang, A.; Molnar, A. A microscale camera using direct Fourier-domain scene capture. *Opt. Lett.* **2011**, *36*, 2949–2951.
19. Wang, A.; Gill, P.; Molnar, A. Fluorescent imaging and localization with angle sensitive pixel arrays in standard CMOS. In Proceedings of the 2010 IEEE Sensors Conference, Waikoloa, HI, USA, 1–4 November 2010; pp. 1706–1709.
20. Levoy, M.; Ng, R.; Adams, A.; Footer, M.; Horowitz, M. Light Field Microscopy. *ACM Trans. Graph.* **2006**, *25*, 924–934.
21. Lee, C.; Molnar, A. Self-quenching, Forward-bias-reset for Single Photon Avalanche Detectors in 1.8 V, 0.18-μm process. In Proceedings of the 2011 IEEE International Symposium on Circuits and Systems (ISCAS), Rio de Janeiro, Brazil, 15–19 May 2011; pp. 2217–2220.
22. Lee, C.; Johnson, B.; Molnar, A. Angle sensitive single photon avalanche diode. *Appl. Phys. Lett.* **2015**, *106*, 231105.
23. Burri, S.; Maruyama, Y.; Michalet, X.; Regazzoni, F.; Bruschini, C.; Charbon, E. Architecture and applications of a high resolution gated SPAD image sensor. *Opt. Express* **2014**, *22*, 17573–17589.
24. Field, R.; Realov, S.; Shepard, K. A 100 fps, Time-Correlated Single-Photon-Counting-Based Fluorescence-Lifetime Imager in 130 nm CMOS. *IEEE J. Solid-State Circuits* **2014**, *49*, 867–880.
25. Veerappan, C.; Richardson, J.; Walker, R.; Li, D.U.; Fishburn, M.; Maruyama, Y.; Stoppa, D.; Borghetti, F.; Gersbach, M.; Henderson, R.; et al. A 160 × 128 single-photon image sensor with on-pixel 55 ps 10 b time-to-digital converter. In Proceedings of the 2011 IEEE International Solid-State Circuits Conference, Digest of Technical Papers (ISSCC), San Francisco, CA, USA, 20–24 February 2011; pp. 312–314.
26. Homulle, H.A.R.; Powolny, F.; Stegehuis, P.L.; Dijkstra, J.; Li, D.U.; Homicsko, K.; Rimoldi, D.; Muehlethaler, K.; Prior, J.O.; Sinisi, R.; et al. Compact solid-state CMOS single-photon detector array for in vivo NIR fluorescence lifetime oncology measurements. *Biomed. Opt. Express* **2016**, *7*, 1797–1814.
27. Dutton, N.A.W.; Gyongy, I.; Parmesan, L.; Henderson, R.K. Single Photon Counting Performance and Noise Analysis of CMOS SPAD-Based Image Sensors. *Sensors* **2016**, *16*, 1122.

28. Leonard, J.; Dumas, N.; Causse, J.P.; Maillot, S.; Giannakopoulou, N.; Barre, S.; Uhring, W. High-throughput time-correlated single photon counting. *Lab Chip* **2014**, *14*, 4338–4343.
29. Harris, C.; Selinger, B. Single-Photon Decay Spectroscopy. II. The Pile-up Problem. *Aust. J. Chem.* **1979**, *32*, 2111–2129.
30. Nie, K.; Wang, X.; Qiao J.; Xu, J. A Full Parallel Event Driven Readout Technique for Area Array SPAD FLIM Image Sensors. *Sensors* **2016**, *16*, 160.
31. Malass, I.; Uhring, W.; Le Normand, J.P.; Dumas, N.; Dadouche, F. Efficiency improvement of high rate integrated time correlated single photon counting systems by incorporating an embedded FIFO. In Proceedings of 2015 the 13th International New Circuits and Systems Conference (NEWCAS), Grenoble, France, 7–10 June 2015; pp. 1–4.
32. Ballew, R.M.; Demas, J.N. An error analysis of the rapid lifetime determination method for the evaluation of single exponential decays. *Anal. Chem.* **1989**, *61*, 30–33.
33. Grant, D.M.; McGinty, J.; McGhee, E.J.; Bunney, T.D.; Owen, D.M.; Talbot, C.B.; Zhang, W.; Kumar, S.; Munro, I.; Lanigan, P.M.; et al. High speed optically sectioned fluorescence lifetime imaging permits study of live cell signaling events. *Opt. Express* **2007**, *15*, 15656–15673.
34. Chan, S.P.; Fuller, Z.J.; Demas, J.N.; DeGraff. B.A. Optimized Gating Scheme for Rapid Lifetime Determinations of Single-Exponential Luminescence Lifetimes. *Anal. Chem.* **2001**, *73*, 4486–4490.
35. Li, D.D.U.; Ameer-Beg, S.; Arlt, J.; Tyndall, D.; Walker, R.; Matthews, D.R.; Visitkul, V.; Richardson, J.; Henderson, R.K. Time-Domain Fluorescence Lifetime Imaging Techniques Suitable for Solid-State Imaging Sensor Arrays. *Sensors* **2012**, *12*, 5650–5669.
36. Villa, F.; Markovic, B.; Bellisai, S.; Bronzi, D.; Tosi, A.; Zappa, F.; Tisa, S.; Durini, D.; Weyers, S.; Paschen, U.; et al. SPAD Smart Pixel for Time-of-Flight and Time-Correlated Single-Photon Counting Measurements. *IEEE Photonics J.* **2012**, *4*, 795–804.
37. Maruyama, Y.; Blacksberg, J.; Charbon, E. A 1024 × 8, 700-ps Time-Gated SPAD Line Sensor for Planetary Surface Exploration with Laser Raman Spectroscopy and LIBS. *IEEE J. Solid-State Circuits* **2014**, *49*, 179–189.
38. De Grauw, C.J.; Gerritsen, H.C. Multiple Time-Gate Module for Fluorescence Lifetime Imaging. *Appl. Spectrosc.* **2001**, *55*, 670–678.
39. Webster, E.; Richardson, J.; Grant, L.; Renshaw, D.; Henderson, R. A Single-Photon Avalanche Diode in 90-nm CMOS Imaging Technology With 44% Photon Detection Efficiency at 690 nm. *IEEE Electron Device Lett.* **2012**, *33*, 694–696.
40. Mizuno, M.; Yamashina, M.; Furuta, K.; Igura, H.; Abiko, H.; Okabe, K.; Ono, A.; Yamada, H. A GHz MOS adaptive pipeline technique using MOS current-mode logic. *IEEE J. Solid-State Circuits* **1996**, *31*, 784–791.
41. Gill, P.R.; Wang, A.; Molnar, A. The In-Crowd Algorithm for Fast Basis Pursuit Denoising. *IEEE Trans. Signal Process.* **2011**, *59*, 4595–4605.
42. Landy, M.; Movshon, J.A. The Plenoptic Function and the Elements of Early Vision. In *Computational Models of Visual Processing*; MIT Press: Cambridge, MA, USA, 1991; pp. 3–20.
43. Maruyama, Y.; Charbon, E. An all-digital, time-gated 128 × 128 SPAD array for on-chip, filter-less fluorescence detection. In Proceedings of 2011 the 16th International Solid-State Sensors, Actuators and Microsystems Conference (TRANSDUCERS), Beijing, China, 5–9 June 2011; pp. 1180–1183.
44. Parmesan, L.; Dutton, N.; Calder, N.; Holmes, A.; Grant, L.; Henderson, R. A 9.8 μm sample and hold time to amplitude converter CMOS SPAD pixel. In Proceedings of 2014 the 44th European Solid State Device Research Conference (ESSDERC), Venice, Italy, 22–26 September 2014; pp. 290–293.
45. Vitali, M.; Bronzi, D.; Krmpot, A.; Nikolic, S.; Schmitt, F.J.; Junghans, C.; Tisa, S.; Friedrich, T.; Vukojevic, V.; Terenius, L.; et al. A Single-Photon Avalanche Camera for Fluorescence Lifetime Imaging Microscopy and Correlation Spectroscopy. *IEEE J. Sel. Top. Quan.* **2014**, *20*, 344–353.

Chapter 4:
Image Reconstruction for Photon-Counting Image Sensors

sensors

MDPI

Article
Vision without the Image

Bo Chen and Pietro Perona *

Computation and Neural Systems, California Institute of Technology, 1200 E California Blvd, Pasadena, CA 91125, USA; bchen3@caltech.edu
* Correspondence: perona@caltech.edu; Tel.: +1-626-395-2084

Academic Editor: Eric R. Fossum
Received: 2 February 2016; Accepted: 4 April 2016; Published: 6 April 2016

Abstract: Novel image sensors transduce the stream of photons directly into asynchronous electrical pulses, rather than forming an image. Classical approaches to vision start from a good quality image and therefore it is tempting to consider image reconstruction as a first step to image analysis. We propose that, instead, one should focus on the task at hand (e.g., detection, tracking or control) and design algorithms that compute the relevant variables (class, position, velocity) directly from the stream of photons. We discuss three examples of such computer vision algorithms and test them on simulated data from photon-counting sensors. Such algorithms work just-in-time, *i.e.*, they complete classification, search and tracking with high accuracy as soon as the information is sufficient, which is typically before there are enough photons to form a high-quality image. We argue that this is particularly useful when the photons are few or expensive, e.g., in astronomy, biological imaging, surveillance and night vision.

Keywords: photon-counting sensors; visual recognition; low-light computer vision

---

1. Introduction

Current computer vision algorithms start with a high-quality image as input. While such images may be acquired almost instantly in a well-lit scene, dark environments demand a significantly longer acquisition time. This long acquisition time is undesirable in many applications that operate in low-light environments: in biological imaging, prolonged exposure could cause health risks [1] or sample bleaching [2]; in autonomous driving, the delay that is imposed by image capture could affect a vehicle's ability to stay on-course and avoid obstacles; in surveillance, long periods of imaging could delay response, as well as produce smeared images. When light is low, the number of photons per pixel is small and images become noisy. Computer vision algorithms are typically not designed to be robust *vis-a-vis* image noise, thus practitioners face an uneasy tradeoff between poor performance and long response times.

Novel sensor technology offers a new perspective on image formation: as soon as a photon is sensed it should be transmitted to the host Central Processing Unit (CPU), rather than wait until a sufficient number of photons has been collected to form good quality image. Thus an image, in the conventional sense, is never formed. Designs and prototypes of photon-counting image sensors, such as the quantum sensors [3], single-photon avalanche detectors [4], quanta image sensors [5,6], and the giga-vision camera [7], have been proposed recently. These sensors are capable of reliably detecting single photons, or a small number of photons. Instead of returning a high-quality image after a long exposure, photon-counting sensors report a stream of photon counts densely sampled in time.

Currently, the dominant use for photon-counting image sensors is image reconstruction [8]: the stream of photon counts is used to synthesize a high-quality image to be used in consumer applications or computer vision. However, the goal of vision is to compute information about the world (class, position, velocity) from the light that reaches the sensor. Thus, reconstructing the image is not a necessary first step. Rather, one should consider computing information directly from the stream of photons [9,10]. This line of thinking requires revisiting the classical image-based paradigm of computer vision, and impacts both the design of novel image sensors and the design of vision algorithms.

Computing directly from the stream of photons presents the advantage that some information may be computed immediately, without waiting for high-quality image to be formed. In other words, information is computed incrementally, enabling the downstream algorithms to trade off reaction times with accuracy. This is particularly appealing in low-light situations were photon arrival times are widely spaced. As the hardware for computation becomes faster, this style of computation will become practical in brighter scenes, especially when response times are crucial (e.g., in vehicle control).

Here we explore three vision applications: classification, search and tracking. In each application, we will propose an algorithm that makes direct use of the stream of photons, rather than an image. We find that each one of these algorithms achieves high accuracy with only a tiny fraction of the photons required for capturing high-quality images. We conclude with a discussion of what was learned.

2. Results

2.1. Simplified Imaging Model

Assume that the scene is stationary and photon arrival times follow a homogeneous Poisson process. Within an interval of length δ_t, the observed photon count X_i at pixel location i is subject to Poisson noise whose mean rate depends on maximum rate $\lambda_{max} \in \mathbb{R}^+$, the true intensity at that pixel $I_i \in [0,1]$ and a dark current rate $\epsilon_{dc} \in [0,1]$ per pixel [8]:

$$P(X_i = k) = Poisson(k; \lambda_{max} \frac{(I_i + \epsilon_{dc})}{(1 + \epsilon_{dc})} \delta_t) \qquad (1)$$

(a model including sensor read noise is described in Section 3.1).

The sensor produces a stream of images $\mathbf{X}_1, \mathbf{X}_2, \ldots$, where $\mathbf{X}_t \in \mathcal{N}^d$ contains the photon counts from d pixel locations from the time interval $[(t-1)\delta_t, t\delta_t]$ (Figure 1a). We use $\mathbf{X}_{1:t}$ to represent the stream of inputs $\{\mathbf{X}_1, \mathbf{X}_2, \ldots, \mathbf{X}_t\}$.

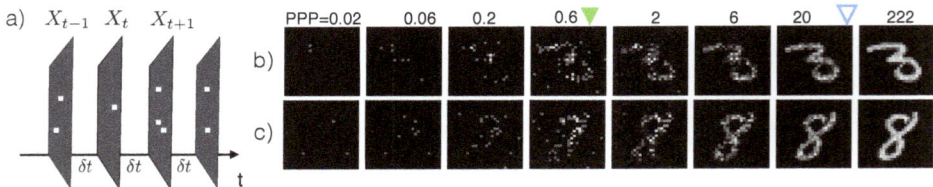

Figure 1. Synthetic low-light images. (**a**) A photon-counting sensor outputs a matrix of photon counts $\mathbf{X}_t$ with a period of δ_t. (**b,c**) Sample synthetic low-light images from the Mixed National Institute of Standards and Technology (MNIST) dataset [11] used in the classification experiments (Section 2.2.2) with increasing average photons per pixel (PPP). PPP is proportional to the exposure time t. Blue hollow arrows indicate the median PPP required for the proposed algorithm (Section 2.2.1) to achieve the same error rate (0.7%) as a model trained and tested using images under normal lighting conditions with about $2^7 \approx 10^4$ PPP. Green solid arrows indicate the median PPP required to to maintain error rates below 1%.

When the illuminance of the environment is constant, the expected number of photons collected by the sensor grows linearly with the exposure time. Hence we use the number of photons per bright pixel (PPP) as a proxy for the exposure time t. PPP $= 1$ means that the a pixel with maximum intensity has collected 1 photon. Additionally, since PPP is linked to the total amount of information content in the image regardless of the illuminance level, we will use PPP when describing the performance of vision algorithms. Figure 1b, c shows two series of inputs $\mathbf{X}_{1:t}$ with increasing PPP.

2.2. Classification

Distinguishing objects of different categories hinges upon the extraction of "features", which are structural regularities in pixel values such as edges, corners, contours, *etc*. For example, the key feature that set apart a handwritten digit "3" from a digit "8" (Figure 1b,c, last column) is the fact that a "3" has open loops and "8" has closed loops—This corresponds to different strokes on the left side of the digit. In normal lighting conditions, these features are fully visible, may be computed by, e.g., convolution with an appropriate kernel, and fed into a classifier to predict the category of the image.

In low light, classification is hard because the features are corrupted by noise. A closed contour may appear broken due to stochastically missing photons. The noise in the features in turn translates to uncertainties on the classification decision. This uncertainty diminishes as the exposure time increases. It is intuitive that a vision algorithm that is designed to compute from a minimal number of photons should keep track of said uncertainties, and dynamically determine the exposure time based on the desired accuracy.

In particular, one wishes to predict the category $Y \in \{1, 2, \ldots, C\}$ of an image based on photon counts $\mathbf{X}_{1:t}$. The predictions must minimize exposure time while being reasonably accurate, *i.e.*,

$$\min \mathbb{E}[T] \qquad\qquad s.t.\ \mathbb{E}[\hat{Y} \neq Y] \leq \gamma \qquad\qquad (2)$$

where T is a random variable denoting the exposure time required to classify an image, $\hat{Y} \in \{1, 2, \ldots, C\}$ is the prediction of the class label, γ is the maximum tolerable misclassification error, and the expectation is taken over all images in a dataset.

2.2.1. Classification Algorithm

In order to make the most efficient use of photons, we first assume that a conditional probabilistic model $P(Y|\mathbf{X}_{1:t})$ is available for any $t \geq 0$ (we will relax this assumption later) and for all possible categories of the input image. An asymptotically optimal algorithm that solves the problem described in Equation (2) is Sequential Probability Ratio Testing (SPRT) [12] (Figure 2a,b):

$$\text{Choose an appropriate error threshold } \theta$$
$$c^* = \underset{c \in \{1,2,\ldots,C\}}{\arg\max}\ P(Y = c|\mathbf{X}_{1:t})$$
$$\begin{cases} \text{report } Y = c^* & \text{if } \log \frac{P(Y=c^*|\mathbf{X}_{1:t})}{P(Y \neq c^*|\mathbf{X}_{1:t})} > \theta \\ \text{increase } t & \text{otherwise} \end{cases} \qquad (3)$$

Essentially, SPRT keeps accumulating photons by increasing exposure time until there is predominant evidence in favor of a particular category. Due to the stochasticity of the photon arrival events and the variability in an object's appearance, the algorithm observes a different stream of photon counts each time. As a result, the exposure time T, and equivalently, the required PPP, are also different each time (see Figure 3).

The accuracy of the algorithm is controlled by the threshold θ. When a decision is made, the declared class c^* satisfies that $\log \frac{P(Y=c^*|\mathbf{X}_{1:t})}{P(Y \neq c^*|\mathbf{X}_{1:t})} > \theta$, which means that class c^* has at least posterior probability $Sigm(\theta) \triangleq \frac{1}{1+e^{-\theta}}$ according to the generative model, and the error rate of SPRT is at most $1 - Sigm(\theta)$. For instance, if the maximum tolerable error rate is 10%, θ should be set so that $1 - Sigm(\theta) = 0.1$, or $\theta \approx 2.2$, while an error rate of 1% would drive θ to 4.6. Since higher thresholds lead to longer exposure times, the threshold serves as a knob to trade off speed *versus* accuracy, and should be set appropriately based on γ (Equation (2)).

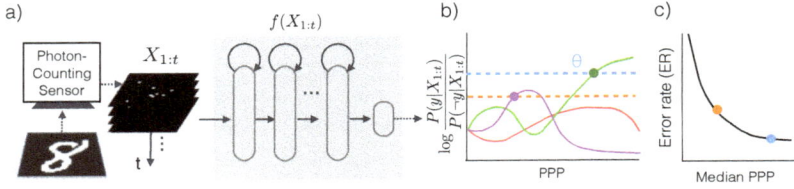

Figure 2. Low-light classification. (**a**) A sensor produces a stream of photon counts $\mathbf{X}_{1:t}$, which is fed into a recurrent neural network to compute class conditional likelihood $f(\mathbf{X}_{1:t}) \approx P(Y|\mathbf{X}_{1:t})$. (**b**) SPRT with three classes. Based on $f(\mathbf{X}_{1:t})$, SPRT compares the class with the highest log likelihood ratio to a threshold θ. Two threshold options (dashed lines) corresponds to different decisions at different times (solid dots). (**c**) The threshold allows the system to traverse the error rate (ER) *versus* PPP curve to find the optimal operating point to minimize the cost function (Equation (2)). We use PPP instead of the exposure time T to measure speed as the former is more closely related to the information content in the photon stream. We use the median PPP instead of the mean because PPP follows a heavy-tailed distribution (Figure 3c) and median is more stable than the mean.

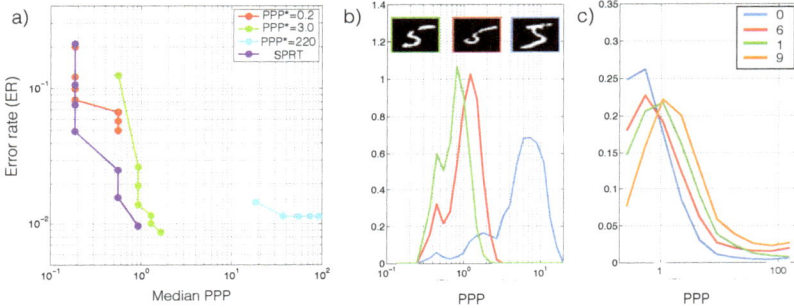

Figure 3. Low-light classification performance. (**a**) Error rate *vs.* PPP tradeoff for the SPRT algorithm (Equation (3)). $PPP^* = x$ denotes a "specialist": A model trained using images only at light level x and tested on other light levels by input normalization (see Section 2.2.2). (**b**) SPRT decision time is stochastic even for the same underlying image. The PPP distribution is plotted separately for multiple images of 5. (**c**) SPRT decision time distribution is category-dependent. Some categories, e.g., "0", are easier (faster decision) than others, say "9".

The assumption that the conditional distribution $P(Y = c|\mathbf{X}_{1:t})$ is known is rather restrictive. Fortunately, the conditional distribution may be directly learned from data. In particular we train a recurrent neural network [13] $f_c(\mathbf{X}_{1:t}) \approx P(Y = c|\mathbf{X}_{1:t})$ to approximate the conditional distribution. This network has a compact representation, and takes advantage of the sparseness of the photon-counts for efficient evaluation. Details of the network may be found in Section 3.2 and [9].

2.2.2. Experiments

We evaluate the low-light classification performance of the SPRT on the MNIST dataset [11], a standard handwritten digits dataset with 10 categories. The images are 28×28 in resolution and in black and white. We simulate the outputs from a photon-counting sensor according to the full noise model (Section 3.1). The images are stationary within the imaging duration. We do not assume a given conditional distribution $P(Y|\mathbf{X}_{1:t})$ but train a recurrent network approximation $f(\mathbf{X}_{1:t})$ from data (Section 3.2).

Recall that classification correctness in each trial and the required exposure time (or PPP) are random variables. We therefore characterize SPRT performance based on the tradeoff between error rates (ER, $\mathbb{E}[\hat{Y} \neq Y]$) and the median PPP in Figure 3a. The tradeoff is generated by sweeping the thresholds $\theta \in [-2.2, 9.2]$. For comparison we tested the performance of models that were trained to classify images from a single PPP. We call these models "specialists" for the corresponding PPP. The specialists are extended to classify images at different light levels by scaling the image to the specialized PPP. To get a sense of the intraclass and interclass PPP variability, we also visualize the PPP histograms for multiple runs of different images in the same class (Figure 3b), and the overall PPP histograms for a few classes (Figure 3c). Lastly, we analyze how SPRT's performance is sensitive to sensor noises in Figure 4. Details of the analysis procedure are found in Section 3.1.

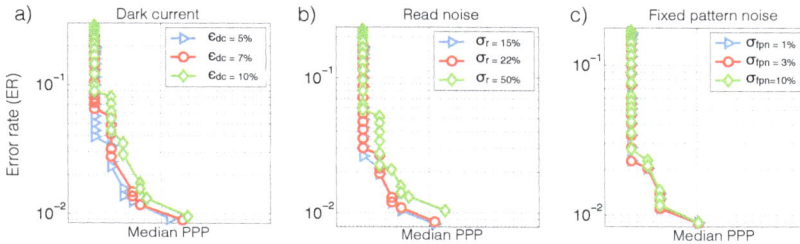

Figure 4. Sensitivity to noise in image classification. Error rate *vs.* PPP tradeoff with different levels of (**a**) dark current ϵ_{dc}, (**b**) read noise and (**c**) fixed pattern noise (see Section 3.1). The default setting uses 3% dark current, 0% read noise and 0% fixed pattern noise.

2.3. Search

Search is a generalization of classification into multiple locations. The task is to identify whether a target object (e.g., keys, a pedestrian, a cell of a particular type) is present in a scene cluttered with distractors (e.g., a messy desk, a busy street at night or a cell culture). Note that despite the multiple candidate positions for a target to appear, we consider search as a binary task, where the two hypotheses are denoted $C = 1$ (target-present) and $C = 0$ (target-absent). We assume for simplicity that at most one target may appear at a time (for multiple targets, see [14]).

The difficulty of search in low-light conditions may be attributed to the following factors. (1) There are multiple objects in the display, and each object is subject to photon count fluctuations. (2) Long range constraints, such as the prior knowledge that at most one target is present in the visual field, must be enforced. (3) Properties of the scene, such as the amount of clutter in the scene and the target and distractor appearance, may be uncertain. For example, we may know that there may be either three or twelve objects in the scene, and intuitively the search strategy for these two scenarios should be drastically different. Therefore, scene properties must be inferred for optimal performance.

We assume that a visual field consists of L non-overlapping locations, out of which M locations may contain an object. M represents the amount of clutter in the scene. The objects are simplified to be oriented bars and the only feature that separates a target from a distractor is the orientation. The orientation at location l is denoted $Y^{(l)}$. The target orientation and the distractor orientation are denoted y_T and y_D, respectively. The scene properties are collected denoted $\phi = \{M, y_T, y_D\}$. The scene properties may be unknown for many search tasks, thus ϕ is a vector of random variables. The variable of interest is $C \in \{0, 1\}$: $C = 1$ iff $\exists l \in \{1, \ldots, L\}, Y^{(l)} = y_T$, (*i.e.*, $C = 1$ iff there exists a location that contains a target).

We also assume that a low-light classifier discussed in Section 2.2 has been developed for classifying bar stimulus: the classifier computes $f_y(\mathbf{X}_{1:t}^{(l)}) \approx P(Y^{(l)} = y|\mathbf{X}_{1:t}^{(l)})$, the probability that the bar orientation at location l is y conditioned only on the local photon counts $\mathbf{X}_{1:t}^{(l)}$.

2.3.1. Search Algorithm

Similar to the low-light classification problem, an asymptotically optimal search algorithm is based on SPRT. The detailed algorithm is [12,14]:

$$\text{Choose two error thresholds } \theta_0 < 0, \theta_1 > 0$$

$$\text{Compute } S(t) \overset{\triangle}{=} \log \frac{P(C=1|\mathbf{X}_{1:t})}{P(C=0|\mathbf{X}_{1:t})}$$

$$\begin{cases} \text{report } C = 1 & \text{if } S(t) > \theta_1 \\ \text{report } C = 0 & \text{if } S(t) < \theta_0 \\ \text{increase } t & \text{otherwise} \end{cases} \tag{4}$$

where $S(t)$ is the log likelihood ratio between the two competing hypotheses, target-present ($C = 1$) and target-absent ($C = 0$). This algorithm is a binary version of the classification algorithm in Equation (3). Similar to Equation (3), the two thresholds θ_0 and θ_1 controls the amount of false reject errors (*i.e.*, declare target-absent when target-present) and false accept errors (*i.e.*, declaring target-present when target-absent).

The key for SPRT is to compute $S(t)$ from photon counts $\mathbf{X}_{1:t}$. The inference procedure may be implemented by two circuits, one infers the scene properties ϕ, and the other computes $S(t)$ (see [14]):

$$S(t) = \log \frac{1}{L} \sum_{l,\phi} R^{(l,\phi)}(\mathbf{X}_{1:t}^{(l)}) P(\phi|\mathbf{X}_{1:t}) \tag{5}$$

where

$$R^{(l,\phi)}(\mathbf{X}_{1:t}^{(l)}) \overset{\triangle}{=} \frac{\sum_y f_y(\mathbf{X}_{1:t}^{(l)}) P(Y^{(l)} = y|\phi, C^{(l)} = 1)}{\sum_y f_y(\mathbf{X}_{1:t}^{(l)}) P(Y^{(l)} = y|\phi, C^{(l)} = 0)} \tag{6}$$

$$P(\phi|\mathbf{X}_{1:t}) \propto P(\phi) \prod_l \sum_y f_y(\mathbf{X}_{1:t}^{(l)}) \frac{P(Y^{(l)} = y|\phi, C^{(l)} = 0)}{P(Y^{(l)} = y)} \tag{7}$$

Therefore, $S(t)$ may be computed by composing the low-light classifiers $f_y(\mathbf{X}_{1:t}^{(l)})$ according to Equations (5)–(7). The probabilities used in Equations (6) and (7), such as $P(Y^{(l)} = y)$ and $P(Y^{(l)} = y|\phi, C = 2)$, may be estimated from past data.

2.3.2. Experiments

We choose a simple setup (Figure 5a) to illustrate how the performance of the search algorithm is affected by scene properties: the amount of clutter M, the target/distractor appearances y_T and y_D, as well the degree of uncertainty associated with them. The setup contains $L = 14$ locations, each occupying a 7×7 area from which the sensor collects photons. The area contains a 3×7-pixel bar with intensity 1 and background pixels with intensity 0. The max emission rate is $\lambda_{max} = 3$ photons/s, and the dark current is 50% (causing the background to emit 1 photon/s). Examples of the lowlight search setup are shown in Figure 6d.

We conduct two experiments, one manipulates the scene complexity M and the other target/distractor appearances. In the first experiment M is either chosen uniformly from $\{3, 6, 12\}$, or fixed at one of the three values (Figure 6a,b): (1) Despite the high dark current noise, a decision may be made quickly with less than 2 photons per pixel. (2) The amount of light required to achieve a given classification error increases as M. (3) Not knowing the complexity further increases the required photon count. (4) Target-absent conditions requires more photons than target-present conditions. In the second experiment the target-distractor appearance difference $\delta y = |y_T - y_D|$ is either chosen uniformly from $\{20°, 30°, 90°\}$ or fixed at one of the three. Figure 6c suggests that target dissimilarity heavily influences the ER-PPP tradeoff, while uncertainty in the target and distractor appearances does not.

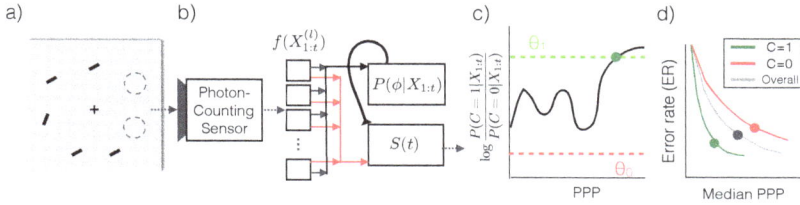

Figure 5. low-light Search. (**a**) Search stimuli with $L = 7$ display locations and 5 objects (oriented bars). The dashed circles are not part of the display but used to indicate empty locations. (**b**) The search algorithm. Local classification results $f(\mathbf{X}_{1:t}^{(l)})$ go through two circuits, one estimates the scene properties $P(\phi|\mathbf{X}_{1:t})$ (Equation (7)) and sends feedback to the other circuit that computes the log likelihood ratio $S(t)$ (Equation (5)). (**c**) SPRT compares $S(t)$ against a pair of thresholds θ_1 and θ_0 to decide whether to declare target-presence or absence, or wait for more evidence (Equation (4)). (**d**) SPRT produces ER *vs.* PPP tradeoffs (sketch) for different conditions.

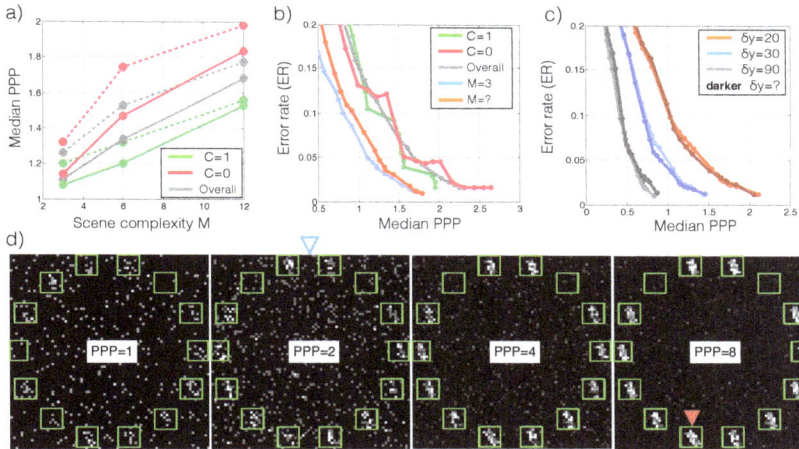

Figure 6. Characteristics of the search algorithm. (**a**) Median PPP required to achieve 5% error rate as a function of scene complexity M. Solid/dashed lines represent search problems where the complexity is known/unknown in advance. (**b**) ER *vs.* PPP tradeoff for various conditions. The first three legends corresond to target-present ($C = 1$), target-absent ($C = 0$) and their average for $M = 12$. The last two corresponds to a simpler image containing three bars with the complexity known ($M = 3$) and unknown ($M = ?$) in advance. (**c**) ER *vs.* PPP tradeoff for different appearance differences $\delta y = |y_T - y_D|$ between the target and the distractor. Darker lines denote the cases where the difference is unknown before hand. (**d**) Examples of lowlight search stimuli as a function of PPP. $L = 14$, $M = 12$, $\delta y = 20°$, target-present ($C = 1$, location indicated by red arrow, same for every PPP). The photon-counting sensors receive inputs from within the green windows. With known complexity the algorithm achieves a 1% error rate with a median PPP of 2 (blue hollow arrow).

2.4. Tracking

Finally, we demonstrate the potential of photon-counting sensors in tracking under low-light conditions. The goal of tracking is to recover time-varying attributes (*i.e.*, position, velocity, pose, *etc.*) of one or multiple moving objects. It is challenging because, unlike classification and search, objects in tracking applications are non-stationary by definition. In low-light environments, as the object transitions from one state to another, it leaves only a transient footprint, in the form of stochastically-sprinkled photons, which is typically insufficient to fully identify the state. Instead, a tracker must postulate the object's dynamics and integrate evidence over time accordingly. The evidence in turn refines the estimates of the dynamics. Due to the self-reinforcing nature of this procedure, the tracker must perform optimal inference to ensure convergence to the true dynamics.

Another challenge that sets low-light tracking apart from regular tracking problems is that the observation likelihood model is not only non-Gaussian, but also often unavailable, as it is commonly the case for realistic images. This renders most Kalman filter algorithms [15] ineffective.

2.4.1. Tracking Algorithm

The tracking algorithm we have designed is a hybrid between the Extended Kalman Filter [15] and the Auxiliary Particle Filter [16]. Let $\mathbf{Z}_t$ denote the state of the object, F the forward dynamics that govern the state transition: $\mathbf{Z}_{t+1} = F(\mathbf{Z}_t)$, which are known and differentiable, and $P_t(Z)$ the posterior distribution over the states at time t: $P_t(Z) \triangleq P(\mathbf{Z}_t = Z | \mathbf{X}_{1:t})$.

We make two assumptions: (1) P_t may be approximated by a multivariate Gaussian distribution; (2) A low-light regressor $f(Z|\mathbf{X}_t) \approx P(\mathbf{Z}_t = Z | \mathbf{X}_t)$ is available to compute a likelihood score of $\mathbf{Z}_t$ given only the snapshot $\mathbf{X}_t$ at time t. $f(Z|\mathbf{X}_t)$ does not have to be normalized. We justify assumption (1) and describe algorithms for realizing assumption (2) in Section 3.3. As we will see in Equation (17), the Poisson noise model (Equation (1)) ensures that $f(Z|\mathbf{X}_t)$ exists and takes a simple form.

Given a prior probability distribution P_0, our goal is to compute the posterior distribution P_t for all t. The tracking algorithm starts with $t = 0$ and repeat the following procedure (Figure 7b).

1. Compute the predictive distribution $P(\mathbf{Z}_{t+1}|\mathbf{Z}_t)$ from P_t
2. Draw K samples Z'_s from $P(\mathbf{Z}_{t+1}|\mathbf{Z}_t)$
3. Observe $\mathbf{X}_{t+1}$ and compute $W_s = f(Z'_s|\mathbf{X}_{t+1})$ (8)
4. Approximate $P_{t+1}(Z')$ as using samples Z'_s weighted by W_s
5. Increase $t = t + 1$

Under the Gaussian assumption for P_t, both steps 1 and 4 may be computed in close-form (Section 3.3). This is in sharp contrast to regular particle filters, which do not assume any parametric form for P_t and accomplish steps 1 and 4 using samples. Empirically we found that the Gaussian assumption is reasonable and often leads to efficient solutions with less variability.

2.4.2. Experiments

We choose the 1D inverted pendulum problem (Figure 7a) that is standard in control theory. A pendulum is mounted via a a a massless pole on a cart. The cart can move horizontally in 1D on a frictionless floor. The pendulum can rotate full circle on a fixed 2D plane perpendicular to the floor. The pendulum is released at time $t = 0$ at an unknown angle $\alpha_0 \in [0, 360°)$ from the vertical line, while the cart is at an unknown horizontal offset $\beta_0 \in \mathbb{R}$. The task is to identify how the angle α_t and the offset β_t change through time from the stream of photon counts $\mathbf{X}_{1:t}$. The state of the pendulum system is $\mathbf{Z}_t = \{\alpha_t, \dot{\alpha}_t, \beta_t, \dot{\beta}_t\}$. The system's forward dynamics is well-known [17].

In our simulations, only the pole of the pendulum is white and everything else is dark. The highest photon emission rate of the scene is λ_{max} and the dark current rate is ϵ_{dc}. We systematically vary λ_{max} and ϵ_{dc} and observe the amount of estimation error in the angle α_t and cart position β_t. See Section 3.4 for the simulation procedure.

We see that (1) estimation errors decrease over time (Figure 8a,b), (2) smaller ϵ_{dc} leads to faster reduction in estimation error on average (Figure 8a,b), and (3) the tracker's *convergence time, i.e.,* the time it takes to achieve a certain level of estimation accuracy, decreases with illuminance (Figure 8c). The time required to satisfy high accuracy requirements (e.g., <1° for α estimation) does NOT follow a simple inverse proportional relationship with illuminance. Instead, the convergence time plateaus, potentially due to the noise in the sampling procedure (Equation (8), step 2).

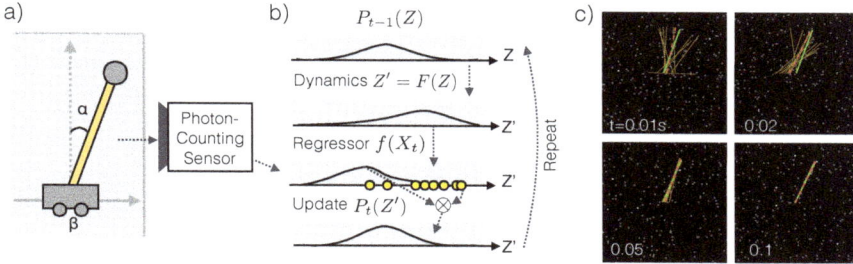

Figure 7. Low-light tracking. (**a**) An illustration of an inverted pendulum with attributes of interest α (pendulum angle) and β (cart location). The pole (yellow) is bright and everything else is dark. (**b**) Tracking algorithm (Equation (8)) iteratively updates the posterior $P_t(Z)$ using new evidence $\mathbf{X}_t$ and a low-light regressor $f(\mathbf{X}_t)$. (**c**) Snapshots of a sample run at exposure times $t = 0.01, 0.02, 0.05$ and 0.1 s. The brightest pixels emit photons at $\lambda_{max} = 10$ photons/s and the dark current $\epsilon_{dc} = 50\%$. The true position of the pendulum pole is shown in green, its estimate in red dashed, and samples from the tracker's posterior in yellow.

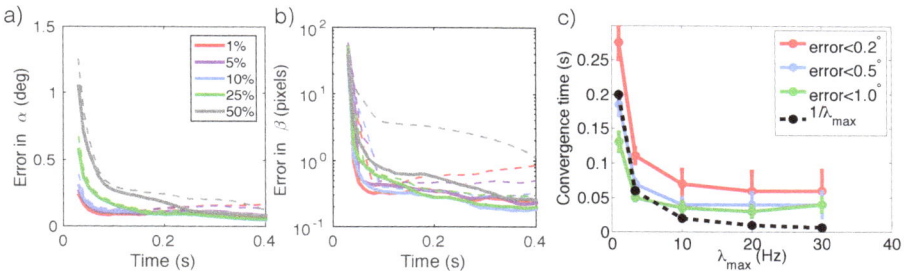

Figure 8. Tracking performance. The average estimation error in (**a**) the pendulum angle α_t and (**b**) in cart position β_t over time as a function of the amount of dark current ϵ_{dc} (color-coded). The photon emission rate is set at $\lambda_{max} = 10$ Hz. Dashed lines shows 1std above the mean. (**c**) The tracker's convergence times for its angle estimates to be within 0.2°, 0.5° and 1°, respectively, of the truth as a function of illuminance. ϵ_{dc} is set at 10%. As a reference "$1/\lambda_{max}$" is inversely proportional to the illuminance and scaled to have roughly the same starting position as the "error < 0.5°" curve.

3. Materials and Methods

3.1. Imaging Model Including Noise Sources

Within an interval δt, the sensor readout x is corrupted by a series of noise sources.

1. The amount of photons N incident on the pixel is subject to Poisson noise (shot noise). The noise level is determined by the true intensity I and the dark current $\epsilon_{dc} \sim \mathcal{N}(0, \sigma_{\epsilon})$.
2. The photon counts are corrupted by an additive Gaussian read noise $\epsilon_r \sim \mathcal{N}(0, \sigma_r)$ and a multiplicative fixed pattern noise $\epsilon_{fpn} \sim \mathcal{N}(0, \sigma_{fpn})$.

$$N \sim Poisson(\cdot|I + \epsilon_{dc}) \tag{9}$$

$$x = max(0, (N + \epsilon_r)(1 + \epsilon_{fpn})) \tag{10}$$

Sensors designed for low light applications (e.g., [5]) have promised low read noise and low fixed pattern noise. Therefore, we focus on modeling the shot noise and dark current, and assume that algorithms have access to N when the algorithms are trained. We then test the algorithm using realistic values of read noise and fixed pattern noise to study robustness against noise (see Figure 4).

3.2. Low-Light Classifier

A low-light classifier $f(\mathbf{X}_{1:t})$ that approximates the conditional distribution $P(Y|\mathbf{X}_{1:t})$ for $Y = \{1, 2, \ldots, C\}$ is developed in [9]. The classifier is a recurrent neural network consisting of multiple layers $h^{(1)}(t), \ldots, h^{(L)}(t)$ where the activation at layer l is:

$$h_j^{(l)}(t) = \sum_{\tau=1}^{t} W_j^{(l)} \mathbf{X}_\tau + b_i t \tag{11}$$

$$= h_j^{(l)}(t-1) + W_j^{(l)} \mathbf{X}_t + b_i \tag{12}$$

where $W_j^{(l)} \in \mathbb{R}^d$ and $b_j^{(l)} \in \mathbb{R}$ are the weights and the biases of the j-th unit. Equation (12) suggests that $h_j^{(l)}(t)$ may be computed incrementally from its old value $h_j^{(l)}(t-1)$.

The hidden units at layer l are organized into non-overlapping groups and pooled. A pooling unit $h_k^{(l)}$ oversees the hidden units at block G_k, and its activation is computed by:

$$m_k^{(l)}(t) = \max(0, \max_{j \in G_k}(h_j^{(l)}(t)) \tag{13}$$

Let $j^*(t) = \arg\max_{j \in G_k}(h_j^{(l)}(t))$ denote the index of the max unit at time $t-1$ ($j^*(t) = 0$ denote the event that the max value is 0). If within time interval δt only a small set $G_k' \subseteq G_k$ of hidden units within group G_k are updated, $m_k^{(l)}(t)$ may also be computed incrementally:

$$m_k^{(l)}(t) = \max_{j \in G_k' \cup j^*(t-1)}(h_j^{(l)}(t)) \tag{14}$$

Both Equations (12) and (14) are critical for an efficient implementation of the classifier. For example, if only a tenth of the units are updated in each layer, the computation time for $f(\mathbf{X}_{1:t})$ may be reduced by a factor of 10.

Finally, the output of the classifier is given by:

$$f_c(\mathbf{X}_{1:t}) = \frac{\exp(m_c^{(L)}(t))}{\sum_{c'} \exp(m_{c'}^{(L)}(t))}, \quad c \in \{1, 2, \ldots, C\} \tag{15}$$

Since $f(\mathbf{X}_{1:t})$ approximates the conditional likelihood $P(Y = c|\mathbf{X}_{1:t})$, the parameters $\{W_j^{(l)}, b_j^{(l)}\}_{j,l}$ of $f(\cdot)$ may be learned by maximum likelihood from a dataset of photon counts. However, it is expensive to keep track of a high number of photon count streams. Fortunately, Equation (11) also suggests that the network's prediction at time t only depends on the cumulative photon counts $\mathbf{S}_t \triangleq \sum_{\tau=1}^{t} \mathbf{X}_\tau$. Therefore, one only needs a dataset of $\{\mathbf{S}_t, t, Y\}$ tuples to perform maximum likelihood learning.

In detail, we simulated a lowlight MNIST dataset $\{\mathbf{S}^{(i)}, PPP^{(i)}, Y^{(i)}\}_i$ where the PPPs are sampled uniformly from $\{0.22, 2.2, 22, 220\}$. Note that we are using PPP instead of the exposure time t for reasons discussed in Section 2.1. At PPP= 220 each image pixel contains around 5 bits of information (log signal-to-noise-ratio ≈ 5). Our implementation uses the MatConvNet package [18] and its default hyper-parameters for training. We train a model with the same connectivity as the LeNet [11] denoted: 784-20-50-500-10. The model contains 784 input units, followed by two convolutional hidden layers with 20 and 50 filters, respectively, of size 5×5. Inputs to a convolutional layer is convolved with the filters, and then pooled over 2×2 non-overlapping windows. After the convolutional layers are a fully connected hidden layer with 500 units and a fully connected softmax layer with 10 output categories. We minimize the negative log likelihood with a $L2$ weight decay:

$$ -\sum_i \log f_{Y^{(i)}}(\mathbf{S}^{(i)}) + \eta \sum_l ||W^{(l)}||_2^2 \tag{16} $$

where $\eta = 0.0005$ is the strength of the weight decay. We use stochastic gradient descent with mini-batches of 100 and train for 60 epochs with learning rate $= 0.001$ and momentum $= 0.9$.

3.3. Tracking Algorithm

3.3.1. Low-Light Regressor

The photon count $\mathbf{X}_t$ at time t depends only on the angle α_t and cart position β_t and not their time derivatives, so our low-light regressor can only predict α_t and β_t: $f(\mathbf{Z}_t|\mathbf{X}_t) = f(\alpha_t, \beta_t|\mathbf{X}_t)$. Since we simulate the scene using a generative model, we can compute the exact form of the low-light regressor. Let $P(\mathbf{I}_t|\alpha_t, \beta_t)$ be the generative model where $\mathbf{I}_t$ is the intensity value of the image, the photon emission rate $\lambda_i(\alpha_t, \beta_t)$ for every pixel may be computed using Equation (1). As a result the log likelihood of observing $\mathbf{X}_t$ is:

$$ \log P(\mathbf{X}_t|\alpha_t, \beta_t) = \log \prod_i Poisson(X_{ti}; \lambda_i(\alpha_t, \beta_t)) = Const. + \mathbf{X}_t^T \log(\lambda(\alpha_t, \beta_t)) - \mathbf{1}^T \lambda(\alpha_t, \beta_t) \tag{17} $$

Note that this likelihood model is linear in the intensity image λ given the parameters α_t and β_t, and *not* linear in terms of the parameters themselves. In addition, the likelihood is Poisson, not Gaussian. Hence Kalman filters are not applicable here.

Using Bayes rule we have:

$$ f(\mathbf{Z}_t|\mathbf{X}_t) = P(\alpha_t, \beta_t|\mathbf{X}_t) \propto P(\mathbf{X}_t|\alpha_t, \beta_t)P(\alpha_t, \beta_t) \tag{18} $$

When a generative model is not available, the regressor may be trained discriminatively on a dataset using maximum likelihood, similar to Section 3.2.

3.3.2. Approximating the Predictive Distribution (Step 1 of Equation (8))

Let $P_t \approx \mathcal{N}(\mu_t, \Sigma_t)$ and F be the dynamics of the inverted pendulum. The predictive distribution may be approximated as a gaussian by linearizing F:

$$ \mathbf{Z}_{t+1}|\mathbf{Z}_t \approx \mathcal{N}(F(\mu_t), (\nabla F|_{\mu_t})\Sigma(\nabla F|_{\mu_t})^T) \tag{19} $$

where $\nabla F|_{\mu_t}$ is the Jacobian of the dynamics F evaluated at the prior mean μ_t.

3.3.3. Approximating the Posterior Distribution (Step 4 of Equation (8))

Given weighted samples $\{Z'_s\}_{s=1}^K$ and normalized weights $\{W_s\}_{s=1}^K$ (*i.e.*, $\sum_s W_s = 1$), the posterior P_{t+1} may be approximated by a Gaussian with: $P_{t+1} \approx \mathcal{N}(\mu_{t+1}, \Sigma_{t+1})$, where

$$\mu_{t+1} = \sum_s W_s Z'_s \tag{20}$$

$$\Sigma_{t+1} = \frac{\sum_s W_s (Z'_s - \mu_{t+1})(Z'_s - \mu_{t+1})^T}{1 - \sum_s W_s^2} \tag{21}$$

3.4. Tracking Experiment

The pole of the pendulum has an intensity of 1 while everything else (background, cart, *etc*) has intensity 0. The pole is 3×30 pixels and the entire scene is 80×80 pixels. The pendulum is half the mass of the cart. The maximum photon emission rate is λ_{max}, and dark current is ϵ_{dc}. The pendulum is released at the origin ($\beta_0 = 0$) with α_0 randomly chosen from $\{10°, 20°, 30°, 40°\}$. Each initial state is simulated 25 times and then aggregated, yielding a total of 100 trials for each condition (*i.e.*, for each pair of λ_{max} and ϵ_{dc}). The distribution on α_0 and β_0 used for simulation is not available to the tracking algorithm, which instead assumes uniform distributions for both.

3.4.1. Internal Noise Due to Sampling

For Figure 7a,b we used $K = 1000$ samples from the predictive distribution (Equation (8), step 2). Repeating the experiment using $K = 300$ samples shows the same trend with larger error bars. One counterintuitive finding is that as the exposure time increases, the estimation error first decreases and then diverges (both higher mean and deviation are visible, Figure 7a,b). The degree of divergence is aggravated by increasing signal-to-noise ratio (or reducing dark current ϵ_{dc}). This may be explained by the internal noise in the tracking algorithm. The algorithm relies on samples from the predictive distribution $P(\mathbf{Z}_{t+1}|\mathbf{Z}_t)$ (Equation (8), step 1) to coincide with states that have a high observation likelihood $f(Z|\mathbf{X}_t)$ (Equation (8), step 3). This coincidence is less likely to happen when the observation likelihood becomes precise, or as the posterior becomes sharply-peaked. Similarly, for Figure 7c we used $K = 1000$, and repeated the experiment with $K = 300$. They obtain exactly the same trends except that smaller K corresponds to a higher plateau for the convergence time (Figure 7c), indicating that sampling noise may be limiting the speed for accurate tracking.

4. Discussion and Conclusions

The advent of photon-counting sensors motivates us to reconsider the prevalent paradigm in computer vision: Rather than first capturing an image and then analyzing it, we should design algorithms that incrementally compute information from the stream of photons that hits the sensor, without any attempt to reconstruct the image. This style of thinking is particularly attractive in low light conditions, where the exposure time required for capturing a high-quality image is prohibitively lengthy.

Photon-counting sensors deliver small increments of the image at short delays and high frequencies. We show that this incremental input could in principle be applied to solve a variety of vision problems with a short exposure time. Algorithms that are inspired by the asymptotically optimal SPRT appear particularly well suited for minimizing photon counts while satisfying a desired accuracy bound.

Our first finding is that useful information may be computed in a short amount of time, well ahead of the integration time that is required for forming (or reconstructing) a high quality image. In Figure 1b,c, we see that a low-light classification algorithm can achieve 1% classification error of handwritten digits before one photon per pixel has been collected. The low-light classifier may be viewed as reconstructing the features (instead of the image), and carrying the uncertainty of the features all the way to classification. This uncertainty is essential in a sequential decision making setting to determine when to stop collecting more photons. In comparison, conventional approaches simply reconstruct the image, and pass it to a classifier trained on high-quality images. The conventional approach suffers from two issues. (1) Since the conventional approach discards the uncertainty information, it is not clear how to determine the required exposure time; and (2) statistics of the reconstruction may be different from that of high-quality images, hence the classifier's performance may not be guaranteed.

Second, algorithms for classification and search from streams of photons are photon-efficient: they stop as soon as a confident decision is made. This efficiency is critical for domains such as astrophysics where each photon is precious [19], and cell imaging applications where the dies that are employed to visualize cell structures are phototoxic [20,21]. As an example of the photon efficiency, Figure 3a shows that at PPP = 1 the low-light classifier based on SPRT can already achieve a better performance than a classifier using PPP = 220. Additionally, contrary to the conventional paradigm that obtains images with a fixed duration, low-light classifiers and search algorithms uses different exposure times depending on the specific photon arrival sequence (Figure 3b) and on the overall classification difficulty of the example (Figure 3b,c).

Third, algorithms become faster when more light is present. For classification and search where the input image is stationary, time is synonymous with the amount of photons. Higher illuminance therefore translates to faster decisions. This simple relationship is useful in that a low-light system trained for classification or search at one illuminance level may be easily applied at another illuminance level. The transition only requires knowing the illuminance level of the new scene, which may be estimated either via an explicit illuminance sensor or from the total photon count across the image [22]. In addition, the ER *vs.* median PPP tradeoff (Figures 3a and 6a–c) is an illuminance-independent characteristic of the algorithm and the task.

Last, the relationship between illuminance and speed is not always simple in tracking. The dynamics governing the object movement/state transition has its own time scale. A regressor $f(\mathbf{X}_t)$ thus has only a finite duration for integrating information before the object moves too far. A tracker relies on accurate prediction of the regressor to postulate the object's next position. An inaccurate prediction due to short exposure time may cause tracking failure. In addition, internal noise in the tracking algorithm (Section 3.4) may cause the speed to plateau after a certain illuminance level. As a result, the relationship between illuminance and convergence time is not a simple inversely proportional relationship, as shown in Figure 8c.

Although we have provided proof-of-concept illustrations of low-light vision applications with photon-counting sensors, many challenges still remain. (1) We are not aware of any hardware specialized at processing streams of photon counts at high speeds. Nonetheless, current Field-programmable Gate Array (FPGA) implementations have achieved over 2000 Hz throughput for classifying images of a similar resolution as those in Section 2.2 [23]. In addition, the low-light classifiers implemented as a recurrent neural network (Section 2.2) can be updated incrementally, *i.e.*, $f(\mathbf{X}_{1:t})$ can be computed from the internal states of $f(\mathbf{X}_{1:t-1})$ and $\mathbf{X}_t$ with sparse updates. The sparseness may be key to expedite computation. (2) We do not yet have datasets collected directly from photon-counting sensors to verify the robustness of the proposed methodology, as many such sensors are still in the making [5–7]. (3) Our noise model (Section 3.1) may be a crude approximation to handle moving objects. For example, we do not model motion induced blur or input disturbances due to camera self-motion. Nonetheless, motion induced blur may not be an issue if the sensor is collecting a single photon at a time, such as in low light and/or high-frequency imaging scenarios. In these scenarios even though the amount of photons is so low that full image reconstruction is difficult, our algorithm can still make correct and full use of the where and when of photon arrivals. This is precisely the advantage of image-free vision.

In conclusion, we propose to integrate computer vision with photon-counting sensors to address the challenges facing low-light vision applications. We should no longer wait for a high-quality image to be formed before executing the algorithm. Novel algorithms and hardware solutions should be developed to operate on streams of photon-counts. These solutions should also sidestep image reconstruction and focus directly on the task at hand.

Acknowledgments: This work was funded by Office of Naval Research (ONR) N00014-10-1-0933 and Gordon and Betty Moore Foundation (Palo Alto , CA, USA).

Author Contributions: B. Chen and P. Perona conceived and designed the experiments. B. Chen performed the experiments and analyzed the data. B. Chen and P. Perona wrote the paper.

Conflicts of Interest: The authors declare no conflict of interest.

Abbreviations

The following abbreviations are used in this manuscript:

CPU	Central Processing Unit
ER	Error rates
FPGA	Field-programmable Gate Arrays
MNIST	The Mixed National Institute of Standards and Technology dataset
PPP	The number of photons per pixel in a low-light image, averaged across all locations and the imaging duration
SPRT	Sequential probability ratio test

References

1. Hall, E.; Brenner, D. Cancer risks from diagnostic radiology. *Cancer* **2014**, *81*, doi:10.1259/bjr/01948454.
2. Stephens, D.J.; Allan, V.J. Light microscopy techniques for live cell imaging. *Science* **2003**, *300*, 82–86.
3. Brida, G.; Genovese, M.; Berchera, I.R. Experimental realization of sub-shot-noise quantum imaging. *Nat. Photonics* **2010**, *4*, 227–230.
4. Zappa, F.; Tisa, S.; Tosi, A.; Cova, S. Principles and features of single-photon avalanche diode arrays. *Sens. Actuators A Phys.* **2007**, *140*, 103–112.
5. Fossum, E. The quanta image sensor (QIS): Concepts and challenges. In Proceedings of the Computational Optical Sensing and Imaging 2011, Toronto, ON, Canada, 10–14 July 2011; doi:10.1364/COSI.2011.JTuE1.
6. Fossum, E.R. Multi-Bit Quanta Image Sensors. In Proceedings of the International Image Sensor Workshop, Vaals, The Netherlands, 8–12 June 2015; pp. 292–295.
7. Sbaiz, L.; Yang, F.; Charbon, E.; Süsstrunk, S.; Vetterli, M. The gigavision camera. In Proceedings of the IEEE International Conference on Acoustics, Speech and Signal Processing, Taipei, Taiwan, 19–24 April 2009; pp. 1093–1096.
8. Morris, P.A.; Aspden, R.S.; Bell, J.E.; Boyd, R.W.; Padgett, M.J. Imaging with a small number of photons. *Nat. Commun.* **2015**, *6*, do:10.1038/ncomms6913.
9. Chen, B.; Perona, P. Scotopic Visual Recognition. In Proceedings of the IEEE International Conference on Computer Vision Workshops, Santiago, Chile, 7–13 December 2015; pp. 8–11.
10. Abu-Naser, A.; Galatsanos, N.P.; Wernick, M.N. Methods to detect objects in photon-limited images. *JOSA A* **2006**, *23*, 272–278.
11. LeCun, Y.; Bottou, L.; Bengio, Y.; Haffner, P. Gradient-based learning applied to document recognition. *IEEE Proc.* **1998**, *86*, 2278–2324.
12. Wald, A. Sequential tests of statistical hypotheses. *Ann. Math. Stat.* **1945**, *16*, 117–186.
13. Elman, J.L. Distributed representations, simple recurrent networks, and grammatical structure. *Mach. Learn.* **1991**, *7*, 195–225.
14. Chen, B.; Perona, P. Speed versus accuracy in visual search: Optimal performance and neural architecture. *J. Vis.* **2015**, *15*, 9–9.
15. Kalman, R.E. A new approach to linear filtering and prediction problems. *J. Fluids Eng.* **1960**, *82*, 35–45.
16. Pitt, M.K.; Shephard, N. Filtering via simulation: Auxiliary particle filters. *J. Amer. Stat. Assoc.* **1999**, *94*, 590–599.
17. Liberzon, D. *Switching in Systems and Control*; Springer Science & Business Media: New York, NY, USA, 2012.
18. Vedaldi, A.; Lenc, K. MatConvNet—Convolutional Neural Networks for MATLAB. Available online: http://arxiv.org/abs/1412.4564 (accessed on 4 April 2016).
19. Martin, D.C.; Chang, D.; Matuszewski, M.; Morrissey, P.; Rahman, S.; Moore, A.; Steidel, C.C. Intergalactic medium emission observations with the Cosmic Web Imager. I. The circum-QSO medium of QSO 1549+19, and evidence for a filamentary gas inflow. *Astrophys. J.* **2014**, *786*, 106–106.

Sensors **2016**, *16*, 484

20. Hoebe, R.; Van Oven, C.; Gadella, T.W.; Dhonukshe, P.; Van Noorden, C.; Manders, E. Controlled light-exposure microscopy reduces photobleaching and phototoxicity in fluorescence live-cell imaging. *Nat. Biotechnol.* **2007**, *25*, 249–253.

21. Ji, N.; Magee, J.C.; Betzig, E. High-speed, low-photodamage nonlinear imaging using passive pulse splitters. *Nat. Methods* **2008**, *5*, 197–202.

22. Cheng, D.; Price, B.; Cohen, S.; Brown, M.S. Effective Learning-Based Illuminant Estimation Using Simple Features. In Proceedings of the IEEE Conference on Computer Vision and Pattern Recognition, Boston, MA, USA, 7–12 June 2015; pp. 1000–1008.

23. Ovtcharov, K.; Ruwase, O.; Kim, J.Y.; Fowers, J.; Strauss, K.; Chung, E.S. Accelerating Deep Convolutional Neural Networks Using Specialized Hardware, Microsoft Research Whitepaper. Available online: http://research.microsoft.com/apps/pubs/?id=240715 (accessed on 4 April 2016).

![sensors logo] *sensors*

MDPI

Review

Three-Dimensional Photon Counting Imaging with Axially Distributed Sensing

Myungjin Cho [1],* and Bahram Javidi [2]

[1] Department of Electrical, Electronic, and Control Engineering,
    Institute of Information Telecommunication Convergence (IITC), Hankyong National University,
    327 Chungang-ro, Anseong-si, Kyonggi-do 456-749, Korea
[2] Electrical and Computer Engineering Department, University of Connecticut, Unit 4157, Storrs, CT 06269,
    USA; bahram@engr.uconn.edu
* Correspondence: mjcho@hknu.ac.kr; Tel.: +82-31-670-5298

Academic Editor: Eric R. Fossum
Received: 25 March 2016; Accepted: 25 July 2016; Published: 28 July 2016

Abstract: In this paper, we review three-dimensional (3D) photon counting imaging with axially distributed sensing. Under severely photon-starved conditions, we have proposed various imaging and algorithmic approaches to reconstruct a scene in 3D, which are not possible by using conventional imaging system due to lack of sufficient number of photons. In this paper, we present an overview of optical sensing and imaging system along with dedicated algorithms for reconstructing 3D scenes by photon counting axially distributed sensing, which may be implemented by moving a single image sensor along its optical axis. To visualize the 3D image, statistical estimation methods and computational reconstruction of axially distributed sensing is applied.

Keywords: axially distributed sensing; photon counting imaging; statistical estimation; Poisson distribution

1. Introduction

Under severely photon-starved conditions, scenes recorded optically may not be properly reconstructed by conventional imaging systems. In many fields, such as noninvasive microscopy, night vision, astronomy, military applications, etc., image acquisition or visualization may be carried out in a low light level environment. Recently, many approaches for three-dimensional (3D) photon counting imaging have been reported [1–6]. For example, 3D information can be recorded and reconstructed under photon-starved conditions with photon counting integral imaging [7–13]. In 3D photon counting imaging, 3D images can be visualized by statistical estimations, such as maximum likelihood estimation [2] and Bayesian approaches [4]. Photon counting detection under such conditions can be modeled using a Poisson distribution since photon events may occur rarely in unit time and space [14]. Using this mathematical photon counting imaging model, 3D visualization and object recognition can be performed under photon-starved conditions. Additionally, optical encryption with improved security level has been accomplished photon counting imaging properties [15–17].

In photon counting imaging, the visual quality of the recorded image or the reconstructed image depends on the number of photons from the scene. To enhance its visual quality, some techniques have been proposed [2–4]. Three-dimensional photon counting imaging captures multiple 2D images from the scenes using a lenslet array or moving camera. The statistical properties of the optical rays, as well as photon counting, are used and the visual quality of the reconstructed image can be enhanced.

To obtain the 3D information with high resolution from the scenes and remove the requirement of lateral parallax of image sensor by integral imaging [11], axially distributed sensing (ADS) may be used [12]. In order to obtain the 3D information, ADS uses only a single camera moving along its optical axis. We show that photon counting with ADS [5] can be used to obtain the 3D information of the scenes under photon-starved conditions. We can obtain the depth map of the 3D objects. Thus, we can create the 3D profile of the object and regenerate elemental images for multiple viewing points for 3D display [18–21].

In this paper, we present an overview of the basic concept of photon counting imaging, and our work on 3D reconstruction using 3D photon counting ADS along with some experiments to illustrate 3D photon counting imaging with ADS.

2. 3D Photon Counting Imaging

2.1. Mathematical Model of Photon Counting Detection

Photon counting detection may be modeled by Poisson distribution because the photon events occur rarely in unit time and space under photon-starved conditions [14]. The photon counting detection fundamental steps are illustrated in Figure 1. For computational simplicity, the image has only one-dimension. Using the following equation, the photon counting image can be constructed [2].

$$\lambda_x = \frac{I_x}{\sum_{x=1}^{N_x} I_x} \tag{1}$$

$$C_x | \lambda_x \sim Poisson\left(N_p \lambda_x\right) \tag{2}$$

where I_x is the light intensity of the image at pixel x, N_x is the total number of pixels in the image, λ_x is the normalized irradiance at pixel x, N_p is the extracted number of photons from the image, C_x is the number of photons at pixel x, respectively. Here, the total energy of λ_x is unity.

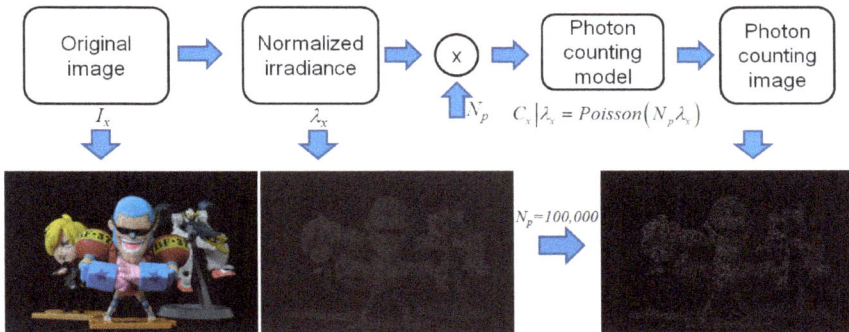

Figure 1. Mathematical model of photon counting detector.

Now, we have 2D photon-limited images, which are generated from Equations (1) and (2). When N_p is very small or the scenes are under severely photon-starved conditions, the image cannot be visualized or recognized. We have shown that passive 3D imaging technique such as integral imaging [1–4] can enhance the visual quality of these photon-limited images and obtain the 3D information. In integral imaging, 3D information can be recorded through a lenslet array or a camera array. Here, multiple 2D images with different perspectives can be acquired. These images are referred to as elemental images. To obtain high lateral and depth resolutions, synthetic aperture integral imaging (SAII) [11], which uses multiple cameras can be used. Then, using computational integral imaging reconstruction (CIIR) [13], 3D images with enhanced visual quality can be reconstructed. In 3D photon counting integral imaging, to estimate 3D information from multiple photon-limited images, statistical estimations such as maximum likelihood estimation (MLE) [2] may be used, as well as computational reconstruction [13].

2.2. Photon Counting Axially Distributed Sensing (ADS)

To remove lateral movement of image sensor required by SAII, axially distributed sensing has been reported [12]. It can capture multiple 2D images with slightly different perspectives by moving single image sensor along its optical axis. Thus, we can adopt this technique to 3D photon counting imaging. Figure 2 shows the basic concept of 3D photon counting imaging with ADS for pickup and reconstruction.

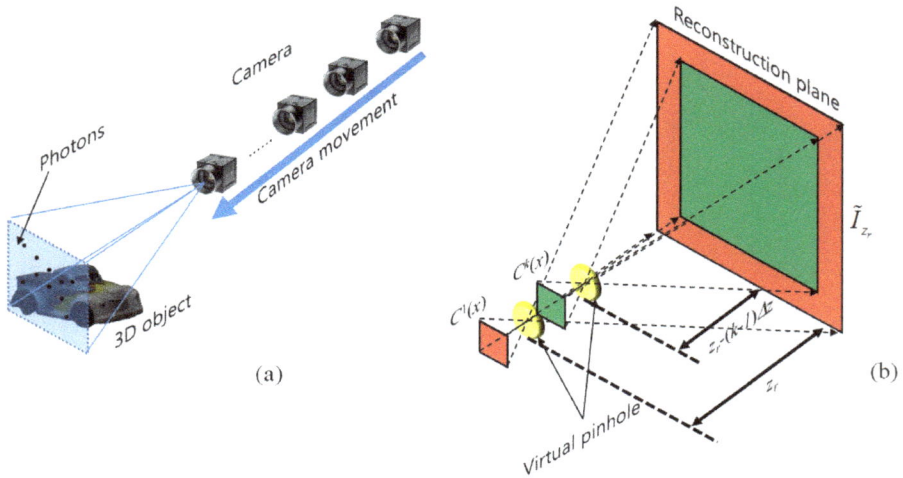

Figure 2. Photon counting imaging with ADS. (**a**) Image sensing; (**b**) Computational reconstruction, k is the number of recorded images by ADS.

In the pickup process, by moving single photon counting camera along its optical axis, multiple 2D images with slightly different perspectives can be recorded as shown in Figure 2a. Computational reconstruction of ADS can be implemented by considering slightly different magnification ratio for each photon-limited image as the following equation [5]:

$$M_k(z_r) = \frac{z_r - k\Delta z}{z_r} \tag{3}$$

where Δz is the moving step for single camera along its optical axis, z_r is the reconstruction depth, and k is the index of the recorded images by ADS, respectively. Since magnification causes the degradation of the reconstructed 3D image quality by the image interpolation method, in this paper, demagnification is used. Then, using MLE process [2], computational reconstruction of photon counting ADS as shown in Figure 2b can be implemented by follows [5]:

$$L(N_p\lambda_k | C_k) = \prod_{k=1}^{K} \frac{(N_p\lambda_k)^{C_k} e^{-N_p\lambda_k}}{C_k!} \tag{4}$$

$$l(N_p\lambda_k | C_k) \propto \sum_{k=1}^{K} [C_k \log(N_p\lambda_k) - N_p\lambda_k] \tag{5}$$

$$\frac{\partial(N_p\lambda_k | C_k)}{\partial \lambda_k} = 0, \quad \hat{\lambda}_k = \frac{C_k}{N_p} \tag{6}$$

$$\hat{I}(x, z_r)_{ADS} = \frac{1}{N_p K} \sum_{k=1}^{K} C_k \left(\frac{x}{M_k(z_r)} \right) \tag{7}$$

where $L(\cdot \mid \cdot)$ and $l(\cdot \mid \cdot)$ are the likelihood function and log-likelihood function, C_k is the kth photon-limited elemental image by ADS and K is the total number of the captured photon-limited elemental images, respectively.

3. Experimental Results

3.1. Photon Counting Imaging with Axially Distributed Sensing

3.1.1. Experimental Setup

Experimental setup for photon counting ADS is illustrated in Figure 3a. In this setup, the focal length of the camera lens is 50 mm. The camera has 1000 (H) × 1000 (V) pixels and axial separation between moving image sensor, Δz, is 2 mm. We used a 3D car model in Figure 3b as the 3D object. Its location is $z_r = 320$ mm. Finally, we recorded 50 multiple images using ADS. In this experiment, we use both non-occluded and occluded 3D objects as shown in Figure 3b,c. The photon counting images are obtained digitally by applying the Poisson model (Equations (1) and (2)) to the digitally captured elemental images.

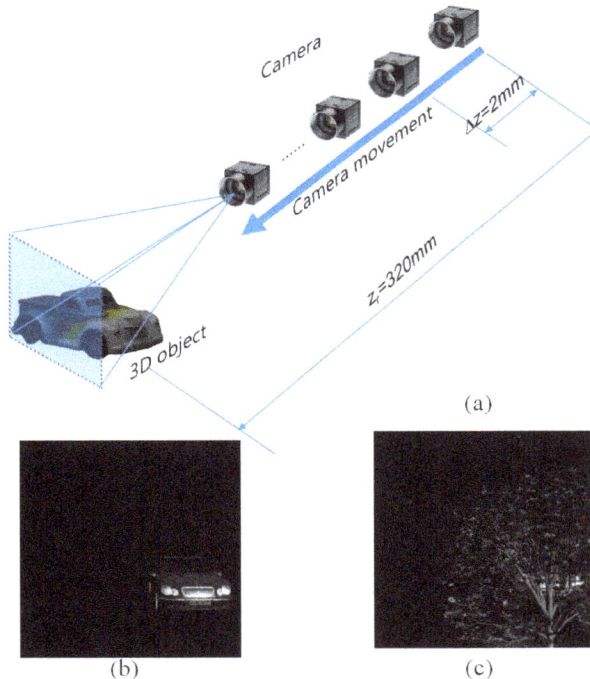

(a)

(b)

(c)

Figure 3. (**a**) Experimental setup for ADS; (**b**) Non-occluded 3D object; (**c**) Partially occluded 3D object.

Figure 4 shows conventional elemental images by ADS for non-occluded 3D object and occluded 3D object. It is noticed that slightly different perspectives between farthest and closest images exist. Thus, using these perspectives and computational reconstruction of ADS, we can reconstruct 3D images.

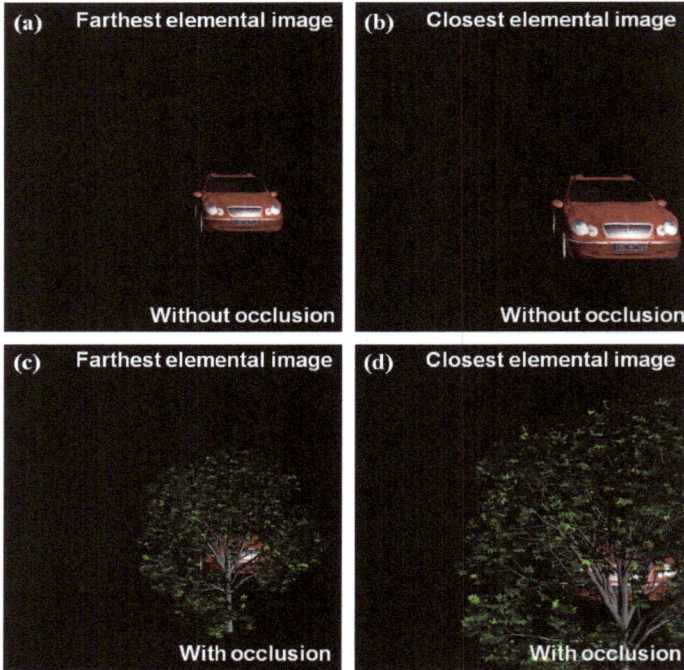

Figure 4. Conventional elemental images with slightly different perspectives by ADS, for: (**a,b**) non-occluded object; (**c,d**) occluded object, respectively.

3.1.2. Results

Figure 5 shows photon-limited elemental images for photon counting ADS of non-occluded and occluded objects at farthest and closest positions, which are obtained by using Equations (1) and (2) applied to the digitally captured images. Since the number of photons is low, that is $N_p = 10,000$ (0.01 photons/pixel), its visual quality is low and the objects are not well recognized. However, using computational reconstruction of ADS, as depicted in Equations (3)–(7), the visual quality of the reconstructed image can be improved, as shown in Figure 6. To evaluate the visual quality of the reconstructed 3D images, we calculate the peak signal to noise ratio (PSNR) between the original reconstructed Three-dimensional images using conventional ADS and the experimental results as shown in Figure 7. The plot has some fluctuations because photons are generated by Poisson random process.

Using multiple images obtained by ADS, we can regenerate the elemental images for 3D multi-view display [18]. Then, the depth map of the 3D objects can be extracted by 3D profilometry [19]. In [19], the extracted depth with contours of equal depth is shown. The errors of the estimated depth may occur due to the specular reflection off of the glossy surface, which departs from the Lambertian assumption. In addition, in [19], the computational reconstruction results of the 3D objects by slicing them in a certain depth range are shown.

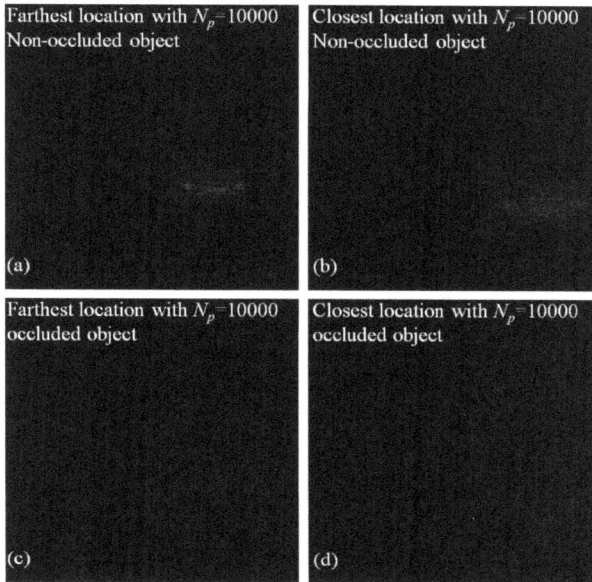

Figure 5. Photon-limited elemental images with N_p = 10,000 of non-occluded object and occluded object at (**a**,**c**) farthest location from camera and (**b**,**d**) closest location from camera.

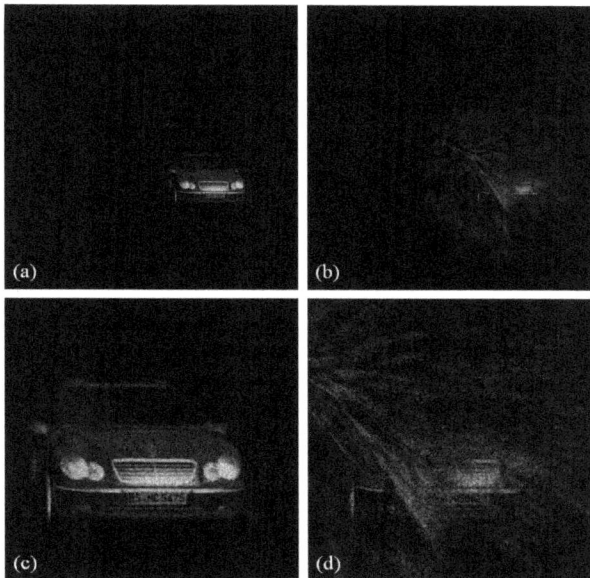

Figure 6. Reconstruction results of photon counting ADS with N_p = 10,000 for (**a**) non-occluded object and (**b**) occluded object. Enlarged part of the reconstructed 3D images for (**c**) non-occluded object and (**d**) occluded object.

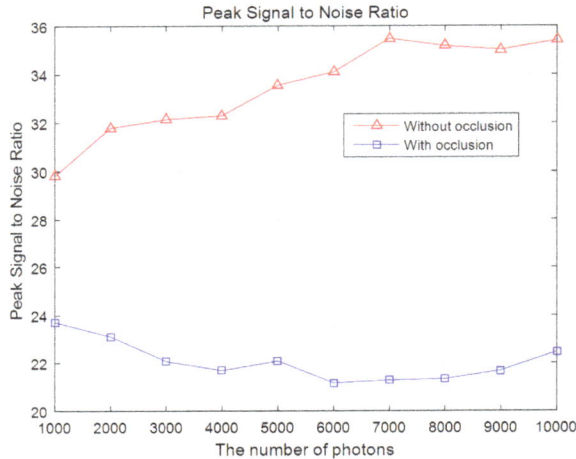

Figure 7. Peak Signal to Noise Ratio (PSNR).

4. Conclusions

In this paper, we have presented an overview of 3D photon counting imaging system under photon-starved conditions using ADS. Photon counting ADS uses camera movement along the optical axis unlike photon counting integral imaging which requires lateral parallax. It can reconstruct 3D images under photon starved conditions, including occluded objects. For real-time photon counting ADS, faster reconstruction algorithm with a Graphic Processing Unit (GPU) may be required.

Acknowledgments: This research was supported in part by Basic Science Research Program through the National Research Foundation of Korea (NRF) funded by the Ministry of Education (NRF-2015R1A2A1A16074936). B. Javidi acknowledges support from NSF under grant # NSF/IIS-1422179.

Author Contributions: M. Cho and B. Javidi conceived and designed the experiments; M. Cho performed the experiments; M. Cho and B. Javidi analyzed the data; M. Cho and B. Javidi wrote the paper.

Conflicts of Interest: The authors declare no conflict of interest.

Abbreviations

The following abbreviations are used in this manuscript:

ADS	Axially distributed sensing
CIIR	Computational integral imaging reconstruction
MLE	Maximum likelihood estimation
SAII	Synthetic aperture integral imaging

References

1. Yeom, S.; Javidi, B.; Watson, E. Photon counting passive 3D image sensing for automatic target recognition. *Opt. Exp.* **2005**, *13*, 9310–9330. [CrossRef]
2. Tavakoli, B.; Javidi, B.; Watson, E. Three dimensional visualization by photon counting computational integral imaging. *Opt. Exp.* **2008**, *16*, 4426–4436. [CrossRef]
3. DaneshPanah, M.; Javidi, B.; Watson, E. Three dimensional object recognition with photon counting imagery in the presence of noise. *Opt. Exp.* **2010**, *18*, 26450–26460. [CrossRef] [PubMed]
4. Jung, J.; Cho, M.; Dey, D.K.; Javidi, B. Three-dimensional photon counting integral imaging using Bayesian estimation. *Opt. Lett.* **2010**, *35*, 1825–1827. [CrossRef] [PubMed]
5. Cho, M.; Javidi, B. Three-dimensional photon counting axially distributed image sensing. *IEEE/OSA J. Disp. Tech.* **2013**, *9*, 56–62. [CrossRef]

6.   Cho, M.; Mahalanobis, A.; Javidi, B. 3D passive photon counting automatic target recognition using advanced correlation filters. *Opt. Lett.* **2011**, *36*, 861–863. [CrossRef] [PubMed]

7.   Lippmann, G. La Photographie Integrale. *Comp. Rend. Acad. Sci.* **1908**, *146*, 446–451. (In French)

8.   Okoshi, T. *Three-Dimensional Imaging Technique*; Academic: New York, NY, USA, 1976.

9.   Okano, F.; Arai, J.; Mitani, K.; Okui, M. Real-time integral imaging based on extremely high resolution video system. *IEEE Proc.* **2006**, *94*, 490–501. [CrossRef]

10. Javidi, B.; Okano, F.; Son, J.Y. *Three-Dimensional Imaging, Visualization, Display*; Springer: New York, NY, USA, 2009.

11. Jang, J.-S.; Javidi, B. Three-dimensional synthetic aperture integral imaging. *Opt. Lett.* **2002**, *27*, 1144–1146. [CrossRef] [PubMed]

12. Schulein, R.; DaneshPanah, M.; Javidi, B. 3D imaging with axially distributed sensing. *Opt. Lett.* **2009**, *34*, 2012–2014. [CrossRef] [PubMed]

13. Hong, S.-H.; Jang, J.-S.; Javidi, B. Three-dimensional volumetric object reconstruction using computational integral imaging. *Opt. Exp.* **2004**, *12*, 483–491. [CrossRef]

14. Goodman, J.W. *Statistical Optics*; Wiley: New York, NY, USA, 1985.

15. Perez-Cabre, E.; Abril, H.C.; Millan, M.; Javidi, B. Photon-counting double-random-phase encoding for secure image verification and retrieval. *J. Opt.* **2012**, *14*. [CrossRef]

16. Pérez-Cabré, E.; Cho, M.; Javidi, B. Information authentication using photon counting double random phase encrypted images. *J. Opt. Lett.* **2010**, *35*, 22–24. [CrossRef] [PubMed]

17. Matoba, O.; Nomura, T.; Perez-Cabre, E.; Millan, M.S.; Javidi, B. Optical techniques for information security. *IEEE J. Proc.* **2009**, *97*, 1128–1148. [CrossRef]

18. Cho, M.; Shin, D. 3D integral imaging display using axially recorded multiple images. *J. Opt. Soc. Korea* **2013**, *17*, 410–414. [CrossRef]

19. Daneshpanah, M.; Javidi, B. Profilometry and optical slicing by passive three dimensional imaging. *Opt. Lett.* **2009**, *34*, 1105–1107. [CrossRef] [PubMed]

20. Pollefeys, M.; Koch, R.; Vergauwen, M.; Deknuydt, B.; Gool, L.V. Three-dimensional scene reconstruction from images. *Proc. SPIE* **2000**, *3958*, 215–226.

21. Saavedra, G.; Martinez-Cuenca, R.; Martinez-Corral, M.; Navarro, H.; Daneshpanah, M.; Javidi, B. Digital slicing of 3D scenes by Fourier filtering of integral images. *Opt. Express* **2008**, *16*, 17154–17160. [CrossRef] [PubMed]

sensors

MDPI

Article

Images from Bits: Non-Iterative Image Reconstruction for Quanta Image Sensors

Stanley H. Chan [1,2,*], Omar A. Elgendy [1] and Xiran Wang [1]

[1] School of Electrical and Computer Engineering, Purdue University, 465 Northwestern Ave,
    West Lafayette, IN 47907, USA; oelgendy@purdue.edu (O.A.E.); wang470@purdue.edu (X.W.)
[2] Department of Statistics, Purdue University, 250 N. University Street, West Lafayette, IN 47907, USA
* Correspondence: stanchan@purdue.edu; Tel.: +1-765-496-0230

Academic Editor: Eric R. Fossum
Received: 8 September 2016; Accepted: 17 November 2016; Published: 22 November 2016

Abstract: A quanta image sensor (QIS) is a class of single-photon imaging devices that measure light intensity using oversampled binary observations. Because of the stochastic nature of the photon arrivals, data acquired by QIS is a massive stream of random binary bits. The goal of image reconstruction is to recover the underlying image from these bits. In this paper, we present a non-iterative image reconstruction algorithm for QIS. Unlike existing reconstruction methods that formulate the problem from an optimization perspective, the new algorithm directly recovers the images through a pair of nonlinear transformations and an off-the-shelf image denoising algorithm. By skipping the usual optimization procedure, we achieve orders of magnitude improvement in speed and even better image reconstruction quality. We validate the new algorithm on synthetic datasets, as well as real videos collected by one-bit single-photon avalanche diode (SPAD) cameras.

Keywords: single-photon image sensor; quanta image sensor (QIS); image reconstruction; quantized Poisson statistics; image denoising; Anscombe Transform; maximum likelihood estimation (MLE)

1. Introduction

1.1. Quanta Image Sensor

Since the birth of charge coupled devices (CCD) in the late 1960s [1] and the complementary metal-oxide-semiconductor (CMOS) active pixel sensors in the early 1990s [2], the pixel pitch of digital image sensors has been continuously shrinking [3]. Shrinking the pixel pitch is intimately linked to the need of increasing image resolution, reducing power consumption and reducing the size and weight of cameras. However, as pixel pitch shrinks, the amount of photon flux detectable by each pixel drops, leading to reduced signal strength. In addition, the maximum number of photoelectrons that can be held in each pixel, known as the full-well capacity, also drops. Small full-well capacity causes reduced maximum signal-to-noise ratio and lowers the dynamic range of an image [4]. Therefore, pushing for smaller pixels, although feasible in the near future, will become a major technological hurdle to new image sensors.

A quanta image sensor (QIS) is a class of solid-state image sensors originally proposed by Eric Fossum as a candidate solution for sub-diffraction-limit pixels. The sensor was first named the digital film sensor [5] and later the quanta image sensor [6–9] (see [10] for a more comprehensive discussion of the history). A similar idea to QIS was developed a few years later by EPFL (École Polytechnique Fédérale de Lausanne, Lausanne, Switzerland), known as the Gigavision camera [11–13]. In the past few years, research groups at the University of Edinburgh (Edinburgh, UK) [14–16], as well as EPFL [17,18] have made new progresses in QIS using binary single-photon avalanche diode (SPAD) cameras. In the industry, Rambus Inc. (Sunnyvale, CA, USA) is developing binary image sensors for high dynamic range imaging [19–21]. For the purpose of this paper, we shall not differentiate these sensors, but refer to them generally as the QIS, because their underlying mathematical principles are similar.

The working principle of QIS is as follows: In CCD and CMOS, one considers each pixel as a "bucket" that collects and integrates photoelectrons. The bucket is partitioned in QIS into thousands of nanoscale cells referred to as "jots". Each jot is capable of detecting a single photon to generate a binary response indicating whether the photon count is above or below a certain threshold q. If the photon count is above q, the sensor outputs a "1"; If the photon count is below q, the sensor outputs a "0". QIS has a very small full-well capacity because it is not designed to accumulate photons. Since the binary response is generated as soon as the number of photons exceeds the threshold, QIS can be operated at very high speed. For example, using single-photon avalanche diodes (SPAD), one can achieve 10k frames per second with a spatial resolution of 320×240 pixels [14] or even 156k frames per second with a spatial resolution of 512×128 pixels [17]. For a higher spatial resolution, Massondian et al. [9] reports a QIS operating at 1000 frames per second for a spatial resolution of 1376×768 pixels.

From a signal processing perspective, the challenge of QIS is the extremely lossy process using binary measurements to acquire the light intensity. In order to compensate for the loss, QIS over-samples the space by using a large number of jots and takes multiple exposures in time. This technique is similar to the classic approach in oversampled analog-to-digital conversions.

1.2. Scope and Contribution

The theme of this paper is about how to reconstruct images from the one-bit quantized measurements. Image reconstruction is a critical component of QIS, for without such an algorithm, we will not be able to form images. However, unlike classical Poisson image recovery problems where solutions are abundant [22–27], the one-bit quantization of QIS makes the problem uniquely challenging, and there is a limited number of existing methods [28–31]. Another challenge we have to overcome is the complexity of the algorithm, which has to be low enough that we can put them on cameras to minimize power consumption, memory consumption and runtime. Numerical optimization algorithms are generally not recommended if they are iterative and require intensive computation for every step.

The main contribution of this paper is a non-iterative image reconstruction algorithm for QIS data. We emphasize the non-iterative nature of the algorithm as it makes the algorithm different from existing methods. The new algorithm is based on a transform-denoise framework. The idea is to apply a variance stabilizing transform known as the Anscombe transform [32] to convert a sum of one-bit quantized Poisson random variables to binomial random variables with equal variances. When variance is stabilized, standard image denoising algorithms can be applied to smooth out the noise in the image. Transform-denoise is a single-pass algorithm with no iteration. Empirically, we find that the new algorithm achieves two orders of magnitude improvement in speed and provides an even better reconstruction result than existing iterative algorithms.

The rest of the paper is organized as follows. First, in Section 2, we present the imaging model of QIS. We discuss how the light intensity of the scene is over-sampled and what statistics do the one-bit quantized measurements have. Unlike existing works, which typically assume a quantization level $q = 1$, we make no assumption about q, except that it is a positive integer. This requires some discussion about the incomplete Gamma function. We also discuss why a simple summation over a local spatial-temporal volume is insufficient to reconstruct images. Section 3 presents the main algorithm. We discuss the concept of variance stabilizing transform, its derivation and its limitations. We also discuss how various image denoising algorithms can be plugged into the framework. In Section 4, we present experimental results. There are two sets of data we will discuss. One is a synthetic dataset in which we can objectively measure the reconstruction quality. The other one is a set of real videos captured by SPAD cameras. We compare the proposed algorithm with existing methods. Proofs of major theorems are given in the Appendix A.

2. QIS Imaging Model

In this section, we provide a quick overview of the QIS imaging model. A similar description of the model was previously discussed in [13,29]. For notational simplicity, we consider one-dimensional signals. The extension to the two-dimensional images is straightforward. Furthermore, we follow [5] by referring to sub-pixels of a QIS as "jots".

The mathematics of QIS is built upon two concepts: (1) a spatial oversampling process to model the acquisition by the sensor; and (2) a quantized Poisson process to model the photon arrivals. The block diagram of the model is summarized in Figure 1.

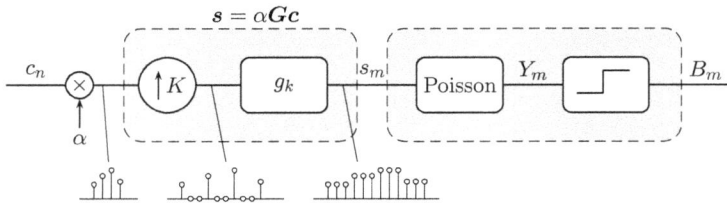

Figure 1. Block diagram of the QIS imaging model. An input signal $c_n \in [0,1]$ is scaled by a constant $\alpha > 0$. The first part of the block diagram is the upsampling ($\uparrow K$) followed by a linear filter $\{g_k\}$. The overall process can be written as $s = \alpha Gc$. The second part of the block diagram is to generate a binary random variable B_m from Poisson random variable Y_m. The example at the bottom shows the case where $K = 3$.

2.1. Oversampling Mechanism

We represent the light intensity of a scene using a digital signal $c = [c_0, c_1, \ldots, c_{N-1}]^T$. To avoid the ambiguity of the scaling, we assume that c_n is normalized so that $c_n \in [0,1]$ for $n = 0, 1, \ldots, N-1$. A fixed and known constant $\alpha > 0$ is multiplied to scale c_n to the proper range.

QIS is a spatial oversampling device. For an N-element signal c, QIS uses $M \gg N$ number of jots to acquire c. Thus, every c_n in the signal c is sampled by M/N jots. The ratio $K \stackrel{\text{def}}{=} M/N$ is called the spatial oversampling factor. To model the oversampling process, we follow [13] by considering an upsampling operator ($\uparrow K$) and a low-pass filter with filtering coefficients $\{g_k\}$, as shown in Figure 1. Since upsampling and low-pass filtering are linear operations, the overall process can be compactly expressed using a matrix-vector multiplication

$$s = \alpha Gc, \tag{1}$$

where $s = [s_0, s_1, \ldots, s_{M-1}]^T$ is the light intensity arriving at those M jots, and the matrix $G \in \mathbb{R}^{M \times N}$ encapsulates the upsampling process and the linear filtering.

There are a variety of choices for the low-pass filter depending on which filter provides a better model of the light intensity. In this paper, we make the following assumption about the low-pass filter.

Assumption 1. *We assume that $\{g_k\}$ is a boxcar function, such that $g_k = 1/K$ for $k = 0, \ldots, K-1$. Consequently, the matrix G is*

$$G = \frac{1}{K} I_{N \times N} \otimes 1_{K \times 1}, \tag{2}$$

where $1_{K \times 1}$ is a K-by-one all one vector, $I_{N \times N}$ is an N-by-N identity matrix and $\otimes$ denotes the Kronecker product.

Intuitively, what Assumption 1 does is to assume that the light intensity is piecewise constant. As will be discussed in Section 3, this is important for us to derive simple algorithms.

2.2. Quantized Poisson Observation

At the surface of the m-th jot, the light intensity s_m generates y_m photons according to the Poisson distribution

$$\mathbb{P}\{Y_m = y_m \;;\; s_m\} = \frac{s_m^{y_m} e^{-s_m}}{y_m!}, \tag{3}$$

where Y_m denotes the Poisson random variable at the m-th jot and y_m denotes the realization of Y_m.

The one-bit quantization of QIS is a truncation of y_m with respect to a threshold $q \in \mathbb{Z}^+$. Precisely, the observed one-bit measurement at the m-th jot is

$$b_m \overset{\text{def}}{=} \begin{cases} 1, & \text{if } y_m \geq q, \\ 0, & \text{if otherwise.} \end{cases} \tag{4}$$

Denoting B_m the random variable of the one-bit measurement at the m-th jot, it follows that the probability of observing $B_m = 1$ is

$$\mathbb{P}\{B_m = 1 \;;\; s_m\} = \mathbb{P}\{Y_m \geq q \;;\; s_m\} = \sum_{k=q}^{\infty} \frac{s_m^k e^{-s_m}}{k!}, \tag{5}$$

and the probability of observing $B_m = 0$ is $\mathbb{P}\{B_m = 0 \;;\; s_m\} = \sum_{k=0}^{q-1} \frac{s_m^k e^{-s_m}}{k!}$.

For general q, keeping track of the sum of exponentials in Equation (5) could be cumbersome. To make our notations simple, we adopt a useful function called the incomplete Gamma function [33], defined as follows.

Definition 1. *The incomplete Gamma function* $\Psi_q : \mathbb{R}^+ \rightarrow [0,1]$ *for a fixed threshold* $q \in \mathbb{Z}^+$ *is defined as*

$$\Psi_q(s) \overset{\text{def}}{=} \frac{1}{\Gamma(q)} \int_s^{\infty} t^{q-1} e^{-t} dt = \sum_{k=0}^{q-1} \frac{s^k e^{-s}}{k!}, \tag{6}$$

where $\Gamma(q) = (q-1)!$ *is the standard Gamma function evaluated at* q.

The incomplete Gamma function is continuous in s, but discrete in q. For every fixed s, $\Psi_q(s)$ is the cumulative distribution of a Poisson random variable evaluated at $q-1$. When q is fixed, Ψ_q is the likelihood function of the random variable B_m. In this paper, we will focus on the latter case where q is fixed so that Ψ_q is a function of s. Since Ψ_q is continuous, the derivative of $\Psi_q(s)$ is available:

Property 1. *The first order derivative of* $\Psi_q(s)$ *is*

$$\frac{d}{ds} \Psi_q(s) = -\frac{e^{-s} s^{q-1}}{\Gamma(q)} \tag{7}$$

With the incomplete Gamma function, we can rewrite Equation (5) as

$$\mathbb{P}\{B_m = 1 \;;\; s_m\} = 1 - \Psi_q(s_m), \quad \text{and} \quad \mathbb{P}\{B_m = 0 \;;\; s_m\} = \Psi_q(s_m). \tag{8}$$

Example 1. *When* $q = 1$, *the incomplete Gamma function becomes* $\Psi_q(s_m) = e^{-s_m}$. *In this case,*

$$\mathbb{P}\{B_m = 1 \;;\; s_m\} = 1 - e^{-s_m} \quad \text{and} \quad \mathbb{P}\{B_m = 0 \;;\; s_m\} = e^{-s_m}.$$

2.3. Image Reconstruction for QIS

We consider a multiple exposure setting. Assuming that the scene is stationary, the measurement we obtain is a sequence of binary bit maps $\{b_{m,t} \mid m = 0, \ldots, M-1, \text{ and } t = 0, \ldots, T-1\}$, where the first index m runs through the M jots spatially, and the second index t runs through the T frames in time. The image reconstruction task is to find the signal $c = [c_0, \ldots, c_{N-1}]^T$ that best explains $\{b_{m,t}\}$. Translating into a probabilistic framework, this can be formulated as maximizing the likelihood:

$$\hat{c} = \underset{c}{\text{argmax}} \ \prod_{t=0}^{T-1}\prod_{m=0}^{M-1} \mathbb{P}[B_{m,t}=1;s_m]^{b_{m,t}}\mathbb{P}[B_{m,t}=0;s_m]^{1-b_{m,t}}, \ \text{ subject to } \ s = \alpha Gc,$$

$$= \underset{c}{\text{argmax}} \ \prod_{t=0}^{T-1}\prod_{m=0}^{M-1} (1-\Psi_q(s_m))^{b_{m,t}}\Psi_q(s_m)^{1-b_{m,t}}, \ \text{ subject to } \ s = \alpha Gc, \tag{9}$$

where the constraint $s = \alpha Gc$ follows from Equation (1). Taking the logarithm on the right-hand side of Equation (9), the maximization becomes

$$\hat{c} = \underset{c}{\text{argmax}} \ \sum_{t=0}^{T-1}\sum_{m=0}^{M-1} \left\{ b_{m,t}\log(1-\Psi_q(s_m)) + (1-b_{m,t})\log \Psi_q(s_m) \right\}, \ \text{ subject to } \ s = \alpha Gc. \tag{10}$$

The optimization in Equation (10) is known as the maximum likelihood estimation (MLE). As shown in [13], the objective function of Equation (10) is concave (for any q), and therefore, the global maximum exists and is unique. When G satisfies Assumption 1, the closed form solution is available, and we will show it in Section 3. However, for general G or if we include additional constraints to enforce the smoothness of the solution, e.g., using total variation [34] or sparsity [35], then numerical algorithms are required. Some existing methods include the gradient descent method [13], Newton method [28] and the alternating direction method of multipliers [29]. These algorithms are iterative, and the computation for each iteration is costly.

3. Non-Iterative Image Reconstruction

In this section, we present a non-iterative image reconstruction algorithm for QIS. There are three components behind the algorithm. The first is the closed-form solution of Equation (10), which is coherent to [13]. The second component is a variance stabilizing transform that transforms the sum of one-bit quantized Poisson random variables to a binomial random variable with equal variances. The transformation is known as the Anscombe transform, named after the English statistician Frank Anscombe (1918–2001). The third idea is the application of an off-the-shelf image denoiser.

3.1. Component 1: Approximate MLE

Because of the piecewise constant property of Assumption 1, the sequence $\{b_{m,t}\}$ can be partitioned into N blocks where each block contains $K \times T$ binary bits. This leads to a decomposition of Equation (10) as

$$\hat{c} = \underset{c}{\text{argmax}} \ \sum_{t=0}^{T-1}\sum_{n=0}^{N-1}\sum_{k=0}^{K-1} \left\{ b_{Kn+k,t}\log\left(1-\Psi_q\left(\frac{\alpha c_n}{K}\right)\right) + (1-b_{Kn+k,t})\log \Psi_q\left(\frac{\alpha c_n}{K}\right) \right\}, \tag{11}$$

where the subsequence $\mathcal{B}_{n,t} \overset{\text{def}}{=} \{b_{Kn,t}, \ldots, b_{Kn+(K-1),t}\}$ denotes the n-th block of the t-th frame. Define

$$S_n \overset{\text{def}}{=} \sum_{t=0}^{T-1}\sum_{k=0}^{K-1} b_{Kn+k,t} \tag{12}$$

be the sum of the bits (i.e., the number of one's) in $\mathcal{B}_{n,t}$. Then, Equation (11) becomes

$$\hat{c} = \operatorname*{argmax}_{c} \sum_{n=0}^{N-1} S_n \log \left(1 - \Psi_q \left(\frac{\alpha c_n}{K}\right)\right) + (L - S_n) \log \Psi_q \left(\frac{\alpha c_n}{K}\right), \tag{13}$$

where $L = KT$. The maximum of Equation (13) is attained when each individual term in the sum attains its maximum. In this case, the closed form solution of Equation (13), for every n, can be derived as in Proposition Equation 1.

Proposition 1. *The solution of Equation (13) is*

$$\hat{c}_n = \frac{K}{\alpha} \Psi_q^{-1} \left(1 - \frac{S_n}{L}\right), \qquad n = 0, \dots, N-1, \tag{14}$$

where $L = KT$, K is the spatial oversampling factor and T is the number of exposures.

Proof. See Appendix A. □

It would be instructive to illustrate Proposition 1 using a figure. Figure 2 shows the case when $T = 1$, i.e., a single exposure, and $K = 16$. The one-bit measurements are first averaged to compute the number of ones within a block of size K. Then, applying the inverse incomplete Gamma function $\Psi_q^{-1}(\cdot)$ and a scaling constant K/α, we obtain the solution $\hat{c}_n$.

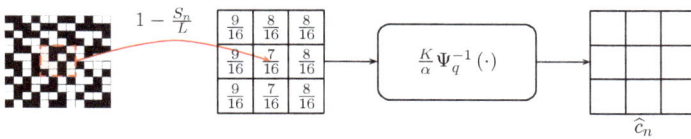

Figure 2. Pictorial interpretation of Proposition 1: Given an array of one-bit measurements (black = 0, white = 1), we compute the number of ones within a block of size K. Then, the solution of the MLE problem in Equation (13) is found by applying an inverse incomplete Gamma function $\Psi_q^{-1}(\cdot)$ and a scaling factor K/α.

Proposition 1 shows why a simple summation S_n/L is inadequate to achieve the desired result, although such summation has been used in [18,36]. By only summing the number of ones, the resulting value S_n/L is the empirical average of these one-bit measurements. Since the probability of drawing a one in QIS follows a quantized Poisson distribution and not a Bernoulli distribution, the nonlinearity due to the quantized Poisson distribution must be taken into account. A comparison of the ground truth image, the summation result and the MLE solution is shown in Figure 3.

As an immediate corollary of Proposition 1, we can simplify the inverse incomplete Gamma function when $q = 1$. This result is sometimes known as the exponential correction function [37].

Corollary 1. *When $q = 1$, the MLE solution of the n-th pixel $\hat{c}_n$ is*

$$\hat{c}_n = -\frac{K}{\alpha} \log \left(1 - \frac{S_n}{L}\right), \qquad n = 0, \dots, N-1, \tag{15}$$

where S_n is the number of ones in the n-th block (See Equation (12)) and $L = KT$ is the total number of pixels in the block.

Proof. The result follows from the fact that $\Psi_q(s) = e^{-s}$ for $q = 1$. Thus, $\Psi_q^{-1}(s) = -\log s$. □

(**a**) Ground truth   (**b**) Simple sum, $\hat{c}_n = S_n/L$ (**c**) $\hat{c}_n = \frac{K}{\alpha}\Psi_q^{-1}\left(1 - \frac{S_n}{L}\right)$

MLE solution

Figure 3. Image reconstruction using synthetic data. In this experiment, we generate one-bit measurements using a ground truth image (**a**) with $\alpha = 160$, $q = 5$, $K = 16$, $T = 1$ (so $L = 16$). The result shown in (**b**) is obtained using the simple summation, whereas the result shown in (**c**) is obtained using the MLE solution. It can be seen that the simple summation has a mismatch in the tone compared to the ground truth.

3.2. Component 2: Anscombe Transform

The MLE solution $\hat{c} = [\hat{c}_0, \ldots, \hat{c}_{N-1}]^T$ computed through Proposition 1 is noisy, as illustrated in Figure 3. The reason is that for a relatively small K and T, the randomness in the one-bit measurement has not yet been eliminated by the summation in S_n. Therefore, in order to improve the image quality, additional steps must be taken to improve the smoothness of the image.

At first glance, this question seems easy because if one wants to mitigate the noise in $\hat{c}$, then directly applying an image denoising algorithm $\mathcal{D}$ to $\hat{c}$ would be sufficient, e.g., Figure 4a. However, a short afterthought will suggest that such an approach is invalid for the following reason. For the majority of image denoising algorithms in the literature, the noise is assumed to be independently and identically distributed (i.i.d.) Gaussian. In other words, the variance of the noise should be spatially invariant. However, the resulting random variable in Figure 2 does not have this property.

Our proposed solution is to apply an image denoiser before the inverse incomplete Gamma function as shown in Figure 4b. Besides the order of denoising and the Gamma function, we also add a pair of nonlinear transforms $\mathcal{T}$ and $\mathcal{T}^{-1}$ before and after the denoiser $\mathcal{D}$. The reasons for these two changes are based on the following observations.

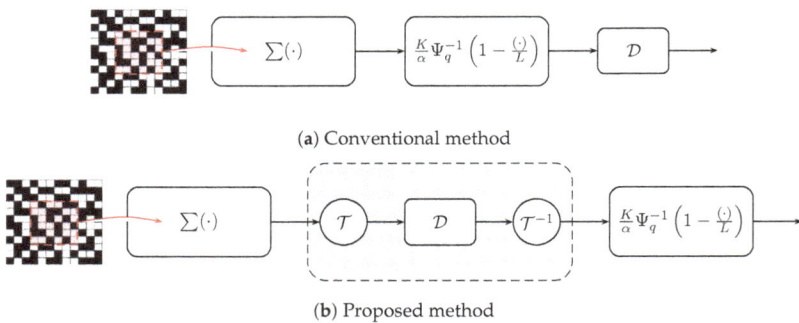

(**a**) Conventional method

(**b**) Proposed method

Figure 4. Two possible ways of improving image smoothness for QIS. (**a**) The conventional approach denoises the image after $\hat{c}_n$ is computed; (**b**) the proposed approach: apply the denoiser before the inverse incomplete Gamma function, together with a pair of Anscombe transforms $\mathcal{T}$. The symbol $\mathcal{D}$ in this figure denotes a generic Gaussian noise image denoiser.

Observation 1. *Under Assumption 1, the random variables* $\{B_{Kn+k,t} \mid k = 0, \ldots, K-1, \text{ and } t = 0, \ldots, T-1\}$ *are i.i.d. Bernoulli of equal probability* $\mathbb{P}[B_{Kn+k,t} = 1] = 1 - \Psi_q\left(\frac{\alpha c_n}{K}\right)$ *for* $k = 0, \ldots, K-1$ *and* $t = 0, \ldots, T-1$.

The proof of Observation 1 follows immediately from Assumption 1 that if $G = (1/K)\boldsymbol{I}_{N \times N} \otimes \boldsymbol{1}_{K \times 1}$, we can divide the M jots into N groups each having $K \times T$ entries. Within the group, the one-bit measurements are all generated from the same pixel c_n.

The consequence of Observation 1 is that for a sequence of i.i.d. Bernoulli random variables, the sum is a Binomial random variable. This is described in Observation 2.

Observation 2. *If* $\{B_{Kn+k,t}\}$ *are i.i.d. Bernoulli random variables with probability* $\mathbb{P}[B_{Kn+k,t} = 1] = 1 - \Psi_q\left(\frac{\alpha c_n}{K}\right)$ *for* $k = 0, \ldots, K-1$ *and* $t = 0, \ldots, T-1$, *then the sum* S_n *defined in Equation (12) is a Binomial random variable with mean and variance:*

$$\mathbb{E}[S_n] = L\left(1 - \Psi_q\left(\frac{\alpha c_n}{K}\right)\right), \qquad \mathrm{Var}[S_n] = L\Psi_q\left(\frac{\alpha c_n}{K}\right)\left(1 - \Psi_q\left(\frac{\alpha c_n}{K}\right)\right).$$

Observation 2 is a classic result in probability. The mean of the Bernoulli random variables is specified by the incomplete Gamma function $\Psi_q\left(\frac{\alpha c_n}{K}\right)$, which approaches one as K increases. Thus, for fixed T, the probability $1 - \Psi_q\left(\frac{\alpha c_n}{K}\right) \to 0$ as $K \to \infty$. When this happens, the binomial random variable S_n can be approximated by a Poisson random variable with mean $L\left(1 - \Psi_q\left(\frac{\alpha c_n}{K}\right)\right)$ [38]. However, as T also grows, the binomial random variable S_n can be further approximated by a Gaussian random variable due to the central limit theorem. Therefore, for a reasonably large K and T, the resulting random variable S_n is approximately Gaussian.

The variance of this approximated Gaussian is, however, not constant. The variance changes across different locations n because $\mathrm{Var}[S_n]$ is a function of c_n. Therefore, if we want to apply a conventional image denoiser (which assumes i.i.d. Gaussian noise) to smooth S_n, we must first make sure that the noise variance is spatially invariant. The technique used to accomplish this goal is called the variance stabilizing transform [39]. In this paper, we use a specific variance stabilizing transform known as the Anscombe transform [32]. Anscombe transform is best known in the image processing literature for Poisson denoising, where one transforms observed Poisson data to approximately Gaussian with equal variance [24]. For binomial random variables S_n, the Anscombe transform and its property are given in Theorem 1.

Theorem 1 (Anscombe transform for binomial random variables). *Let* S_n *be a binomial random variable with parameters* (L, p_n), *where* $p_n = 1 - \Psi_q\left(\frac{\alpha c_n}{K}\right)$ *and* $L = KT$. *Define the Anscombe transform of* S_n *as a function* $\mathcal{T} : \{0, \ldots, L\} \to \mathbb{R}$, *such that*

$$Z_n = \mathcal{T}(S_n) \overset{\text{def}}{=} \sqrt{L + \frac{1}{2}} \sin^{-1}\left(\sqrt{\frac{S_n + \frac{3}{8}}{L + \frac{3}{4}}}\right). \tag{16}$$

Then, the variance of Z_n *is* $\mathrm{Var}[Z_n] = \frac{1}{4} + \mathcal{O}(L^{-2})$ *for all* n.

Proof. The proof of Theorem 1 is given in the Appendix A. It is a simplified version of a technical report by Brown et al. [40]. The original paper by Anscombe [32] also contains a sketch of the proof. However, the sketch is rather brief, and we believe that a complete derivation would make this paper self-contained. □

The implication of Theorem 1 is that regardless of the location n, the transformed random variable Z_n has a constant variance $\frac{1}{4}$ when L is large. Therefore, the noise variance is now location independent, and hence, a standard i.i.d. Gaussian denoiser can be used.

Example 2. *To provide readers a demonstration of the effectiveness of Theorem 1, we consider a checkerboard image of $N = 64$ pixels with intensity levels $c_0, \ldots, c_{N-1}$. The n-th pixel c_n generates $K = 100$ binary quantized Poisson measurements $\{B_{Kn}, \ldots, B_{Kn+(K-1)}\}$ using $\alpha = 100$, $q = 1$, $T = 1$ (so $L = 100$). From each of these K measurements, we sum to obtain a binomial random variable $S_n = \sum_{k=0}^{K-1} B_{Kn+k}$. We then compute the variance of $\text{Var}[S_n]$ and $\text{Var}[\mathcal{T}(S_n)]$ using 10^4 independent Monte Carlo trials. The results are shown in Figure 5, where we observe that $\text{Var}[S_n]$ varies with the location n, and $\text{Var}[\mathcal{T}(S_n)]$ is nearly constant for all n.*

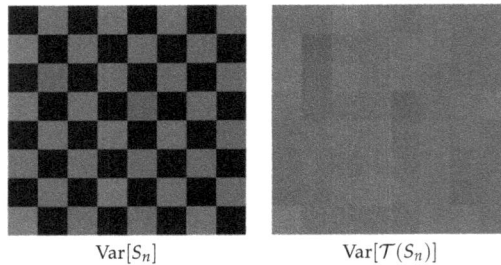

$$\text{Var}[S_n] \qquad\qquad \text{Var}[\mathcal{T}(S_n)]$$

Figure 5. Illustration of Anscombe transform. Both sub-figures contain $N = 64$ (8×8) pixels $c_0, \ldots, c_{N-1}$. For each pixel, we generate 100 binary Poisson measurements and sum to obtain binomial random variables $S_0, \ldots, S_{N-1}$. We then calculate the variance of each S_n. Note the constant variance after the Anscombe transform.

Remark 1. *The inverse Anscombe transform is*

$$S_n = \mathcal{T}^{-1}(Z_n) = \left(L + \frac{3}{4}\right) \sin^2\left(\frac{Z_n}{\sqrt{L + \frac{1}{2}}}\right) - \frac{3}{8}, \tag{17}$$

which we call the algebraic inverse. Another possible inverse of the Anscombe transform is the asymptotic unbiased inverse [32], defined as

$$S_n = \mathcal{T}^{-1}_{\text{unbias}}(Z_n) = \left(1 + \frac{1}{2L}\right)^{-1}\left[\left(L + \frac{3}{4}\right)\sin^2\left(\frac{Z_n}{\sqrt{L + \frac{1}{2}}}\right) - \frac{1}{8}\right]. \tag{18}$$

The performance of the unbiased inverse is typically better for low noise (large L), whereas the algebraic inverse is better for high noise (small L). This is consistent with the Poisson denoising literature. (e.g., [24]).

Example 3. *Table 1 shows the peak signal to noise ratio (PSNR) values of the reconstructed images using the algebraic inverse and the asymptotic unbiased inverse. In this experiment, we consider 10 standard images commonly used in the image processing literature: Baboon, Barbara, Boat, Bridge, Couple, Hill, House, Lena, Man and Peppers. The sizes of the images are either 256×256 or 512×512. For each image, we set $T = 1$, $q = 1$ and $\alpha = K$ and vary $K = \{1, 4, 9, 16, 25, 36, 49, 64\}$. The results in Table 1 indicate that $\mathcal{T}^{-1}_{\text{unbias}}$ is consistently better than $\mathcal{T}^{-1}$ for $K > 1$, although the difference diminishes as K grows.*

Table 1. PSNR values using algebraic inverse $\mathcal{T}^{-1}$ and asymptotic unbiased inverse $\mathcal{T}^{-1}_{\text{unbias}}$. The results are averaged over 10 standard images. In this experiment, we set $T = 1$, $q = 1$ and $\alpha = K$.

K	1	4	9	16	25	36	49	64
$\mathcal{T}^{-1}$	20.51	23.08	25.00	26.47	27.49	28.40	29.09	29.71
$\mathcal{T}^{-1}_{\text{unbias}}$	19.43	23.64	25.30	26.62	27.57	28.45	29.12	29.73

3.3. Component 3: Image Denoiser

An important feature of the proposed algorithm is that it can take any off-the-shelf image denoiser for the operator $\mathcal{D}$. Here, by image denoiser, we meant an image denoising algorithm designed to remove i.i.d. Gaussian noise from an observed image.

Image denoising is an important research topic by its own. In the following, we provide a few popular image denoising algorithms.

- Total variation denoising [34]: Total variation denoising was originally proposed by Rudin, Osher and Fatemi [34], although other researchers had proposed similar methods around the same time [41]. Total variation denoising formulates the denoising problem as an optimization problem with a total variation regularization. Total variation denoising can be performed very efficiently using the alternating direction method of multipliers (ADMM), e.g., [42–44].
- Bilateral filter [45]: The bilateral filter is a nonlinear filter that denoises the image using a weighted average operator. The weights in a bilateral filter are the Euclidean distance between the intensity values of two pixels, plus the spatial distance between the two pixels. A Gaussian kernel is typically employed for these distances to ensure proper decaying of the weights. Bilateral filters are extremely popular in computer graphics for applications, such as detail enhancement. Various fast implementations of bilateral filters are available, e.g., [46,47].
- Non-local Means [48]: non-local means (NLM) was proposed by Buades et al. [48] and, also, an independent work of Awante and Whitaker [49]. Non-local means (NLM) is an extension of the bilateral filter where the Euclidean distance is computed from a small patch instead of a pixel. Experimentally, it has been widely agreed that such patch-based approaches are very effective for image denoising. Fast NLM implementations are now available [50–52].
- BM3D [53]: 3D block matching (BM3D) follows the same idea of non-local means by considering patches. However, instead of computing the weighted average, BM3D groups similar patches to form a 3D stack. By applying a 3D Fourier transform (or any other frequency domain transforms, e.g., discrete cosine transform), the commonality of the patches will demonstrate a group sparse behavior in the transformed domain. Thus, by applying a threshold in the transformed domain, one can remove the noise very effectively. BM3D is broadly regarded as a benchmark of today's image denoising algorithm.

The factors to consider in choosing an image denoiser are typically the complexity and quality. Low complexity algorithms, such as bilateral filter, are fast, but the denoising ability is limited. High end algorithms, such as BM3D and non-local means, produce very good images, but require much computation. The trade-off between complexity and performance is a choice of the user.

Readers at this point may perhaps ask a question: What will happen if we apply a denoiser after the MLE solution, like the one shown in the block diagram in Figure 4a? As we have explained in the Anscombe transform section, this will lead to a suboptimal result because the noise is not i.i.d. Gaussian. To illustrate the difference in terms of performance, we show in Figure 6 a comparison between applying image denoising using the two block diagrams shown in Figure 4. The denoiser we use in this experiment is BM3D. The metric we use to evaluate the performance is the peak signal to noise ratio (PSNR), which will be defined formally in Section 4. In short, a large PSNR value is equivalent to a low mean squared error comparing the estimated image and the ground truth image. The results are shown in Figure 6. Although denoising after the MLE (the conventional idea) generates some reasonable images, the PSNR values are indeed significantly lower than the proposed Anscombe approach. This is not surprising, because the denoiser tends to oversmooth the dark regions and undersmooth the bright regions due to the signal-dependent noise levels.

| Ground truth | Sum of *K* binary observations 17.07 dB | Denoise after the MLE solution (Conventional) 25.85 dB | Denoise using Anscombe Transform (Proposed) 29.96 dB |

Figure 6. Comparison between image denoising after the MLE solution and using the proposed Anscombe transform. The denoiser we use in this experiment is 3D block matching (BM3D) [53]. The binary observations are generated using the configurations $\alpha = 160$, $q = 5$, $K = 16$, $T = 1$. The values shown are the peak signal to noise ratio (PSNR).

3.4. Related Work in the Literature

The proposed algorithm belongs to a family of methods we call the transform-denoise methods. The idea of transform-denoise is similar to what we do here: transform the random variable using a variance stabilizing transform, then denoise using an off-the-shelf image denoiser. Among the existing transform-denoise methods, perhaps the most notable work is the one by Makitalo and Foi [24], where they considered the optimal inverse of the Anscombe transform for the case of Poisson–Gaussian random variables. A more recent work by the same research group [27] showed that it is possible to boost the denoising performance by applying the transform-denoise iteratively. We should also mention the work by Foi [54], which considered the modeling and transformation for clipped noisy images. The problem setting of that work is for conventional sensors. However, the underlying principle using the transform-denoise approach is similar to that of QIS.

The approximate MLE solution in Section 3.1 is based on the piecewise constant assumption (Assumption 1). Under this assumption, summing of the Bernoulli random variables can be thought of as performing a "binning" of the pixels. Binning is a common technique in restoring images from Poisson noise, especially when the signal-to-noise ratio is low [23,25,26]. Binning can also be applied together with transform-denoise, e.g., in [27], to achieve improved results. For QIS, the result of binning is different from that of the Poisson noise, for the sum of QIS bits leads to a binomial random variables, whereas the sum of Poisson noise leads to a Poisson random variable.

4. Experimental Results

In this section, we provide further experimental results to evaluate the proposed algorithm.

4.1. Synthetic Data

In this experiment, we consider 100 natural images downloaded from the Berkeley Segmentation database [55]. The input resolution of these images is 481×321 (or 321×481), and all images are converted to a gray-scale image with values in the range $[0, 1]$. To generate the synthetic data, for each image, we consider an oversampling factor of four along the horizontal and the vertical directions (so $K = 16$). The sensor gain is set as $\alpha = 16$, and the threshold level is $q = 1$ to simulate a single photon sensor that triggers at one photon. We generate one-bit observations using a quantized Poisson statistics and use the proposed algorithm to reconstruct. As a comparison, we also test the MLE solution, i.e., a summation followed by the inverse incomplete Gamma transform (see Theorem 1), and an ADMM algorithm using a total variation regularization [29,31]. For the proposed algorithm, we use BM3D as the image denoiser.

We report two results in this experiment. First, we consider the peak signal to noise ratio (PSNR) as an evaluation metric. The PSNR of an estimated image $\hat{c}$ compared to the ground truth image c^* is defined as

$$\text{PSNR} = -20 \log_{10} \left(\frac{\|\hat{c} - c^*\|^2}{N} \right),$$ (19)

where N is the number of pixels in c. Typically, higher PSNR values imply better image reconstruction quality. The PSNR values of these 100 images are shown in Figure 7. In this figure, we observe that the proposed algorithm is better than the MLE solution and the ADMM solution by 10.20 dB and 2.75 dB, respectively, which are very substantial amounts from a reconstruction perspective.

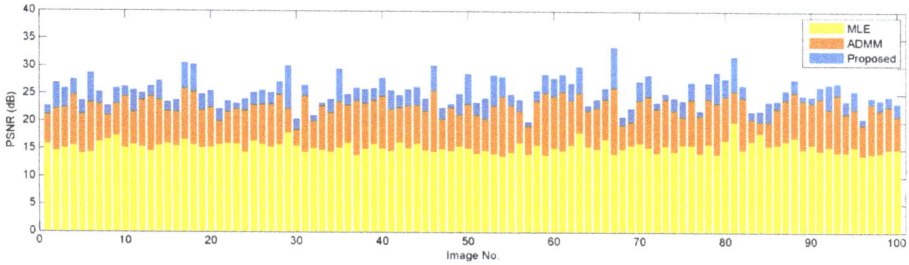

Figure 7. PSNR comparison of various image reconstruction algorithms on the Berkeley Segmentation database [55]. In this experiment, we fix $q = 1$, $\alpha = 16$, and $K = 16$. The proposed algorithm uses BM3D [53] as the image denoiser.

Apart form the PSNR values, we also report the runtimes of the algorithms. The runtimes of the algorithms are recorded by running the methods on the same machine and the same platform, which is an Intel i7-6700 3.4-GHz desktop with Windows 7/MATLAB 2014. As shown in Figure 8, the runtime of the proposed algorithm is approximately two orders of magnitude (100×) faster than the ADMM algorithm.

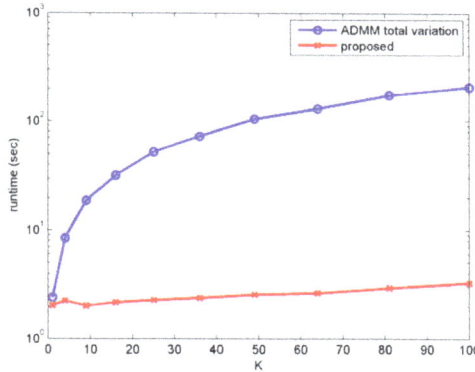

Figure 8. Runtime comparison of the proposed algorithm and the alternating direction method of multipliers (ADMM) algorithm [31].

As for the influence of the oversampling factor K on the reconstruction quality, we show in Figure 9 the reconstructed results using $K = 4, 16, 64$ and the binary one-bit measurements. While it is clear from the figure that the reconstructed image improves as K increases, we observe that most of the visual content has been recovered even at $K = 4$.

| One-bit measurement | $K = 4$ | $K = 16$ | $K = 64$ |

Figure 9. Influence of the oversampling factor K on the image reconstruction quality. In this experiment, we set $\alpha = K$, $q = 1$. $T = 1$.

4.2. Real Data

In this experiment, we consider two single-photon avalanche diode (SPAD) cameras for capturing high speed videos. The first camera is a CMOS SPAD-based image sensor developed by Dutton et al. [14–16]. This camera has a resolution of 320×240, with a frame rate of 10k frames per second. The second camera is the SwissSPAD camera developed by Burri et al. [17,18]. This camera has a resolution of 512×128, with frame rate of 156k frames per second. Both cameras capture one-bit measurements from a scene containing a stationary background with a rapidly moving foreground. Our experimental goal is to test if the proposed algorithm can resolve the spatial content with minimal trade-off in the temporal resolution.

There are several points of this experiment on which we should comment. First, since the spatial resolution of these two cameras is relatively small (as compared to the synthetic case), we do not assume any spatial oversampling, i.e., $K = 1$. Instead, we use T temporal frames to reconstruct one output image. To ensure smooth transitions across adjacent frames, we use a temporally-sliding window as we progress to the next output image. Second, since these real videos do not have a ground truth, we can only compare the quality of the resulting images visually.

We first look at the results of the SPAD camera by Dutton et al. [14–16]. Figure 10 shows several snapshots of the "fan" sequence and the "milk" sequence. To generate this result, we run the proposed algorithm using $T = 16$ frames in a sliding window mode. That is, we use Frames 1 to 16 to recover Frame 1 and Frames 2 to 17 to recover Frame 2, etc. The quantization level q is set as $q = 1$, and the sensor gain α is adjusted to produce the best visual quality. For the "fan" sequence, we set $\alpha = 4$, and for the "milk" sequence, we set $\alpha = 0.75$. As we can see from the figures, the proposed algorithm recovers most of the content from the scene, even revealing the textures of the milk in the scene.

As for the SwissSPAD camera, we consider a video sequence "oscilloscope" captured at a frame rate of 156k frames per second. The goal is to track the sinusoid shown on the oscilloscope's screen. We consider four values of $T = 4, 16, 64$ and 256 for the number of frames. The results are shown in Figure 11. As one may expect, when $T = 256$, the image quality improves because we are effectively summing 256 frames to reconstruct one output frame. However, since we are summing the 256 frames, the temporal resolution is severely distorted. In particular, the trace of the sinusoid signal disappears because of the strong averaging effect. When we reduce the number of frames to $T = 16$, we observe that the trace of the signal can be observed clearly. The same image obtained by the MLE (i.e., simple summation) is still highly noisy.

(a) "Fan" sequence   (b) "Milk" sequence

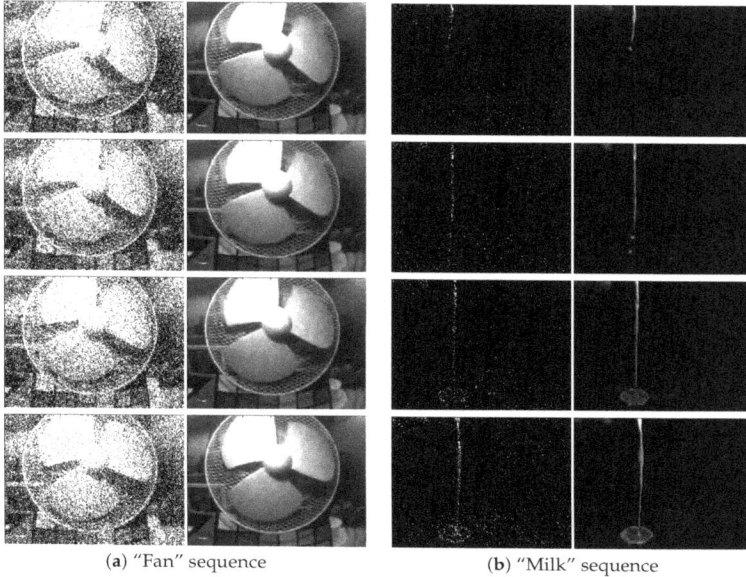

Figure 10. Image reconstruction of two real video sequences captured using a 320×240 single-photon avalanche diode (SPAD) camera running at 10k frames per second [14–16]. In this experiment, we use $T = 16$ frames to construct one output frame. In both columns, the left are the raw one-bit measurements, and the right are the recovered images using the proposed algorithm.

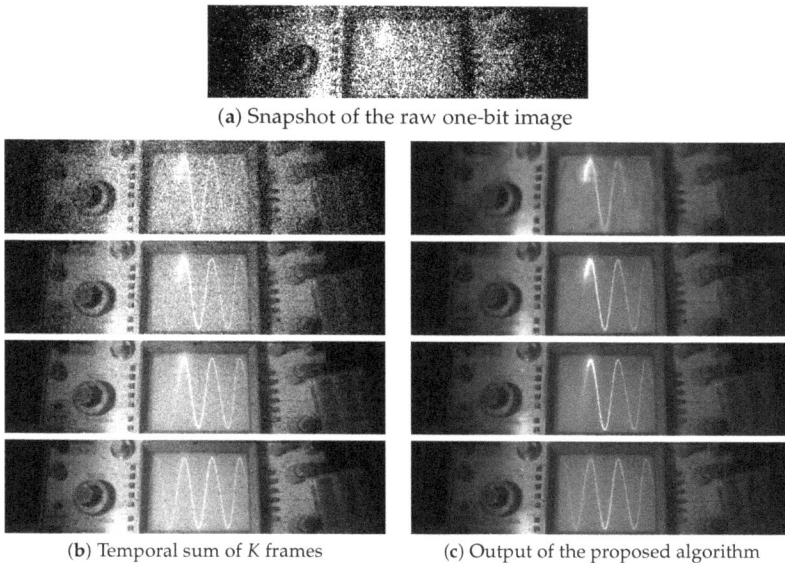

(a) Snapshot of the raw one-bit image

(b) Temporal sum of K frames   (c) Output of the proposed algorithm

Figure 11. Image reconstruction of real video sequences captured using the 512×128 SwissSPAD camera running at 156k frames per second [17,18]. (a) is a snapshot of the raw one-bit image. (b) shows the result of summing $T = 4, 16, 64, 256$ temporal frames with $K = 1$. (c) shows the corresponding results using the proposed algorithm.

5. Conclusions

We present a new image reconstruction algorithm to recover images from one-bit quantized Poisson measurements. Different from existing algorithms that are mostly iterative, the new algorithm is non-iterative. The algorithm consists of three key components: (1) an approximation to the standard maximum likelihood estimation formulation that allows us to decouple the dependency of pixels; (2) a nonlinear transform known as the Anscombe transform that converts a sum of one-bit quantized Poisson random variables to a Gaussian random variable with equal variance; (3) an off-the-shelf image denoising algorithm that performs the smoothing. Experimental results confirm the performance of the proposed algorithm. The algorithm demonstrates two orders of magnitude improvement in speed compared to existing iterative methods and shows several dBs of improvement in terms of the peak signal to noise ratio (PSNR) as a metric of image quality.

Acknowledgments: We thank Guest Editor Eric R. Fossum for inviting us to submit this work. We thank Neale Dutton and his team for sharing the "fan" and the "milk" sequences collected by their SPAD camera. We thank Edoardo Charbon and his team for sharing the "oscilloscope" sequence collected by the SwissSPAD camera. We would like to give special thanks to the anonymous reviewers, who provided us tremendously helpful feedback on this paper.

Author Contributions: S.H.C.initiated and developed the algorithm. S.H.C. and O.A.E. conducted the analysis of the Anscombe transform. S.H.C. and X.W. analyzed the image denoising algorithms. S.H.C. designed the experiments. O.A.E. and X.W. conducted the experiments. S.H.C. wrote the paper.

Conflicts of Interest: The authors declare no conflict of interest.

Abbreviations

The following abbreviations are used in this manuscript:

CCD	Charge coupled device
CMOS	Complementary metal-oxide-semiconductor
QIS	Quanta image sensors
MLE	Maximum likelihood estimation
ADMM	Alternating direction method of multipliers
SPAD	Single-photon avalanche diode
PSNR	Peak signal to noise ratio
BM3D	3D block matching
i.i.d.	Independently and identically distributed

Appendix A

Appendix A.1. Proof of Proposition 1

Proof. For notational simplicity, we drop the subscript n. Then, the optimization problem is

$$\hat{c} = \underset{c}{\operatorname{argmax}} \; S \log \left(1 - \Psi_q \left(\frac{\alpha c}{K} \right) \right) + (L - S) \log \Psi_q \left(\frac{\alpha c}{K} \right).$$

Taking the first order derivative with respect to c and setting to zero yields (with the help of Property 1):

$$\left(\frac{S}{1 - \Psi_q \left(\frac{\alpha c}{K} \right)} \right) \left(\frac{\alpha}{K} \frac{e^{-\frac{\alpha c}{K}} \left(\frac{\alpha c}{K} \right)^{q-1}}{\Gamma(q)} \right) + \left(\frac{L - S}{\Psi_q \left(\frac{\alpha c}{K} \right)} \right) \left(-\frac{\alpha}{K} \frac{e^{-\frac{\alpha c}{K}} \left(\frac{\alpha c}{K} \right)^{q-1}}{\Gamma(q)} \right) = 0.$$

Rearranging the terms leads to the desired result that

$$\Psi_q\left(\frac{\alpha c}{K}\right) = 1 - \frac{S}{L}.$$

☐

Appendix A.2. Proof of Theorem 1

Proof. For notational simplicity, we drop the subscript n. Our goal is to show that if $X \sim \mathrm{Binomial}(L, p)$, then the transformed variable:

$$T(X) = \sqrt{L + \frac{1}{2}} \sin^{-1}\left(\sqrt{\frac{X + \frac{3}{8}}{L + \frac{3}{4}}}\right) \tag{A1}$$

has a variance $\mathrm{Var}[T(X)] = \frac{1}{4} + \mathcal{O}(L^{-2})$. To this end, we first consider the function $\mathcal{Q}$, such that

$$\mathcal{Q}(X) = T(X) - \sqrt{L + \frac{1}{2}} \sin^{-1}\sqrt{p}.$$

Since $\mathrm{Var}[\mathcal{Q}(X) + c] = \mathrm{Var}[\mathcal{Q}(X)]$ for any c, by letting $c = -\sqrt{L + \frac{1}{2}} \sin^{-1}\sqrt{p}$, we observe that showing Proposition 1 is equivalent to showing $\mathrm{Var}[\mathcal{Q}(X)] = \frac{1}{4} + \mathcal{O}(L^{-2})$.

To show the desired result, we note that for any α and β, the arcsin function has the property that

$$\sin^{-1}\alpha - \sin^{-1}\beta = \sin^{-1}\left(\alpha\sqrt{1 - \beta^2} - \beta\sqrt{1 - \alpha^2}\right).$$

Define $F \stackrel{\text{def}}{=} \frac{X + \frac{3}{8}}{L + \frac{3}{4}}$, and substitute $\alpha = \sqrt{F}$, $\beta = \sqrt{p}$; it follows that

$$\mathcal{Q}(X) = \sqrt{L + \frac{1}{2}} \sin^{-1}\left(\sqrt{(1 - p)F} - \sqrt{p(1 - F)}\right). \tag{A2}$$

There are two terms in this equation. The first term $\sqrt{L + \frac{1}{2}}$ can be expanded (using Taylor expansion) to its first second order as

$$\sqrt{L + \frac{1}{2}} = \sqrt{L}\left(1 + \frac{1}{2L}\right)^{\frac{1}{2}} = \sqrt{L}\left(1 + \frac{1}{4L} + \mathcal{O}(L^{-2})\right).$$

The arcsin function can be expanded to its second order as

$$\sin^{-1}W = W + \frac{W^3}{6} + \frac{3W^5}{40} + \frac{5W^7}{112} + \dots,$$

for $W = \sqrt{(1 - p)F} - \sqrt{p(1 - F)}$.

We next consider the standardized binomial random variable by defining

$$Y \stackrel{\text{def}}{=} \frac{X - Lp}{\sqrt{Lp(1 - p)}}. \tag{A3}$$

Then, by Lemma A1, it follows that

$$W = \sqrt{(1-p)F} - \sqrt{p(1-F)}$$
$$= \frac{Y}{2\sqrt{L}} + \frac{(2p-1)(2Y^2-3)}{16L\sqrt{p(1-p)}} + \frac{-16Y^3p^2 + 16Y^3p - 6Y^3 + 9Y}{96L^{\frac{3}{2}}p(1-p)} + \mathcal{O}(L^{-2}).$$

Therefore,

$$\mathcal{Q}(X) = \sqrt{L}\left(1 + \frac{1}{4L} + \mathcal{O}(L^{-2})\right)\left(W + \frac{W^3}{6} + \mathcal{O}(W^5)\right)$$
$$= a_0 + a_1 Y + a_2 Y^2 + a_3 Y^3 + \mathcal{O}(Y^5),$$

where

$$a_0 = -\frac{3(2p-1)}{16\sqrt{Lp(1-p)}}, \quad a_1 = \frac{1}{2} + \frac{1}{8L} - \frac{3}{32Lp(1-p)}$$
$$a_2 = \frac{2p-1}{8\sqrt{Lp(1-p)}}, \quad a_3 = \frac{16p^2-16p+6}{96Lp(1-p)}.$$

Since the first four moments of Y are

$$\mathbb{E}[Y] = 0, \quad \mathbb{E}[Y^2] = 1, \quad \mathbb{E}[Y^3] = -\frac{2p-1}{\sqrt{Lp(1-p)}}, \quad \mathbb{E}[Y^4] = 3 + \frac{1-6p(1-p)}{Lp(1-p)},$$

we conclude that

$$\text{Var}[\mathcal{Q}(X)] = a_1^2 \text{Var}[Y] + a_2^2 \text{Var}[Y^2] + 2a_1 a_2 \text{Var}[Y^3] + 2a_1 a_3 \text{Var}(Y^4)$$
$$= a_1^2 - a_2^2 + 2a_1 a_2 \mathbb{E}[Y^3] + (2a_1 a_3 + a_2^2)\mathbb{E}[Y^4] = \frac{1}{4} + \mathcal{O}(L^{-2}).$$

□

Lemma A1. *Let* $F = \frac{X+\frac{3}{8}}{L+\frac{3}{4}}$ *and* $Y = \frac{X-Lp}{\sqrt{Lp(1-p)}}$. *It holds that*

$$\sqrt{(1-p)F} = \sqrt{p(1-p)} + \frac{(1-p)Y}{2\sqrt{L}} - \frac{\sqrt{1-p}(6p+2(1-p)Y^2-3)}{16L\sqrt{p}}$$
$$- \frac{(1-p)Y(6p-2(1-p)Y^2+3)}{32pL^{\frac{3}{2}}} + \mathcal{O}(L^{-2}). \tag{A4}$$

$$\sqrt{p(1-F)} = \sqrt{p(1-p)} - \frac{pY}{2\sqrt{L}} - \frac{\sqrt{p}(-6p+2pY^2+3)}{16L\sqrt{1-p}}$$
$$- \frac{pY(6p+2pY^2-9)}{32(1-p)L^{\frac{3}{2}}} + \mathcal{O}(L^{-2}). \tag{A5}$$

Proof. Note that $Y = \frac{X-Lp}{\sqrt{Lp(1-p)}}$ is equivalent to $X = Y\sqrt{Lp(1-p)} + Lp$. Thus, F can be expressed in terms of Y as

$$F = \frac{\left(Y\sqrt{Lp(1-p)} + Lp\right) + \frac{3}{8}}{L + \frac{3}{4}} = \left(Y\sqrt{\frac{p(1-p)}{L}} + p + \frac{3}{8L}\right)\left(1 + \frac{3}{4L}\right)^{-1}.$$

For large L, we have $\frac{3}{4L} \ll 1$. Thus, by expanding $\left(1 + \frac{3}{4L}\right)^{-1}$, we have

$$F = \left(Y\sqrt{\frac{p(1-p)}{L}} + p + \frac{3}{8L}\right)\left(1 - \frac{3}{4L} + \mathcal{O}(L^{-2})\right) = p(1 + E_1),$$

where

$$E_1 = \sqrt{\frac{1-p}{p}}\frac{Y}{\sqrt{L}} - \frac{\frac{3}{4} - \frac{3}{8p}}{L} - \sqrt{\frac{p}{1-p}}\frac{3Y}{4L^{\frac{3}{2}}} + \mathcal{O}(L^{-2}).$$

By expanding $\sqrt{1 + E_1}$, we arrive at

$$\sqrt{F} = \sqrt{p}\sqrt{1 + E_1} = \sqrt{p}\left(1 + \frac{E_1}{2} - \frac{E_1^2}{8} + \frac{E_1^3}{16} + \mathcal{O}(E_1^4)\right).$$

Multiplying both sides by $\sqrt{1-p}$ and substituting for E_1 yields

$$\sqrt{(1-p)F} = \sqrt{p(1-p)} + \frac{(1-p)Y}{2\sqrt{L}} - \frac{\sqrt{1-p}(6p + 2(1-p)Y^2 - 3)}{16L\sqrt{p}}$$

$$- \frac{(1-p)Y(6p - 2(1-p)Y^2 + 3)}{32pL^{\frac{3}{2}}} + \mathcal{O}(L^{-2}).$$

The proof of the second equality can be done by expressing $1 - F$ in terms of Y as

$$1 - F = \left(-Y\sqrt{\frac{p(1-p)}{L}} + (1-p) + \frac{3}{8L}\right)\left(1 + \frac{3}{4L}\right)^{-1}.$$

$$= \left(-Y\sqrt{\frac{p(1-p)}{L}} + (1-p) + \frac{3}{8L}\right)\left(1 - \frac{3}{4L} + \mathcal{O}(L^{-2})\right) = (1-p)(1 + E_2),$$

where

$$E_2 = -\sqrt{\frac{p}{1-p}}\frac{Y}{\sqrt{L}} - \frac{\frac{3}{4} - \frac{3}{8(1-p)}}{L} + \sqrt{\frac{1-p}{p}}\frac{3Y}{4L^{\frac{3}{2}}} + \mathcal{O}(L^{-2}).$$

By expanding $\sqrt{1 + E_2}$, we arrive at

$$\sqrt{1 - F} = \sqrt{1 - p}\sqrt{1 + E_2} = \sqrt{1 - p}\left(1 + \frac{E_2}{2} - \frac{E_2^2}{8} + \frac{E_2^3}{16} + \mathcal{O}(E_2^4)\right).$$

Multiplying both sides by $\sqrt{p}$ and substituting for E_2 yields

$$\sqrt{p(1 - F)} = \sqrt{p(1-p)} - \frac{pY}{2\sqrt{L}} - \frac{\sqrt{p}(-6p + 2pY^2 + 3)}{16L\sqrt{1-p}}$$

$$- \frac{pY(6p + 2pY^2 - 9)}{32(1-p)L^{\frac{3}{2}}} + \mathcal{O}(L^{-2}).$$

□

References

1. Press Release of Nobel Prize 2009. Available online: http://www.nobelprize.org/nobel_prizes/physics/laureates/2009/press.html (accessed on 18 November 2016).
2. Fossum, E.R. Active Pixel Sensors: Are CCD's Dinosaurs? *Proc. SPIE* **1993**, *1900*, 2–14.

3. Clark, R.N. Digital Camera Reviews and Sensor Performance Summary. Available online: http://www. clarkvision.com/articles/digital.sensor.performance.summary (accessed on 18 November 2016).
4. Fossum, E.R. Some Thoughts on Future Digital Still Cameras. In *Image Sensors Signal Processing for Digital Still Cameras*; Nakamura, J., Ed.; CRC Press: Boca Raton, FL, USA, 2005; Chapter 11, pp. 305–314.
5. Fossum, E.R. What to do with sub-diffraction-limit (SDL) pixels?—A proposal for a gigapixel digital film sensor (DFS). In Proceedings of the IEEE Workshop Charge-Coupled Devices and Advanced Image Sensors, Nagano, Japan, 9–11 June 2005; pp. 214–217.
6. Fossum, E.R. The Quanta Image Sensor (QIS): Concepts and Challenges. In *OSA Technical Digest (CD), Paper JTuE1, Proceedings of the OSA Topical Mtg on Computational Optical Sensing and Imaging, Toronto, ON, Canada, 10–14 July 2011*; Optical Society of America: Washington, DC, USA, 2011.
7. Masoodian, S.; Song, Y.; Hondongwa, D.; Ma, J.; Odame, K.; Fossum, E.R. Early research progress on quanta image sensors. In Proceedings of the International Image Sensor Workshop (IISW), Snowbird, UT, USA, 12–16 June 2013.
8. Ma, J.; Fossum, E.R. Quanta image sensor jot with sub 0.3 e− r.m.s. read noise and photon counting capability. *IEEE Electron Device Lett.* **2015**, *36*, 926–928.
9. Masoodian, S.; Rao, A.; Ma, J.; Odame, K.; Fossum, E.R. A 2.5 pJ/b binary image sensor as a pathfinder for quanta image sensors. *IEEE Trans. Electron Devices* **2016**, *63*, 100–105.
10. Fossum, E.R.; Ma, J.; Masoodian, S. Quanta image sensor: Concepts and progress. *Proc. SPIE Adv. Photon Count. Tech. X* **2016**, *9858*, 985804.
11. Sbaiz, L.; Yang, F.; Charbon, E.; Susstrunk, S.; Vetterli, M. The gigavision camera. In Proceedings of the 2009 IEEE International Conference on Acoustics, Speech and Signal Processing (ICASSP), Taipei, Taiwan, 19–24 April 2009; pp. 1093–1096.
12. Yang, F.; Sbaiz, L.; Charbon, E.; Süsstrunk, S.; Vetterli, M. On pixel detection threshold in the gigavision camera. *Proc. SPIE* **2010**, *7537*, doi:10.1117/12.840015.
13. Yang, F.; Lu, Y.M.; Sbaiz, L.; Vetterli, M. Bits from photons: Oversampled image acquisition using binary Poisson statistics. *IEEE Trans. Image Process.* **2012**, *21*, 1421–1436.
14. Dutton, N.A.W.; Parmesan, L.; Holmes, A.J.; Grant, L.A.; Henderson, R.K. 320 × 240 oversampled digital single photon counting image sensor. In Proceedings of the IEEE Symposium VLSI Circuits Digest of Technical Papers, Honolulu, HI, USA, 9–13 June 2014; pp. 1–2.
15. Dutton, N.A.W.; Gyongy, I.; Parmesan, L.; Henderson, R.K. Single photon counting performance and noise analysis of CMOS SPAD-based image sensors. *Sensors* **2016**, *16*, 1122.
16. Dutton, N.A.W.; Gyongy, I.; Parmesan, L.; Gnecchi, S.; Calder, N.; Rae, B.R.; Pellegrini, S.; Grant, L.A.; Henderson, R.K. A SPAD-based QVGA image sensor for single-photon counting and quanta imaging. *IEEE Trans. Electron Devices* **2016**, *63*, 189–196.
17. Burri, S.; Maruyama, Y.; Michalet, X.; Regazzoni, F.; Bruschini, C.; Charbon, E. Architecture and applications of a high resolution gated SPAD image sensor. *Opt. Express* **2014**, *22*, 17573–17589.
18. Antolovic, I.M.; Burri, S.; Hoebe, R.A.; Maruyama, Y.; Bruschini, C.; Charbon, E. Photon-counting arrays for time-resolved imaging. *Sensors* **2016**, *16*, 1005.
19. Vogelsang, T.; Stork, D.G. High-dynamic-range binary pixel processing using non-destructive reads and variable oversampling and thresholds. In Proceedings of the 2012 IEEE Sensors, Taipei, Taiwan, 28–31 October 2012; pp. 1–4.
20. Vogelsang, T.; Guidash, M.; Xue, S. Overcoming the full well capacity limit: High dynamic range imaging using multi-bit temporal oversampling and conditional reset. In Proceedings of the International Image Sensor Workshop, Snowbird, UT, USA, 16 June 2013.
21. Vogelsang, T.; Stork, D.G.; Guidash, M. Hardware validated unified model of multibit temporally and spatially oversampled image sensor with conditional reset. *J. Electron. Imaging* **2014**, *23*, 013021.
22. Figueiredo, M.A.T.; Bioucas-Dias, J.M. Restoration of Poissonian images using alternating direction optimization. *IEEE Trans. Image Process.* **2010**, *19*, 3133–3145.
23. Harmany, Z.T.; Marcia, R.F.; Willet, R.M. This is SPIRAL-TAP: Sparse Poisson intensity reconstruction algorithms: Theory and practice. *IEEE Trans. Image Process.* **2011**, *21*, 1084–1096.
24. Makitalo, M.; Foi, A. Optimal inversion of the generalized Anscombe transformation for Poisson–Gaussian noise. *IEEE Trans. Image Process.* **2013**, *22*, 91–103.

25. Salmon, J.; Harmany, Z.; Deledalle, C.; Willet, R. Poisson noise reduction with non-local PCA. *J. Math Imaging Vis.* **2014**, *48*, 279–294.

26. Rond, A.; Giryes, R.; Elad, M. Poisson inverse problems by the Plug-and-Play scheme. *J. Visual Commun. Image Represent.* **2016**, *41*, 96–108.

27. Azzari, L.; Foi, A. Variance stabilization for noisy+estimate combination in iterative Poisson denoising. *IEEE Signal Process. Lett.* **2016**, *23*, 1086–1090.

28. Yang, F.; Lu, Y.M.; Sbaiz, L.; Vetterli, M. An optimal algorithm for reconstructing images from binary measurements. *Proc. SPIE* **2010**, *7533*, 75330K.

29. Chan, S.H.; Lu, Y.M. Efficient image reconstruction for gigapixel quantum image sensors. In Proceedings of the 2014 IEEE Global Conference on Signal Information Processing (GlobalSIP), Atlanta, GA, USA, 3–5 December 2014; pp. 312–316.

30. Remez, T.; Litany, O.; Bronstein, A. A picture is worth a billion bits: Real-time image reconstruction from dense binary threshold pixels. In Prroceedings of the 2016 IEEE International Conference on Computational Photography (ICCP), Evanston, IL, USA, 13–15 May 2016; pp. 1–9.

31. Elgendy, O.; Chan, S.H. Image reconstruction and threshold design for quanta image sensors. In Proceedings of the 2016 IEEE International Conference on Image Processing (ICIP), Phoenix, AZ, USA, 25–28 September 2016; pp. 978–982.

32. Anscombe, F.J. The transformation of Poisson, binomial and negative-binomial data. *Biometrika* **1948**, *35*, 246–254.

33. Abramowitz, M.; Stegun, I.A. *Handbook of Mathematical Functions with Formulas, Graphs, and Mathematical Tables*; Dover Publications: New York, NY, USA, 1965.

34. Rudin, L.; Osher, S.; Fatemi, E. Nonlinear total variation based noise removal algorithms. *Physica D* **1992**, *60*, 259–268.

35. Elad, M. *Sparse and Redundant Representations*; Springer: New York, NY, USA, 2010.

36. Fossum, E.R.; Ma, J.; Masoodian, S.; Anzagira, L.; Zizza, R. The quanta image sensor: Every photon counts. *Sensors* **2016**, *16*, 1260.

37. Antolovic, I.M.; Burri, S.; Bruschini, C.; Hoebe, R.; Charbon, E. Nonuniformity analysis of a 65-kpixel CMOS SPAD imager. *IEEE Trans. Electron Devices* **2016**, *63*, 57–64.

38. Leon-Garcia, A. *Probability, Statistics, and Random Processes for Electrical Engineering*; Pearson Prentice Hall: Upper Saddle River, NJ, USA, 2008.

39. Wasserman, L. *All of Nonparametric Statistics*; Springer: New York, NY, USA, 2006.

40. Brown, L.; Cai, T.; DasGupta, A. On Selecting a Transformation: With Applications. Available online: http://www.stat.purdue.edu/~dasgupta/vst.pdf (accessed on 18 November 2016).

41. Sauer, K.; Bouman, C.A. Bayesian estimation of transmission tomograms using segmentation based optimization. *IEEE Trans. Nucl. Sci.* **1992**, *39*, 1144–1152.

42. Afonso, M.; Bioucas-Dias, J.; Figueiredo, M. Fast image recovery using variable splitting and constrained optimization. *IEEE Trans. Image Process.* **2010**, *19*, 2345–2356.

43. Chan, S.H.; Khoshabeh, R.; Gibson, K.B.; Gill, P.E.; Nguyen, T.Q. An augmented Lagrangian method for total variation video restoration. *IEEE Trans. Image Process.* **2011**, *20*, 3097–3111.

44. Wang, Y.; Yang, Y.; Yin, W.; Zhang, Y. A new alternating minimization algorithm for total variation image reconstruction. *SIAM J. Imaging Sci.* **2008**, *1*, 248–272.

45. Tomasi, C.; Manduchi, R. Bilateral filtering for gray and color images. In Proceedings of the Sixth International Conference on Computer Vision, Bombay, India, 4–7 January 1998; pp. 839–846.

46. Paris, S.; Durand, F. A fast approximation of the bilateral filter using a signal processing approach. *Int. J. Comput. Vis.* **2009**, *81*, 24–52.

47. Chaudhury, K.N.; Sage, D.; Unser, M. Fast $\mathcal{O}(1)$ bilateral filtering using trigonometric range kernels. *IEEE Trans. Image Process.* **2011**, *20*, 3376–3382.

48. Buades, A.; Coll, B.; Morel, J.M. A non-local algorithm for image denoising. In Proceedings of the IEEE Computer Society Conference on Computer Vision and Pattern Recognition (CVPR), San Diego, CA, USA, 20–26 June 2005; Volume 2, pp. 60–65.

49. Awate, S.P.; Whitaker, R.T. Higher-order image statistics for unsupervised, information-theoretic, adaptive, image filtering. In Proceedings of the IEEE Computer Society Conference on Computer Vision and Pattern Recognition (CVPR), San Diego, CA, USA, 20–26 June 2005; Volume 2, pp. 44–51.

50. Adams, A.; Baek, J.; Davis, M.A. Fast high-dimensional filtering using the permutohedral lattice. *Comput. Graph. Forum* **2010**, *29*, 753–762.

51. Chan, S.H.; Zickler, T.; Lu, Y.M. Monte Carlo non-local means: Random sampling for large-scale image filtering. *IEEE Trans. Image Process.* **2014**, *23*, 3711–3725.

52. Gastal, E.; Oliveira, M. Adaptive manifolds for real-time high-dimensional filtering. *ACM Trans. Graph.* **2012**, *31*, 33.

53. Dabov, K.; Foi, A.; Katkovnik, V.; Egiazarian, K. Image denoising by sparse 3D transform-domain collaborative filtering. *IEEE Trans. Image Process.* **2007**, *16*, 2080–2095.

54. Foi, A. Clipped noisy images: Heteroskedastic modeling and practical denoising. *Signal Process.* **2009**, *89*, 2609–2629.

55. Martin, D.; Fowlkes, C.; Tal, D.; Malik, J. A database of human segmented natural images and its application to evaluating segmentation algorithms and measuring ecological statistics. In Proceedings of the Eighth IEEE International Conference on Computer Vision (ICCV), Vancouver, BC, Canada, 9–12 July 2001; Volume 2, pp. 416–423.

MDPI AG

St. Alban-Anlage 66

4052 Basel, Switzerland

Tel. +41 61 683 77 34

Fax +41 61 302 89 18

http://www.mdpi.com

Sensors Editorial Office

E-mail: sensors@mdpi.com

http://www.mdpi.com/journal/sensors

www.ingramcontent.com/pod-product-compliance
Lightning Source LLC
Chambersburg PA
CBHW051709210326
41597CB00032B/5422